D1486637

The Analysis of Covariance and Alternatives

Bradley E. Huitema

Professor of Psychology
Western Michigan University

A Wiley-Interscience Publication

JOHN WILEY & SONS, New York · Chichester · Brisbane · Toronto

Copyright © 1980 by John Wiley & Sons, Inc.

All rights reserved. Published simultaneously in Canada.

Reproduction or translation of any part of this work
beyond that permitted by Sections 107 or 108 of the
1976 United States Copyright Act without the permission
of the copyright owner is unlawful. Requests for
permission or further information should be addressed to
the Permissions Department, John Wiley & Sons, Inc.

Library of Congress Cataloging in Publication Data:

Huitema, Bradley E   1938–
    The analysis of covariance and alternatives.

    "A Wiley-Interscience publication."
    Includes index.
    1. Analysis of covariance.   I. Title

QA279.H83   519.5′352   80-11319
ISBN 0-471-42044-1

Printed in the United States of America

10 9 8 7 6 5 4 3 2 1

To Craig, Laura, and
my Parents

Preface

The purpose of this book is to provide graduate students and research workers in the behavioral and biological sciences with an applied and comprehensive treatment of the analysis of covariance (ANCOVA) and alternative procedures. Even though covariance analysis was introduced to research workers many years ago (Fisher 1932), it remains one of the least understood and most misused of all statistical methods. Even well-qualified researchers often apply ANCOVA in situations where such analysis leads to invalid conclusions and fail to employ it when it is appropriate. Perhaps the widespread misuse of this procedure can be traced to the lack of emphasis in the standard statistics texts on the design and interpretation problems that go along with ANCOVA. Often the emphasis is on computational complexities and mathematics rather than logical considerations and assumptions. Moreover, the existing treatments of ANCOVA (which are generally found at the end of most analysis of variance texts) provide the reader with little or no guidance concerning what to do when the assumptions underlying the analysis are not met. For example, the typical treatment of ANCOVA includes the statement that the analysis is invalid if the regression slopes are not homogeneous. Period. Researchers confronted with data yielding heterogeneous slopes want to know (1) what the consequences are if they go ahead and employ ANCOVA regardless and (2) whether some other procedure is available that is more appropriate in this situation. These questions and similar ones frequently arise in reference to real data; fortunately, most of them have relatively clear answers.

The book is broken down into three major parts. Part 1 includes a brief review of statistical inference through simple analysis of variance and regression. Simple analysis of covariance is then developed through the use of a simple diagrammatic approach to illustrate the basic rationale. The general linear regression approach to analysis of variance (ANOVA) and ANCOVA is then introduced. This approach is also employed in subsequent sections dealing with complex ANCOVA. The beauty of the general linear regression approach is that it allows us to both (1) clarify the relationship between correlation, regression, and analysis of variance procedures and (2) concentrate on rationale rather than computational difficulties because all basic computations can be carried out with any standard multiple regression computer program. The remainder of Part 1 deals with mutliple comparison procedures, assumptions and, perhaps most importantly, design and interpretation problems.

Part 2 covers complex varieties of ANCOVA, including procedures for dealing with multiple covariates, nonlinearity, multiple factors, and multiple dependent variables. Alternatives to standard parametric ANCOVA are presented in Part 3. The first chapter in this part is on rank ANCOVA—an alternative useful when the assumptions of normality and homogeneity of variance are encountered with unequal sample sizes. Next the Johnson–Neyman technique is described as the appropriate alternative to ANCOVA when regression slopes are not homogeneous. The last two chapters attend to the problem of analyzing data from nonequivalent-group quasi experiments. Since ANCOVA is frequently inappropriately applied in this situation, two promising alternatives are presented—true-score ANCOVA and standardized change-score analysis.

The first part of the book includes a review of elementary statistics; however, it is hoped that all readers have an understanding of basic statistical inference through analysis of variance and regression. Most chapters require almost no knowledge of mathematics, but there are a few exceptions in some sections of Chapters 8, 11, 13, and 15.

Many people have contributed to the development of this book. Michael R. Stoline was of great help in the construction of the Bonferroni F tables, in providing a critical review of the first half of the manuscript, and in providing encouragement throughout the project. The thinking of David A. Kenny was very helpful in my attempts to conceptualize the problems in the analysis of nonequivalent-group designs. He also provided helpful criticism of procedures that I had proposed in earlier drafts of Chapter 15. Charles S. Reichardt also provided a very thorough and useful review of this chapter. Lee J. Cronbach and David R. Rogosa provided helpful comments on Chapter 13. Among many others who provided feedback and checked computational examples, special thanks go to Suzanne M. Girman, Robert W. High, and Michael F. Masters. Brian T. Mitchell provided essential help in producing the Bonferroni tables.

I am indebted to the literary executor of the late Sir. Ronald A. Fisher, F.R.S., Cambridge; to Frank Yates, F.S.R., Rothamsted; and to Oliver and Boyd Ltd., Edinburgh, for permission to reprint Tables A.1, A.2, A.3, and A.7 from their book *Statistical Tables for Biological, Agricultural and Medical Research*.

I am also indebted to J. L. Bryant, A. S. Paulson, and the editor of *Biometrika* for permission to reprint Table A.4; to E. S. Pearson and H. O. Hartley, editors of *Biometrika Tables for Statisticians*, Vol. 1; and to *Biometrika* trustees for permission to reprint Table A.5.

BRADLEY E. HUITEMA

Kalamazoo, Michigan
August 1980

Contents

PART ONE
SIMPLE ANALYSIS OF COVARIANCE

CHAPTER ONE
Overview of Some Basic Statistical Considerations

1.1 DESCRIPTIVE VERSUS INFERENTIAL STATISTICS

Statistical methods are often subsumed under the general headings "descriptive" and "inferential." The emphasis in most behavioral and biological science statistics books (including this one) is on the inferential rather than the descriptive aspects of statistics. This emphasis is understandable because descriptive statistics such as means, variances, standard deviations, and correlation coefficients can be described and explained in relatively few chapters. The foundations of inferential statistics generally require longer explanations, a higher level of abstract thinking, and more complex formulas. Unfortunately, even though the inferential aspects of statistics are important, it seems that researchers frequently ignore descriptive statistics. It is not unusual, especially in the behavioral sciences, to hear research workers describe the outcome of their studies in terms of probability values or statements of "significant" and "nonsignificant" with little or no mention of the size of the mean difference or any other appropriate descriptive statistic.

Although it is possible to describe the outcome of a research project in many ways, the orientation of this book is primarily toward means and mean differences. Before any hypothesis tests are run and before any confidence intervals are constructed, we suggest that a question be asked as to whether the obtained mean difference is of any practical or theoretical importance. It is often possible for a researcher to state values of mean differences that fall into one of the following categories: (1) trivial or unimportant, (2) of questionable importance, or (3) of definite practical or theoretical importance.

If the obtained difference is clearly unimportant, there seems to be little justification for reporting the results of significance tests. If the obtained mean difference is not trivial, significance tests provide useful information concerning the degree of confidence that can be placed in the results. This does not mean that significance tests are invalid when applied to trivial

3

differences. Rather, it seems that it is unnecessary to ask whether a trivial difference is statistically significant (as is often the case with large sample sizes). Probably few researchers care whether a trivial difference is real.

If the nontrivial difference is statistically significant, it may be useful to provide a descriptive measure of the extent to which the observed sample behavior is under experimental control; that is, it may be useful to determine what proportion of the total variability in the study is accounted for by treatment or group differences. This question can be answered with use of the analysis of variance (ANOVA) by forming the ratio of the between group over the total sum of squares (SS). This is sometimes called the *correlation ratio*.*

$$\text{Correlation ratio} = \frac{\text{SS between groups}}{\text{SS total}} = \text{proportion of total variability on dependent variable accounted for by independent variable}$$

If t rather than ANOVA is computed for a two-group experiment, the correlation ratio can be obtained from the following formula:

$$\frac{t_{\text{obt}}^2}{t_{\text{obt}}^2 + (N-2)}$$

where t_{obt} is the obtained t statistic and N is the total number of subjects in both groups combined.

It is possible to have a nontrivial and statistically significant mean difference but a relatively small correlation ratio. An advantage of reporting the correlation ratio in addition to mean differences, significance tests, and/or confidence intervals is that it provides a description of the effectiveness of the treatments in relative rather than absolute units. If the correlation ratio is 0.40, we can see that 40% of the total variability in the experiment is under experimental control; 60% of the total variability is due to sources other than treatment or group differences. If *all* variability in the experiment is due to treatments, the correlation ratio is 1.0; in this case the experimenter has complete control. The correlation ratio provides an especially useful description of treatment effects if the dependent variable is not a well-known measure. Mean differences sometimes convey little information to readers of research reports if the characteristics of the variable on which the means are based are not well known. The interpretation of the correlation ratio as the proportion of variability on the dependent variable accounted for by the independent variable remains the same for all dependent variables.

---

*Statistics texts are inconsistent in the definition of the correlation ratio. Hays (1973) and Ferguson (1976), for example, define it as we have, whereas McNemar (1969), Lewis (1960), and others use the square root of the quantity employed here.

Statistics similar to the correlation ratio are described in Chapters 3 and 4 for analysis of covariance problems and in Chapter 15 for standardized change-score analysis. Comments on the interpretation of the correlation ratio and its variants can be found in Gaito (1958) and Levin (1967).

1.2 HYPOTHESIS TESTS AND CONFIDENCE INTERVALS

Research workers generally employ inferential statistics in dealing with the problem of generalizing the results of sample data to the populations from which the subjects were selected. If we (1) randomly select N mentally retarded subjects from a Michigan institution, (2) randomly assign these subjects to treatments 1 and 2, (3) carry out the experiment, and (4) obtain an outcome measure Y on each subject, a useful descriptive measure of the differential effectiveness of the two treatments is the difference between the two sample means $\bar{Y}_1$ and $\bar{Y}_2$. (This difference is an unbiased estimate of the difference between population means and thus should be viewed as an inferential as well as a descriptive measure.) If it turns out that the mean difference is of practical importance, the investigator may want to make a statement about the difference between the unknown population means μ_1 and μ_2. Population mean μ_1 is the mean score that would be obtained if treatment 1 were administered to *all* mentally retarded subjects from the Michigan institution. Population mean μ_2 is the mean score that would be

Table 1.1 Computation and Interpretation of Significance Tests and Confidence Intervals

Independent Sample t Test Using $\alpha = .05$	95% Confidence Interval
Computation	Computation

$$\frac{\bar{Y}_1 - \bar{Y}_2}{\sqrt{(\Sigma y_1^2 + \Sigma y_2^2)/(n_1 + n_2 - 2)\left(\dfrac{1}{n_1} + \dfrac{1}{n_2}\right)}} = \frac{\bar{Y}_1 - \bar{Y}_2}{s_{\bar{Y}_1 - \bar{Y}_2}} = t \qquad (\bar{Y}_1 - \bar{Y}_2) \pm (t_{(.05, N-2)})(s_{\bar{Y}_1 - \bar{Y}_2})$$

where Σy_1^2 and Σy_1^2 are within-group sums of squares of deviation scores (i.e., $(Y - \bar{Y})^2 = y^2$), n_1 and n_2 are sample sizes associated with groups 1 and 2, obtained t value (t_{obt}) is compared with $t_{(.05, N-2)}$—critical value of t statistic for $\alpha = .05$ (two-tailed) based on N (the total number of subjects) minus 2 degrees of freedom, and $s_{\bar{Y}_1 - \bar{Y}_2}$ is estimated standard error of difference

Interpretation	Interpretation
If $\lvert t_{obt}\rvert$ exceeds $t_{(.05, N-2)}$, reject H_0: $\mu_1 = \mu_2$ at .05 level; otherwise, retain H_0: $\mu_1 = \mu_2$	Probability is .95 that interval $(\bar{Y}_1 - \bar{Y}_2) \pm (t_{(.05, N-2)})(s_{\bar{Y}_1 - \bar{Y}_2})$ will span population difference $\mu_1 - \mu_2$

obtained if treatment 2 were administered to *all* mentally retarded subjects from the Michigan institution. The investigator may decide to test the hypothesis that $\mu_1 = \mu_2$ with a *t* test or an analysis of variance *F* test. If the obtained *t* or *F* statistic exceeds the critical value, the null hypothesis is rejected. It is concluded that the difference between the population means is not zero. If the obtained *t* or *F* statistic does not exceed the critical value, the null hypothesis is retained and it is concluded that there is insufficient information to state that the difference is not zero. A more informative approach is to construct a confidence interval on the mean difference rather than to test the hypothesis that the mean difference is zero. A comparison of the significance test and confidence interval procedures is shown in Table 1.1.

Example

Suppose that a two-group experiment employing 12 subjects is run and the following results are obtained:

$$\overline{Y}_1 = 40$$

$$\overline{Y}_2 = 35$$

$$s_{\overline{Y}_1 - \overline{Y}_2} = 2$$

Mean difference: $\qquad \overline{Y}_1 - \overline{Y}_2 = 40 - 35 = 5$

Hypothesis test: $\qquad\qquad t_{obt} = \frac{5}{2} = 2.5$

$$t_{(.05, 10)} = 2.228$$

Since $2.5 > 2.228$, reject H_0: $\mu_1 = \mu_2$.

Confidence interval: $\qquad 5 \pm (2.228)(2) = (0.544, 9.456)$

Correlation ratio: $\qquad \dfrac{(2.5)^2}{(2.5)^2 + (12 - 2)} = 0.38$

The mean difference of 5 points is, in the opinion of the experimenter, of practical or theoretical importance. The hypothesis that the true population means are equal is rejected on the basis of the *t* test. The confidence interval also leads the experimenter to conclude that the population means are not equal because zero is not included between the upper and lower limits of the interval. Suppose the experimenter also wants to know whether the population difference is likely to be 0.5, 10, or 15 points. The hypothesis test answers none of these questions because the null hypothesis employed in this particular example stated that the population means are equal. A quick glance at the confidence interval answers the experimenter's additional questions. Since the confidence interval does not span 0.5, 10, or 15, it is concluded that none of these values represents the population mean difference. The correlation ratio suggests that approximately 38% of the total variability in the experiment is accounted for by the treatments; 62% is due to other sources of variability.

1.3 ELEMENTARY DECISION THEORY

The concepts of Type I error, Type II error, and power are central to elementary decision theory.

Type I Error

It was pointed out in Section 1.2 that the decision to reject the null hypothesis

$$H_0: \mu_1 = \mu_2$$

is made when the obtained test statistic t_{obt} or ANOVA F_{obt} exceeds the critical (i.e., tabled) value of the t or F statistic. If the obtained statistic is less than the critical value, the null hypothesis is retained. This decision strategy will not always lead to the correct conclusion. If the null hypothesis is true, the population means μ_1 and μ_2 are equal, but we do not expect each sample mean on the basis of a random sample from population 1 or 2 to be equal to the corresponding population mean. We expect the mean of an infinite number of means, each based on a random sample, to equal the population mean, but we know that the individual sample means will generally be somewhat different from the population values as a result of sampling fluctuation. Since each sample mean is subject to sampling fluctuation, it follows that the difference between sample means is also subject to sampling fluctuation. This suggests that differences between sample means will sometimes be large even though the populations from which the subjects were selected have exactly the same mean. We employ significance tests to help decide whether an obtained difference between two sample means is due to sampling fluctuation or to treatment effects. If we reject the null hypothesis when the null hypothesis is true, a Type I error is committed. That is, because of random fluctuation, the difference between means will sometimes be large enough to result in the decision to reject the null hypothesis even though there is no difference between population means. The probability of making a Type I error is known as *alpha* (α). The researcher can control α by selecting the desired probability value in the tables of the critical values of the test statistics.

Type II Error

If the difference between sample means is not large enough to yield an obtained t or ANOVA F that exceeds the critical value, the null hypothesis is retained. This is a decision error of the second kind if the null hypothesis is actually false. That is, a Type II error is committed when the experimenter fails to reject a false null hypothesis. The probability of making a Type II error is known as *beta* (β).

Power

The power of a statistical test is the probability of rejecting the null hypothesis when it should be rejected. In other words, power is the probability of correctly rejecting a false null hypothesis. Power is equal to $1 - \beta$ (i.e., one minus the probability of making a Type II error). Unlike α, which is easily determined by the experimenter, power (and β) is somewhat difficult to estimate. Power is a function of (1) the size of the population effect (e.g., mean difference), (2) the size of the random error variance, (3) the sample size, and (4) the α level employed. Power increases as the population effect, sample size, and α increase and as error variance decreases. Charts and tables for estimating power can be found in most advanced statistics books such as Winer (1971) and in Cohen's (1977) book, which is devoted to the topic of power.

1.4   GENERALIZATION OF RESULTS

It is necessary to consider the sampling procedure employed in the experimental design to specify the population to which the results can be generalized. Three types of selection are described in the paragraphs that follow.

Case 1—Random Selection and Random Assignment.   The experimental design described in Section 1.3 involved the random selection of N subjects from a defined population and the random assignment of these subjects to two treatment conditions. The results of this experiment can be generalized to the population from which the subjects were randomly selected.

Case 2—Accessible Selection and Random Assignment.   If the subjects are not randomly selected from a defined population but are simply accessible to the experimenter, it is still appropriate to randomly assign these subjects to treatment conditions and to apply conventional inferential procedures. Although the tests and/or confidence intervals are appropriate, the generalization of results is more ambiguous than with Case 1. The experimenter can only state that the results can be generalized to a population of subjects who have characteristics like those who were included in the sample. The generalization in Case 1 situations is based on principles of statistical inference, whereas Case 2 generalization is based on logical considerations and speculation concerning the similarity of the relevant characteristics of the sample and the population to which an experimenter hopes to generalize. Although Case 1 designs are superior from the point of view of generalization, they are almost always impractical; Case 2 designs are the rule rather than the exception in most behavioral science experiments.

Case 3—Assignment of Treatments to Intact Groups (Quasi Experiment).   If subjects are neither randomly selected nor randomly assigned to treatment conditions, the use of conventional statistical procedures is generally questionable. When treatments are applied to intact groups, it is appropriate to employ analysis procedures specially designed for this situation (see Chapters 14 and 15). The generalization of results is limited, as with Case 2, to subjects having characteristics (perhaps unknown) similar to those associated with the subjects included in the study.

1.5   EXPERIMENTAL DESIGNS

The simple randomized design mentioned in the previous sections involved one independent variable and two groups. Most books that cover ANOVA procedures describe randomized designs involving multiple groups and multiple independent variables. These designs are briefly reviewed in the remainder of this section.

Multiple-group One-factor Designs

If an experimenter is interested in comparing the mean effects of three or more treatments applied to groups formed through random assignment of subjects, the two-group t test is no longer appropriate. The analysis of variance F test is generally applied in this case to test the following overall null hypothesis:

$$H_0: \mu_1 = \mu_2 = \cdots = \mu_J$$

If this hypothesis concerning the equality of J population means is rejected on the basis of the overall ANOVA F test, subsequent multiple comparison tests or simultaneous confidence intervals are computed to evaluate differences between specific pairs of means. A brief review of the rationale underlying the analysis of variance F test is provided in Chapter 2.

Multiple-factor Designs

The next step in complexity beyond the multiple-group one-factor design is the two-factor design. Each factor (i.e., independent variable) is a collection of two or more treatment levels. A 2×3 two-factor design has two levels on the first factor and three levels on the second factor. For example, an experimenter might be interested in the effects of different levels of music loudness and drug type on the galvanic skin response under certain environmental conditions. If two loudness levels of music (50 db and 100 db) and three types of drug are employed, the design is said to have two levels on factor A (the first factor) and three levels on factor B (the second factor). The number of groups (or treatment combinations or cells) associated with this design is $(2)(3)=6$. The available subjects are randomly assigned to the six cells. An advantage of employing both factors in one experiment rather than employing each factor in a separate experiment is that the interaction effects of the two factors can be analyzed. The results of a two-factor analysis of variance include an F test on the effects of the first factor, an F test on the effects of the second factor, and an F test on the interaction of the two factors. Suppose we have summarized the following data:

		Drug Type (Factor B)			
		B_1	B_2	B_3	
Music Loudness (Factor A)	100 db A_1	Cell mean $=20$	Cell mean $=30$	Cell mean $=60$	36.67
	50 db A_2	Cell mean $=25$	Cell mean $=35$	Cell mean $=45$	35.00
		22.5	32.5	52.5	

The F test on the first factor (factor A—music loudness) tests the statistical significance of the difference between the two marginal row means 36.67 and 35.00. The F test on the second factor (factor B—drug type) tests the statistical significance of the difference between the three marginal column means 22.5, 32.5, and 52.5. The F test on the interaction of factors A and B tests the inconsistency in the effects of A at the three levels of B. The results of two-factor designs are generally plotted as shown in Figure 1.1.

The magnitude of the dependent variable is on the ordinate, levels of factor B are on the abscissa, and cell means are plotted in the graph. Descriptively, we can see that the effects of A are not consistent at all three levels of B because lines A_1 and A_2 are not parallel; the interaction F test helps us decide whether this inconsistency is due to chance.

In this example both factors are of experimental interest. Sometimes one of the two factors is included in the design as a nuisance factor. A nuisance factor is a variable that is believed to affect scores on the dependent variable

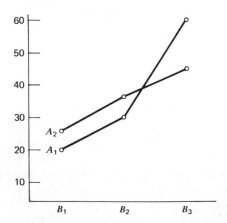

Fig. 1.1 Results of two-factor designs.

but is of no experimental interest. Suppose we are interested in the effects of three methods of teaching reading. Students from a single classroom are randomly assigned to the three treatment conditions. If the students in the classroom are heterogeneous with respect to reading ability, we can expect much within group variability. Since the effect of large within-group variability is to reduce the power of the ANOVA F test, we may find relatively large differences among the means but nonsignificant results associated with the F test. This will occur because the variability in reading ability found within groups is treated as random fluctuation under the one-factor analysis of variance (ANOVA) model. Variability among subjects in reading ability is masking the treatment effect in this case. Some method of at least partially controlling this nuisance variable (reading ability) should be employed to increase the power of the analysis.

One method of controlling nuisance variability involves selection of only those subjects with the same ability level (as measured by a pretest) for participation in the experiment. Obviously, if all subjects included in the experiment have the same reading ability level, we have eliminated reading ability as a nuisance variable. A problem with this method is that it is usually very difficult to find a sufficient number of subjects falling at the same level on the pretest measure.

A more frequently used method of dealing with nuisance variability is to employ the nuisance variable as a factor in the design. If pretest reading ability scores are available before the experimental groups are formed, we can form high, medium, and low reading ability groups. Then subjects from each level can be randomly assigned to the three treatment conditions. The analysis is carried out as an ordinary two-factor design by using reading ability level as factor A and method of teaching reading as factor B. We will have little interest in the outcome of the F test on factor A because we know a priori that differences in this factor exist. We have employed ability level as a factor in the design to remove variability due to this factor from the estimate of error variability. The advantage of this design over the one-factor design is that the test on treatments (method of teaching reading) is more powerful (or, correspondingly, the simultaneous confidence intervals are narrower) and that it is possible to determine whether the treatment effects are consistent for subjects classified as low, medium, or high on reading ability.

A third method of controlling nuisance variability is through statistical control through the use of the analysis of covariance (ANCOVA). This procedure has certain practical and statistical advantages over the two procedures just described. These advantages are described in subsequent chapters.

Other Experimental Designs

Many complex experimental designs such as randomized block, split-plot, Latin-square, and higher-order factorial designs are described in most texts on the analysis of variance. Unfortunately, these designs, as well as the simpler one- and two-factor designs, are often difficult to implement in real-world settings. The major problem with these conventional designs is that random assignment of subjects or treatment orders is required. Randomization eliminates many interpretation problems and ambiguities, but data are often collected from groups of subjects not formed through randomization. Some statisticians suggest that these data be deposited in the nearest circular file. It turns out, however, that the statistician does not usually have the final word. Once data have been collected, they *will* be analyzed—by someone. Whereas some studies do not employ randomization simply because it does not occur to the researchers to do so, there are many research areas in which randomization is impossible to carry out. Regardless of the reason for the use of intact groups of subjects, it is important to remember that such designs are far less easily interpreted than are randomized designs. There are, however, recently developed procedures that appear to yield meaningful analyses of nonrandomized studies. Some of these analyses are described in Chapters 14 and 15, which cover the topic of nonequivalent-group studies.

1.6   SUMMARY

Both descriptive and inferential statistics are relevant to the task of describing the outcome of a research study and to generalizing the outcome to other subjects. The major emphasis in evaluating data should be on descriptive statistics. If the descriptive results reveal effects that appear to be of theoretical or practical importance, the issue of statistical significance is worthwhile considering. Although the reporting of results in terms of statistical significance is conventional practice, confidence intervals can be claimed to be much more informative. The correlation ratio is a descriptive measure that is sometimes a useful addition to the conventional descriptive and inferential methods of evaluating the outcome of a study. The essentials of elementary decision theory include the concepts of Type I error, Type II error, and power. Among the goals of a well-designed experiment are low probability of Type I error (α), low probability of Type II error (β), and high power ($1 - \beta$). More important than these goals is the extent to which the experimental design provides unambiguous information. The use of random assignment is strongly recommended as a method of establishing groups that can yield meaningful information concerning treatment effects. Comparisons of groups formed through other procedures make analyses and clear statements of treatment effects more difficult. Complex experimental designs are employed to yield greater information and greater power than are simple designs.

CHAPTER TWO

Review of Models Related to Analysis of Covariance

2.1 INTRODUCTION

The analysis of covariance (ANCOVA) model represents an integration of the analysis of variance (ANOVA) and the analysis of variance of regression (ANOVAR) models. Similar to ANOVA, ANCOVA is often used to test the null hypothesis that two or more sample means were obtained from populations with the same mean. The basic advantages of ANCOVA over ANOVA are (1) generally greater power and (2) reduction in bias caused by differences between groups that exist before experimental treatments are administered.

When the design involves the random assignment of subjects to treatments, the increase in power is the major payoff in selecting the analysis of covariance. That is, the size of the error term is smaller with the use of ANCOVA rather than ANOVA if certain conditions are met. At the same time, the ANCOVA procedure includes an adjustment of the treatment effect that reduces bias that may be caused by pretreatment differences between groups. This type of bias is generally small if subjects are appropriately randomly assigned to treatments. If randomization is not employed and a nonequivalent-group design appears to be useful, the differences between groups before treatments may be large and hence the adjustment of the treatment effects and the corresponding means will generally be large. Even though it has become common practice to employ nonequivalent group designs, these designs should be avoided if at all possible; serious interpretation problems are associated with their usage. Regardless of the design, the ANCOVA F test concerns the null hypothesis that two or more adjusted population means are equal. (Both ANOVA and ANCOVA test the same null hypothesis H_0: $\mu_1 = \mu_2 = \cdots = \mu_J$ if a randomized design is involved because adjusted and unadjusted population means are the same in this case.) The appropriate interpretation of this test and the associated adjusted means requires a thorough knowledge of the design employed in the collection of the

13

data. Design and interpretation problems are discussed further in Chapters 6 and 7.

Suppose that an experimenter is interested in comparing the effects of two drugs on scores obtained on a complex psychomotor test. Subjects are randomly assigned to the two treatments, treatments are applied, and the psychomotor test data are obtained. A simple two-group ANOVA is computed, and a nonsignificant F value is obtained. The experimenter then points out that there is a great deal of age variability within the two groups and that age is well known to be highly related to scores on the psychomotor test employed in the experiment. Analysis of covariance is then computed using age as the so-called covariate. This analysis statistically controls or, more correctly, partitions the effect of the covariate from the relationship between the treatments and the dependent variable. In the present example age is partitioned from the relationship between drug type and psychomotor test scores. The ANCOVA F test reveals a statistically significant treatment (drug) effect. The reason for the contradictory results of the two analyses can most easily be understood by inspecting the distributions associated with each analysis (see Figure 2.1).

The major distinction between the two analyses is that the analysis of variance is based on unconditional Y scores, shown as distributions A and B, whereas the analysis of covariance is based on conditional $Y|X$ scores, distributions C and D. That is, the dependent variable scores or test scores (Y) are conditional on the covariate or age scores (X). More simply, the unconditional distributions are distributions of scores around their respective group means. The conditional distributions reflect the distributions of scores

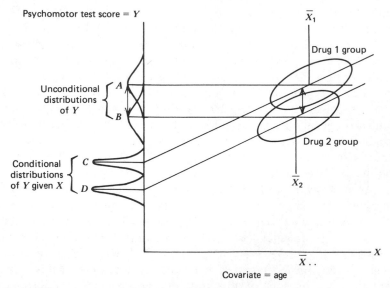

Fig. 2.1 Distribution of unconditional and conditional scores.

around regression lines. It should be obvious that the effect of the smaller within-group variance associated with the conditional distributions is an increase in the power of the analysis. Notice that the unconditional distributions overlap more than do the conditional distributions.

The second reason for employing ANCOVA rather than ANOVA—that of bias reduction—is also illustrated in Figure 2.1. Examine the positions of the two scatter diagrams. The diagram for drug 1 scores is slightly to the right of the drug 2 diagram; correspondingly, the mean age score for drug 1 subjects is slightly above the mean age score for drug 2 subjects. Since age is positively correlated with test performance, we would expect the mean test score to be slightly higher for those subjects assigned to receive drug 1 simply because they are slightly older. Thus even if treatments are not applied, we would predict that the group with the higher mean age will also have superior performance on the psychomotor test; in other words, a slight amount of bias is present because the older group is favored from the start. The analysis of covariance adjusts for this bias by slightly decreasing the mean of group 1 and slightly increasing the mean of group 2. This is because the adjustment procedure changes each treatment mean to the level we would expect it to have if all treatments (drug groups in this example) had the same covariate (age) mean. More specifically, the adjusted means are *adjusted to the level that would be expected if all group covariate means were equal to the grand covariate mean $\overline{X}..$* (i.e., the mean of all subjects in the experiment on the covariate). This causes the difference between the adjusted means of distributions C and D to be smaller than the difference between the unadjusted means of distributions A and B. In this particular example the difference between the adjusted means is less than the difference between the unadjusted means, but the smaller difference will be associated with a higher F value because the error term of the ANCOVA will be much smaller than the error term of the ANOVA. Again notice that distributions C and D overlap less than do distributions A and B *even though the mean difference between C and D is smaller than the mean difference between A and B.*

Since the details of the analysis of covariance can most easily be explained by building on the rationale of the analysis of variance and the analysis of variance of regression, a short review of the two simpler analyses is presented in Section 2.2.

2.2 REVIEW OF ANOVA AND ANOVAR

The purpose of the fixed-effects one-way ANOVA is to test the hypothesis that J population means are equal. The purpose of the simple ANOVAR is to test the hypothesis that the population regression slope of Y on X is equal to zero. Hence ANOVA is generally employed in experimental studies in which tests on differences among means are of primary interest, whereas regression analysis is employed when the regression equation relating Y to some predictor variable X is to be analyzed.

Both ANOVA and ANOVAR involve partitioning the total variance on the dependent variable. Under the ANOVA model the total variance is partitioned into (1) between-group (also called *among-group* or *treatment*) variance and (2) within-group (or error) variance, that is,

$$\sigma_t^2 = \sigma_b^2 + \sigma_w^2$$

where the subscripts t, b, and w represent total, between, and within, respectively.

In working with actual data the population parameters are not, of course, available, and σ_t^2, σ_b^2, and σ_w^2 must be estimated from sample data. Three variance estimators may be obtained by:

1. Computing total sum of squares.
2. Partitioning the total sum of squares into between-group and within-group sum of squares.
3. Dividing each sum of squares by the appropriate degrees of freedom.

$$\frac{SS_t}{N-1} = \hat{\sigma}_t^2$$

$$\frac{SS_b}{J-1} = \hat{\sigma}_b^2$$

$$\frac{SS_w}{N-J} = \hat{\sigma}_w^2$$

where N is the number of subjects, J is the number of treatment groups, and $\hat{\sigma}^2$ is the estimator of the population variance. When the hypothesis H_0: $\mu_1 = \mu_2 = \cdots = \mu_J$ is valid, all three variance estimators ($\hat{\sigma}_t^2$, $\hat{\sigma}_b^2$, and $\hat{\sigma}_w^2$) yield unbiased estimates of the population variance. Only two of the three variance estimators [generally termed *mean squares* (MS) in the ANOVA summary table] are used in testing the null hypothesis. The between-group estimate (MS between) is divided by the within-group estimate (MS within) to obtain the F statistic. If the samples or groups in the experiment have been drawn from populations in which $\mu_1 = \mu_2 = \cdots = \mu_J$, we should expect the F value obtained in the experiment to be near* 1.0 because the numerator and denominator mean squares both yield unbiased estimates of the same parameter. If the null hypothesis is not valid (i.e., if there are differences among the means of the various treatment populations), the between-group MS has an expected value that is greater than the mean square within, and the obtained F will reflect this effect by being large. The obtained F statistic is compared with the tabled value of F (for some specified α level) to determine whether

---

*The exact expectation of F is $\nu_2/(\nu_2-2)$, where ν_2 is the denominator degrees of freedom.

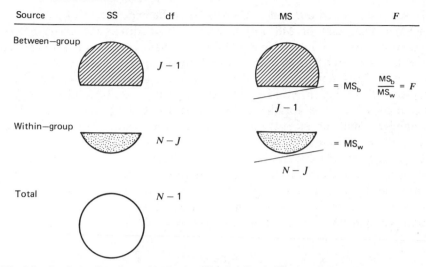

Source	SS	df	MS	F

Between—group $J-1$

$= MS_b$ $\dfrac{MS_b}{MS_w} = F$

Within—group $N-J$

$= MS_w$

Total $N-1$

Fig. 2.2   Analysis of variance (single-classification fixed-effects model) partition and F test.

the obtained F is larger than would be expected by chance alone. The partitioning of the total sum of squares is shown in Figure 2.2.

Analysis of variance of regression differs from the simple ANOVA in that the total variance is partitioned into regression and residual components rather than between and within components. As with the analysis of variance, the population parameters are estimated from the available sample data. A simple regression partition is described in Figure 2.3.

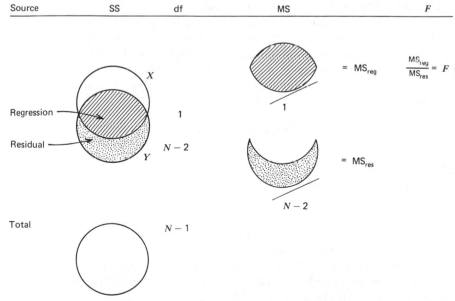

Source	SS	df	MS	F

Regression 1

$= MS_{reg}$ $\dfrac{MS_{reg}}{MS_{res}} = F$

Residual $N-2$

$= MS_{res}$

Total $N-1$

Fig. 2.3   Simple regression partition and F test.

It was just pointed out under ANOVA that the expectation of MS between is greater than that of MS within when a true difference exists between population means. Similarly, in ANOVAR the MS regression has an expected value that is greater than the MS residual if the population regression slope or coefficient β_1 is nonzero. Hence the statistical significance of the sample regression coefficient b_1 is evaluated by comparing the obtained F value with the tabled value of F for the α level specified by the investigator. If the obtained F equals or exceeds the critical or tabled value, the null hypothesis H_0: $\beta_1 = 0.0$ is rejected.

It can be seen in Figure 2.2 that information concerning continuous X variables is not included in the single-classification analysis of variance. Only between-group and within-group variabilities are analyzed. Between-group variance is, under this ANOVA model, treated as the only source of systematic (nonerror) variation. Formally, the ANOVA model is written

$$Y_{ij} = \mu + \alpha_j + \varepsilon_{ij}$$

where

Y_{ij} = dependent variable score of ith individual in jth treatment group

μ = population mean common to all observations

α_j = effect of treatment j (a constant associated with all individuals in treatment j)

ε_{ij} = error component associated with ith individual in treatment j

Notice that all variability in Y not accounted for by treatment effects is treated as error; this is true even though much of the error variability may be predictable from other variables not included in the model.

The ANOVAR model, in contrast to the ANOVA model, may be written to disregard between-group differences as a source of systematic variability. Using this approach, variance on the dependent variable is partitioned into (1) what is accounted for with the use of X and a linear rule and (2) what is not accounted for with the use of X and a linear rule. The model may be written

$$Y_i = \beta_0 + \beta_1 X_i + \varepsilon_i$$

where

Y_i = dependent variable score for individual i

β_0 = Y intercept

β_1 = linear regression coefficient of Y on X

X_i = predictor score for individual i

ε_i = error component associated with individual i

This model, unlike the ANOVA model, does not include information concerning treatment group membership. Notice the absence of the j subscript in the various terms of the model.

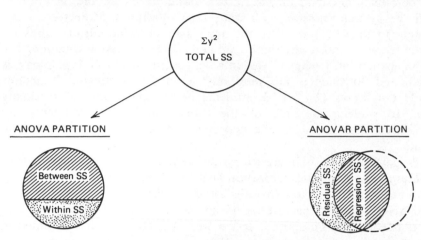

Fig. 2.4   Analysis of variance and analysis of variance of regression partitions.

In summary, ANOVA and ANOVAR both involve partitioning total variability into systematic and error components; the former treats between-group variability as the systematic component, whereas the latter treats variability accounted for by regression as the systematic component of the partition. The two partitions are summarized in Figure 2.4.

ANOVA vs. ANOVAR: A Comparative Example

Two investigators are interested in academic achievement. Investigator A is primarily concerned with the effects of three different types of study objective on student achievement in freshmen biology. The three types of objectives are:

1. *General*—Students are told to know and understand everything in the text.
2. *Specific*—Students are provided with a clear specification of the terms and concepts they are expected to master and of the testing format.
3. *Specific with study time allocations*—The amount of time that should be spent on each topic is provided in addition to specific objectives that describe the type of behavior expected on examinations.

The dependent variable is the biology achievement test.

Investigator B is interested in predicting student achievement using a recently developed aptitude test. This investigator is not particularly concerned with differences in the mean level of achievement resulting from the use of different types of study objective, but rather with making a general statement of the predictive efficiency of the test across a population of freshmen biology students of which one-third have been exposed to general study objectives, one-third to specific behavioral objectives, and one-third to specific behavioral objectives with study time allocations. More specifically, he wants to know (1) whether the regression of achievement scores on aptitude scores is statistically significant and (2) the regression equation.

A population of freshmen students scheduled to enroll in biology is defined, and 30 students are randomly selected. Investigator B obtains aptitude test scores (X) for all students before investigator A randomly assigns 10 students to each of the three treatments. Treatments are administered, and scores on the dependent variable are obtained for all students.

Investigator A, who is interested in differences among treatment means, computes a simple single-classification analysis of variance on the 30 biology achievement test (Y) scores. Investigator B, on the other hand, computes a simple ANOVAR since his interest lies in the regression of biology achievement test scores (Y) on aptitude test scores (X). The data are presented in Table 2.1.

Scores on the dependent variable (Y) are, of course, the same in both analyses. The data are arranged somewhat differently because group membership (i.e., treatment group 1, 2, or 3) is considered in ANOVA but not in ANOVAR. On the other hand, the X variable (aptitude) is included in ANOVAR. Hence it can be seen that both types of analysis can be used with a given set of dependent variable scores to answer different questions. The results of the two analyses appear in Figure 2.5.

The analysis of variance of regression reveals an F value of 18.3, which clearly exceeds the tabled value of F, which is 13.50 for $\alpha = 0.001$ and degrees of freedom (df) 1 and 28. Since F obtained exceeds $F_{(0.001, 1, 28)}$ the null hypothesis H_0: $\beta_1 = 0.0$ is rejected, and investigator B concludes that biology achievement scores can be predicted from aptitude test scores with less error using the regression equation than would result from predicting the overall mean on the biology achievement test.

The sample intercept b_0 and slope b_1 are 9.413 and 0.519, respectively. Biology achievement scores from future students can be predicted by obtaining aptitude test scores and entering the scores in the prediction equation

$$\hat{Y} = 9.413 + 0.519(X)$$

Investigator A tentatively concludes that the differences among the sample means of the three treatment groups are the result of chance fluctuation because the ANOVA-obtained F value of 1.60 is less than the critical value, which is 3.35 using $\alpha = 0.05$. At this point the two investigators get together and discuss the results of the experiment. Investigator B examines the results of his colleague's ANOVA and suggests that a more powerful technique, one that controls for aptitude differences among subjects, might reveal significant treatment effects; ANCOVA is one such technique.

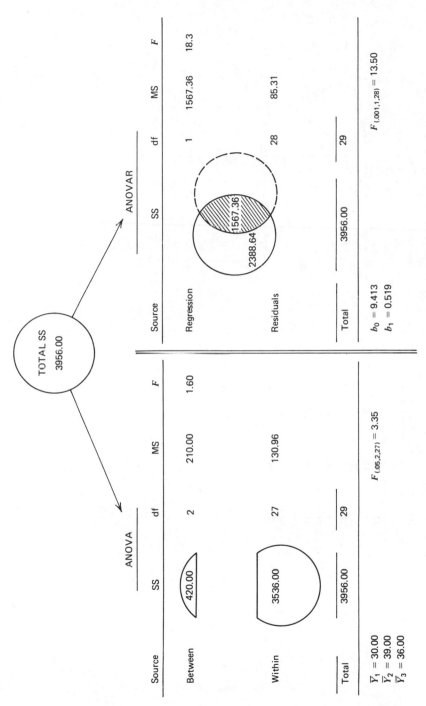

Fig. 2.5 Analysis of variance and analysis of variance of regression for behavioral objectives data.

Table 2.1 Data for Behavioral Objectives Study

A. Data Layout for Analysis of Variance

Treatment (Type of Objective)

1	2	3
15	20	14
19	34	20
21	28	30
27	35	32
35	42	34
39	44	42
23	46	40
38	47	38
33	40	54
50	54	56

B. Data Layout for Analysis of Variance of Regression

Aptitude Test score (X)	Biology Achievement Score (Y)
29	15
49	19
48	21
35	27
53	35
47	39
46	23
74	38
72	33
67	50
22	20
24	34
49	28
46	35
52	42
43	44
64	46
61	47
55	40
54	54
33	14
45	20
35	30
39	32
36	34
48	42
63	40
57	38
56	54
78	56

C. Computation of Quantities Required for ANOVA and ANOVAR

ANOVA

1. Computation of total sum of squares

$$\sum Y_t^2 - \frac{(\sum Y_t)^2}{N} = \sum y_t^2 = SS_t$$

$$40,706 - \frac{(1050)^2}{30} = 3956.00$$

2. Computation of between-group sum of squares

$$\frac{(\sum Y_1)^2}{n_1} + \frac{(\sum Y_2)^2}{n_2} + \cdots + \frac{(\sum Y_J)^2}{n_J} - \frac{(\sum Y_t)^2}{N} = SS_b$$

$$\frac{(300)^2}{10} + \frac{(390)^2}{10} + \frac{(360)^2}{10} - \frac{(1050)^2}{30} = 420.00$$

3. Computation of within-group sum of squares (by subtraction)

$$SS_t - SS_b = SS_w$$

$$3956.00 - 420.00 = 3536.00$$

4. Computation of group means

$$\bar{Y}_1 = \frac{\sum Y_1}{n_1} = \frac{300}{10} = 30.00$$

$$\bar{Y}_2 = \frac{\sum Y_2}{n_2} = \frac{390}{10} = 39.00$$

$$\bar{Y}_3 = \frac{\sum Y_3}{n_3} = \frac{360}{10} = 36.00$$

ANOVAR

1. Computation of total sum of squares on y (same as under ANOVA)

2. Computation of regression sum of squares

$$\frac{\{\sum XY_t - [(\sum X_t)(\sum Y_t)/N]\}^2}{\sum X_t^2 - [(\sum X_t)^2/N]} = \frac{(\sum xy_t)^2}{\sum x_t^2} = SS\ reg$$

$$\frac{\{54,822 - [(1480)(1050)/30]\}^2}{78,840 - (2,190,400/30)} = \frac{9132484}{5826.67} = 1567.36$$

3. Computation of residual sum of squares (by subtraction)

$$SS_t - SS\ reg = SS\ res$$

$$3956.00 - 1567.36 = 2388.64$$

4. Computation of slope and intercept

$$Slope\ (b_1) = \frac{\sum XY_t - [(\sum X_t)(\sum Y_t)/N]}{\sum X_t^2 - \frac{(\sum X_t)^2}{N}} = \frac{\sum xy_t}{\sum x_t^2}$$

$$\frac{54,822 - [(1480)(1050)/30]}{78,840 - \frac{2,190,400}{30}} = \frac{3022}{5826.67} = 0.519$$

Intercept $(b_0) = \bar{Y}.. - b_1 \bar{X}..$

$$35 - 0.519(49.33) = 9.413*;$$

regression equation: $\hat{Y} = 9.413 + 0.519(X)$

*This value was computed using far more decimal places than are shown on the left side of the equation; this practice is followed in many subsequent examples.

23

2.3 SUMMARY

The purpose of covariance analysis is to estimate and test differences among J adjusted population means. Two basic advantages of covariance analysis over the analysis of variance are bias reduction and greater precision or power. The way in which these two advantages come about can be understood if the essentials of ANOVA and ANOVAR are kept in mind.

The analysis of variance and the analysis of variance of regression are two seemingly different approaches to data analysis. Whereas ANOVA is generally used to test the equality of J population means, ANOVAR is used to test the hypothesis of zero population slope. Both analyses involve the partitioning of the total sum of squares into two sources of variability. Under the ANOVA model the total sum of squares is partitioned into the between-group sum of squares and the within-group sum of squares. Likewise, under the regression model, the total sum of squares is partitioned into the regression sum of squares and the residual sum of squares. An F ratio is formed in ANOVA by dividing the MS between groups by the MS within groups. Since the former may contain one source of variability (due to differences between population means) that cannot affect the latter, large values of F are attributed to differences between population means. In ANOVAR the MS regression may contain variability that is due to a nonzero population slope. The MS residual does not contain this source of variability. Hence the F ratio based on MS regression over MS residual will tend to be large when the population slope is nonzero. The manner in which ANOVA and ANOVAR concepts are integrated into one model is described in Chapter 3.

CHAPTER THREE
One- factor Analysis of Covariance

3.1 ANALYSIS OF COVARIANCE MODEL

It has been pointed out that the analysis of variance (ANOVA) model treats only between-group variance as systematic variance and that the analysis of variance of regression (ANOVAR) model treats only variance accounted for by regression as systematic. The analysis of covariance (ANCOVA) model treats *both* between-group and regression variance as systematic (nonerror) components. The mathematical model for the analysis of covariance is

$$Y_{ij} = \mu + \alpha_j + \beta_1 \left(X_{ij} - \bar{X}.. \right) + \varepsilon_{ij}$$

where
  Y_{ij} = dependent variable score of ith individual in jth group
  μ = population mean (on dependent variable) common to all observations
  α_j = effect of treatment j (a constant associated with all individuals in treatment j)
  β_1 = linear regression coefficient of Y on X
  X_{ij} = covariate score for ith individual in jth group
  $\bar{X}..$ = mean of all individuals on covariate
  ε_{ij} = error component associated with ith individual in jth group
A comparison of the models for ANOVA, ANOVAR, and ANCOVA and illustrations of the general type of partitioning involved for each are presented in Figure 3.1.

It can be seen in Figure 3.1 that when the regression term is added to the ANOVA model the result is the ANCOVA model. If the assumptions for the analysis of covariance are met for a given set of data, the error term will be smaller than in ANOVA because much of the within-group variability will be accounted for and removed by the regression of the dependent variable on the covariate. Since the result of the smaller error term is an increase in power, it is quite possible that data analyzed by using ANCOVA will yield highly significant results where ANOVA yields nonsignificant results. The

25

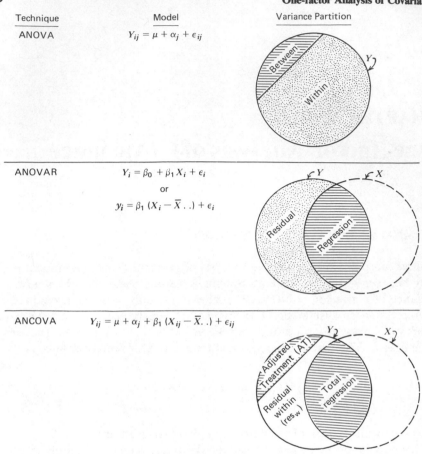

Technique	Model	Variance Partition
ANOVA	$Y_{ij} = \mu + \alpha_j + \epsilon_{ij}$	
ANOVAR	$Y_i = \beta_0 + \beta_1 X_i + \epsilon_i$ or $y_i = \beta_1 (X_i - \overline{X}..) + \epsilon_i$	
ANCOVA	$Y_{ij} = \mu + \alpha_j + \beta_1 (X_{ij} - \overline{X}..) + \epsilon_{ij}$	

Fig. 3.1 Comparison of ANOVA, ANOVAR, and ANCOVA.

opposite is possible under certain conditions, but unlikely. The justification for these statements should become clear as the computation and rationale for the analysis are explained.

3.2 COMPUTATION AND RATIONALE

The starting point for ANCOVA is exactly the same as for ANOVAR or ANOVA; the total sum of squares ($\sum y_t^2$) is computed. In the experiment described previously (in Section 2.2) three treatments were applied, and the purpose of the analysis was to detect mean treatment effects if present. The total sum of squares can be viewed as containing variability in academic achievement (Y) resulting from:

1. Treatment effects (effects of being exposed to different types of behavioral objectives) that are independent of test X.

2. Differences in achievement among subjects that can be predicted from test X.
3. Differences among subjects that are not due to treatment effects and cannot be predicted from test X (i.e., error).

After the total sum of squares is computed, the second step is to remove the sum of squares due to the regression of Y on X from the total sum of squares. That is, those differences in achievement (Y) that can be predicted by using the test (X) are subtracted from the total variability on Y. These and the remaining ANCOVA steps are illustrated in the remainder of this section, using n_1, n_2, and n_3 as the number of subjects in groups 1, 2, and 3 respectively; N as the total number of subjects in the experiment; X as a raw score on the covariate; x as a deviation score (i.e., $X - \bar{X}$) on the covariate, Y as a raw score on the dependent variable, and y as a deviation score on the dependent variable.

Step 1. Computation of total sum of squares.

$$\text{Total SS} = \Sigma y_t^2 = \Sigma Y_t^2 - \frac{(\Sigma Y_t)^2}{N}$$

The total sum of squares includes SS due to (1) treatment effects independent of X, (2) differences in achievement predictable from test X (i.e., variability accounted for by regressing Y on X), and (3) differences among subjects that are not due to treatment effects and cannot be predicted from test X (i.e., error).

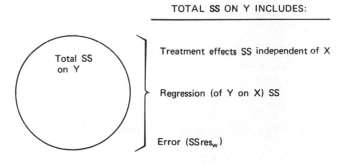

TOTAL SS ON Y INCLUDES:

Total SS on Y

Treatment effects SS independent of X

Regression (of Y on X) SS

Error (SSres$_w$)

Step 2: Computation of total residuals.

$$\text{Total residual SS (SSres}_t) = \Sigma y_t^2 - \frac{(\Sigma xy_t)^2}{\Sigma x_t^2}$$

$$= \left[\Sigma Y_t^2 - \frac{(\Sigma Y_t)^2}{N} \right] - \frac{\left[\Sigma XY_t - \frac{(\Sigma X_t)(\Sigma Y_t)}{N} \right]^2}{\Sigma X_t^2 - \frac{(\Sigma X_t)^2}{N}}$$

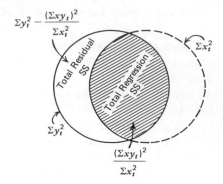

The left-hand circle represents total SS on Y and the right-hand circle, total SS on X. The shaded portion represents variability on Y predictable from X. If the shaded portion is removed from the total left-hand circle, the remaining part is the total residual SS. Since the total SS contains treatment effects independent of X, variability due to the regression of Y on X, and error, it follows that the total residual SS contains only variability due to treatment effects independent of X and error because the regression SS has been removed; that is:

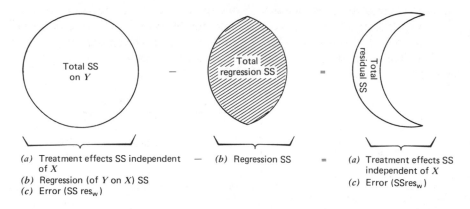

(a) Treatment effects SS independent   —   (b) Regression SS   =   (a) Treatment effects SS
    of X independent of X
(b) Regression (of Y on X) SS   (c) Error (SSres$_w$)
(c) Error (SS res$_w$)

Step 3. Computation of within-group sum of squares.

Within-group $SS = \Sigma y_w^2 = \Sigma y_1^2 + \Sigma y_2^2 + \Sigma y_3^2$

$$= \left[\Sigma Y_1^2 - \frac{(\Sigma Y_1)^2}{n_1} \right] + \left[\Sigma Y_2^2 - \frac{(\Sigma Y_2)^2}{n_2} \right] + \left[\Sigma Y_3^2 - \frac{(\Sigma Y_3)^2}{n_3} \right]$$

The within-group sum of squares may be obtained by computing Σy^2 for each group and then pooling the results of the three separate computations. The within-group SS is not, of course, influenced by treatment or between-group differences. The within-group sum of squares includes SS due to differences predictable from X and differences not predictable from X (error).

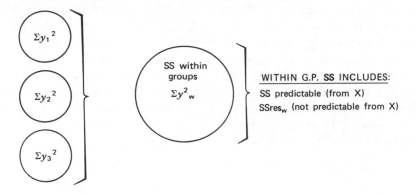

The within-group deviation cross products and sums of squares on X required in the next step are computed using the following formula:

$$\Sigma xy_w = \Sigma xy_1 + \Sigma xy_2 + \Sigma xy_3$$

$$= \left[\Sigma XY_1 - \frac{(\Sigma X_1)(\Sigma Y_1)}{n_1} \right] + \left[\Sigma XY_2 - \frac{(\Sigma X_2)(\Sigma Y_2)}{n_2} \right]$$

$$+ \left[\Sigma XY_3 - \frac{(\Sigma X_3)(\Sigma Y_3)}{n_3} \right]$$

and

$$\Sigma x_w^2 = \Sigma x_1^2 + \Sigma x_2^2 + \Sigma x_3^2$$

$$= \left[\Sigma X_1^2 - \frac{(\Sigma X_1)^2}{n_1} \right] + \left[\Sigma X_2^2 - \frac{(\Sigma X_2)^2}{n_2} \right] + \left[\Sigma X_3^2 - \frac{(\Sigma X_3)^2}{n_3} \right]$$

Step 4. Computation of within-group residual SS (or error SS). By subtracting SS due to predictable differences among subjects within groups (sometimes called *within-group regression SS*) from SS within groups, we obtain residual

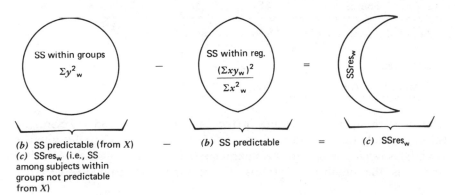

sum of squares within (i.e., SSres$_w$), which is used as the error SS in ANCOVA. Differences contributing to this sum are not predictable from X (using a linear rule) and are not accounted for by treatment differences.

Step 5. Computation of adjusted-treatment effects (AT). By subtracting the SS residual within (step 4) from the total residual SS (step 2), the adjusted-treatment SS is obtained. This quantity was described as the sum of squares due to "treatment effects independent of X" in the previous steps. It is more conventional, however, to refer to this portion of the partition as the adjusted-treatment sum of squares.

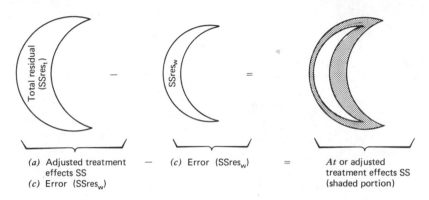

(a) Adjusted treatment   —   (c) Error (SSres$_w$)   = At or adjusted
    effects SS   treatment effects SS
(c) Error (SSres$_w$)   (shaded portion)

Step 6. Computation of F ratio. Step 5 involves the partitioning of the total residual SS into the sum of squares residual within (i.e., SSres$_w$) and the adjusted treatment SS. The latter two correspond directly to within- and between-group (or treatment) SS in a simple analysis of variance. Thus the F ratio can be obtained by dividing mean square adjusted treatment (MSAT) by mean square error (MSres$_w$). Degrees of freedom are computed in essentially the same way as in ANOVA except that an additional degree of freedom is lost from the error MS for each covariate (or covariate polynomial) employed in the analysis. The ANCOVA summary table has the following form:

Source	SS	df	MS	F
Adjusted treatment (AT)	SSAT	$J-1$	$\dfrac{\text{SSAT}}{J-1}$	$\dfrac{\text{MSAT}}{\text{MSres}_w}$
Error (res$_w$)	SSres$_w$	$N-J-C$	$\dfrac{\text{SSres}_w}{N-J-C}$	
Total residual (res$_t$)	SSres$_t$	$N-1-C$		

The sums of squares appearing in the first three rows (SSAT, SSres$_w$, and SSres$_t$) are obtained in steps 5, 4, and 2, respectively; C is the number of covariates (which is one in this chapter); N is the total number of subjects; and J is the number of groups. If we form the ratio SSAT/SSres$_t$, a descriptive measure of the proportion of the variability explained by the treatments when the effects of the covariate are controlled statistically is obtained.

3.3 ADJUSTED MEANS

An adjusted mean is a predicted score. It is the mean dependent variable score that would be expected or predicted for a specified group of subjects if the covariate mean for this group were the same as the grand covariate mean. When subjects are assigned to treatments, the various samples usually will not have *exactly* the same covariate mean even though a randomization procedure may have been employed. Since these group mean differences on the covariate are likely to exist and the covariate is (if properly chosen) highly related to the dependent variable, the investigator may wonder whether the dependent variable mean for one group is higher than for another group because differences between the groups existed *before* the experimental treatments were administered. In the long run (i.e., across many such experiments) the group means will be equal (before treatment administration) on the covariate, the dependent variable, and all other variables if random assignment to treatments has been employed. In a particular experiment, however, there will likely be *some* difference among sample means due to sampling fluctuation.

Suppose a randomized two-group experiment with equal sample sizes is conducted to investigate the effects of two different types of desensitization treatment of snake phobia. A behavioral avoidance test is administered before treatment and again after treatment. The pretreatment scores are employed as the covariate and the posttreatment scores are employed as the dependent variable. Let us say that the covariate means for the two groups are 25 and 30; this five-point difference has occurred even though randomization was employed in establishing the treatment groups. After the treatments have been administered and the dependent variable means are available, it would be reasonable for the investigator to wonder what the two treatment means on the dependent variable would be if the two groups would have had *exactly* the same covariate means rather than covariate means that differ by five points. We attempt to answer this question by computing the adjusted means. In the present example the adjusted means are the predicted dependent variable means that we would expect to occur under the two treatments if the covariate means for both groups were 27.5, which is the covariate grand mean. Very simply, we are answering a "what if" question. What would the dependent variable means be if the covariate means were exactly 27.5 (i.e., equal to the covariate grand mean) rather than 25 and 30? If the assumptions underlying ANCOVA are met, the adjusted means provide a precise answer to this question. A discussion of the assumptions is provided in Chapter 6.

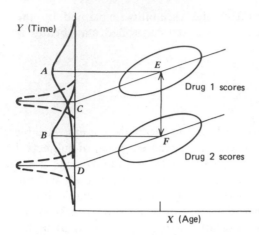

Fig. 3.2 Unconditional and conditional means and intercepts.

It has been stated that the purpose of ANCOVA is to test the null hypothesis that two or more adjusted population means are equal. Alternatively, we could state that the purpose is to test the equality of two or more regression intercepts. Under the assumption of parallel regression lines, the difference between intercepts must be equal to the difference between adjusted means. This equality can be seen in Figure 3.2.

Notice that the difference between points C and D (the intercepts) is exactly the same as the difference between points E and F (the adjusted means). It also turns out that the difference between points A and B (the unadjusted means) is equal to the difference between the adjusted means in *this particular example*. Only when the covariate means for all groups are the same, or the slope b_w is zero, will the unadjusted means be identical to the adjusted means.

Adjusted means should be reported as a standard part of any covariance analysis because inferential statements concerning the results of ANCOVA refer to differences among adjusted means (or intercepts). It is critical that the adjusted means be computed and closely inspected before the results of a covariance analysis are reported because it is often difficult to determine the relative size of adjusted means by simply inspecting the unadjusted means. This point should become clear when the details of the adjustment procedure are described.

The formula for the computation of adjusted means is

$$\overline{Y}_j - b_w\left(\overline{X}_j - \overline{X}..\right) = \overline{Y}_{j\,\text{adj}}$$

where

$\overline{Y}_{j\,\text{adj}} = $ adjusted treatment mean for jth treatment group

$\overline{Y}_j = $ unadjusted treatment mean for jth treatment group

$b_w = $ pooled within-group regression coefficient

$\overline{X}_j = $ covariate mean for jth treatment group

$\overline{X}.. = $ grand covariate mean

Example

A two-group ANCOVA is computed, and the following data are obtained.

Treatment 1	Treatment 2
$\bar{X}_1 = 50$	$\bar{X}_2 = 10$
$\bar{Y}_1 = 120$	$\bar{Y}_2 = 80$

$$\bar{X}.. = 30$$
$$b_w = 0.70$$

The adjusted means for the two treatment groups are

$$120 - [0.70(50 - 30)] = 120 - 14$$
$$= 106$$
$$= \bar{Y}_{1\,adj}$$

and

$$80 - [0.70(10 - 30)] = 80 - [-14]$$
$$= 94$$
$$= \bar{Y}_{2\,adj}$$

An examination of the adjustment formula indicates that the magnitude of the adjustment is a function of (1) the difference between the treatment group covariate mean $\bar{X}_j$ and the grand covariate mean $\bar{X}..$ and (2) the pooled within-group regression coefficient.

The effects of varying the degree and direction of the difference between covariate means can be seen in Figure 3.3, where each situation is drawn with the same slope b_w and the same difference between unadjusted means. Situation I may be used as a reference against which the others may be compared. Notice that the adjusted and unadjusted differences are the same in this situation. Situation IIa indicates what happens to the adjusted means when the group with the higher unadjusted mean also has a somewhat higher covariate mean—the difference becomes smaller. Situation IIb indicates that the adjustment procedure may result in no difference between adjusted means, and situation IIc indicates that adjustment can actually reverse the order of the adjusted means relative to the unadjusted means. That is, the unadjusted mean for treatment 1 is higher than the unadjusted treatment mean 2, but adjusted treatment mean 1 is lower than adjusted mean 2. Situation III diagrams indicate that if the treatment 1 covariate mean is lower than the treatment 2 covariate mean, the difference between the adjusted means will increase as the $\bar{X}_1 - \bar{X}_2$ difference increases.

The general rule to keep in mind is that "winners lose and losers win." That is, the group with the highest covariate mean will lose the most (i.e., the

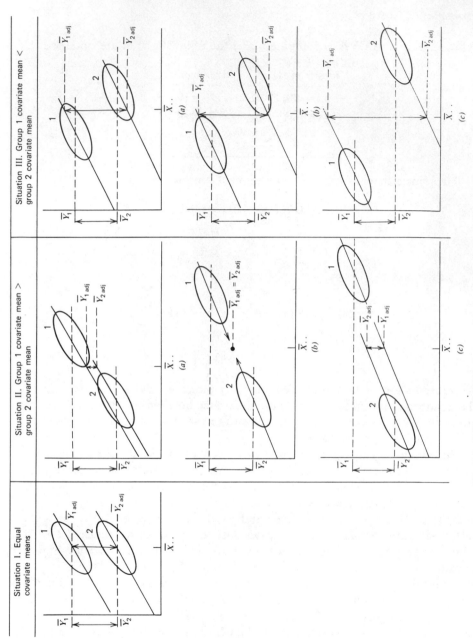

Fig. 3.3 Effects of different covariate means on adjusted means.

downward adjustment will be the greatest), and the group with the lowest covariate mean will gain the most. This will be true as long as the slope b_w is positive.

Suppose that the following data have been obtained in an experiment:

	$\bar{X}$	$\bar{Y}$		
Group 1	30	50	$n_1 = 30$	
Group 2	20	40	$n_2 = 30$	$b_w = 0.80$

Since group 1 has the higher covariate mean we know that the group 1 adjusted mean will be lower than 50 and that the group 2 adjusted mean will be higher than 40. Using the adjustment formula, we find:

		$\bar{Y}_{adj}$
$50 - 0.80(30 - 25) =$ Group 1		46
$40 - 0.80(20 - 25) =$ Group 2		44

If the covariate means in the experiment were equal, we would find the following situation:

	$\bar{Y}$	$=$	$\bar{Y}_{adj}$
Group 1	50		50
Group 2	40		40

because

$$50 - 0.80(0) = 50$$
$$40 - 0.80(0) = 40$$

On the other hand, if the difference between covariate means were 20 points, we would find

	$\bar{Y}$	$\bar{Y}_{adj}$
Group 1	50	42
Group 2	40	48

because

$$50 - 0.80(10) = 42$$
$$40 - 0.80(-10) = 48$$

The effect of the slope on the adjusted means is illustrated in Figure 3.4. Situation *a* in Figure 3.4 indicates that no adjustment is involved, even though the slope is quite steep, when the covariate means are equal. Situation *b* indicates that no adjustment is possible when the slope is equal to 0.0 regardless of the difference between covariate means. Situation *c* involves a combination of the slope of *a* with the covariate mean difference of *b*. A large adjustment is evident in this situation. Situation *d* indicates that it is possible to have no difference between unadjusted means and yet a relatively large difference between adjusted means when the slope is steep and a substantial difference exists between covariate means.

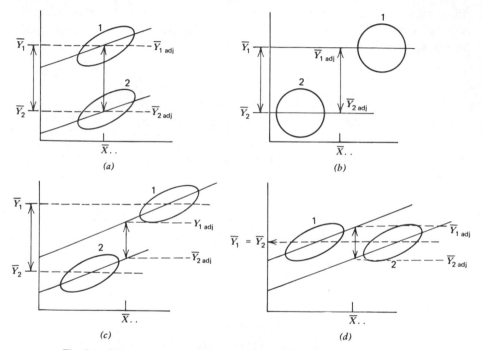

Fig. 3.4 Effects of varying slope b_w and covariate means on adjusted means.

Sample sizes have been equal in the adjustment examples described up to this point. If the absolute size of the adjusted means rather than the difference between them is of primary interest, the relative number of subjects in the groups should be considered in the interpretation. When the number of subjects is the same in the two group situation, the magnitude of the adjustment is the same for both groups. For example:

	$\bar{X}$	$\bar{Y}$	$\bar{Y}_{adj}$
Group 1	5	10	11.25
Group 2	10	10	8.75

$$\Sigma X_1 = 250 \qquad n_1 = 50 \qquad \bar{Y}_{1\,adj} = 10 - [0.5(5 - 7.5)] = 11.25$$
$$\Sigma X_2 = 500 \qquad n_2 = 50$$
$$\Sigma X_t = 750 \qquad b_w = 0.5 \qquad \bar{Y}_{2\,adj} = 10 - [0.5(10 - 7.5)] = 8.75$$
$$\bar{X}.. = 7.5$$

Notice that the $\bar{Y}$ values have been adjusted for both groups by the same number of units.

When the number of subjects in the groups is not the same, the grand mean $\bar{X}..$ is more heavily weighted by the group with the larger n, and, as the following contrived example shows, the Y means of the two groups are adjusted by different amounts:

	$\bar{X}$	$\bar{Y}$	$\bar{Y}_{adj}$
Group 1	5	10	10.05
Group 2	10	10	7.55

$$\Sigma X_1 = 490 \qquad n_1 = 98 \qquad \bar{Y}_{1\,adj} = 10 - [0.5(5 - 5.1)] = 10.05$$
$$\Sigma X_2 = 20 \qquad n_2 = 2$$
$$\Sigma X_t = 510 \qquad b_w = 0.5 \qquad \bar{Y}_{2\,adj} = 10 - [0.5(10 - 5.1)] = 7.55$$
$$\bar{X}.. = 5.1$$

The two previous examples of adjustment indicate once again that it is possible for adjusted treatment means to differ when the unadjusted values are identical. Hence it should be clear that the adjusted means must be computed to discover the values that are being compared in the ANCOVA.

A more extreme example of the necessity for computing the adjusted means is as follows:

	$\bar{X}$	$\bar{Y}$	Y_{adj}
Group 1	115	90	78
Group 2	85	85	97

$$n_1 = n_2 \qquad \bar{Y}_{1\,adj} = 90 - [0.8(115 - 100)] = 78$$
$$b_w = 0.8$$
$$\bar{Y}_{2\,adj} = 85 - [0.8(85 - 100)] = 97$$
$$\bar{X}.. = 100$$

If nothing beyond the unadjusted Y means and the ANCOVA F value were to be inspected, it would be easy to misinterpret the results completely because the order of the means would be reversed in the adjusted and unadjusted situations. Thus it can be seen that the adjusted means can be smaller than, larger than, in the same rank order as, or in a different rank order from, the corresponding unadjusted means. It is also true that the ANCOVA F value can be smaller, larger, or equal to the ANOVA F value for given sets of data. This does not mean, however, that a small difference between adjusted means is necessarily associated with a small F value. The use of ANCOVA often results in a dramatic reduction in the size of the error term (relative to the ANOVA error term), and a small difference between adjusted means may be associated with relatively large F statistic. In most experiments the ANCOVA F exceeds the ANOVA F. Procedures for comparing differences between specific pairs of adjusted means are discussed in Chapter 5.

Some interesting problems in the interpretation of adjusted means in agricultural research can be found in Smith (1957). Interpretation problems associated with adjusted means based on nonequivalent groups (i.e., groups not formed through random assignment) are described in Chapters 14 and 15.

3.4 NUMERICAL EXAMPLE

Analysis of covariance is employed in the analysis of the behavioral objectives study previously described. Here we employ the aptitude test scores (X) as the covariate and the type of behavioral objectives as the independent variable. The data presented in Table 2.1 for ANOVA and ANOVAR are rearranged for ANCOVA in Table 3.1.

Step 1. Computation of total sum of squares. This step and step 2 were previously carried out when the simple regression analysis of Y on X was

Table 3.1. Data Layout for Analysis of Covariance
Treatment (Type of Objective)

1		2		3	
X	Y	X	Y	X	Y
29	15	22	20	33	14
49	19	24	34	45	20
48	21	49	28	35	30
35	27	46	35	39	32
53	35	52	42	36	34
47	39	43	44	48	42
46	23	64	46	63	40
74	38	61	47	57	38
72	33	55	40	56	54
67	50	54	54	78	56

performed.

$$\text{Total SS} = \Sigma y_t^2$$

$$= \Sigma Y_t^2 - \frac{(\Sigma Y_t)^2}{N}$$

$$= 40,706 - \frac{1,102,500}{30}$$

$$= 3956.00$$

This quantity contains variability predictable from the aptitude test (X), variability due to differences produced by the three types of behavioral objective, and error (i.e., differences unrelated to either the aptitude test or the treatments).

Step 2. Computation of total residuals.

$$\text{SSres}_t = \Sigma y_t^2 - \frac{(\Sigma xy_t)^2}{(\Sigma x_t^2)}$$

$$= \Sigma y_t^2 \text{ (from step 1)} - \frac{\left[\Sigma XY_t - \frac{(\Sigma X_t)(\Sigma Y_t)}{N}\right]^2}{\Sigma X_t^2 - \frac{(\Sigma X_t)^2}{N}}$$

$$= 3956.00 - \frac{\left[54,822 - \frac{(1480)(1050)}{30}\right]^2}{78,840 - \frac{2,190,400}{30}}$$

$$= 3956.00 - 1567.36 = 2388.64$$

Since the portion of the total variability in the biology achievement test scores that is predictable from the aptitude test scores is the total regression sum of squares (i.e., 1567.36), the remaining sum of squares (total residuals) contains variability due to treatment effects (different types of behavioral objective) and variability not due to different treatments. To determine how much of the total residual variability is due to treatments and how much is not due to treatments, we perform steps 3, 4, and 5.

Step 3. Computation of within-group sum of squares. We must compute the quantity Σy^2 for each treatment group separately and then add the results to obtain the within-group sum of squares.

$$\Sigma y_1^2 = \Sigma Y_1^2 - \frac{(\Sigma Y_1)^2}{n_1} = 10{,}064 - \frac{90{,}000}{10} = 1064.00$$

$$\Sigma y_2^2 = \Sigma Y_2^2 - \frac{(\Sigma Y_2)^2}{n_2} = 16{,}106 - \frac{152{,}100}{10} = 896.00$$

$$\Sigma y_3^2 = \Sigma Y_3^2 - \frac{(\Sigma Y_3)^2}{n_3} = 14{,}536 - \frac{129{,}600}{10} = 1576.00$$

$$3536.00 = \Sigma y_w^2$$

The within-group sum of squares includes variability predictable from the aptitude test (X) and variability not predictable from X. Notice that between-group or treatment variability is of no concern here because the computations were carried out separately for each group; this, of course, eliminates the possibility of any variability due to treatments affecting the within group sum of squares.

Step 4. Computation of within-group residual sum of squares. Since the within-group sum of squares contains variability predictable from the aptitude test, it is necessary to remove this predictable variability to obtain the amount of variability in the total residuals that is unrelated to both the treatments and the aptitude test, that is, the within-group residual sum of squares. It is necessary to obtain Σxy and Σx^2 for each group independently.

$$\Sigma xy_1 = \Sigma XY_1 - \frac{(\Sigma X_1)(\Sigma Y_1)}{n_1} = 16{,}603 - \frac{(520)(300)}{10} = 1003.00$$

$$\Sigma xy_2 = \Sigma XY_2 - \frac{(\Sigma X_2)(\Sigma Y_2)}{n_2} = 19{,}241 - \frac{(470)(390)}{10} = 911.00$$

$$\Sigma xy_3 = \Sigma XY_3 - \frac{(\Sigma X_3)(\Sigma Y_3)}{n_3} = 18{,}978 - \frac{(490)(360)}{10} = 1338.00$$

$$3252.00 = \Sigma xy_w$$

$$\Sigma x_1^2 = \Sigma X_1^2 - \frac{(\Sigma X_1)^2}{n_1} = 29{,}054 - \frac{270{,}400}{10} = 2014.00$$

$$\Sigma x_2^2 = \Sigma X_2^2 - \frac{(\Sigma X_2)^2}{n_2} = 23{,}888 - \frac{220{,}900}{10} = 1798.00$$

$$\Sigma x_3^2 = \Sigma X_3^2 - \frac{(\Sigma X_3)^2}{n_3} = 25{,}898 - \frac{240{,}100}{10} = 1888.00$$

$$5700.00 = \Sigma x_w^2$$

The within-group regression coefficient (to be used later in adjusting the means) is

$$b_w = \frac{\Sigma xy_w}{\Sigma x_w^2} = \frac{3252}{5700} = 0.5705$$

and the within-group sum of squares regression is

$$\frac{(\Sigma xy_w)^2}{\Sigma x_w^2} = \frac{(3252)^2}{5700} = 1855.35 = SSreg_w$$

By subtraction of within regression from the previously computed within-group sum of squares, we obtain the within-group residual sum of squares:

$$\Sigma y_w^2 - \frac{(\Sigma xy_w)^2}{\Sigma x_w^2} = 3536 - 1855.35 = 1680.65 = SSres_w$$

We now know how much of the total residual sum of squares is error variability unrelated to both treatments and aptitude test scores.

Step 5. Computation of treatment effects. By subtracting the within-group residual sum of squares from the total residual sum of squares, we obtain the adjusted treatment (AT) sum of squares. This quantity represents differences among treatment groups that are not predictable from the aptitude test scores.

$$\begin{array}{l} \text{Total residuals (step 2)} = 2388.64 = SSres_t \\ - \text{Within residuals (step 4)} = 1680.65 = SSres_w \\ \hline \text{Adjusted treatment effect (AT)} = 707.99 = SSAT \end{array}$$

Step 6. Computation of F ratio.

Source	SS	df	MS	F
Adjusted treatment (AT)	707.99	2	354.00	5.48
Error (res$_w$)	1680.65	26	64.64	
Total residuals (res$_t$)	2388.64	28		

The study involves one covariate, three groups, and 30 subjects. The degrees of freedom are $J-1$ or 2 for treatments, $N-J-C$ or $30-3-1=26$ for error, and $N-1-C$ for total residuals. The obtained F is then compared with the critical value of F with 2 and 26 degrees of freedom; $F_{(.05, 2, 26)}$ is 3.37, and the null hypothesis of no treatment effect is rejected at the .05 level.

A summary of the computations involved in the ANCOVA and the adjustment of means is presented in Table 3.2. Since it may be pointless to compare

Table 3.2 Summary of Computations for ANCOVA

	Total	Group 1	Group 2	Group 3	Within
Sums of Squares on Y	$\Sigma y_t^2 = 3956.00$	$\Sigma y_1^2 = 1064.00$	$\Sigma y_2^2 = 896.00$	$\Sigma y_3^2 = 1576.00$	$\Sigma y_w^2 = 3536.00$
Sums of Squares on X	$\Sigma x_t^2 = 5826.67$	$\Sigma x_1^2 = 2014.00$	$\Sigma x_2^2 = 1798.00$	$\Sigma x_3^2 = 1888.00$	$\Sigma x_w^2 = 5700.00$
Sums of Cross Products	$\Sigma xy_t = 3022.00$	$\Sigma xy_1 = 1003.00$	$\Sigma xy_2 = 911.00$	$\Sigma xy_3 = 1338.00$	$\Sigma xy_w = 3252.00$
Regression Weights	$\dfrac{\Sigma xy_t}{\Sigma x_t^2} = 0.5187 = b_t$	$\dfrac{\Sigma xy_1}{\Sigma x_1^2} = 0.4980 = b^{(\text{group 1})}$	$\dfrac{\Sigma xy_2}{\Sigma x_2^2} = 0.5067 = b^{(\text{group 2})}$	$\dfrac{\Sigma xy_3}{\Sigma x_3^2} = 0.7087 = b^{(\text{group 3})}$	$\dfrac{\Sigma xy_w}{\Sigma x_w^2} = 0.5705 = b_w$
Sums-of-squares Regression	$\dfrac{(\Sigma xy_t)^2}{\Sigma x_t^2} = 1567.36 = \text{SSreg}_t$	$\dfrac{(\Sigma xy_1)^2}{\Sigma x_1^2} = 499.51 = \text{SSreg}_1$	$\dfrac{(\Sigma xy_2)^2}{\Sigma x_2^2} = 461.58$	$\dfrac{(\Sigma xy_3)^2}{\Sigma x_3^2} = 948.22$	$\dfrac{(\Sigma xy_w)^2}{\Sigma x_w^2} = 1855.35 = \text{SSreg}_w$
Sums-of-squares Residuals	$\Sigma y_t^2 - \dfrac{(\Sigma xy_t)^2}{\Sigma x_t^2} = 2388.64 = \text{SSres}_t$	$\Sigma y_1^2 - \dfrac{(\Sigma xy_1)^2}{\Sigma x_1^2} = \boxed{564.49}$	$\Sigma y_2^2 - \dfrac{(\Sigma xy_2)^2}{\Sigma x_2^2} = \boxed{434.42}$	$\Sigma y_3^2 - \dfrac{(\Sigma xy_3)^2}{\Sigma x_3^2} = \boxed{627.78}$	$\Sigma y_w^2 - \dfrac{(\Sigma xy_w)^2}{\Sigma x_w^2} = 1680.65 = \text{SSres}_w$

$1626.69 = \text{SSres}_i$

adjusted means if slopes are not homogeneous, the homogeneity of regression test should be considered as a necessary adjunct to the ANCOVA summary. The details of this test are presented in the next section.

The ANCOVA steps based on quantities presented in Table 3.2 are:

1. Total SS $= 3956.00$
2. Total residual SS $= 2388.64$
3. Within SS $= 3536.00$
4. Within residual SS $= 1680.65 = SSres_w$
5. Adjusted treatment SS $= 707.99$ (SSAT)
6. F test:

Source	SS	df	MS	F
Adjusted treatment (AT)	707.99	2	354.00	5.48
Residual within (res$_w$)	1680.65	26	64.64	
Total residuals	2388.64	28		

$$\text{Critical value} = F_{(.05, 2, 26)} = 3.37$$

The adjusted means are:

Group 1. $30 - 0.57(52 - 49.33) = 28.48 = \overline{Y}_{1\,adj}$

Group 2. $39 - 0.57(47 - 49.33) = 40.33 = \overline{Y}_{2\,adj}$

Group 3. $36 - 0.57(49 - 49.33) = 36.19 = \overline{Y}_{3\,adj}$

The proportion of adjusted variability on Y accounted for by adjusted treatment effects is:

$$SSAT/SSres_t = 707.99/2388.64 = 0.30$$

3.5   TESTING HOMOGENEITY OF REGRESSION SLOPES

An assumption underlying the correct usage of ANCOVA is that the population regression slopes associated with the treatment populations are equal. The consequences of violating this assumption are described in detail in Chapter 6. Briefly, the problem is that the adjusted means are inadequate descriptive measures of the outcome of a study if the size of the treatment effect on Y (i.e., the vertical distance between the regression lines) is not the same at different levels of the covariate X. If the slopes are heterogeneous, the treatment effects are not the same at different levels of the covariate; consequently, the adjusted means can be misleading because they do not convey this important information. When the slopes are homogeneous, the adjusted means are adequate descriptive measures because the treatment effects are the same at different levels of the covariate. A method of testing the assumption of homogeneous regression slopes is presented in this section. When the results of this test suggest that the slopes are heterogeneous, the procedures described in Chapter 13 should be employed.

If the slopes for the treatment populations in an experiment are equal, that is, $\beta_1^{(\text{group } 1)} = \beta_1^{(\text{group } 2)} = \cdots = \beta_1^{(\text{group } J)}$, the best way of estimating the value of this common slope from the samples is by computing an average of the sample b_1 values. In the behavioral objectives study the sample regression slopes for the three treatment groups are

$$b_1^{(\text{group } 1)} = 0.4980$$

$$b_1^{(\text{group } 2)} = 0.5067$$

$$b_1^{(\text{group } 3)} = 0.7087$$

It can be seen in Table 3.2 that these values were obtained by dividing Σxy by Σx^2 for each group. If the three separate values Σxy_1, Σxy_2, and Σxy_3 are summed to obtain Σxy_w and divided by the sum of Σx_1^2, Σx_2^2, and Σx_3^2, which is Σx_w^2, we have the pooled within-group regression slope b_w, which is the weighted average of the three separate within-group slopes. (For consistency, the symbol for the pooled within-group slope should be $b_1^{(w)}$ rather than b_w; the latter, however, is employed consistently in the ANCOVA literature.) The slope b_w is our best estimate of the population slope β_1, which is the slope common to all treatment populations. As long as $\beta_1^{(\text{group } 1)} = \beta_1^{(\text{group } 2)} = \beta_1^{(\text{group } 3)} = \beta_1$, the estimate b_w is a useful statistic to employ. But if the population slopes are not equal, it no longer makes sense to obtain an average or pooled slope b_w to estimate a single population parameter because the separate sample values are not all estimates of the same parameter.

Now the problem is to decide whether the treatment populations have the same slope. More specifically, in terms of the example problem, the question is, "Are the three sample values 0.4980, 0.5067, and 0.7087 all estimates of a single parameter β_1 or are the differences among these values so large that it is unlikely that they have come from populations with the same slope?"

If we have reason to believe that the sample values differ more than would be expected on the basis of sampling fluctuation, we may conclude that the population slopes are not equal. The homogeneity of regression F test is designed to answer the question of the equality of the population slopes. The null hypothesis associated with this test is

$$H_0\colon \beta_1^{(\text{group } 1)} = \beta_1^{(\text{group } 2)} = \cdots = \beta_1^{(\text{group } J)}$$

The steps involved in the computation of the test are given in the following paragraphs.

Steps 1 and 2. Computation of within-group sum of squares and within-group residual sum of squares (SSres_w). Steps 1 and 2 were described and carried out in Section 3.4 as the third and fourth steps of ANCOVA. The example data summary values are repeated here.

$$\sum y_w^2 = 3536.00$$

$$\frac{(\sum xy_w)^2}{\sum x_w^2} = 1855.35$$

$$\sum y_w^2 - \frac{(\sum xy_w)^2}{\sum x_w^2} = 1680.65 = \text{SSres}_w$$

Step 3.   Computation of individual sum of squares residual (SSres_i).   The third step involves computation of the sum of squares residual for each treatment group separately and then pooling these residuals to obtain the pooled individual residual sum of squares (SSres_i). The difference in the computation of SSres_w and SSres_i to keep in mind is that SSres_w involves computing only one residual sum of squares whereas SSres_i involves computation of a separate residual sum of squares for each group.

$$\sum y_1^2 - \frac{(\sum xy_1)^2}{\sum x_1^2} = \text{SS residual for group } 1 = 564.49$$

$$\sum y_2^2 - \frac{(\sum xy_2)^2}{\sum x_2^2} = \text{SS residual for group } 2 = 434.42$$

$$\sum y_3^2 - \frac{(\sum xy_3)^2}{\sum x_3^2} = \text{SS residual for group } 3 = 627.78$$

$$1626.69 = \text{SSres}_i$$

Step 4.   Computation of heterogeneity of slopes sum of squares.   The discrepancy between SSres_w and SSres_i reflects the extent to which the individual regression slopes are different from the pooled within-group slope b_w; hence the heterogeneity of slopes SS is simply SSres_w − SSres_i. The rationale for this computation is straightforward. Notice that SSres_i is less than SSres_w for the example data. It turns out that SSres_i can never be larger than SSres_w, just as, in an ordinary ANOVA, the sum of squares within can never be larger than the sum-of-squares total. There is only one explanation for SSres_w being larger than SSres_i—the individual within-group slopes must be different. The heterogeneous slope case is illustrated in Figure 3.5. Within-group SS residual around the average least-squares regression line b_w will be equal to the residuals around the individual within-group slopes if and only if $b_1^{(\text{group 1})} = b_1^{(\text{group 2})} = \cdots = b_1^{(\text{group J})}$. Obviously, when the sample slopes are all equal, the sum of squares for the heterogeneity of the slopes is zero, because

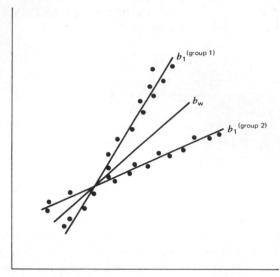

Fig. 3.5 Heterogeneous regression slopes.

when the individual slopes are the same they are also equal to b_w, and $SSres_i$ then must equal $SSres_w$. When the individual within-group slopes differ, the single slope b_w cannot have residuals as small as those around the separate slopes. And when large differences between individual slopes exist, $SSres_w$ is much larger than $SSres_i$. Notice that this is the situation in Figure 3.5; the vertical distance between an observed score and the regression line is much larger for b_w than for the individual lines. A single regression sope simply cannot fit different samples of data as well as can a separate slope for each sample, unless there are no differences among the slopes. The heterogeneity of slopes SS for the example data is $(1680.65 - 1626.69) = 53.96$.

Step 5. Computation of F ratio. The summary table for the F test is as follows:

Source	SS	df	MS	F
Heterogeneity of slopes		$J-1$	$SShet/J-1$	$MShet/MSres_i$
Individual residual (res_i)		$N-(J2)$	$SSres_i/N-(J2)$	
Within residual (res_w)		$N-J-1$		

If the obtained F is equal to or greater than $F_{[(\alpha, J-1, N-(J2)]}$, the null hypothesis $\beta_1^{(group\ 1)} = \beta_1^{(group\ 2)} = \cdots = \beta_1^{(group\ J)}$ is rejected. For the example data, the summary of the analysis is:

Source	SS	df	MS	F
Heterogeneity of slopes	53.96	2	26.98	0.40
Individual residual (res_i)	1626.69	24	67.78	
Within residual (res_w)	1680.65	26		

The critical value using $\alpha = .05$ is $F_{(.05, 2, 24)} = 3.40$; hence the null hypothesis is not rejected, and the investigator can conclude that the ANCOVA assumption of homogeneous regression slopes is met. The individual slopes 0.4980, 0.5067, and 0.7087 appear to differ because of sampling fluctuation only.

3.6 SUMMARY

Performance on a response variable is conceptualized in three different ways under the ANOVA, ANOVAR, and ANCOVA models. Under the ANOVA model the total variability in an experiment is viewed as a function of treatment effects (mean differences) and random fluctuation. Under the ANOVAR model the total variability is viewed as a function of the level of performance on a predictor variable and random fluctuation. Under the ANCOVA model the total variability is viewed as a function of performance on a predictor variable (or covariate), treatment effects that are independent of the predictor variable, and random fluctuation. Hence the ANCOVA model is an integration of the ANOVA and ANOVAR models.

Since one purpose of ANCOVA is to estimate and test differences among adjusted means, it is very important to recognize the factors that affect the adjustment. The relative size of the samples, the pooled within-group regression coefficient, and the mean difference on the covariate all play a part in the adjustment process. There is generally little difference between adjusted and unadjusted means when randomized group experiments are employed. This is because small differences on the covariate can generally be expected with these designs. When nonrandomized designs are employed, the mean differences on the covariate and hence the degree of adjustment may be large.

The interpretation of ANCOVA and the associated adjusted means relies very heavily on the assumption of homogeneous regression slopes for the various treatment groups. If this assumption is not met, the ANCOVA F test and the adjustment process can lead to highly misleading results. For this reason the homogeneity of slopes test should be carried out whenever ANCOVA is employed.

CHAPTER FOUR

Analysis of Covariance through Linear Regression

4.1 INTRODUCTION

The similarity among analysis of variance (ANOVA), analysis of variance of regression (ANOVAR), and analysis of covariance (ANCOVA) models was described in Chapters 2 and 3. Since these analyses are all based on linear models, the reader familiar with multiple linear regression will not be surprised to discover that analysis of variance and analysis of covariance problems can be computed with any typical multiple regression computer program.

4.2 SIMPLE ANALYSIS OF VARIANCE THROUGH LINEAR REGRESSION

Simple (one-way) analysis of variance problems can be computed through regression analysis by regressing the dependent variable scores (Y) on so-called dummy variable(s). The dummy variables, which are used to identify group membership, are easily constructed. Suppose we have a simple two-group analysis of variance problem with five subjects in each group. The 10 dependent variable scores are regressed on a column of dummy scores arranged as follows:

	D Dummy Variable	Y Dependent Variable	
Group 1 dummy variable scores	$\begin{cases} 1 \\ 1 \\ 1 \\ 1 \\ 1 \end{cases}$	$\begin{cases} Y_1 \\ Y_2 \\ Y_3 \\ Y_4 \\ Y_5 \end{cases}$	Group 1 dependent variable scores
Group 2 dummy variable scores	$\begin{cases} 0 \\ 0 \\ 0 \\ 0 \\ 0 \end{cases}$	$\begin{cases} Y_6 \\ Y_7 \\ Y_8 \\ Y_9 \\ Y_{10} \end{cases}$	Group 2 dependent variable scores

If a subject belongs to the first treatment group, he or she is assigned a dummy variable score of one. If a subject does not belong to the first treatment group, he or she is assigned a dummy variable score of zero. Hence the column of dummy scores simply indicates whether a subject belongs to the first treatment group. The analysis involves nothing more than regressing the dependent variable scores on the dummy variable scores. The output of the typical regression program will provide the correlation between the two variables, the regression equation, a test of the significance of the regression slope, and/or an ANOVAR.

If the computer output does not include ANOVAR or any other test statistics, the correlation between the dummy variable(s) and the dependent variable provides enough information to carry out the analysis of variance. This is true regardless of the number of treatment groups.

In the two-group case the simple correlation between the dummy variable and the dependent variable (r_{yd}) can be tested for significance as follows:

$$\frac{r_{yd}^2/1}{\left(1-r_{yd}^2\right)/N-2} = F$$

$$\text{Critical value} = F_{(\alpha,1,N-2)}$$

This F test is a test of the significance of the point-biserial correlation between the dummy variable and the dependent variable. It turns out that this test is equivalent to an ANOVA F test on the difference between the means of the two treatments, which, in turn, is equivalent to the independent sample t test.

Since the F test of the statistical significance of the regression slope, the F test for the point-biserial correlation coefficient, the ANOVA F test on the difference between two means, and the independent sample t test on the difference between two means are all equivalent tests, it follows that any one of these tests can be substituted for any other one. Likewise, any of these test statistics can be converted into a point-biserial correlation coefficient as follows:

$$\sqrt{\frac{\text{ANOVAR } F}{\text{ANOVAR } F + (N-2)}} = \sqrt{\frac{\text{Point biserial } F}{\text{Point biserial } F + (N-2)}}$$

$$= \sqrt{\frac{\text{ANOVA } F}{\text{ANOVA } F + (N-2)}}$$

$$= \sqrt{\frac{t^2}{t^2 + (N-2)}}$$

$$= r_{\text{point biserial}}$$

It may be helpful to consider why these various tests are equivalent. Suppose the data presented in Table 4.1 were collected in a study of the effectiveness of vigorous exercise in the reduction of angina pain in patients who have had one heart attack. The experimental group is exposed to a carefully monitored exercise routine that involves maintaining the heart rate at 140 for 25 min every day for 12 months. The control group is composed of subjects who continue their normal living patterns without an exercise program. Subjects are assigned to the two conditions using a table of random numbers. (The very small sample sizes employed here are not recommended in practice.) The dependent variable is the number of angina attacks experienced by the subjects during the last 2 months of the program.

Table 4.1 Comparison of Significance Tests on Slope, Point-biserial Correlation Coefficient, and Mean Difference

Data:	D Dummy Variable	Y Dependent Variable
	1	3
	1	4
	1	3
	1	2
	1	3
	0	3
	0	12
	0	14
	0	7
	0	9

Analyses:

1. ANOVAR on regression slope

Source	SS	df	MS	F
Regression	90	1	90.0	9.474
Residual	76	8	9.5	
Total	166	9		

$$b_0 = 9.00$$
$$b_1 = -6.00$$

Regression equation: $\hat{Y} = 9.00 - 6.00(D)$

2. Point-biserial correlation F test

$$r_{\text{point biserial}} = r_{yd} = 0.73632$$

$$\frac{r_{yd}^2/1}{1 - r_{yd}^2/N - 2} = \frac{0.54217}{(1 - .54217)/10 - 2} = 9.474 = F_{\text{obt}}$$

3. ANOVA F test

Source	SS	df	MS	F
Between	90	1	90.0	9.474
Within	76	8	9.5	
Total	166	9		

$$\bar{Y}_1 = 3$$
$$\bar{Y}_2 = 9$$

4. Independent sample t test

$$\frac{\bar{Y}_1 - \bar{Y}_2}{\sqrt{\frac{\sum y_1^2 + \sum y_2^2}{n_1 + n_2 - 2}[(1/n_1) + (1/n_2)]}} = \frac{3 - 9}{\sqrt{[(2 + 74)/(5 + 5 - 2)](\frac{1}{5} + \frac{1}{5})}} = -3.078$$

and $\quad t_{\text{obt}}^2 = 9.474 = F_{\text{obt}}$

The purpose of illustrating the relationships among the tests on the slope, the point-biserial correlation coefficient, and the mean difference is to give the reader some feel for the appropriateness of employing correlation and regression procedures in testing mean differences. It is likely, however, that questions remain concerning *why* the regression of the dependent variable on a dummy variable gives the same F as a conventional ANOVA F test. The answer to this question is fairly simple if one of the central concepts in regression analysis, the least-squares criterion, is kept in mind.

Recall that when Y is regressed on X (or D, as we have labeled it here) the intercept (b_0) and the slope (b_1) are fit to the data in such a way that the sum of the squared discrepancies between the observed and the predicted scores is minimum; that is,

$$\sum_1^N (Y_i - \hat{Y}_i)^2 \quad \text{is minimum}$$

where $\hat{Y}_i = b_0 + b_1 D_i$ and D_i is one or zero, depending on whether the ith observation falls in group 1 or group 2. There is no way that the sum of the squared discrepancies or errors can be less if knowledge of values of D and a linear prediction rule are used. This sum of squared errors is generally called the *sum of squares residuals* in the ANOVAR summary table. Now think about the quantity called the *sum of squares within* in the ANOVA. The sum of squares within is also based on a least-squares criterion.

Suppose you are assigned the task of selecting *one* value for *each* of two groups of scores; each value must have the property that the sum of the squared differences between the value selected and the actual scores is minimum. The value you should select for each group is the mean of that group. Any value other than the group mean (in the case of each group) will yield a sum-of-squared-difference scores that is larger than the sum obtained by using the group mean.

Now if (1) the sum-of-squares residuals in a regression analysis is the minimum sum of squared differences between Y and $\hat{Y}$, where $\hat{Y}$ is based on information on the predictor variable D which indicates group membership and (2) if the sample means provide the minimum sum of squares within each group in a two-group ANOVA, what values do you think the regression equation will predict for each group when Y is regressed on the dummy variable D?

The mean of treatment group 1 is the predicted value when $D=1$, and the mean of the second treatment group is the predicted value when $D=0$; in other words,

$$\overline{Y}_1 = b_0 + b_1(1)$$
$$\overline{Y}_2 = b_0 + b_1(0)$$

Notice that these formulas do in fact yield the means for the data in Table 4.1, where it can be seen that the prediction equation is

$$\hat{Y} = 9 - 6(D)$$

Since the value of D for the first group is one, the predicted value is

$$\hat{Y} = 9 - 6(1) = 3$$

Hence $\overline{Y}_1 = 3$. Likewise, the predicted value associated with the second group which has a D value of zero is

$$\hat{Y}=9-6(0)=9$$

Hence $\bar{Y}_2=9$. Clearly, the regression of Y on D yields the same information as the ANOVA. The sum-of-squares residual in the ANOVAR is equivalent to the within-group sum of squares in the ANOVA; the sum-of-squares regression in the ANOVAR is equivalent to the sum of squares between groups in the ANOVA. Since the degrees of freedom are the same for these two analyses it follows that the F also must be the same. If the equivalence of these two analyses is understood, it is not difficult to grasp why the F test for the point-biserial correlation coefficient is also equivalent.

Recall from elementary correlation analysis that the point-biserial correlation is used to provide a measure of the relationship between a continuous variable and a dichotomous variable. Also recall that a squared correlation coefficient is called a *coefficient of determination*. In the analysis of the data in Table 4.1 the point-biserial correlation between the dependent variable (number of pain attacks, a continuous variable) and the dummy variable D (group membership, a dichotomous variable) is .736. The square of this value is approximately .54, which is the coefficient of determination. This is interpreted as the proportion of the variability in pain frequency that is accounted for on the basis of knowledge of group membership. This notion of the proportion of variability in Y, explained on the basis of information on group membership D, can be easily understood if the concept of prediction error is kept in mind.

Suppose we know nothing about which treatment group is associated with the 10 dependent variable scores in Table 4.1. Next, suppose that the list of 10 scores is lost but we know that the grand mean of the 10 lost scores is 6. If our task is to guess the score associated with each one of the 10 subjects, our best guess for each is the grand mean 6. This guess is the "best" in a least-squares sense. There is no *single* value that we can guess that will provide a smaller sum of squared differences between the value guessed (predicted) and the actual score. This sum of squared prediction errors is, of course, the total sum of squares in ANOVA and ANOVAR. In the example the total sum of squares is

$$\sum_{1}^{10}\left(Y_i-\bar{Y}..\right)^2=166$$

Hence if we predict that each of the 10 subjects has a score of 6 and then later locate the 10 actual obtained scores, we will discover that the total sum of squares of the obtained scores is the same as the sum of the squared prediction errors we have made by predicting the grand mean for each subject.

At this stage we might ask how much better our prediction would be if we had knowledge of the treatment to which each subject had been exposed. If the information concerning the treatment exposure is available, we simply include that information in a regression analysis in the form of the group membership dummy variable D.

If there is a difference between the two sample means, the sample slope b_1 will not be zero, and the value predicted by the regression equation will not be the grand mean; rather, the predicted value will be the mean of the treatment group to which an individual belongs. As long as the treatment means are different, the sum of the prediction errors or residuals will be smaller if the values predicted by the regression equation (i.e., the group means) rather than the grand mean are used.

This difference between the total sum of squares and the residual sum of squares in the regression analysis yields the regression sum of squares; that is,

$$\text{SStotal} - \text{SSres} = \text{SSreg}$$

Since $\text{SStotal} = \text{SSreg} + \text{SSres}$, it follows that the ratio $\text{SSreg}/\text{SStotal}$ is *the proportion of Y that is accounted for by group membership D*. This ratio, then, is interpreted in the same way as the squared point-biserial correlation coefficient; the two are equivalent:

$$\frac{\text{SSreg}}{\text{SStotal}} = r^2_{\text{point biserial}}$$

Alternatively, it turns out that

$$\left(1 - r^2_{\text{point biserial}}\right)\text{SS Total} = \text{SS}_{\text{residual}}$$

$$\left(r^2_{\text{point biserial}}\right)\text{SS Total} = \text{SS}_{\text{regression}}$$

Hence it can be seen that the ANOVA, the ANOVAR, and the point-biserial correlation F tests are all equivalent. It was also pointed out that the independent sample t test yields the same information as do the F tests mentioned earlier. Since most elementary statistics texts contain a proof that $t^2 = \text{ANOVA } F$ in the case of two groups, this correspondence is not pursued here.

ANOVA through Regression with Three or More Groups

If more than two groups are involved, the basic changes in the analysis procedure described earlier are that (1) there are more dummy variables, and (2) the correlation between the dummy variables and the dependent variable is a multiple correlation rather than a simple correlation. The number of dummy variables, regardless of the number of treatments, is $J-1$. Hence the number of dummy variables in a three-group problem is two. The data layout for a three-group analysis of variance through multiple regression is provided in Table 4.2 for the behavioral objectives study mentioned in Chapter 2.

Table 4.2   Data Layout for ANOVA through Multiple Regression Analysis (Behavioral Objectives Data)

D_1	D_2	Y	
1	0	15	
1	0	19	
1	0	21	
1	0	27	
1	0	35	Group 1
1	0	39	
1	0	23	
1	0	38	
1	0	33	
1	0	50	
0	1	20	
0	1	34	
0	1	28	
0	1	35	
0	1	42	
0	1	44	Group 2
0	1	46	
0	1	47	
0	1	40	
0	1	54	
0	0	14	
0	0	20	
0	0	30	
0	0	32	
0	0	34	Group 3
0	0	42	
0	0	40	
0	0	38	
0	0	54	
0	0	56	

The steps followed in arranging the data were:

Step 1.   All dependent variable scores were entered under the Y column heading.

Step 2.   The number of dummy variables required was computed $(3-1=2)$, and column headings D_1 and D_2 were entered next to the Y column heading.

Step 3.   A "one" was entered in column D_1 for each dependent variable score associated with the first treatment group; a "zero" was entered in this column for each dependent variable score not associated with the first treatment group.

Step 4. A "one" was entered in column D_2 for each dependent variable score associated with the second treatment group; a "zero" was entered in this column for each dependent variable score not associated with the second treatment group.

Notice that the first 10 rows in the D_1 column contain "ones." This is because the first 10 subjects belong to the first treatment group. Since the next 20 subjects do not belong to the first treatment group, they are each assigned a "zero." The first 10 rows in column D_2 are "zeros" because the first 10 subjects do not belong to the second treatment group. Subjects 11 through 20 are assigned "ones" in column D_2 because they do belong to the second treatment group. Hence the "ones" in column D_1 indicate which subjects belong to treatment 1. The "ones" in column D_2 indicate which subjects belong to treatment group 2. Obviously, if a subject has "zeros" in both columns D_1 and D_2, he belongs to group 3.

Each dummy variable is treated as a predictor or independent variable in a multiple regression analysis; that is, the dependent variable scores are regressed on the dummy variables. The output of the typical multiple regression program will contain, as a minimum, the multiple correlation coefficient (R) and the multiple regression equation. The analysis of variance can be carried out by employing the following general formula for the ANOVA F test:

$$\frac{R^2_{yd_1, d_2, \ldots, d_{J-1}}/m}{\left(1 - R^2_{yd_1, d_2, \ldots, d_{J-1}}\right)/N - m - 1} = F$$

$$\text{Critical value} = F_{(\alpha, m, N - m - 1)}$$

where

$R_{yd_1, d_2, \ldots, d_{J-1}}$ = multiple correlation coefficient (more correctly, the multiple point-biserial correlation coefficient) between the dummy variables and the dependent variable

m = number of dummy variables

N = total number of subjects

The rationale for this formula is quite straightforward. Recall that the basic sum-of-squares partition in the analysis of variance is between group + within group = total. The squared multiple R provides the proportion of the total variability in Y that is accounted for by the dummy variables. This turns out to be equivalent to the proportion of the total variability that is between-group variability. It follows that $1 - R^2$ must provide the proportion of the total variability that is within-group or error variability. The F ratio then becomes

$$\frac{R^2/\text{between-group degrees of freedom}}{(1 - R^2)/\text{within-group degrees of freedom}} = \frac{R^2/J - 1}{(1 - R^2)/N - J}$$

$$= \frac{R^2/m}{(1 - R^2)/N - m - 1}$$

$$= F$$

If we multiply the total sum of squares (SST) by R^2 and $1 - R^2$, we obtain the between- and within-group sum of squares, respectively. If these sums of squares are computed, the equivalence of the conventionally computed ANOVA and ANOVA through multiple regression may become more obvious. The ANOVA summary using this approach is tabulated as follows.

Source	SS	df	MS	F
Between	$R^2(\text{SST})$	$J-1$	$\dfrac{R^2(\text{SST})}{J-1}$	$\dfrac{\text{MS}_b}{\text{MS}_w}$
Within	$(1-R^2)\text{SST}$	$N-J$	$\dfrac{(1-R^2)\text{SST}}{N-J}$	
Total	$(1)\text{SST}$	$N-1$		

In addition to the multiple correlation coefficient, the multiple regression prediction equation is generally provided by the computer output. This equation is generally written as

$$\hat{Y} = b_0 + b_1 X_1 + b_2 X_2 + \cdots + b_m X_m$$

where
$\hat{Y}$ = predicted score,
b_0 = multiple regression intercept,
$b_1 - b_m$ = partial regression coefficients associated with $X_1 - X_m$
$X_1 - X_m$ = predictor or independent variable scores
Since we have labeled the dummy variable columns with Ds rather than Xs, the equation can be written as

$$\hat{Y} = b_0 + b_1 d_1 + b_2 d_2 + \cdots + b_{J-1} d_{J-1}$$

where
$\hat{Y}$ = predicted score (which turns out to be mean of treatment identified by dummy variables),
b_0 = multiple regression intercept,
$b_1 - b_{J-1}$ = partial regression coefficients associated with first through $J-1$th dummy variables
$d_1 - d_{J-1} = J - 1$ dummy variables required to identify group membership
The information contained in the multiple regression equation can be employed to compute the sample means as follows:

$$\overline{Y}_1 = b_0 + b_1$$
$$\overline{Y}_2 = b_0 + b_2$$
$$\cdot \quad \cdot \quad \cdot$$
$$\cdot \quad \cdot \quad \cdot$$
$$\cdot \quad \cdot \quad \cdot$$
$$\overline{Y}_{J-1} = b_0 + b_{J-1}$$
$$\overline{Y}_J = b_0$$

The rationale for this multiple regression approach leading to the same results in multiple-group experiments as are obtained with a conventional analysis of variance is the same as was described previously for two-group experiments. Recall that in the two-group case the dependent variable is regressed on one dummy variable that indicates membership in one of the two groups. The residual sum of squares associated with this simple regression is equivalent to the within-group sum of squares. Likewise, when three or more groups are involved, the dependent variable is regressed on the collection of dummy variables that is required to indicate group membership.

Keep in mind that the parameter estimates in the multiple regression equation (i.e., the intercept and the partial regression coefficients) are computed in such a way that the sum of the squared prediction errors is a minimum; that is,

$$\sum_{1}^{N}(Y_i - \hat{Y}_i)^2 \text{ is a minimum}$$

where Y_i is the observed or actual dependent variable score for the ith subject, $\hat{Y}_i$ is the score predicted for the ith subject using the multiple regression equation, and N is the total number of subjects in the experiment.

Just as the regression of Y on a single dummy variable in the two-group case yields a prediction equation that predicts the mean of the group to which a subject belongs, the multiple regression of Y on a collection of dummy variables yields a multiple regression equation that predicts the mean of the group to which a subject belongs. The only difference between the two-group case and the case of more than two groups is that the former involves one dummy variable and simple regression where the latter involves more than one dummy variable and multiple regression. Since the least-squares criterion is associated with both simple and multiple regression, it makes sense that group means are predicted regardless of the number of groups in an experiment. This is true because there is no single value other than the mean associated with a given group that will meet the least-squares criterion. Another parallel between the case where $J = 2$ and the case where J is greater than two can be seen in the correlation values associated with the simple and multiple regressions.

Just as r^2 in the two-group case yields the proportion of the total variability in Y accounted for by knowledge of group membership, the coefficient of multiple determination (R^2) yields the proportion of the total variability in Y accounted for by knowledge of group membership in the case of more than two groups. Similarly, the product

$$r^2(\text{SST}) = \text{between-group SS for 2-group experiments}$$

and the product

$$R^2(\text{SST}) = \text{between or among group SS for experiments with} > 2 \text{ groups}$$

The regression analysis on the data presented in Table 4.2 results in the following:

$$R = .32583$$
$$R^2 = .10617$$
$$b_0 = 36$$
$$b_1 = -6$$
$$b_2 = 3$$

The regression intercept (b_0) and the partial regression coefficients (b_1 and b_2) provide the constants required in the computation of the predicted scores. As was just mentioned, the predicted score for any subject will be equal to the mean of the group to which the subject belongs. Hence the predicted score (or group mean) for any subject is obtained by entering the multiple regression equation

$$\hat{Y} = b_0 + b_1 d_1 + b_2 d_2 = 36 - 6d_1 + 3d_2$$

Since each subject in the experiment has a d_1 score and a d_2 score, we simply enter the equation with these scores to obtain the predicted score for that subject (or to obtain the mean of the group to which the subject belongs). For example, the first subject in the first group has a d_1 score of one and a d_2 score of zero. This means that the predicted score for this subject, or any subject in the first group, is

$$b_0 + b_1(1) + b_2(0) = 36 - 6(1) + 3(0) = 30$$

This reduces to

$$b_0 + b_1 = 36 - 6 = 30$$

Hence the mean of the first group is 30.

Since subjects in the second group have $d_1 = 0$ and $d_2 = 1$, the predicted score is

$$b_0 + b_1(0) + b_2(1) = 36 - 6(0) + 3(1) = 39$$

This reduces to

$$b_0 + b_2 = 36 + 3 = 39$$

The mean of the second group is 39.

The dummy scores for subjects in group 3 are $d_1 = 0$ and $d_2 = 0$. The predicted score is

$$b_0 + b_1(0) + b_2(0) = 36 - 6(0) + 3(0) = 36.$$

This reduces to

$$b_0 = 36$$

Thus it can be seen that the three group means are

$$\overline{Y}_1 = b_0 + b_1$$

$$\overline{Y}_2 = b_0 + b_2$$

$$\overline{Y}_3 = b_0$$

Table 4.3   Actual, Predicted, Deviation, and Squared Deviation Scores for Behavioral Objectives Data

Subject	Actual Y	Predicted $\hat{Y}$	Deviation $Y - \hat{Y}$	(Deviation)^2 $(Y - \hat{Y})^2$		
1	15	30	−15	225		
2	19	30	−11	121		
3	21	30	− 9	81		
4	27	30	− 3	9		
5	35	30	5	25	1064	
6	39	30	9	81		
7	23	30	− 7	49		
8	38	30	8	64		
9	33	30	3	9		
10	50	30	20	400		
11	20	39	−19	361		
12	34	39	− 5	25		
13	28	39	−11	121		
14	35	39	− 4	16		
15	42	39	3	9	896	3536
16	44	39	5	25		
17	46	39	7	49		
18	47	39	8	64		
19	40	39	1	1		
20	54	39	15	225		
21	14	36	−22	484		
22	20	36	−16	256		
23	30	36	− 6	36		
24	32	36	− 4	16		
25	34	36	− 2	4	1576	
26	42	36	6	36		
27	40	36	4	16		
28	38	36	2	4		
29	54	36	18	324		
30	56	36	20	400		

The predicted scores ($\hat{Y}$), actual scores (Y), prediction errors ($Y - \hat{Y}$), and squared prediction errors ($Y - \hat{Y})^2$ for all subjects are presented in Table 4.3. Notice that the sum of the squared deviation scores is exactly the same as the within group sum of squares that was computed using a conventional analysis of variance procedure in Chapter 2.

The analysis is summarized as follows:

Source	SS	df	MS	F
Between-group	(0.10617)3956 = 420.00	2	210.00	1.60
Within-group	(1 − 0.10617)3956 = 3536.00	27	130.96	
Total	3956.00			

or

$$\frac{0.10617/2}{(1-0.10617)/27} = 1.60 = F$$

4.3   ANALYSIS OF COVARIANCE THROUGH LINEAR REGRESSION

Once the regression approach to ANOVA problems is mastered, the analysis of covariance can be easily conceptualized as a slight extension of the same approach. As the models in Table 4.4 indicate, ANCOVA differs from ANOVA only in that it contains a term for the regression of Y on the covariate (X). This means that we can carry out the analysis of covariance by using a regression program by regressing the Y scores on both the dummy variable scores (necessary to designate group membership or treatments) and the covariate scores. The coefficient of multiple determination ($R^2_{yd_1,\ldots,d_{J-1},x}$) resulting from regressing Y on dummy variables and the covariate yields the proportion of the total variability accounted for by *both* group membership and covariate scores. Then Y is regressed on the covariate *alone* to obtain the proportion of the total variability accounted for by X (i.e., r^2_{yx}). The difference between the two coefficients of determination indicates the unique contribution of the dummy variables to the first computed coefficient.

This procedure has been applied in the analysis of the behavioral objectives data described previously. The results appear in the right-hand column of Table 4.4 in two forms. The F values obtained under the two forms of the analysis presented in this table are the same as those obtained using the traditional solution (see Table 3.2). Table 4.4 also contains the data matrices and summaries for the ANOVA on Y and the ANOVAR. The reader should carefully compare the similarities and differences between the data matrices and computations associated with the three models.

The Y columns are, of course, the same under the three models since the purpose of this example is to illustrate how a given set of dependent variable scores is analyzed under ANOVA, ANOVAR, and ANCOVA. The first two

Table 4.4 ANOVA, ANOVAR, and ANCOVA as General Linear Models

			Model					
ANOVA $Y_{ij}=\mu+\alpha_j+\epsilon_{ij}$			ANOVAR $Y_i=\beta_0+\beta_1\bar{X}_i+\epsilon_i$		ANCOVA $Y_{ij}=\mu+\alpha_j+\beta(X_{ij}-\bar{X}_{..})+\epsilon_{ij}$			
D_1	D_2	Y	X	Y	D_1	D_2	X	Y
			Data Layout for Example Problem					
1	0	15	29	15	1	0	29	15
1	0	19	49	19	1	0	49	19
1	0	21	48	21	1	0	48	21
1	0	27	35	27	1	0	35	27
1	0	35	53	35	1	0	53	35
1	0	39	47	39	1	0	47	39
1	0	23	46	23	1	0	46	23
1	0	38	74	38	1	0	74	38
1	0	33	72	33	1	0	72	33
0	1	50	67	50	0	1	67	50
0	1	20	22	20	0	1	22	20
0	1	34	24	34	0	1	24	34
0	1	28	49	28	0	1	49	28
0	1	35	46	35	0	1	46	35
0	1	42	52	42	0	1	52	42
0	1	44	43	44	0	1	43	44
0	1	46	64	46	0	1	64	46
0	1	47	61	47	0	1	61	47
0	0	40	55	40	0	0	55	40
0	0	54	54	54	0	0	54	54
0	0	14	33	14	0	0	33	14
0	0	20	45	20	0	0	45	20
0	0	30	35	30	0	0	35	30
0	0	32	39	32	0	0	39	32
0	0	34	36	34	0	0	36	34
0	0	42	48	42	0	0	48	42
0	0	40	63	40	0	0	63	40
0	0	38	57	38	0	0	57	38
0	0	54	56	54	0	0	56	54
0	0	56	78	56	0	0	78	56

ANOVA

Source	SS	df	MS	F
Between	$=(R^2_{yd_1,d_2})SST$ $=(0.10165)3956$ $=420.00$	2	210.00	1.60
Within	$=(1-R^2_{yd_1,d_2})SST$ $=(1-0.10165)3956$ $=3536.00$	27	130.96	
Total	$=3956.00$	29		

$$\frac{R^2_{yd_1,d_2}/2}{1-R^2_{yd_1,d_2}/27} = \frac{0.0531}{0.0331} = 1.60 = F$$

ANOVAR

Source	SS	df	MS	F
Regression	$=(r^2_{yx})SST$ $=(0.396195)3956$ $=1567.36$	1	1567.36	18.37
Residuals	$=(1-r^2_{yx})SST$ $=(1-0.396195)3956$ $=2388.64$	28	85.31	
Total	$=3956.00$	29		

Abbreviated Summary

$$\frac{r^2_{yx}/1}{1-r^2_{yx}/28} = \frac{0.3962}{0.02156} = 18.37 = F$$

ANCOVA

Source	SS	df	MS	F
AT	$=(R^2_{yd_1,d_2.x}-r^2_{yx})SST$ $=(0.575164-.396198)3956$ $=707.99$	2	354.00	5.48
Res_w	$=(1-R^2_{yd_1,d_2.x})SST$ $=(1-0.575164)3956$ $=1680.65$	26	64.64	
Res_t	$=(1-r^2_{yx})SST$ $=(0.603802)3956$ $=2388.64$	28		

$$\frac{R^2_{yd_1,d_2.x}-r^2_{yx}/J-1}{1-R^2_{yd_1,d_2.x}/N-J-1} = \frac{0.089483}{0.016340} = 5.48 = F$$

columns under the ANOVA are the dummy variables, which have already been explained (Section 4.2). Regressing Y on the dummy variables results in the ANOVA on Y. This ANOVA is not of particular interest here, but the dummy variables are. Notice that these dummy variables again appear under the ANCOVA model. Also notice that the X column under the ANOVAR is identical to the X column under the ANCOVA model.

The regression approach to ANCOVA on these data proceeds as follows:

1. Y is regressed on X and r_{yx}^2 is obtained. In this example

$$r_{yx}^2 = .396198$$

2. Y is regressed on the two dummy variables and X.

$$R_{yd_1, d_2, x}^2 = .575164$$

3. The value r_{yx}^2 is subtracted from $R_{yd_1, d_2, x}^2$. Since r_{yx}^2 is the proportion of the total variability in Y that is accounted for by the covariate and $R_{yd_1, d_2, x}^2$ is the proportion accounted for by both the covariate and the independent variable (i.e., the dummy variables), the difference between these two coefficients of determination must reflect the proportion of the total variability that is uniquely accounted for by the independent variable. Hence $0.575164 - 0.396198 = 0.178966$, which is the AT (i.e., adjusted treatment) effect expressed as a proportion. The sum of squares for adjusted treatments (SSAT) can be obtained by multiplying the difference $(R_{yd_1, d_2, x}^2 - r_{yx}^2)$ times the total sum of squares (SST). In this case $(0.178966)3956 = 707.99$.

4. The proportion of the total variability in Y that is not explained by the independent variable or the covariate is $1 - R_{yd_1, d_2, x}^2$. We can obtain the within-group residual sum of squares (SSres$_w$) by multiplying $(1 - R_{yd_1, d_2, x}^2)$ times the total sum of squares. Hence $(0.424836)3956 = 1680.65$.

5. The sum of the AT sum of squares and the within-group residual sum of squares is the total residual sum of squares. As a computational check, the SSres$_t$ can also be computed by multiplying $1 - r_{yx}^2$ times the total sum of squares (SST). In the example $707.99 + 1680.65 = 2388.64$ and $(0.603802)3956 = 2388.64$.

As is the case with ANOVA and ANOVAR, results of ANCOVA are generally presented in conventional summary tables even though regression procedures are employed. It is possible, of course, to compute the F ratio by using only R^2 coefficients and degrees of freedom as shown at the bottom of Table 4.4. Since some multiple regression programs do not provide sums of squares, the R^2 form of summary is sometimes the only simple choice. Thus the general form for the summary of the simple (one factor–one covariate)

ANCOVA is:

Source	SS	df	MS	F
Adjusted treatment	$(R^2_{yd_1,\ldots,d_{J-1},x} - r^2_{yx})$SST	$J-1$	MSAT	MSAT/MSres$_w$
Within residual	$(1 - R^2_{yd_1,\ldots,d_{J-1},x})$SST	$N-J-1$	MSres$_w$	
Total residual	$(1 - r^2_{yx})$SST	$N-2$		

or simply

$$\frac{\left(R^2_{yd_1,\ldots,d_{J-1},x} - r^2_{yx}\right)/J-1}{\left(1 - R^2_{yd_1,\ldots,d_{J-1},x}\right)/N-J-1} = F$$

where

$R^2_{yd_1,\ldots,d_{J-1},x}$ = coefficient of multiple determination obtained by regressing the dependent variable on all group membership dummy variables and covariate

r^2_{yx} = coefficient of determination obtained by regressing the dependent variable on covariate

J = number of groups

N = total number of subjects and obtained F is evaluated with $F_{(\alpha, J-1, N-J-1)}$

4.4 COMPUTATION OF ADJUSTED MEANS

The equation associated with the regression of Y on the dummy variables and the covariate provides the information required for the computation of the adjusted means. The adjusted mean for any group is $b_0 + b(d_1) + \cdots + b(d_{J-1}) + b(\bar{X}..)$.

For the example data, the regression of Y on D_1, D_2 and X results in the following regression intercept and weights:

$$b_0 = 8.0442099$$

$$b_1 = -7.71158$$

$$b_2 = 4.14105$$

$$b_3 = 0.570526$$

The grand mean on X is 49.33. Hence $\bar{Y}_{j\,adj} = 8.0442099 - 7.71158(d_1) + 4.14105(d_2) + 0.570526(\bar{X}..)$. The adjusted means for the three treatment groups are

$$\bar{Y}_{I\,adj} = 8.0442099 - 7.71158(1) + 4.14105(0) + 0.570526(49.33) = 28.48$$

$$\bar{Y}_{II\,adj} = 8.0442099 - 7.71158(0) + 4.14105(1) + 0.570526(49.33) = 40.33$$

$$\bar{Y}_{III\,adj} = 8.0442099 - 7.71158(0) + 4.14105(0) + 0.570526(49.33) = 36.19$$

These adjusted means are, of course, the same as those presented in Table 3.2. Similarly, we find the ratio

$$\frac{R^2_{yd_1,\ldots,d_{J-1},x} - r^2_{yx}}{1 - r^2_{yx}} = \frac{0.178966}{0.603802} = 0.30$$

is the same as the ratio of SSAT/SSres$_t$ in the same table.

4.5 SIMILARITY OF ANCOVA TO PART AND PARTIAL CORRELATION METHODS

The reader familiar with multiple correlation analysis will recognize the similarities among ANCOVA, multiple, part, and partial correlation procedures. The ANCOVA procedure described in the previous section is the same as the procedure for testing the difference between simple and multiple correlation coefficients based on the same sample. It follows, then, that ANCOVA can be conceptualized as a variety of part or partial correlation. That is, for a two-group case, the ANCOVA F test is the same as the F test of the significance of the partial correlation $r_{yx_2 \cdot x_1}$. If X_2 is viewed as the dummy variable, we can rewrite the partial correlation between the dependent variable and the dummy variable, holding constant the covariate (X) as $r_{yd \cdot x_1}$. An even more straightforward analogy is drawn between ANCOVA and the part correlation $r_{y(d \cdot x)}$ as shown in Figure 4.1.

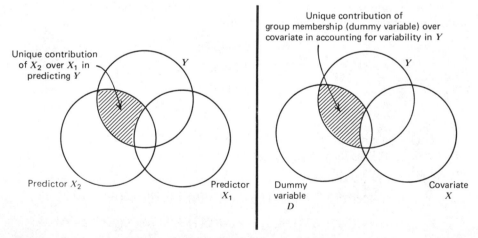

Fig. 4.1 Comparison of partitioning for conventional part correlation and analysis of covariance through multiple regression.

Hence in the two-group situation (1) the conventional ANCOVA F test, (2) the F test of significance of the difference between $R_{yd,x}$ and r_{yx}, (3) the F test of the significance of the difference between the partial $r_{yd\cdot x}$ and zero, and (4) the F test of the significance of the difference between the part $r_{y(d\cdot x)}$ and zero, are all equivalent.

When more than two groups are involved, the number of dummy variables will be greater than one, and the following F tests are equivalent:

1. ANCOVA F.
2. F test of difference between multiple $R_{yd_1,\ldots,d_{J-1},x}$ and r_{yx}.
3. F test of significance of multiple partial correlation $R_{yd_1,\ldots,d_{J-1}\cdot x}$.
4. F test of significance of multiple part correlation $R_{y(d_1,\ldots,d_{J-1}\cdot x)}$.

Since all these tests are equivalent it seems reasonable to learn only one. It will be seen in Chapter 8 that the F test of the difference between the multiple correlation coefficient $R_{yd_1,\ldots,d_{J-1},x}$ and the simple correlation coefficient r_{yx} that has been employed in this chapter, that is.

$$\frac{\left(R^2_{yd_1,\ldots,d_{J-1},x} - r^2_{yx}\right)/J - 1}{\left(1 - R^2_{yd_1,\ldots,d_{J-1},x}\right)/N - J - 1} = F$$

is easily generalized to the case in which multiple covariates are employed.

4.6 HOMOGENEITY OF REGRESSION TEST THROUGH GENERAL LINEAR REGRESSION

The design matrix of Table 4.4 allows us to compute ANOVA on X, ANOVA on Y, ANOVAR, and ANCOVA through the general linear regression approach. The homogeneity of regression analysis is not included in this list. This analysis can be accomplished through a simple extension of the ANCOVA procedure just described. An expanded version of the ANCOVA design matrix of Table 4.4 is presented in Table 4.5. The additional columns are labeled D_1X and D_2X. These columns are simply the products of the values of the associated columns. The first subject in the matrix has a D score of 1 and an X score of 29. Hence his D_1X score $= 1(29) = 29$. His D_2 score is zero; therefore, his D_2X score is zero. Design matrices for simple ANCOVA (one factor and one covariate) problems containing up to seven groups are provided in Appendix B.

After the design matrix is filled in, Y is regressed on all other columns. The coefficient of multiple determination $R^2_{yd_1,\ldots,d_{J-1},x,d_1x,\ldots,d_{J-1}x}$ resulting from this regression gives us the proportion of the total variability in Y that is explained by the dummy variables, the covariate, and the interaction of dummy variables with the covariate. It turns out that the interaction of

Table 4.5 Design Matrix for ANCOVA and Homogeneity of Regression Test

D_1	D_2	X	D_1X	D_2X	Y
1	0	29	29	0	15
1	0	49	49	0	19
1	0	48	48	0	21
1	0	35	35	0	27
1	0	53	53	0	35
1	0	47	47	0	39
1	0	46	46	0	23
1	0	74	74	0	38
1	0	72	72	0	33
1	0	67	67	0	50
0	1	22	0	22	20
0	1	24	0	24	34
0	1	49	0	49	28
0	1	46	0	46	35
0	1	52	0	52	42
0	1	43	0	43	44
0	1	64	0	64	46
0	1	61	0	61	47
0	1	55	0	55	40
0	1	54	0	54	54
0	0	33	0	0	14
0	0	45	0	0	20
0	0	35	0	0	30
0	0	39	0	0	32
0	0	36	0	0	34
0	0	48	0	0	42
0	0	63	0	0	40
0	0	57	0	0	38
0	0	56	0	0	54
0	0	78	0	0	56

dummy variables with the covariate is the heterogeneity of the regression slopes. Thus the difference between (1) the R^2 based on dummy variables, the covariate, and the interaction of dummy variables with the covariate and (2) the R^2 based on only dummy variables and the covariate must reflect the extent to which there are heterogeneous regression slopes.

The homogeneity of regression F test is

$$\frac{\left(R^2_{yd_1,\ldots,d_{J-1},x,d_1x,\ldots,d_{J-1}x} - R^2_{yd_1,\ldots,d_{J-1},x}\right)/J-1}{\left(1 - R^2_{yd_1,\ldots,d_{J-1},x,d_1x,\ldots,d_{J-1}x}\right)/N-[2J]} = F$$

where $R^2_{yd_1,\ldots,d_{J-1},x,d_1x,\ldots,d_{J-1}x}$ is the coefficient of multiple determination obtained by regressing the dependent variable on the group membership dummy variables, the covariate, and the products of group membership dummy variables times the covariate. Henceforth this coefficient is written $R^2_{yD,X,DX}$. The coefficient of multiple determination $R^2_{yd_1,\ldots,d_{J-1},x}$ is obtained by regressing the dependent variable on the group membership dummy variables and the covariate. Henceforth this coefficient is written $R^2_{yD,X}$.

The total number of subjects is N, the number of groups is J, and the obtained F is evaluated with $F_{(\alpha, J-1, N-[2J])}$. For the example data, the required R^2 values are $R^2_{yD,X,DX} = R^2_{yd_1,d_2,x,d_1x,d_2x} = 0.588800$ and $R^2_{yD,X} = R^2_{yd_1,d_2,x} = 0.575164$. The difference is 0.01364, and the test is

$$\frac{0.01364/2}{0.4112/24} = \frac{0.00682}{0.01713} = 0.398 = F$$

The obtained F is less than the critical value of $F_{(.05, 2, 24)}$, and the null hypothesis is retained; the regression slopes are considered homogeneous. Notice that this result is in perfect agreement with the result obtained in Section 3.5 that was based on the conventional computation procedure.

4.7 SUMMARY

The analysis of covariance, the analysis of variance, and simple regression analysis are all special cases of the general linear model. Multiple regression computer programs can be conveniently employed to carry out the analysis of covariance and the homogeneity of regression slopes tests. It is necessary to construct dummy variables if this multiple regression approach is to be utilized.

Dummy variables are employed to identify the group to which an individual belongs. Since group membership is the independent variable in an analysis of variance problem, it is possible to carry out the ANOVA F test by regressing the dependent variable on the dummy variables. The conventional F test associated with such a regression analysis is equivalent to a conventional ANOVA F test. Likewise, with covariance analysis, the regression approach can be followed.

The dependent variable is regressed on both the dummy variables and the covariate as the first step in ANCOVA. One result of this regression is an estimate of the proportion of the variability that is accounted for by group membership and the covariate combined. The next step is to regress the dependent variable on the covariate alone. An estimate of the proportion of the variability in the dependent variable that is accounted for by just the covariate is obtained from this regression. The difference between the proportion found in the first regression (on the dummy variables *and* the covariate) and the proportion found in the second regression (on the covariate only) yields the proportion of the total variability that can be attributed to the treatment groups independent of the covariate. The F test on this difference in proportions is equivalent to the ANCOVA F test on adjusted means.

The homogeneity of regression slopes test can also be computed through a regression approach. This involves the regression of the dependent variable on the dummy variables, the covariate, and the products of the dummy variables and the covariate. The difference between the proportion of the total variability that is accounted for by this regression and the proportion that is accounted for by regressing on the dummy variables and the covariate yields the proportion accounted for by covariate–treatment interaction. This turns out to be the proportion due to heterogeneity of regression slopes. Hence the F test on this proportion is equivalent to the conventional homogeneity of regression F test.

CHAPTER FIVE

Multiple Comparison Tests and Confidence Intervals

5.1 INTRODUCTION

If only two groups are involved in an analysis of covariance problem, the ANCOVA F test, the adjusted means, the confidence interval, and the homogeneity of regression slopes F test are generally considered the essentials of a complete analysis. Investigations dealing with three or more groups, however, generally require additional analysis.

Recall the nature of the null hypothesis associated with the analysis of covariance. In the two-group case the hypothesis H_0: $\mu_{1\,adj} = \mu_{2\,adj}$ is rejected if the obtained F equals or exceeds the critical value of F. We conclude that the adjusted sample means were obtained from populations having different adjusted means, that is, $\mu_{1\,adj} \neq \mu_{2\,adj}$. More specifically, we conclude that $\mu_{1\,adj} > \mu_{2\,adj}$ or that $\mu_{1\,adj} < \mu_{2\,adj}$, depending on the relative size of the sample values $\overline{Y}_{1\,adj}$ and $\overline{Y}_{2\,adj}$. When more than two groups are involved, the number of possible alternative hypotheses is great.

Suppose three groups are involved and that the ANCOVA F is significant and hence the null hypothesis H_0: $\mu_{1\,adj} = \mu_{2\,adj} = \mu_{3\,adj}$ is rejected. The alternative hypothesis that the adjusted population means are not all equal is accepted. The F test does not, however, indicate where the differences among the adjusted population means lie. It only indicates that there is at least one difference among the three adjusted population means. Any one of the following six conditions could exist:

$$\mu_{1\,adj} > \mu_{2\,adj} > \mu_{3\,adj}$$
$$\mu_{1\,adj} < \mu_{2\,adj} < \mu_{3\,adj}$$
$$\mu_{1\,adj} > \mu_{3\,adj} > \mu_{2\,adj}$$
$$\mu_{1\,adj} < \mu_{3\,adj} < \mu_{2\,adj}$$
$$\mu_{2\,adj} > \mu_{1\,adj} > \mu_{3\,adj}$$
$$\mu_{2\,adj} < \mu_{1\,adj} < \mu_{3\,adj}$$

This is not a complete list of alternative hypotheses because only those situations in which all populations differ have been included. Two of the three populations may have the same adjusted mean that differs from the adjusted mean of the third population, for example, $\mu_{1\ adj} = \mu_{2\ adj} > \mu_{3\ adj}$.

Clearly, there are many possible hypotheses that can be considered when the overall null hypothesis is rejected. Some type of procedure is needed to differentiate among these various alternative hypotheses. Several different "multiple comparison" procedures are available for this purpose. Four different multiple comparison procedures are presented in this chapter, the choice of which should be dictated by the type of comparisons of greatest interest to the researcher. Before examining the various comparison procedures, two methods of writing contrasts or comparisons are mentioned.

5.2 TWO METHODS OF EXPRESSING PAIRWISE AND COMPLEX COMPARISONS

Two basic types of comparison in a single-factor experiment are pairwise (or simple) and complex. Only two treatments are considered in a pairwise comparison, whereas three or more treatments are considered in complex comparisons. An example should clarify the distinction.

Suppose we have run a four-group experiment and have obtained the following adjusted means:

$$\overline{Y}_{1\ adj} = 4$$

$$\overline{Y}_{2\ adj} = 8$$

$$\overline{Y}_{3\ adj} = 7$$

$$\overline{Y}_{4\ adj} = 12$$

If we make all possible pairwise or simple comparisons, we simply look at the differences between all pairs of adjusted means in the experiment. The number of pairwise comparisons in an experiment with J groups is $J(J-1)/2$. Hence our four-group experiment has six possible pairwise comparisons. A simple way to present these comparisons is as follows:

Comparison Groups	Mean Difference Between Adjusted Sample Means	Sample Contrast $\hat{\psi}$
1,2	$4-8$	-4
1,3	$4-7$	-3
1,4	$4-12$	-8
2,3	$8-7$	1
2,4	$8-12$	-4
3,4	$7-12$	-5

Each difference between a pair of adjusted sample means is a sample contrast. Such a difference is often given the label $\hat{\psi}$; the hat indicates that the difference is an estimated population difference that is based on sample data. The population contrast that is estimated by $\hat{\psi}$ is ψ. Since the adjusted sample means for the first two treatment groups are 4 and 8, respectively, the sample contrast is $4-8=-4=\hat{\psi}$.

This approach is very straightforward since each $\hat{\psi}$ is simply the difference between a pair of adjusted means. There are certain multiple comparison procedures, however, that require a more complicated method of writing the contrasts. Rather than simply computing the difference between adjusted means, a linear expression can be formed by using so-called contrast coefficients.

If the linear expression approach is used to form the sample contrast between the first two treatments in the example just described, the contrast coefficients associated with these two treatments are 1 and -1, respectively. The linear expression is

$$1(4)-1(8)+0(7)+0(12)=-4=\hat{\psi}$$

Notice that the third and fourth treatments do not affect this contrast because the contrast coefficients associated with these treatments are zero. Two other aspects of this expression should be noted. First, the contrast is the same whether it is written $4-8=-4$ or $1(4)-1(8)+0(7)+0(12)=-4$. Second, the sum of the contrast coefficients is zero. That is, the sum of 1, -1, 0, and 0 is zero. When the linear expression is formed, the sum of the contrast coefficients must be zero. The general form of the linear expression for a sample contrast is

$$c_1\left(\overline{Y}_{1\ \text{adj}}\right)+c_2\left(\overline{Y}_{2\ \text{adj}}\right)+\cdots+c_J\left(\overline{Y}_{J\ \text{adj}}\right)=\hat{\psi}$$

Correspondingly, the linear expression for a population contrast is

$$c_1(\mu_{1\ \text{adj}})+c_2(\mu_{2\ \text{adj}})+\cdots+c_J(\mu_{J\ \text{adj}})=\psi$$

In both cases the c values are the contrast coefficients.

When the linear expression approach is employed to obtain the six possible pairwise contrasts previously described, the contrasts are written

$$1(4)-1(8)+0(7)+0(12)=-4$$
$$1(4)+0(8)-1(7)+0(12)=-3$$
$$1(4)+0(8)+0(7)-1(12)=-8$$
$$0(4)+1(8)-1(7)+0(12)=1$$
$$0(4)+1(8)+0(7)-1(12)=-4$$
$$0(4)+0(8)+1(7)-1(12)=-5$$

Note that the sum of the contrast coefficients is zero in each row.

The four-group example just described yielded six pairwise comparisons. Each pairwise comparison involved only two adjusted sample means. Since the contrasts or comparisons never included more than two adjusted means, it is appropriate to describe such pairwise contrasts as simple contrasts. Complex contrasts involve three or more adjusted means.

Suppose there is good reason to compare the average of the first three treatments with the mean of the fourth treatment in the data described previously. That is, the investigator may want to test the hypothesis described as follows:

$$H_0: \frac{\mu_{1\,adj} + \mu_{2\,adj} + \mu_{3\,adj}}{3} - \mu_{4\,adj} = 0 = \psi$$

If this contrast is written as a linear expression, we have

$$\tfrac{1}{3}(\mu_{1\,adj}) + \tfrac{1}{3}(\mu_{2\,adj}) + \tfrac{1}{3}(\mu_{3\,adj}) - 1(\mu_{4\,adj}) = 0 = \psi$$

The corresponding sample contrast is

$$\frac{\overline{Y}_{1\,adj} + \overline{Y}_{2\,adj} + \overline{Y}_{3\,adj}}{3} - \overline{Y}_{4\,adj} = \hat{\psi}$$

or, using the linear expression,

$$\tfrac{1}{3}(\overline{Y}_{1\,adj}) + \tfrac{1}{3}(\overline{Y}_{2\,adj}) + \tfrac{1}{3}(\overline{Y}_{3\,adj}) - 1(\overline{Y}_{4\,adj}) = \hat{\psi}$$

On substitution of the previously described sample values, the sample contrast is

$$\frac{4+8+7}{3} - 12 = -5.67 = \hat{\psi}$$

or

$$\tfrac{1}{3}(4) + \tfrac{1}{3}(8) + \tfrac{1}{3}(7) - 1(12) = -5.67 = \hat{\psi}$$

There are many other complex comparisons that may be of interest. For example, the average of the first and second treatments minus the average of the third and fourth treatments:

$$\frac{4+8}{2} - \frac{7+12}{2} \quad \text{or} \quad \tfrac{1}{2}(4) + \tfrac{1}{2}(8) - \tfrac{1}{2}(7) - \tfrac{1}{2}(12) = -3.5 = \hat{\psi}$$

Another contrast that might be of interest is the difference between the average of the first and second treatments and the third treatment. The linear expression for this sample contrast is

$$\tfrac{1}{2}(4) + \tfrac{1}{2}(8) - 1(7) + 0(12) = -1 = \hat{\psi}$$

The sum of the contrast coefficients in this expression, as with all other examples, is zero. The value of $\hat{\psi}$ is interpreted as an unbiased estimate of the corresponding population contrast ψ.

In most cases pairwise comparisons are of greatest interest because they are easily interpreted and meaningful. Complex comparisons, on the other hand, are frequently difficult to interpret in a straightforward manner. This is because, in many cases, it simply does not make sense to average the effects of two or more distinctly different treatments. There are situations, however, where a common characteristic runs through a collection of treatments; complex comparisons may be useful in this case. If, for example, a study is undertaken to evaluate the effects of (1) marihuana, (2) hashish, (3) tetra-hydrocannabinol, and (4) lysergic acid diethylamide on visual perception, it might be reasonable to compare the average of the first three treatments with the fourth treatment because the first three are all cannabis derivatives.

5.3   CONCEPTS OF ERROR RATE

When an experiment involves only two groups, there is only one comparison. The comparison is the difference between the two means in the case of the analysis of variance; with the analysis of covariance the comparison is the difference between the two adjusted means. The probability of making a Type I error (rejecting a true null hypothesis) is, of course, known as *alpha* (α). Since two-group experiments have only one comparison it is obvious that k two-group experiments will have k comparisons. Hence if the null hypothesis is true in 100 independent two-group experiments and a significance test using $\alpha = .05$ is performed on each experiment, it should be expected that five of the comparisons will be falsely declared significant. In this situation the conceptual unit for the significance level is the individual comparison. In this type of experiment (two groups) the notion of a Type I error refers to the *individual comparison*. Whenever a significance test is performed on the one comparison involved in a two-group experiment, the error rate is *per comparison*. In other words, the level of α associated with the significance test refers to the probability that an individual comparison will be incorrectly declared significant.

When three or more groups are employed in an experiment, the interpretation of the level of α is *not* as was just described for the two-group case. The interpretation is not the same because an experiment with three or more groups has more than one comparison. In this case the experiment involves a collection of comparisons rather than a single comparison; likewise, the conceptual unit for a Type I error involves the whole collection of comparisons rather than a single comparison. In this case the level of α associated with an ANOVA or ANCOVA F test does *not* refer to the probability that a specific comparison will be incorrectly declared significant. Rather, it refers to the probability that *one or more* of all possible comparisons will be incorrectly declared significant. Since the collection of all possible comparisons defines the experiment, this probability of Type I error is known as the *experimentwise* error rate. This distinction between the error rate per comparison and the error rate experimentwise is an important one to keep in mind when interpreting ANOVA and ANCOVA F tests and the associated multiple comparison procedures.

5.4 OVERVIEW OF FOUR MULTIPLE COMPARISON PROCEDURES

Four multiple comparison procedures are presented in the remainder of this chapter: the Fisher protected LSD (least significant difference), the Bryant–Paulson generalization of Tukey's HSD (honestly significant difference), the Dunn–Bonferroni procedure, and the Scheffé procedure.

The choice among these procedures should be based on the type of comparisons of greatest interest to the investigator and whether simultaneous confidence intervals are desired. General recommendations are as follows:

1. Use the protected LSD procedure when the comparisons of interest are limited to hypothesis tests on any or all pairwise comparisons and simultaneous confidence intervals are not desired.
2. Use the Bryant–Paulson generalization of Tukey's HSD procedure when most or all pairwise comparisons are of interest and simultaneous confidence intervals are desired.
3. Use the Dunn–Bonferroni procedure for hypothesis tests and simultaneous confidence intervals on a small number of planned pairwise and/or complex (i,j) comparisons.
4. Use the Scheffé procedure when a large number of planned or unplanned complex comparisons of any type is of interest.

Notice that recommendations 2 through 4 contain the ambiguous words "most," "small," and "large". This is unavoidable because the best choice (in terms of power or width of simultaneous confidence intervals) among the last three procedures is a function of several factors, including the number of groups, the number of comparisons, and the degrees of freedom. Fortunately, it is quite acceptable to compute simultaneous confidence intervals by using more than one procedure and to then select the procedure that yields the shortest intervals. This approach presumes, of course, that the decision concerning which contrasts are of interest has been made a priori.

Interpretation of Multiple Comparison Hypothesis Tests

The distinction between per comparison and experimentwise error rates was described previously when the interpretation of the ANCOVA F in the two-group case was compared with the interpretation in the case of three or more groups. The error rate associated with multiple comparison hypothesis tests is also an experimentwise rate. There is a slight difference, however, between the definition of the experimentwise error rate in the case of the

ANCOVA F test and the definition of the error rate associated with several of the multiple comparison procedures.

With the protected LSD, the Bryant–Paulson generalization of Tukey's HSD, and the Dunn–Bonferroni procedure, the probability of making one or more Type I errors in the set of contrasts is equal to *or less than* α when the overall F test is used to decide whether multiple comparison tests are to be carried out. It is quite possible for the ANCOVA F to be statistically significant but to then find that no contrasts are significant using protected LSD, Bryant–Paulson, or Dunn–Bonferroni tests. This apparent inconsistency can occur because the F test and these three multiple comparison procedures are based on different sampling distributions and different conceptualizations of the experiment.

The experiment is conceptualized as all possible contrasts (which is an infinite number when there are at least three groups) under the ANCOVA F model. The experiment is viewed as a collection of all *pairwise* comparisons with protected LSD and Bryant–Paulson tests. If the F test is significant but no pairwise contrasts are significant when these tests are used, it can be concluded that some complex comparison is significant.

It is also possible that the overall F is significant but none of the (i,j) contrasts are significant. An (i,j) contrast is of the form where the mean of i means (where i may equal 1) is compared with the mean of j different means (where j may equal 1). The number of (i,j) contrasts is finite. For example, the total number of (i,j) contrasts in a three-group experiment is six; these contrasts are

$$\mu_{1\ adj} - \mu_{2\ adj} = \psi_1$$

$$\mu_{1\ adj} - \mu_{3\ adj} = \psi_2$$

$$\mu_{2\ adj} - \mu_{3\ adj} = \psi_3$$

$$\mu_{1\ adj} - \frac{\mu_{2\ adj} + \mu_{3\ adj}}{2} = \psi_4$$

$$\frac{\mu_{1\ adj} + \mu_{2\ adj}}{2} - \mu_{3\ adj} = \psi_5$$

$$\frac{\mu_{1\ adj} + \mu_{3\ adj}}{2} - \mu_{2\ adj} = \psi_6$$

The linear expression approach to writing these contrasts yields the following:

$$1(\mu_{1\ adj}) - 1(\mu_{2\ adj}) + 0(\mu_{3\ adj}) = \psi_1$$

$$1(\mu_{1\ adj}) + 0(\mu_{2\ adj}) - 1(\mu_{3\ adj}) = \psi_2$$

$$0(\mu_{1\ adj}) + 1(\mu_{2\ adj}) - 1(\mu_{3\ adj}) = \psi_3$$

$$1(\mu_{1\ adj}) - 0.5(\mu_{2\ adj}) - 0.5(\mu_{3\ adj}) = \psi_4$$

$$0.5(\mu_{1\ adj}) + 0.5(\mu_{2\ adj}) - 1(\mu_{3\ adj}) = \psi_5$$

$$0.5(\mu_{1\ adj}) - 1(\mu_{2\ adj}) + 0.5(\mu_{3\ adj}) = \psi_6$$

The Dunn–Bonferroni and Scheffé procedures can be employed to test these contrasts. If none of these tests is statistically significant but the overall F test is, there is no inconsistency. The set of all (i,j) contrasts is not the set of all possible contrasts. Since the collection of all (i,j) contrasts might appear to exhaust all contrasts that the reader can think of, it is reasonable to question the existence of an infinite number of possible contrasts in this three-group case. But consider the following contrast:

$$1(\mu_{1\ adj}) - 0.3(\mu_{2\ adj}) - 0.7(\mu_{3\ adj}) = \psi$$

This is not an (i,j) contrast since $\mu_{2\ adj}$ and $\mu_{3\ adj}$ are not given equal weight; that is, the mean of these two adjusted population means has not been compared with $\mu_{1\ adj}$. It can be seen, then, that there are many possible contrasts other than the (i,j) contrasts. It may seem absurd to think about testing the significance of a contrast defined as anything other than an (i,j) contrast. Surprisingly, the ANOVA and ANCOVA F tests are almost always tests of the significance of some non-(i,j) contrast!

In essence, the overall ANCOVA F test is a test of the significance of the optimum linear combination of the adjusted population means. The linear combination is optimum in the sense that there is no set of contrast coefficients that will result in a larger contrast. In any ANOVA or ANCOVA problem it is possible to identify the exact linear expression that is associated with the F test. Hence the ANCOVA F test is a test of the statistical significance of a particular identifiable linear combination of adjusted means. But the mere fact that there is no linear combination (or set of coefficients) that will yield a larger estimated contrast does not imply that the contrast identified is of any particular interest.

Since many researchers do not understand that ANOVA and ANCOVA F tests are associated with a linear combination that is dictated by the nature of the data, there is much head scratching when the overall F is significant but the multiple comparison tests identify no significant differences. If the F test is significant, there is a significant contrast that can be identified. There is no reason to expect, however, that the contrast that led to the significant F is a pairwise contrast or even an (i,j) contrast.

Since the protected LSD, the Bryant–Paulson, and the Dunn–Bonferroni procedures differ from the F test in the implied definition of the experiment it is quite possible for no significant contrasts to be found with these procedures in an experiment with a significant F statistic. Another related explanation for this possible outcome is that the sampling distributions associated with these tests are not the F distribution. The protected LSD employs the t distribution, the Bryant–Paulson employs the generalized studentized range distribution, and the Dunn–Bonferroni employs the Bonferroni t distribution.

Since the F test may reject the null hypothesis concerning the equality of the adjusted population means in cases where none of these multiple comparison tests yields significant values, the frequency of false rejections will be slightly less with these multiple comparison procedures than with the F test. Hence whereas the probability of a Type I error experimentwise is *equal* to α with the ANCOVA F, it is equal to or less than α with these three multiple comparison procedures.

These distinctions between F and the first three multiple comparison tests in terms of the definition of the experiment and the sampling distributions involved do not apply to Scheffé's procedure. That is, the overall F and the Scheffé procedure involve the same conceptualization of the experiment, and both are based on the F distribution. If an investigator were content to always define an experiment as a collection of *all possible* comparisons there would be need for only this method. The rationale underlying the Scheffé multiple comparison procedure, like the overall F test, is based on the notion that an experiment is a collection of all possible comparisons. Since the definition of the experiment is that of all possible contrasts with both the overall F test and the Scheffé procedure it should be no surprise that the F distribution is employed with the Scheffé's method. Hence the Scheffé method is the only multiple comparison procedure that is tied directly to the overall F test. If the F test is significant, some linear contrast is definitely significant. Why, then, if the F and Scheffé tests are related to each other logically and mathematically, are other procedures generally recommended instead? The answer is power.

There is a trade-off of generality for power that should be considered in the selection of a multiple comparison procedure. If a researcher is interested in all possible contrasts (which is unlikely), the Scheffé procedure is the only logical choice. But it should be pointed out that the use of this procedure for a small number of contrasts is a very conservative approach. Since all possible contrasts can be tested with this method, the power associated with each test in the set of a small number of contrasts is generally very low. In fact, pairwise contrasts are frequently not significant with this procedure when F is significant. This situation is more likely to develop with this procedure than with any of the other multiple comparison procedures.

The investigator who is interested only in tests on pairwise or (i,j) comparisons does not need the capability of making tests on all possible contrasts and should, therefore, employ a procedure designed for the comparisons of interest. But a caveat is in order concerning how the decision is made on what is "interesting." This decision concerning relevant contrasts should be made *before* the data are analyzed. Obviously, if an investigator were to compute a large number of comparisons, inspect each comparison, and then decide that the only comparison of interest is the largest one (because it is large), it would be inappropriate to employ an analysis procedure that is based on the rationale that this one contrast was the only one planned. There are statisticians who are so concerned with this problem that they always recommend the use of the Scheffé procedure. We agree that it would be very suspicious if a researcher carried out an experiment with several groups and then analyzed only one mean difference as if it were planned. But usually all pairwise comparisons (not all possible comparisons) are the comparisons of interest.

An alternative way of dealing with the multiple comparison problem is through the use of simultaneous confidence intervals rather than multiple comparison tests. Essentially the same issues are involved with both approaches.

Simultaneous Confidence Intervals

The distinction previously made between an experiment defined as a simple contrast and an experiment defined as a collection of contrasts is also important when confidence intervals are constructed. In the case of two-group ANCOVA experiments, there is interest in constructing intervals that will contain the population difference between adjusted means. If, for example, 95% confidence intervals are constructed in each of 100 independent two-group experiments, the true difference between adjusted means can be expected to be contained in all but five of the 100 intervals. In other words, in the long run 95% of the true differences will be contained in such an interval. It should be kept in mind that the interpretation of confidence intervals is not affected by whether the null hypothesis is true. That is, there is no reason to run hypothesis tests to decide whether to construct confidence intervals. If a confidence interval contains zero, the corresponding hypothesis test will not result in the rejection of the null hypothesis. If the confidence interval does not contain zero, the corresponding hypothesis test will result in the rejection of the null hypothesis. In any event, when the confidence interval approach is employed with *two-group* experiments, the population difference will be interpreted to be contained in the intervals $100(1 - \alpha)\%$ of the time. If the experiment contains *more* than two groups and confidence intervals are constructed by using one of the multiple comparison procedures, the interpretation of the intervals is not the same as in the two-group case.

When three or more groups are involved, the experiment contains more than one comparison and, of course, more than one confidence interval. In this case the collection of confidence intervals in an experiment is the set of *simultaneous* confidence intervals. As is the case with multiple comparison hypothesis tests, the experiment is viewed as a collection of comparisons, and the interpretation of simultaneous confidence intervals is tied to this collection. Suppose a three-group experiment is carried out and confidence intervals are constructed on the three pairwise contrasts by using one of the procedures described in this chapter. If 95% simultaneous confidence intervals are constructed, the three population contrasts can be expected to be contained in the three simultaneous confidence intervals; the probability is at least .95 that all three true population contrasts (i.e., $\mu_{1\ adj} - \mu_{2\ adj}$, $\mu_{1\ adj} - \mu_{3\ adj}$, and $\mu_{2\ adj} - \mu_{3\ adj}$) are contained within the intervals. The probability statement (i.e., .95 in this case) refers to the whole collection of intervals; it does not refer to a separate statement for each interval. It is *not* correct to state that the probability of $\mu_{1\ adj} - \mu_{2\ adj}$, for example, falling in the associated interval is .95. Actually, the probability of any one of the population differences falling within the corresponding interval is *greater* than .95.

Table 5.1 Summary of Multiple Comparison Procedures and Situations to Which They Apply

Type of Comparison	Protected LSD	Bryant–Paulson	Dunn–Bonferroni	Scheffé
All pairwise—hypothesis tests only	X (F must be significant)			
All pairwise—hypothesis tests and simultaneous confidence intervals		X (F test not required)		
Small number of planned pairwise comparisons—hypothesis tests and simultaneous confidence intervals			X (F test not required)	
Small number of planned complex or a mixture of planned pairwise and complex comparisons—hypothesis tests and simultaneous confidence intervals			X (F test not required)	
Large number of comparisons of any type—hypothesis tests and simultaneous confidence intervals				X (F test not required, but no comparison will be significant if F is not significant)

In summary, two-group experiments contain one contrast. The probability of making a Type I error in a two-group experiment defines the *per comparison* error rate:

$$\text{Per comparison error rate} = \frac{\text{Number of two-group experiments in which difference is falsely declared significant}}{\text{Total number of two-group experiments}}$$

Confidence intervals associated with two-group experiments are individual confidence intervals. Experiments with three or more groups are generally analyzed by using multiple comparison procedures. The probability of making at least one false rejection of the null hypothesis in an experiment with these tests is termed the *experimentwise* error rate:

$$\text{Experimentwise error rate} = \frac{\text{Number of experiments in which at least one difference is falsely declared significant}}{\text{Total number of experiments}}$$

Confidence intervals associated with experiments containing more than one contrast are termed *simultaneous confidence intervals* because the probability statement refers to the whole collection of intervals rather than to each interval individually. Three of the four multiple comparison procedures described can be used to establish simultaneous confidence intervals. The Bryant-Paulson generalization of the Tukey HSD procedure should generally be used for pairwise contrasts. Simultaneous confidence intervals cannot be obtained through the protected LSD method. The Dunn–Bonferroni procedure is recommended for most situations where a small number of (i,j) contrasts is of interest. The Scheffé approach can be employed with any possible contrast or set of contrasts. The best choice among these procedures can generally be made by simply selecting the one that is associated with the type of contrasts of interest. It is acceptable, however, to select the procedure that produces the shortest simultaneous confidence intervals—as long as the contrasts of interest have been selected before observing the data. A summary of the situations to which the four multiple comparison procedures apply is provided in Table 5.1. Computational details are provided in the following sections.

5.5 TESTS ON ALL PAIRWISE COMPARISONS: FISHER'S PROTECTED LSD PROCEDURE

Probably the most frequently encountered multiple comparison situation is the one in which the researcher is interested in all pairwise comparisons. Many multiple comparison procedures can be used in this situation, but,

R. A. Fisher's protected LSD procedure is generally superior (in terms of power) to other methods that have become popular in recent years. A large-scale Monte Carlo study by Carmer and Swanson (1973) clearly reveals the advantages of this test relative to several competing tests. This study demonstrated that even though the protected LSD test is distinctly more powerful than the Tukey HSD test, for example, the probability of committing a Type I error is approximately the same for both procedures. Similar results were obtained by Bernhardson (1975). These studies involved ANOVA but can be expected to generalize to ANCOVA.

Two steps precede the computation of any multiple comparison procedure used in ANCOVA: (1) the analysis of covariance and (2) the analysis of variance on the *covariate*. Quantities from both analyses are employed in the multiple comparison formulas. The ANCOVA and the ANOVA on the covariate for the behavioral objectives study described in previous chapters are tabulated as follows for reference throughout the remainder of this chapter.

Analysis of Covariance (Data from Table 3.1)

Source	SS	df	MS	F
Adjusted treatment	707.99	2	354.00	5.48
Residual within	1680.65	26	64.64	
Total residuals	2388.64	28		
Critical value $= F_{(.05, 2, 26)} = 3.37$				

The adjusted means are

$$\text{Group } 1 = 28.48 = \bar{Y}_{1 \text{ adj}}$$

$$\text{Group } 2 = 40.33 = \bar{Y}_{2 \text{ adj}}$$

$$\text{Group } 3 = 36.19 = \bar{Y}_{3 \text{ adj}}$$

Analysis of Variance on Covariate (Data from Table 3.1)

Source	SS	df	MS	F
Between	126.67	2	63.33	.30
Within	5700.00	27	211.11	
Total	5826.67	29		
Critical value $= F_{(.05, 2, 27)} = 3.35$				

The covariate means are

$$\text{Group } 1 = 52.00 = \bar{X}_1$$

$$\text{Group } 2 = 47.00 = \bar{X}_2$$

$$\text{Group } 3 = 49.00 = \bar{X}_3$$

Protected LSD Test Procedure.

Stage 1. Perform the ANCOVA F test at level α. If this test is nonsignificant, retain the hypothesis that $\mu_{1\,adj} = \mu_{2\,adj} = \cdots = \mu_{J\,adj}$. If this test is significant, proceed to stage 2.

Stage 2. The LSD tests on each pairwise difference between adjusted means are carried out at the same α as was employed during stage 1.

The error term for evaluating the difference between two adjusted means, say, i and j, is computed by using the following formula:

$$s_{\overline{Y}_{i\,adj} - \overline{Y}_{j\,adj}} = \sqrt{\text{MSres}_w \left[\left[\frac{1}{n_i} + \frac{1}{n_j} \right] + \frac{(\overline{X}_i - \overline{X}_j)^2}{\text{SS}_{w_x}} \right]}$$

where

  MSres_w = analysis of covariance mean square error term obtained from the ANCOVA summary table

  n_i, n_j = sample sizes for the ith and jth groups, respectively

  $\overline{X}_i, \overline{X}_j$ = covariate means for ith and jth groups, respectively

  SS_{w_x} = sum of squares within groups on X variable; this is obtained from summary table for ANOVA on covariate

This formula, which is appropriate for both randomized and nonrandomized studies, can also be employed in two-group experiments to construct confidence intervals using

$$\left(\overline{Y}_{\text{I}\,adj} - \overline{Y}_{\text{II}\,adj} \right) \pm s_{\overline{Y}_{\text{I}\,adj} - \overline{Y}_{\text{II}\,adj}} \left[t_{(\alpha, N-3)} \right]$$

When the difference between the two adjusted means $(\overline{Y}_{i\,adj} - \overline{Y}_{j\,adj})$ is divided by the LSD error term, the resulting test statistic is evaluated with the tabled values of the ordinary t distribution by using $N - J - 1$ degrees of freedom:

$$\frac{\overline{Y}_{i\,adj} - \overline{Y}_{j\,adj}}{s_{\overline{Y}_{i\,adj} - \overline{Y}_{j\,adj}}} = t.$$

It is important to keep in mind that these pairwise t statistics are calculated *only* if the initial ANCOVA F test is significant at level α. This is the reason that the procedure is called the "protected" LSD.

Interpretation of Protected LSD Tests. Recall the meaning of the rejected null hypothesis in analysis of covariance problems. If three or more groups are involved, the rejection of the hypothesis $H_0: \mu_{1\,adj} = \mu_{2\,adj} = \cdots = \mu_{J\,adj}$ leads us to conclude that some linear combination of adjusted means results in a population contrast that is not zero. The level of α associated with the ANCOVA F test is the probability that the observed differences among the sample adjusted means would occur if the population adjusted means were equal. Each significant LSD test leads to the conclusion that the associated pairwise difference between adjusted population means is not zero. The researcher can conclude that the probability of obtaining one or more significant differences (using LSD) in the experiment by chance alone is not greater than α.

Numerical Example of Protected LSD Tests.   The analysis of covariance on the behavioral objectives data described previously yields a significant F; it is appropriate to carry out the LSD tests. The test on the difference between adjusted means 1 and 2 is

$$\frac{28.48 - 40.33}{\sqrt{64.64\left[\left(\frac{1}{10} + \frac{1}{10}\right) + \frac{(52-47)^2}{5700}\right]}} = \frac{-11.85}{3.64} = -3.26 = t_{obt}$$

The obtained t is 3.26 (ignoring the sign), and the critical value against which this obtained value is compared is $t_{(.05, 26)} = 2.056$. The difference between the two adjusted means is significant at the .05 level.
    The test on the difference between adjusted means 1 and 3 is

$$\frac{28.48 - 36.19}{\sqrt{64.64\left[\left(\frac{1}{10} + \frac{1}{10}\right) + \frac{(52-49)^2}{5700}\right]}} = \frac{-7.71}{3.61} = -2.14 = t_{obt}$$

Since $|2.14|$ is greater than 2.056, the null hypothesis H_0: $\mu_{1\ adj} = \mu_{3\ adj}$ is rejected at the .05 level.
    The test on the difference between adjusted means 2 and 3 is

$$\frac{40.33 - 36.19}{\sqrt{64.64\left[\left(\frac{1}{10} + \frac{1}{10}\right) + \frac{(47-49)^2}{5700}\right]}} = \frac{4.14}{3.60} = 1.15 = t_{obt}$$

The null hypothesis H_0: $\mu_{2\ adj} = \mu_{3\ adj}$ is retained.
    Three pairwise comparisons have been computed and two of these were found to be significant using $\alpha = .05$. The probability that one or more of the

comparisons has incorrectly been declared significant because of random fluctuation is less than .05.

5.6   ALL PAIRWISE COMPARISONS: BRYANT–PAULSON GENERALIZATION OF TUKEY'S HSD

The Bryant–Paulson generalization of Tukey's HSD procedure (BPT for short) is appropriate for all pairwise comparisons. This procedure is different from the conventional Tukey procedure, which has been inappropriately recommended for ANCOVA in many of the standard references (e.g., Kirk 1968, Winer 1971). Scheffé (1959) has pointed out the problem in using the Tukey procedure with ANCOVA. The Bryant–Paulson procedure (Bryant and Paulson 1976) was derived under the assumption that the covariate (or covariates) is a random variable. As is discussed in Chapter 6, this assumption is realistic for virtually all applications of ANCOVA. (If the covariate is a fixed variable, the Tukey procedure is appropriate.) The distribution associated with this procedure is known as the *generalized studentized range distribution*. The critical values of this distribution are contained in Table A.4.

Unlike the protected LSD procedure, which is for hypothesis tests *only*, the BPT procedure can be employed for both hypothesis tests *and* simultaneous confidence intervals. Since our preference is to report the results of research in terms of simultaneous confidence intervals rather than hypothesis tests, and since pairwise comparisons are generally of greatest interest, this procedure is strongly recommended.

It is typical to employ the overall F test as a preliminary test before carrying out any type of multiple comparison procedure. It is not essential, however, that this convention be followed with the BPT procedure. That is, the overall hypothesis of the equality of the J population adjusted means can be tested using either the ANCOVA F test or, alternatively, the BPT test on the largest difference between adjusted means. In most cases the two procedures will yield the same conclusion. Bryant and Paulson (1976) have shown that for certain combinations of sample size and patterns of differences among population means, their procedure is more powerful than the ANCOVA F test. However, if the results of a great deal of research on the power characteristics of ANOVA F generalize to ANCOVA F (and we believe that they do), the F approach should be uniformly most powerful. That is, when all combinations of sample size, number of treatment groups, and patterns of differences among population means are considered, the F should, overall, have higher power than competing procedures.

Since most researchers are more concerned with specific pairwise comparisons than with overall hypotheses, and since confidence intervals provide more information than do hypothesis tests, we recommend that emphasis be placed on simultaneous confidence intervals based on the BPT procedure rather than on the ANCOVA F test.

Computation Procedure for Bryant–Paulson Generalization of Tukey's HSD: Tests and Simultaneous Confidence Intervals

All pairwise comparisons between adjusted means can be tested by using

$$\frac{\overline{Y}_{i\,\text{adj}} - \overline{Y}_{j\,\text{adj}}}{\sqrt{\text{MSres}_{\text{w}}\left[1+\left(\text{MS}_{b_x}/\text{SS}_{w_x}\right)\right]/n}} = Q_p$$

where

$\overline{Y}_{i\,\text{adj}}, \overline{Y}_{j\,\text{adj}}$ = any two adjusted means

$\qquad n$ = sample size of each group

MSres_{w} = error term obtained from ANCOVA summary table

$\quad\text{MS}_{b_x}$ = mean square between groups obtained from summary table for ANOVA on covariate

$\quad\text{SS}_{w_x}$ = sum of squares within groups obtained from summary table for ANOVA on covariate

$\qquad Q_p$ = generalized studentized range statistic

The critical value against which the obtained Q_p is compared is found in Table A.4. The critical value is denoted $Q_{p_{(\alpha:\ C,J,N-J-C)}}$, where α is the (nondirectional) experimentwise error rate, C is the number of covariates (one in this chapter), and J is the number of groups.

The 95% simultaneous confidence intervals around all pairwise sample mean differences on the true population mean differences are computed by using

$$\left(\overline{Y}_{i\,\text{adj}} - \overline{Y}_{j\,\text{adj}}\right) \pm Q_{p_{(.05:\ 1,J,N-J-1)}}\sqrt{\text{MSres}_{\text{w}}\left[1+\frac{\text{MS}_{b_x}}{\text{SS}_{w_x}}\right]/n}$$

Numerical Example of BPT Tests and Simultaneous Confidence Intervals

1. Compute (a) ANOVA on the covariate and (b) covariate means (see Section 5.5).
2. Compute (a) ANCOVA and (b) adjusted means (see Section 5.5).
3. Suppose that we are interested in all the pairwise comparisons in the experiment. Since the adjusted means are 28.48, 40.33, and 36.19 the sample differences are:

$$28.48 - 40.33 = -11.85$$
$$28.48 - 36.19 = -7.71$$
$$40.33 - 36.19 = 4.14$$

The test of the $H_0: \mu_{1 \text{ adj}} = \mu_{2 \text{ adj}}$ is

$$\frac{-11.85}{\sqrt{64.64[1 + (63.33/5700)]}/10} = \frac{-11.85}{2.5565}$$

$$= -4.64$$

$$= Q_{p_{\text{obt}}}$$

The critical value is $Q_{p_{(.05;\ 1,3,36)}}$ or 3.59 (see Table A.4), and we reject the null hypothesis. (*Note*: The sign of the obtained Q_p is disregarded.)

Since the error term (i.e., 2.5565) is the same for all pairwise comparisons, the tests of the hypotheses $H_0: \mu_{1 \text{ adj}} = \mu_{2 \text{ adj}}$ and $H_0: \mu_{2 \text{ adj}} = \mu_{3 \text{ adj}}$ are simply:

$$\frac{-7.71}{2.5565} = 3.016 = Q_{p_{\text{obt}}}$$

and

$$\frac{4.14}{2.5565} = 1.62 = Q_{p_{\text{obt}}}.$$

We conclude that the only significant difference is the difference between treatments 1 and 2 because the obtained Q_p values for the other comparisons are below the critical value 3.59.

The set of simultaneous 95% confidence intervals around $\overline{Y}_{1 \text{ adj}} - \overline{Y}_{2 \text{ adj}}$, $\overline{Y}_{1 \text{ adj}} - \overline{Y}_{3 \text{ adj}}$, and $\overline{Y}_{2 \text{ adj}} - \overline{Y}_{3 \text{ adj}}$ on $\mu_{1 \text{ adj}} - \mu_{2 \text{ adj}}$, $\mu_{1 \text{ adj}} - \mu_{3 \text{ adj}}$ and $\mu_{2 \text{ adj}} - \mu_{3 \text{ adj}}$, respectively, are:

1. $-11.85 \pm (2.5565)(3.59) = -11.85 \pm 9.18 = (-21.03, -2.67)$
2. $-7.71 \pm (2.5565)(3.59) = -7.71 \pm 9.18 = (-16.89, 1.47)$
3. $4.14 \pm (2.5565)(3.59) = 4.14 \pm 9.18 = (-5.04, 13.22)$

All pairwise confidence intervals have been computed, and the statement that all intervals contain the true population difference between adjusted means can be made with 95% confidence.

Unequal Sample Sizes. If the sample sizes are not the same for each group, substitute the harmonic mean of the sample sizes involved in each comparison in place of n in the computational formula. No other modifications are required. For example, if the sample sizes are 8, 12, and 15 in a three-group experiment, use $[2/(\frac{1}{8} + \frac{1}{12})] = 9.6$ (in the comparison of groups 1 and 2) where n is called for in the previously described formula for the BPT tests and confidence intervals. The use of the harmonic mean has precedent (Kramer 1956) with Tukey's test in the ANOVA case. Recent work by Dunnett (1979) suggests that the use of the harmonic mean in the unequal sample size case maintains the Type I error very near the tabled value even though the Tukey procedure is often not recommended unless equal sample sizes are available.

Since the Tukey procedure can be conceptualized as a special case of the BPT procedure without covariates, it is likely that Dunnett's results generalize to the BPT procedure for ANCOVA.

Nonrandomized Studies. If the data have been obtained from a randomized experiment, the formulas presented earlier in this section are appropriate. If the data have been obtained from a biased assignment or nonrandomized study (discussed in Chapters 6, 7, 14, and 15), the following formula is recommended:

$$\frac{\overline{Y}_{i\,\text{adj}} - \overline{Y}_{j\,\text{adj}}}{\sqrt{\text{MSres}_w\left\{(2/n) + \left[(\overline{X}_i - \overline{X}_j)^2/\text{SS}_{w_x}\right]\right\}/2}} = Q_{p_\text{obt}}$$

The error term must be computed separately for each contrast with this formula because the difference between the ith and jth group means on the covariate (i.e., $\overline{X}_i - \overline{X}_j$) will generally be different for each contrast. The obtained value of Q_p is compared with the critical value of the generalized studentized range statistic $Q_{p_{(\alpha,\,1,J,N-J-1)}}$.

5.7 PLANNED PAIRWISE AND COMPLEX COMPARISONS: DUNN–BONFERRONI PROCEDURE

Occasionally an experiment is designed in which the experimenter has interest in a limited number of pairwise and/or complex comparisons. Suppose that an experimenter is interested a priori in the following three comparisons in a six-group experiment:

$$
\begin{array}{cccccc}
\text{I} & \text{II} & \text{III} & \text{IV} & \text{V} & \text{VI} \\
\underline{\lfloor\quad\rfloor} & & \underline{\lfloor\quad\rfloor} & & \underline{\lfloor\quad\rfloor} &
\end{array}
$$

The decision to make these comparisons should be made *before* the data are collected. The Dunn–Bonferroni procedure is appropriate for planned pairwise and complex comparisons rather than for comparisons that are suggested by the outcome of the experiment.

The advantages of the Dunn–Bonferroni procedure over the BPT procedure are: (1) there is greater power when the number of comparisons relative to the total number of possible pairwise comparisons is small, and (2) complex comparisons can be tested. If an investigator should decide to make a large number of planned pairwise comparisons, say, almost as many as the total number of pairwise comparisons, the power advantage of the Dunn–Bonferroni procedure will be lost. It is only when the number of comparisons is relatively small that the Dunn–Bonferroni is superior to the BPT for pairwise comparisons. Before concluding that the Dunn–Bonferroni procedure is the most appropriate method for a relatively small number of pairwise comparisons, the protected LSD procedure must be considered again. It turns out that as long as the ANCOVA F is significant, no hypothesis testing method is superior in terms of power to the protected LSD test for pairwise comparisons. Hence even if the number of groups is huge and the number of planned contrasts is only 2, the protected LSD is more sensitive than the Dunn–Bonferroni. However, simultaneous confidence intervals cannot be computed with the use of the protected LSD approach.

Computation Procedure for Dunn–Bonferroni Tests and Simultaneous Confidence Intervals

The formula for carrying out the Dunn–Bonferroni (DB) test is

$$\frac{c_1\left(\overline{Y}_{1\,\text{adj}}\right)+c_2\left(\overline{Y}_{2\,\text{adj}}\right)+\cdots+c_J\left(\overline{Y}_{J\,\text{adj}}\right)}{\sqrt{\text{MSres}_w\left[1+\dfrac{\text{MS}_{b_x}}{\text{SS}_{w_x}}\right]\left[\dfrac{(c_1)^2}{n_1}+\dfrac{(c_2)^2}{n_2}+\cdots+\dfrac{(c_J)^2}{n_J}\right]}}=t\text{DB}$$

where:

$c_1, c_2, \ldots c_j$ = contrast coefficients for comparisons of interest
$\overline{Y}_{1\,\text{adj}}, \ldots, \overline{Y}_{J\,\text{adj}}$ = adjusted means
$n_1, \ldots, n_J$ = sample sizes associated with groups $1, \ldots, J$
MSres_w = error term obtained from the ANCOVA summary table
MS_{b_x} = mean square between groups obtained from summary table for ANOVA on covariate
SS_{w_x} = sum of squares within groups obtained from summary table for ANOVA on covariate

The critical value against which the absolute value of the obtained tDB is compared is $t_{\text{DB}(\alpha,\,C',\,N-J-1)}$ found in Table A.6. [Dunn (1961) was the first to tabulate the critical values of the Bonferroni t; hence we have labeled this the Dunn–Bonferroni statistic.] It is important to note that the second term in the subscript (i.e., C') is the number of planned comparisons—not the number of groups or the number of covariates. The third term is, of course, the degrees of freedom associated with the ANCOVA error term.

The $100(1-\alpha)\%$ confidence interval for any planned comparison is constructed by using

$$c_1\left(\overline{Y}_{1\,\text{adj}}\right)+c_2\left(\overline{Y}_{2\,\text{adj}}\right)+\cdots+c_J\left(\overline{Y}_{J\,\text{adj}}\right)$$

$$\pm\,t_{\text{DB}(\alpha,\,C',\,N-J-1)}\sqrt{\text{MSres}_w\left[1+\dfrac{\text{MS}_{b_x}}{\text{SS}_{w_x}}\right]\left[\dfrac{(c_1)^2}{n_1}+\dfrac{(c_2)^2}{n_2}+\cdots+\dfrac{(c_J)^2}{n_J}\right]}$$

Numerical Example of Dunn–Bonferroni Tests and Simultaneous Confidence Intervals. If the investigator conducting the behavioral objectives study *planned* to compare the treatment 1 adjusted mean with the treatment 2 adjusted mean and the treatment 2 adjusted mean with the treatment 3 adjusted mean, two of three possible pairwise comparisons are involved. Since the ANCOVA F is significant the protected LSD procedure is the best choice as a testing procedure; but if confidence intervals are desired, the choice between the BPT and the Dunn–Bonferroni procedures should be made before the results are reported. It can be seen in the computational steps that follow that the width of the confidence intervals associated with the two planned contrasts are narrower with the Dunn–Bonferroni procedure than with the BPT procedure.

1. The ANCOVA and adjusted means are computed (see Section 5.5).
2. The ANOVA on the covariate is computed (see Section 5.5).
3. The differences $\bar{Y}_{1\ adj} - \bar{Y}_{2\ adj}$ and $\bar{Y}_{2\ adj} - \bar{Y}_{3\ adj}$ are tested $(\alpha = .05)$ as follows:

 a.

$$\frac{1(28.48) - 1(40.33) + 0(36.19)}{\sqrt{64.64[1 + (63.33/5700)][(1)^2/10 + (-1)^2/10 + (0)^2/10]}}$$

$$= \frac{-11.85}{\sqrt{(65.36)(0.2)}}$$

$$= \frac{-11.85}{3.62} = -3.27 = t_{DB_{obt}}$$

 b.

$$\frac{0(28.48) + 1(40.33) - 1(36.19)}{\sqrt{64.64[1 + (63.33/5700)][(0)^2/10 + (1)^2/10 + (-1)^2/10]}}$$

$$= \frac{4.14}{\sqrt{(65.36)(0.2)}}$$

$$= \frac{4.14}{3.62} = 1.14 = t_{DB_{obt}}$$

The critical value against which the obtained t_{DB} values are compared is $t_{DB_{(\alpha,2,26)}}$ or 2.38 for $\alpha = .05$.

4. Simultaneous confidence intervals are as follows:

 a. The confidence interval around

$$1(\bar{Y}_{1\ adj}) - 1(\bar{Y}_{2\ adj}) + 0(\bar{Y}_{3\ adj}) \quad \text{on} \quad 1(\mu_{1\ adj}) - 1(\mu_{2\ adj}) + 0(\mu_{3\ adj})$$

$$\text{is} \quad -11.85 \pm 2.38(3.62) = (-20.47, -3.23)$$

 b. The confidence interval around

$$0(\bar{Y}_{1\ adj}) + 1(\bar{Y}_{2\ adj}) - 1(\bar{Y}_{3\ adj}) \quad \text{on} \quad 0(\mu_{1\ adj}) + 1(\mu_{2\ adj}) - 1(\mu_{3\ adj})$$

$$\text{is} \quad 4.14 \pm 2.38(3.62) = (-4.48, 12.76)$$

The intervals for these two contrasts are $(-21.03, -2.67)$ and $(-5.04, 13.32)$ if the BPT procedure is used and $(-20.47, -3.23)$ and $(-4.48, 12.76)$ if the Dunn–Bonferroni procedure is used. Since the intervals are shorter with the Dunn–Bonferroni procedure, these results should be reported. This approach of computing the intervals through both procedures and reporting the shorter ones is acceptable practice as long as the decision concerning the number of contrasts of interest is made *before* the data are examined.

Nonrandomized Studies. If the data have been obtained from a randomized experiment, the formulas presented earlier in this section are appropriate. If the data have been obtained from a biased assignment or nonrandomized study, the following formula should be employed instead:

$$\frac{c_1 \overline{Y}_{1\,\text{adj}} + c_2 \overline{Y}_{2\,\text{adj}} + \cdots + c_J \overline{Y}_{J\,\text{adj}}}{\sqrt{\text{MSres}_w \left[\dfrac{c_1^2}{n_1} + \dfrac{c_2^2}{n_2} + \cdots + \dfrac{c_j^2}{n_J} + \left[\left(c_1 \overline{X}_1 + c_2 \overline{X}_2 + \cdots c_J \overline{X}_J \right)^2 / \text{SS}_{w_x} \right] \right]}} = t_{\text{DB}}$$

Notice that the error term is computed separately for each contrast when this formula is employed. The critical value is $t_{\text{DB}_{(\alpha, C', N-J-1)}}$, which is the same value used with the randomized experiment formula.

5.8 ANY OR ALL COMPARISONS: SCHEFFÉ PROCEDURE

The Scheffé procedure is the most general of the available procedures for making multiple comparisons. Although it can be used for any type or number of comparisons, it is recommended for only certain types of comparison. The Scheffé approach is most useful when (1) the experimenter discovers comparisons he would like to test after examining the outcome of the experiment and (2) a very large number of pairwise and/or complex comparisons are planned. The protected LSD and BPT procedures are much more powerful for all pairwise comparisons, and the Dunn–Bonferroni procedure is more powerful for a limited number of planned complex comparisons.

If the experimenter is interested in testing most of the possible (i,j) contrasts, the Scheffé procedure should be used. If the experimenter plans to make a small number of complex contrasts, the Dunn–Bonferroni procedure should be used. The choice between the Scheffé and Dunn–Bonferroni procedures for complex comparisons depends on a number of factors, including the number of comparisons the investigator plans to make, the number of groups involved, and the degrees of freedom associated with the ANCOVA MSres_w. The choice between the two can easily be made by selecting the one with the smaller critical value. Keep in mind that the Scheffé procedure is appropriate regardless of the number of complex comparisons. Only when the contrasts are planned is the Dunn–Bonferroni considered.

The Scheffé test can be used to test any possible contrast; if the ANCOVA F is significant, there is some contrast [not necessarily an (i,j) contrast] that is significant if the Scheffé procedure is used. If the ANCOVA F is not significant, there is no contrast that will be declared significant by using the Scheffé procedure. The view that the experiment is a collection of all possible comparisons is consistent with the ANCOVA F test but not with any other multiple comparison tests described in this chapter. In a sense, the Scheffé test is the only multiple comparison procedure that is "correct" from the point of view of the F-test framework. This, of course, does not mean that the other procedures are inappropriate. It simply means that the Scheffé approach is appropriate if the experimenter is interested in a very large number of contrasts. When the experiment is viewed as a collection of all pairwise contrasts or a small number of planned contrasts, other procedures are more appropriate because they are more likely to detect true differences when they exist.

The Scheffé procedure has one advantage over other multiple comparison procedures; it is an exact method for both equal and unequal sample sizes (Halperin and Greenhouse 1958, Scheffé 1959). The BPT method is also exact if sample sizes are equal.

Computation Procedure for Scheffé Tests

The test statistic is

$$
\frac{c_1\left(\overline{Y}_{1\,\mathrm{adj}}\right)+c_2\left(\overline{Y}_{2\,\mathrm{adj}}\right)+\cdots+c_J\left(\overline{Y}_{J\,\mathrm{adj}}\right)}{\sqrt{\mathrm{MSres_w}\left[1+(\mathrm{MS}_{b_x}/\mathrm{SS}_{w_x})\right]\left\{\left[(c_1)^2/n_1\right]+\left[(c_2)^2/n_2\right]+\cdots+\left[(c_J)^2/n_J\right]\right\}}}
$$
$$= F'$$

where

$c_1, c_2, \ldots c_J =$ contrast coefficients for comparison of interest

$\overline{Y}_{1\,\mathrm{adj}}, \overline{Y}_{2\,\mathrm{adj}}, \ldots, \overline{Y}_{J\,\mathrm{adj}} =$ adjusted means

$n_1, n_2, \ldots, n_J =$ sample sizes associated with groups $1, 2, \ldots, J$

$\mathrm{MSres_w} =$ error term obtained from ANCOVA summary table

$\mathrm{MS}_{b_x} =$ mean square between groups obtained from summary table for ANOVA on covariate

$\mathrm{SS}_{w_x} =$ sum of squares within groups obtained from summary table for ANOVA on covariate

The critical value against which the obtained F' is compared is

$$\sqrt{(J-1)F_{(\alpha, J-1, N-J-1)}}$$

Simultaneous Confidence Intervals Based on Scheffé Procedure

The simultaneous 95% confidence intervals are computed by using the following formula for each comparison:

$$\left[c_1\left(\overline{Y}_{1\,adj} \right) + c_2\left(\overline{Y}_{2\,adj} \right) + \cdots + c_J\left(\overline{Y}_{J\,adj} \right) \right] \pm \sqrt{(J-1)\left[F_{(.05,\,J-1,\,N-J-1)} \right]}$$

$$\sqrt{MSres_w\left[1 + (MS_{bx}/SS_{w_x}) \right]\left[\frac{(c_1)^2}{n_1} + \frac{(c_2)^2}{n_2} + \cdots + \frac{(c_J)^2}{n_J} \right]}$$

Numerical Example of Scheffé Tests and Simultaneous Confidence Intervals

Suppose that the six (i,j) comparisons in the behavioral objectives study are of interest. The computational steps are as follows:

1. Compute (a) ANOVA on X and (b) covariate means (see Section 5.5).
2. Compute (a) ANCOVA and (b) adjusted means (see Section 5.5).
3. Compute the contrast estimates

$$1(28.48) - 1(40.33) + 0(36.19) = -11.85 = \hat{\psi}_1$$
$$1(28.48) + 0(40.33) - 1(36.19) = -7.71 = \hat{\psi}_2$$
$$0(28.48) + 1(40.33) - 1(36.19) = 4.14 = \hat{\psi}_3$$
$$1(28.48) - 0.5(40.33) - 0.5(36.19) = -9.78 = \hat{\psi}_4$$
$$-0.5(28.48) + 1(40.33) - 0.5(36.19) = 7.995 = \hat{\psi}_5$$
$$-0.5(28.48) - 0.5(40.33) + 1(36.19) = 1.785 = \hat{\psi}_6$$

The tests on the contrast estimates are

$$\frac{-11.85}{\sqrt{64.64\left[1 + (63.33/5700) \right]\left[(1)^2/10 + (-1)^2/10 + (0)^2/10 \right]}} = \frac{-11.85}{3.62}$$
$$= -3.28$$

$$\frac{-7.71}{\sqrt{64.64\left[1 + (63.33/5700) \right]\left[(1)^2/10 + (0)^2/10 + (-1)^2/10 \right]}} = \frac{-7.71}{3.62}$$
$$= -2.13$$

$$\frac{4.14}{\sqrt{64.64\left[1 + (63.33/5700) \right]\left[(0)^2/10 + (1)^2/10 + (-1)^2/10 \right]}} = \frac{4.14}{3.62}$$
$$= 1.14$$

$$\frac{-9.78}{\sqrt{64.64[1+(63.33/5700)][(1)^2/10+(-0.5)^2/10+(-0.5)^2/10]}} = \frac{-9.78}{3.13}$$

$$= -3.12$$

$$\frac{7.995}{\sqrt{64.64[1+(63.33/5700)][(-0.5)^2/10+(1)^2/10+(-0.5)^2/10]}} = \frac{7.995}{3.13}$$

$$= 2.55$$

$$\frac{1.785}{\sqrt{64.64[1+(63.33/5700)][(-0.5)^2/10+(-0.5)^2/10+(1)^2/10]}} = \frac{1.785}{3.13}$$

$$= 0.57$$

The absolute value of each obtained F' is compared with the critical value of F', which is

$$\sqrt{2[F_{(.05,2,26)}]} = \sqrt{2[3.37]} = 2.60$$

The hypotheses

$$\mu_{1\ adj} - \mu_{2\ adj} = 0.0$$

$$\mu_{1\ adj} - \frac{\mu_{2\ adj} + \mu_{3\ adj}}{2} = 0.0$$

are rejected at the .05 level because the tests associated with $\hat{\psi}_1$ and $\hat{\psi}_4$ yield absolute values of F'_{obt} that exceed the critical value of F' (i.e., 3.28 and 3.12 are greater than 2.60).

4. The 95% simultaneous confidence intervals on ψ_1, ψ_2, ψ_3, ψ_4, ψ_5, and ψ_6 around the corresponding contrast estimates $\hat{\psi}_1$, $\hat{\psi}_2$, $\hat{\psi}_3$, $\hat{\psi}_4$, $\hat{\psi}_5$, and $\hat{\psi}_6$ are

$\hat{\psi}$	Confidence Interval
$-11.85 \pm (2.60)(3.62)$	$= (-21.26, -2.44)$
$-7.71 \pm (2.60)(3.62)$	$= (-17.12, 1.70)$
$4.14 \pm (2.60)(3.62)$	$= (-5.27, 13.55)$
$-9.78 \pm (2.60)(3.13)$	$= (-17.92, -1.64)$
$7.995 \pm (2.60)(3.13)$	$= (-0.14, 16.13)$
$1.785 \pm (2.60)(3.13)$	$= (-6.35, 9.92)$

The probability is .95 that the entire set of confidence intervals constructed according to the Scheffé procedure will cover all of the population contrasts they estimate. Alternatively, the probability is .05 that one or more of the population contrasts will not be covered by such a set of intervals.

Nonrandomized Studies

The nonrandomized study formula presented at the end of Section 5.7 in regard to the Dunn–Bonferroni procedure is also appropriate for the Scheffé procedure. The difference between the two procedures is that a different critical value is employed. Rather than using the critical value of the Dunn–Bonferroni t statistic, the Scheffé procedure uses the critical value of F', which is

$$\sqrt{(J-1)F_{(\alpha, J-1, N-J-1)}}$$

5.9 OTHER MULTIPLE COMPARISON PROCEDURES

There are several multiple comparison procedures that are frequently employed with ANOVA that have not been described in this chapter. Some of these procedures, including the Newman–Keuls (Newman 1939, Keuls 1952) test and the Tukey (Tukey 1949) test based on allowances, can be carried out for ANCOVA problems by straightforward modifications of formulas and by substituting the critical values of the Bryant–Paulson generalized studentized range statistic for the studentized range statistic.

Two other popular procedures are the Duncan (1955) multiple range test and Dunnett's (Dunnett 1955; Dunnett and Sabel 1954) procedure. The extensions of these procedures to the ANCOVA case have been derived mathematically (Bryant and Paulson 1976, Paulson 1979), but the critical values have not yet been tabulated. Those interested in these procedures can expect to see Paulson's tabulations published in the statistical literature in the near future.

5.10 SUMMARY

The ANCOVA F test is only a starting point in the analysis of studies with three or more groups. To analyze differences between particular pairs of adjusted means (pairwise comparisons) or differences based on complex linear combinations of adjusted means (complex comparisons), multiple comparison procedures are required. Four different multiple comparison procedures were described: Fisher's protected LSD, the Bryant–Paulson generalization of Tukey's HSD, the Dunn–Bonferroni procedure, and Scheffé's method. Fisher's protected LSD was suggested for hypothesis tests on pairwise comparisons. The Bryant–Paulson approach was suggested for both hypothesis tests and simultaneous confidence intervals on pairwise comparisons. The Dunn–Bonferroni procedure was recommended for hypothesis tests and simultaneous confidence intervals on a small number of comparisons of any type (simple or complex) that have been planned before the data are examined. The Scheffé method was suggested for both hypothesis tests and simultaneous confidence intervals on any comparisons that are suggested by the data or for a large number of planned pairwise and/or complex comparisons. The Bryant–Paulson procedure was recommended as the procedure of greatest utility for most researchers because it provides the narrowest simultaneous confidence intervals on the type of comparisons of most general interest (i.e., all pairwise comparisons).

The correct interpretation of multiple comparison procedures is based on the concept of experimentwise error. The level of α associated with a multiple comparison procedure sets the experimentwise error rate. This rate is the probability (approximate with some procedures) that one or more of the comparisons associated with a multiple comparison procedure will be incorrectly declared significant. Simultaneous confidence intervals are the confidence intervals associated with three of the four multiple comparison procedures. The interpretation of the simultaneous confidence intervals is parallel to the interpretation of the experimentwise error rate. If, for example, 95% simultaneous confidence intervals are employed, the interpretation is that, in 95 of 100 such sets, all population differences between adjusted means will be included in the set of intervals.

CHAPTER SIX
Assumptions

6.1 INTRODUCTION

An examination of the linear model $Y_{ij} = \mu + \alpha_j + \beta_1(X_{ij} - \bar{X}..) + \varepsilon_{ij}$ underlying the one-factor ANCOVA makes it clear that dependent variable scores are conceptualized as the sum of the terms of the model. That is, each observation (Y_{ij}) is viewed as the sum of four independent components: (1) the grand mean of all dependent variable observations (μ), (2) the effect of the jth treatment ($\alpha_j = \mu_j - \mu$), (3) the effect of the covariate [$\beta_1(X_{ij} - \bar{X}..)$], and (4) the error ($\varepsilon_{ij}$) that is the deviation of the observation from the adjusted or conditional mean of the treatment group of which it is a member ($Y_{ij} - \mu_{j\ adj}$). Some assumptions concerning the nature of the data follow directly from an elementary understanding of the model. Other assumptions are not so obvious. An understanding of the specific assumptions and the consequences of their violation is important for the researcher who attempts to evaluate the fit of his data to the model. The hypothesis of no treatment effect can be appropriately tested and interpreted with the fixed effects analysis of covariance when the following assumptions are met:

1. Randomization.
2. Homogeneity of within-group regressions.
3. Statistical independence of covariate and treatment.
4. Fixed covariate values that are error free.
5. Linearity of within-group regressions.
6. Normality of conditional Y scores.
7. Homogeneity of variance of conditional Y scores.
8. Fixed treatment levels.

This list is arranged in an approximate rank order of the importance of the assumptions to the interpretation of the results of covariance analysis. Each assumption is explained in the remainder of this chapter.

98

6.2 RANDOMIZATION*

Randomization is involved in true experiments when (1) subjects are randomly selected from some defined population and (2) when the subjects are randomly and independently assigned to the treatment groups. As with other inferential techniques, the random selection of subjects from a defined population can be disregarded if the experimenter has little concern about the specific population to which the results can be generalized. The random assignment of subjects to treatments is, however, essential for a valid interpretation of F tests and confidence intervals associated with the analysis of covariance in experimental studies.

There are two reasons why the random assignment of subjects is important. First, this procedure results in groups of subjects having the same expected values on *all* characteristics. In other words, this is *the* method of forming equivalent groups. Other procedures such as matching are no substitute for random assignment. The second reason for employing random assignment concerns the distribution of the error (ε_{ij}) in the ANCOVA model. It is assumed that the errors are independent. If subjects are randomly assigned to treatments, it is more likely that the errors will be independent than if subjects are assigned according to a nonrandom procedure. The random assignment to treatments does not, however, guarantee that the errors will be independent. For example, if all subjects within a given randomly formed treatment group are exposed to the treatment at the same time rather than independently, the errors may be correlated because certain extraneous variables may affect many or all subjects within the group and exert a dependency among the responses. Unfortunately, it is very difficult to know when this type of dependence is a problem. An excellent treatment of this issue can be found in Peckham et al. (1969).

Testing the Assumption of Randomization

Tests for randomization are not employed in deciding whether subjects have "really" been randomly assigned to treatments. Obviously, the experimenter should know whether he has in fact employed a procedure resulting in random assignment. Occasionally, it can be argued that nonrandomly formed groups have been formed through an assignment procedure that is random in effect. If the experimenter is willing to state, on the basis of thorough information concerning the subjects, that the groups are not systematically different in terms of characteristics related to the dependent variable, statistical tests may be justified. One way of supporting the contention that the

---

*The "randomization model" discussed by some mathematical statisticians differs from the "normal theory model" described in this text. The distinction between the two models is not discussed because the asymptotic distributions of the tests under the two models are the same (Robinson 1973).

groups are not systematically different is to carry out an ANOVA on the covariate using a liberal α level such as .10 or .25. If the null hypothesis is not rejected, the researcher has evidence that the groups are equivalent on X. This test is not completely satisfactory, however, because the groups may differ on many relevant variables even though the ANOVA on the covariate is not statistically significant.

Subjects should not be matched on the covariate as a means of eliminating group differences. Obviously, this procedure can guarantee a nonsignificant ANOVA on the covariate, but it is also likely to lead to a regression artifact. If the matched (on X) subjects have been selected from populations having different covariate means, the expected mean difference on the dependent variable is *not* zero. Hence both ANOVA and ANCOVA F tests on the dependent variable will be biased. That is, differences may frequently be detected when treatment effects do not exist, or treatment effects that do exist may frequently not be identified. The problems of matching and the regression artifact are related to the issue of the reliability of the covariate. This topic is a very important one to which we return later in this chapter and again in several other chapters. Important works on the problems associated with matching and the regression artifact are those by Campbell and Boruch (1975), Campbell and Erlebacher (1970), McNemar (1940), and Thorndike (1942).

The second randomization issue, that of the independence of the errors, ε_{ij}, is not easily tested. The basic issue here is deciding whether the subjects within treatment groups are responding independently of each other. This is important because dependence can have drastic effects on the F test. Suppose two methods of teaching calculus are being evaluated, and a two-group randomized ANCOVA design with 20 subjects in each group is employed by using a pretest as the covariate and a final exam as the dependent variable. If 12 of the 20 students in the first group are copying all their answers from one of the remaining eight students, several problems arise. First, it should be clear that there will not be 40 subjects responding independently of each other. One student is completely controlling the responses of 12 other students. The degrees of freedom for error in this design would normally be $40 - 2 - 1$ (i.e., $N - J - 1$), but the formula for computing degrees of freedom is based on the assumption that there are 40 independent pieces of information on the dependent variable. Actually, there are only $40 - 12 = 28$ independent responses. The first group is contributing only eight independent scores. Hence the errors (ε_{ij}) are not independent.

Consequences of Violating the Randomization Assumption

It has been pointed out that two problems are likely to occur when randomization is not employed: (1) the equivalence of the populations from which the samples have been obtained is suspect, and (2) the assumption of independent errors is less likely to be met if the subjects are not independently assigned and treated. The consequences of these two problems are discussed in the remainder of this section.

If the subjects have *not* been randomly assigned to treatments, the ANCOVA F test and the adjusted means are likely to be biased. If the subjects *have* been randomly assigned to treatments, and even if the ANOVA on the covariate is statistically significant, the ANCOVA F test and the adjusted means will *not* be biased. The situation in which ANOVA on X is significant even though treatments have been randomly assigned is sometimes described as "unhappy randomization" and can be expected to occur $\alpha(100)$ percent of the time. That is, since significant results are expected to occur $\alpha(100)$ percent of the time when the null hypothesis is true, it is not unheard of for significant differences on the covariate to occur when random assignment has been employed. Indeed, this is a situation in which ANCOVA clearly yields less ambiguous results concerning treatment effects than does ANOVA.

It should be pointed out that there are certain nonrandomized designs (described in Chapter 7) that are unambiguously analyzed by using ANCOVA, but the typical "observational study" to which ANCOVA is often applied does not qualify. In this situation ANCOVA cannot be counted on to adjust the means in such a way that they can be interpreted in the same way as with a randomized design. Although ANCOVA is often employed in this situation, a covariate can't be expected to eliminate the bias that goes along with comparing means from populations that probably differ in many ways.

With randomized designs the difference between adjusted means is an unbiased estimate of what the difference between group means would be if each group had a covariate mean equal to the grand covariate mean. With observational studies (where selection into groups is based on unknown characteristics) the difference between adjusted means is *not* generally an unbiased estimate of what the mean treatment difference would be if each group had a covariate mean equal to the grand covariate mean. The reasons for this bias and difference in interpretation in these two cases is not obvious or simple. Chapters 7, 14, and 15 deal with this issue of attempting to obtain unbiased estimates of treatment effects when random assignment is impossible.

The second major consequence of not employing randomization is possible bias in the F test resulting from dependence of scores. If random assignment is carried out but the observations of the various subjects are not independent of each other, the true probability of making a Type I error may be greater than the nominal value. If the dependence among the scores is positive, the error variance is likely to be underestimated and the degrees of freedom will be fewer than those specified in the computation procedure for ANCOVA. Both of these effects of dependence generally contribute to an increase in the true α.

The extent to which the true α exceeds the nominal α is difficult to compute, but it is clear that dependence can have a large effect. The true α could easily be five or six times the nominal alpha.

Since dependence is an important concern it is reasonable to attempt to identify its existence. Unfortunately, there is no definitive test of the independence assumption for most experimental designs. In the case of randomized-group designs, about all that can be recommended is that experimenters employ all the knowledge they have concerning subject behavior, experimental procedures, and the environment surrounding the experiment to judge independence. If it is clear that dependence is a problem, it is reasonable to reconsider the definition of the sample and the sampling unit. In the case of the two-group experiment described previously, there were 40 subjects but only 28 observations that appeared to be independent. It would be reasonable in this case to omit the 12 redundant observations from the analysis since they provide no new information and can be expected to bias the F test. The experimenter's task is to identify the sampling units that are free to vary; the analysis should then be performed on these units.

The issues of independence become somewhat more complex when designs other than randomized group designs are considered. The repeated measure designs, for example, have the problem of the dependence of (1) several responses from the same subjects and (2) observations from different subjects. Fortunately, the former problem can be identified much more easily than the latter. Most current textbooks on experimental design describe a test of the assumption of "compound symmetry" that is directly related to the independence assumption. We do not consider the details of this test, but the interested reader is referred to Winer (1971). Keep in mind that this test is relevant to only one of the two independence problems in repeated measure designs.

There is one class of design in which the independence of the observations can be thoroughly investigated: interrupted time-series designs. These designs involve many observations across time from a single subject or sampling unit of some kind. Since only one subject or sampling unit is involved the problem of dependence with these designs is one of serial dependence (i.e., dependence of responses across time) rather than dependence among different subjects. Box and Tiao (1965, 1975) and Glass et al. (1975) are important references on interrupted time-series experiments. Box and Jenkins (1970) is the definitive reference on the identification of dependence in time-series data.

6.3 HOMOGENEITY OF WITHIN-GROUP REGRESSION SLOPES

It is assumed that the regression slopes associated with the various treatment groups are the same. The reason for this becomes clear when the β_1 coefficient in the ANCOVA model and the sample b_w coefficient in the adjustment formula are examined. Recall that b_w is the pooled within-group estimate of the population parameter β_1. The fact that β_1 has no superscript to identify different treatments indicates that one common slope is assumed to fit the data. It makes sense to pool the data from the different groups to obtain the single estimate (b_w) only if there is a single parameter (β_1). If there were different population slopes associated with the different treatments, the pooled estimate (b_w) would not be an appropriate estimate of the different population values.

Testing the Assumption of Homogeneity of Regression

A test of the null hypothesis

$$H_0: \beta_1^{(\text{group 1})} = \beta_1^{(\text{group 2})} = \cdots = \beta_1^{(\text{group } J)}$$

is found in Sections 3.5 and 4.6. If this hypothesis is rejected, the conclusion is that the population slopes associated with the various treatments are not the same. This means that the conventional ANCOVA model does not fit the data. Another model should be considered. Some alternative models are described in the chapter on the Johnson–Neyman technique. Reasons for considering another model are described next.

Consequences of Violating the Homogeneity of Regression Assumption

A treatment–slope interaction is said to exist when the slopes are not parallel. Suppose a two group experiment employing one covariate yields the slopes shown in Figure 6.1.

Since the mean Y scores and the mean X scores are the same for both treatments it is obvious that the mean adjusted scores are the same. It would not, however, be wise to interpret the nonsignificant ANCOVA by simply stating that the null hypothesis was retained. It is true that those individuals at the mean on X do not differ on Y from group to group. But it is true *only*

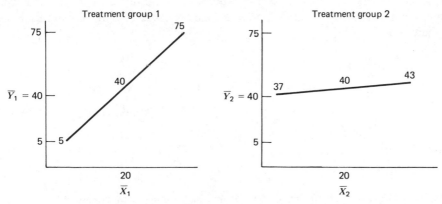

Fig. 6.1 Two heterogeneous regression slopes.

for those individuals with X scores falling at the grand mean $(\bar{X}..)$ on the covariate. Notice that the Y scores are quite different for individuals in the two groups when we look only at those with high or low covariate scores. A complete verbal description of the situation must include the statement that the effect of the treatments depends on the level of X. Subjects with high X scores appear to perform better under treatment 1 than under treatment 2, low X subjects appear to perform better under treatment 2 than under treatment 1, and average X subjects do not appear to differ in performance under the two treatments. Since different interpretations are associated with different levels of X, the use of covariance analysis with heterogeneous within-group slopes should be questioned.

When the slopes of groups 1 and 2 described in Figure 6.1 are plotted on the same diagram, as in Figure 6.2a, the heterogeneous regression slopes problem can be seen to be equivalent to the problem associated with the interpretation of main effects in a two-factor ANOVA with interaction present. An appropriate procedure for dealing with the two-factor ANOVA is to (a) test for $A \times B$ interaction, and if interaction is present, (b) ignore the main effect tests, and (c) test for simple main effects. Similarly, with ANCOVA, an appropriate procedure is to (1) test for heterogeneity of regression, and if heterogeneity is present, (2) ignore the adjusted treatment effects test, and (3) compute and report separate regression slopes for all groups and/or employ the Johnson–Neyman procedure (described in Chapter 13).

In addition to the logical problem of comparing adjusted means when slope heterogeneity is present is the statistical problem of the distribution of the F statistic. The effects of heterogeneous regression slopes on the ANCOVA F test have been the subject of a number of Monte Carlo investigations (Robinson 1969; Peckham 1970; McClaran 1972; Hamilton 1974, 1976; Kocker 1974). The basic conclusion of these studies is that discrepancies in regression slopes have a conservative effect on the ANCOVA F test as long as sample sizes are equal and error variances are homogeneous. That is, when homogeneity of slopes was not present, the probability of Type I error was lower than the nominal value. The rationale for these findings is essentially as follows.

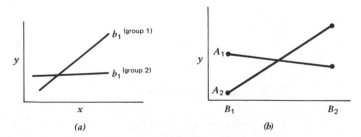

Fig. 6.2 Comparison of one-factor ANCOVA with heterogeneous slopes and two-factor ANOVA with interaction: (a) one-factor ANCOVA; (b) two-factor ANOVA.

Recall that the error sum of squares in ANCOVA ($SSres_w$) is the pooled within-group sum of squared deviations around the pooled within-group slope, b_w. If the sample slopes are not homogeneous, the sum of squares around the pooled slope b_w will be larger than would be obtained if the sum of squares around the individual slopes were computed. Hence as the degree of heterogeneity of slopes increases, the slope b_w will become an increasingly less adequate fit of the data points. That is, the error sum of squares (i.e., squared deviations around b_w) must increase as the degree of slope heterogeneity increases. It should make sense, then, that heterogeneity of slopes will yield smaller ANCOVA F values (smaller than with homogeneous slopes) because the error mean square (the denominator of the F ratio) is an *overestimate* of the population conditional variance. Small F ratios are, of course, associated with large probability values. This is consistent with the general findings of the studies mentioned previously and with the mathematical conclusions reached by Atiqullah (1964).

One study (Hollingsworth 1976) contradicts this general finding. Hollingsworth concluded that the probability of Type I error increases with slope heterogeneity; this conclusion is based on the assumption that the differences between the covariate means are not zero. In randomized studies, however, the expected mean difference between groups is equal to zero for all variables (i.e., dependent variables and covariates) unless the treatment affects the covariate. Hence only in nonrandomized observational studies and randomized experiments where the treatment affects the covariate is it likely that slope heterogeneity will increase Type I error.

In summary, the two consequences of violating the assumption of homogeneity of regression leads to (1) difficulty in interpreting the meaning of a retained null hypothesis (i.e., uncertainty as to whether overall mean effects are masking treatment differences associated with specific levels of X) and (2) biased F tests. Since a complete interpretation of results requires information concerning the slopes, it is suggested that the data be plotted and the test for the homogeneity of regression be carried out routinely in any situation where the analysis of covariance is employed. When the slopes are clearly heterogeneous, the investigator will probably be more interested in examining the regression functions for each treatment group separately than in ANCOVA. This problem and appropriate methods of analysis are discussed in Chapter 13.

6.4   STATISTICAL INDEPENDENCE OF COVARIATE AND TREATMENT

The statistical independence of the covariate and the treatment is not a mathematical assumption underlying the ANCOVA model. It is, however, a condition that should exist for the most unambiguous interpretation of an experimental study. In a randomized two-group experiment, for example, the adjusted mean difference can be interpreted as an unbiased estimate of what the mean difference on Y would be if both groups had exactly the same covariate mean.

The mean difference on X will generally be small in this case (since random assignment has been employed) as long as the treatment has not affected the covariate. Since the mean difference on X is due to sampling fluctuation, the interpretation of the adjusted mean difference as the difference that would be expected if both groups had the same covariate mean presents no logical problem. That is, the adjusted difference can be interpreted as the mean difference that would be expected if a matching design had been employed and if subjects with X scores equal to $\overline{X}..$ had been randomly assigned to treatments.

When the treatment affects the covariate, the results are interpreted in essentially the same way as when the treatment is independent of the covariate, but two problems may arise. First, treatments may produce covariate means that, when averaged, yield a grand covariate mean that has no counterpart in reality. There may be no subjects in existence who do (or even could have) covariate scores equal to the grand covariate mean.

Suppose an investigation is being carried out to evaluate the effects of two different on-the-job methods of assembling a recently designed complex electronic component. Employees are randomly assigned to two groups, and one of the two experimental assembly methods is assigned to each group; the dependent variable is the number of hours of training required to produce a component with no defects. The assembly tasks are performed in various areas of the plant; each area is subjected to different levels of uncontrolled noise from various sources that are unrelated to the assembly task. It is believed that the performance of the employees on the assembly task is affected by the noise.

Since the noise level is not the same at the various work sites associated with the different employees, it could affect the mean difference between the two assembly methods. In an attempt to remove the possible contamination effect of noise from the experimental comparison, noise levels associated with each employee's work site are measured and used as the covariate in ANCOVA. If the treatments (methods of assembly) do not affect the noise level, ANCOVA can be expected to provide an unbiased estimate of what the treatment effect would be if all employees were subjected to the same $(\overline{X}..)$ noise level.

If, however, the characteristics of the methods of assembly of the components constitute the *cause* of the noise, the treatment would be causing the variability in the covariate. In this case it would be questionable to employ noise as the covariate, because, as can be seen in Figure 6.3, to remove variability in Y due to noise would be removing part of the treatment effect. That is, the noise is a part of the treatment package in the sense that it is caused by the assembly procedure.

In this case the hypothesis of equal adjusted population means has little meaning. This is because one of the essential elements of the treatments and

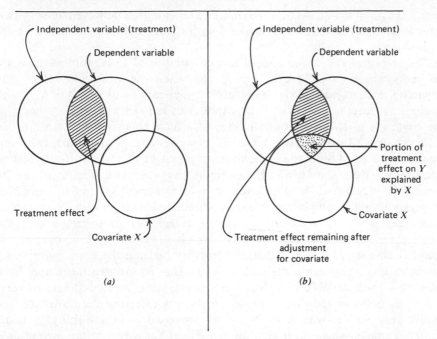

Fig. 6.3   One consequence of treatment affecting covariate: (*a*) treatment effect uncorrelated with covariate; (*b*) treatment effect correlated with covariate.

the covariate overlap. If it is impossible to conceptualize having two treatments without different noise levels, the ANCOVA is comparing two treatments that do not exist in reality.

Suppose the noise levels for the two assembly methods are as follows:

Assembly Method	Mean	Range
1	20 db	10 db
2	90 db	10 db

The grand mean on X is 55, but neither method is associated with a noise level near 55. It is meaningless to interpret the outcome of ANCOVA; it provides an answer to a question that has no relevance. No one cares about what the estimated difference in mean performance would be for subjects who are exposed to a noise level of 55 db. It is not useful to try to estimate this difference because, first, neither method 1 nor method 2 is associated with a 55-db noise level. The research question (viz., "What is the difference in performance under the two methods?") is not answered because one aspect of each treatment is the noise level. Second, even if someone thought it would be useful to know the difference given a noise level of 55 db, there are no observations in the experiment anywhere near 55 db and hence no data to provide an answer. The ANCOVA could be computed, but it is not clear why an investigator would want to estimate and interpret adjusted means that refer to levels of X that are associated with treatments and subjects that are nonexistent.

The preceding discussion should not be interpreted to mean that there are no cases where ANCOVA is appropriate when the treatment affects the covariate. Indeed, one of the more useful applications of ANCOVA is with studies that employ each of several response measures (in several analyses) as the covariate to clarify how the treatment affects behavior. For example, in studying the effects of two types of exercise on resting heart rate and frequency of psychosomatic complaints, it might be of interest to use resting heart rate as the covariate and frequency of psychosomatic complaints as the dependent variable. If the difference between treatment groups on frequency of psychosomatic complaints is largely eliminated when resting heart rate is employed as the covariate, it would be reasonable to *speculate* that the treatment effect on the dependent variable is mediated by the covariate. It couldn't be *concluded* that X causes Y to vary because there are many other variables that could cause variability on Y. That is, the treatment may have affected a third variable that causes both complaints and heart rate to vary.

At this point in the discussion of treatments affecting the covariate, the reader may wonder why ANCOVA was dismissed as nonmeaningful in the earlier example (where noise was the covariate) but useful in the second case (where heart rate was the covariate). This is because the covariate (noise level) was an integral part of the treatment in the first case, but this was not true in the second case. It is reasonable to employ resting heart rate as a covariate because the treatments do not limit the possible resting heart rate values; the treatments did directly control noise level.

Suppose the mean resting heart rates for the exercise groups 1 and 2 were 75 and 85, respectively. It is not difficult to argue that an experiment could be carried out by randomly assigning subjects (each having a resting heart rate of 80) to two different exercise programs. Recall that it did not make sense to consider assigning subjects associated with 55-db noise levels to treatments in the other experiment.

A second possible problem in the interpretation of ANCOVA when the covariate is affected by the treatment concerns the reliability of the covariate. Suppose that the mean difference on the number of psychosomatic complaints (Y) is not explained by heart rate (X) in the previously described experiment. That is, the ANCOVA F is significant (or the confidence interval does not contain zero). It can be appropriately concluded that the adjusted difference is independent of the variability on X *as measured in this experiment*. This is not the same thing, however, as saying that the adjusted mean difference is independent of the basic characteristic X. If this distinction is not understood, the results of ANCOVA are likely to be misinterpreted. In most cases the researcher wants to know what the adjusted Y difference is when true scores (perfectly reliable measures of the basic characteristic) rather than actual observed scores on X are held constant. Conventional

ANCOVA does not generally answer this question if the treatment affects the covariate in a randomized design or when nonequivalent observational groups are employed. Hence the difference between adjusted means is usually a biased estimate of the treatment effect when these designs are employed. This problem is discussed further in Section 6.5 and in Chapter 14. It is recommended that true-score ANCOVA (see Chapter 14) be employed when the treatment affects the covariate in randomized designs (Overall and Woodward 1977). This recommendation holds when it makes sense to investigate treatment effects for a population of subjects falling at the grand mean on X.

Testing the Assumption of Independence of Treatment and Covariate

It is possible to test the assumption of the independence of the treatment and the covariate by performing an ANOVA on the covariate. In a randomized experiment in which the covariate has been measured before the treatments are administered, there is no reason to report the result of such a test since the treatment can't have an effect in this case. If the covariate is measured during or after the treatment, the ANOVA on the covariate should be carried out because it provides information on the effect of the treatments.

For most nonrandomized studies, the ANOVA on the covariate should be carried out regardless of whether treatments have been applied. If treatments have not been applied but the ANOVA F is significant, this is a warning that the ANCOVA F and adjusted means are almost certain to be biased as reflections of the treatment effects. If the ANOVA on the covariate is not significant, the ANCOVA F and adjusted means are still subject to problems of bias, but the degree of bias, although unknown, may not be large. Unfortunately, it is impossible to know.

There is one experimental design, the biased assignment design (described in the next chapter), that is not a randomized design, but it is not subject to the problems of bias in the ANCOVA F and adjusted means. If a nonrandomized design other than the biased assignment design is employed and the covariate is measured after treatments are administered, the ANOVA on the covariate as well as the ANOVA and the ANCOVA on Y will be essentially uninterpretable because treatment effects and pretreatment differences among the populations will be confounded.

Summary of Consequences of Violating the Assumption of Independence of Treatment and Covariate

If the treatment affects the covariate in a randomized group experiment, the hypothesis tested is not, in general, the same with ANCOVA as with ANOVA. That is, the adjusted and unadjusted population means are not generally the same when the population covariate means are different. The covariate may either remove or introduce differences on Y that may be misinterpreted as reflections of treatment effects. Also, the adjusted mean difference may be impossible to interpret meaningfully because such a difference may be the predicted mean difference for nonexistent treatment conditions and subjects.

6.5   FIXED COVARIATE VALUES MEASURED WITHOUT ERROR

The interpretation of any type of statistical analysis is related to the nature of the measurement model underlying the variables employed. For example, the values of the covariate can be conceptualized as being (1) fixed with no measurement error, (2) fixed with measurement error, (3) random with no measurement error, or (4) random with measurement error. If all the population values of X about which inferences are going to be made are included in the sample, the covariate can be conceptualized as a fixed variable. In this case additional samples drawn from the same population will include the same values of X. Hence if the covariate is an attitude scale and the X values obtained (in an unreasonably small sample) are 27, 36, 71, and 93, the covariate would be considered to be fixed if (1) the values of X in future samples drawn from the same population include only 27, 36, 71, and 93, and (2) the experimenter is interested in generalizing the results to a population having these four values. This, of course, is not very realistic. If the range of possible scores on the attitude scale is zero through 100, for example, it is very unlikely that future samples will include only the four values obtained in the first sample. It is also unlikely that the experimenter would be interested in limiting the generalization of ANCOVA to a population having covariate values of 27, 36, 71, and 93.

Future samples will almost certainly contain different values of the covariate. Since subjects are randomly sampled from the population, it is realistic to view the X values in a given experiment as a random sample of X values from a collection of X population values that range (in this example) from zero to 100. One reason, then, why future samples will generally have different values of X is because X is a random variable.

Another issue to be considered in the conceptualization of the covariate is measurement error. Almost all characteristics are imperfectly measured. If additional measures of the covariate (either repeated measures with the same measuring instrument or alternative measures of the same characteristic) are employed with the same subjects who originally obtained the X scores of 27, 36, 71, and 93, it is unlikely, because of errors in measurement, that exactly the same values will be obtained once again. Hence "measurement error" refers to imperfections in the measurement process employed to collect data; it does not refer to sampling error. This problem of measurement error or unreliability is one of the central issues in psychometric theory. Nunnally (1978) should be consulted for an excellent discussion of these topics.

Since the covariate is likely to be a random variable in most studies (virtually all studies in the behavioral and social sciences) and since measurement error is almost always present, it is realistic to view X as a random variable measured with error. Interestingly, the analysis of covariance was derived under the assumption that the X variable is fixed and measured without error. We see later that whereas violations of this assumption are not important with all applications of ANCOVA, there are some cases in which serious interpretation errors will result.

Consequences of Violating the Assumption of Fixed Covariate Values

The consequences of violating the assumption of fixed error-free X values depend on the nature of the violation and the type of research design employed. The effects of three types of violation are presented in this section.

Case 1.   If the covariate is fixed but measured with error, the exact consequences in terms of bias in F or adjusted means are not known. The problem has been investigated, however, by Calkins (1974) with the use of Monte Carlo computer simulation procedures. His conclusions are that (1) there is little discrepancy between the actual and nominal Type I error rate, and (2) the estimates of the population slope and intercept (or adjusted mean) are not seriously affected when the covariate is fixed but measured with error. An expected effect of measurement error is lower power, but Calkins did not report results on this. Keep in mind that this case (X fixed but measured with error) is not the typical case encountered in most research applications.

Case 2.   The next case to be considered is where the X variable is random but measured without error; this has been investigated by Calkins and Jennings (1972) and Calkins (1974) with the use of Monte Carlo procedures. Their results indicate that as long as the sample size is at least 12, the use of random variables (without error) rather than fixed variables has little consequence of practical concern for the researcher. This outcome applies to both the ANCOVA and the homogeneity of regression test. The conclusion that ANCOVA is appropriate with a random covariate is consistent with the suggestions of Scheffé (1959, pp. 195–196) and Winer (1971, p. 765) that X need not be a fixed variable.

Case 3.   Behavioral science applications of ANCOVA generally involve the use of covariates that are considered random and measured with error. Unfortunately, this case can present serious problems of interpretation. When measurement error is present, the obtained ANCOVA F may be either too high, too low, or correct. In general, when a true randomized experiment is involved and subjects are randomly assigned to treatment groups, the reliability of the covariate is not critical; the sensitivity (power) of the analysis is, however, lowered. It is when the ANCOVA is used to adjust means in nonrandomized studies that the measurement error problem is of *great* concern. The following example, originally presented by Lord (1960), suggests the nature of the problem.

Suppose that we have two populations with different means on both X and Y. Also suppose that the reliability of X is perfect and that the common slope obtained by regressing Y on X is 1.0. Now draw a sample from each population and plot the data. The data in Figure 6.4a illustrate what such a plot would look like. Notice that the group means on the covariate, as well as on the dependent variable, are quite different. Keep in mind that the mean difference on the covariate is not a chance difference; these sample data are from populations having different covariate means. This is clearly not a randomized group experiment. Rather, it is an observational study in which there is interest in employing ANCOVA to reduce bias in the difference between the Y means that can be accounted for by the covariate X.

There is nothing unreasonable about attempting to remove bias in Y that can be accounted for by the covariate. Indeed, in Figure 6.4a it can be seen that the adjustment has worked perfectly in the sense that there are no differences on Y that cannot be predicted from X. This is as it should be if X and Y are both measures of the same characteristic. But keep in mind that the reliability of X was unrealistically set at 1.0; that is, measurement is perfectly reliable.

Let's now take the data of Figure 6.4a and introduce measurement error into the covariate. Since error has been added to the X variable, the scores are spread horizontally around the regression line and the correlation between X and Y can no longer be perfect; thus regression slopes must change as indicated in Figure 6.4b.

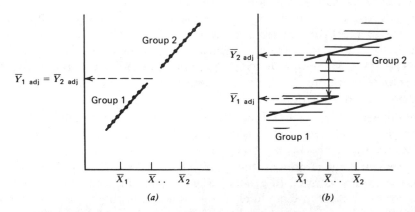

Fig. 6.4 Effect of measurement error in X on adjusted means based on samples from two different populations. (a) ANCOVA two-group situation with no error in X. (b) ANCOVA two-group situation with error introduced into X.

The extent to which measurement error affects the slope can be seen in the following formula:

$$\hat{\beta}_e = \beta_p \left(\frac{\sigma_{tr}^2}{\sigma_{obs}^2} \right)$$

where

$\hat{\beta}_e$ = expected value of slope when measurement error is present in covariate

β_p = population slope based on a perfectly reliable covariate

σ_{tr}^2 = true-score variance in X (i.e., variance of X scores that, hypothetically, have been measured without error)

σ_{obs}^2 = observed variance of X (i.e., σ_{tr}^2 + variance due to measurement error = σ_{obs}^2).

If there is no measurement error, the true-score variance and the observed score variance are the same, which means that, in this case,

$$\hat{\beta}_e = \beta_p(1) = \beta_p$$

When measurement error is present, the term $(\sigma_{tr}^2/\sigma_{obs}^2)$ will be less than one. Therefore, the slope associated with scores that contain measurement error is less steep than the slope based on X measured without error. [See Cochran (1968) and Fuller and Hidiroglou (1978) for additional details on the effects of measurement error.] Since measurement error is virtually always present, the regression slopes in ANCOVA are virtually always underestimated relative to what they would be with a perfectly reliable covariate. Hence the major consequence of carrying out ANCOVA in the case of observational studies is that the adjusted mean difference might be misinterpreted. This can be seen by comparing the two situations in Figure 6.4.

Notice the difference between the slopes under the error and no-error situations and the effect this difference has on the adjusted means. Without error in X the adjusted Y means are identical; with error in X the adjusted means are quite different. (Introducing error into Y does not affect the slope.) It is important to note that this error–no-error outcome discrepancy has occurred even without change in the mean scores on X or Y in the two situations. The point of the example is that *if the covariate means are not equal, the difference between the adjusted means is partly a function of the reliability of the covariate.*

Figure 6.5 shows that the difference between adjusted means can be too small *or even in the wrong direction* (relative to what it would be with perfect measurement) when the covariate contains measurement error. Notice in Figure 6.5a that the group 1 adjusted mean exceeds the group 2 adjusted mean when true scores (scores measured without error) are available as the covariate. In this case of groups from two different populations, neither the X nor the Y means are equal. Group 1 is superior to group 2, but if X is

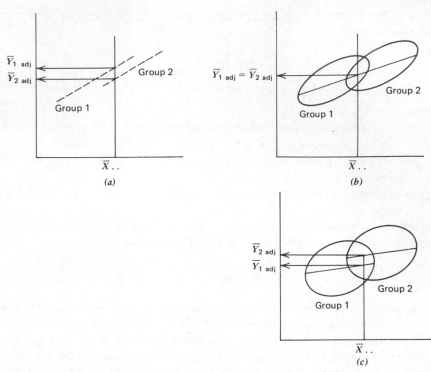

Fig. 6.5 Effects of varying degrees of measurement error on adjusted means. (*a*) Situation with no error in X and $\overline{Y}_{1\,adj} > \overline{Y}_{2\,adj}$. (*b*) Situation with moderate amount of error introduced into X that results in the conclusion that the adjusted means are equal. (*c*) Situation with large amount of error introduced into X that results in the conclusion that $\overline{Y}_{2\,adj} > \overline{Y}_{1\,adj}$.

measured with moderate measurement error, as is depicted in Figure 6.5*b*, the adjusted means will be equal even though there actually is a group effect. In Figure 6.5*c* the covariate is measured with greater error than in Figure 6.5*b* and the adjusted mean difference leads to the conclusion that group 2 is superior to group 1.

Several comments are in order before the reader gets the impression that ANCOVA always yields misleading results. First, the problem of "incorrect" adjusted means just described is not a problem with all research designs. Remember that the design here was one in which samples were drawn from two *different* populations. It was a simple observational study where X was employed to remove bias in the comparison of the two groups on Y. The hypothetical situations were designed in such a way that the true (or completely reliable) X variable completely accounted for bias in Y, but the unreliable X variable did not account for all bias in the comparison of the two groups on Y. Actually, the ANCOVA performed what it was supposed to do in both cases; some methodologists (e.g., Bock 1969, Meehl 1970) argue that the problem is not with the analysis but with the interpretation.

Recall that one way of interpreting an adjusted mean difference is to state that it is the difference on Y that would be expected if the groups all had the same mean scores on X. This statement holds for the case of unreliable as well as reliable X scores and for both randomized and observational studies. But the insidious problem here is that the mathematical expectation for the adjusted mean difference, under the hypothesis of *no* treatment effects, is *not* always zero!

When subjects in the various groups are not from the same population, the expected adjusted mean differences are generally *not* zero when there are no treatment effects. Hence a significant ANCOVA F and the associated adjusted means should *not* be interpreted as a reflection of treatment effects when the groups have not been selected from a single population. This interpretation problem is not peculiar to ANCOVA. As is pointed out in Chapter 7, the problem is a general one; it is probably inappropriate to damn ANCOVA (as many have done) when the same problem is associated with even the simplest matching designs and analyses—ANCOVA is innocent; measurement error and nonequivalent group studies are culpable. But there is some hope for useful analyses of the nonequivalent group design; Chapters 14 and 15 describe several of these possibilities. Designs other than randomized group and nonequivalent group studies exist and can be very useful. Analysis of covariance is an ideal method for analyzing several designs that are described in Chapter 7.

The reliability of the covariate is not a major consideration with most designs that are based on subjects selected from a single population. The effect of measurement error with these designs is only one of a reduction in power. Since measurement error attenuates the correlation between X and Y, the error term in ANCOVA may not be much smaller than with ANOVA when measurement error is great. But bias is *not* introduced to the adjusted means by measurement error with these designs as long as the treatment does not affect the covariate. Since the error mean square is likely to be somewhat smaller with ANCOVA (even when reliability of X is not high) than with ANOVA and since some adjustment of the means will occur as long as $b_w \neq 0.0$ (and covariate mean differences exist), there is no reason to avoid the use of covariance analysis. This is true regardless of the reliability of the covariate as long as (1) it is known that the subjects have been randomly assigned to treatments from a single population or assigned on the basis of the covariate score (an approach described in Chapter 7), and (2) the treatment does not affect the covariate.

6.6 LINEARITY OF WITHIN-GROUP REGRESSIONS

It is assumed that the relationship between the covariate scores and the dependent variable scores is linear for each treatment group.

Testing the Assumption of Linearity

Several procedures are available for testing linearity in conventional regression problems. Useful references on these procedures are Draper and Smith (1966) and Neter and Wasserman (1974). These linearity tests, although very useful in many situations involving regression analysis, are not presented here. A brief description of a procedure for identifying nonlinearity by comparing linear and nonlinear ANCOVA models is presented in Chapter 9.

Consequences of Violating the Linearity Assumption

Recall from elementary correlation and regression analysis that if a linear model is employed where a nonlinear model fits the data, the degree of relationship between X and Y is underestimated. That is, the simple correlation will be much smaller than the correlation ratio or the index of correlation if the degree of nonlinearity is great. Correspondingly, in ANCOVA the reduction of the total and within-group sums of squares after adjustment for X will be too small if the linear model is inappropriate. The most obvious consequence of the underadjustment is that the utility of the covariate will be diminished because the adjustment of the means (if any) and the gain in power of ANCOVA over ANOVA depends on the degree of linear relationship between the covariate and the dependent variable.

A mathematical treatment of the bias problem resulting from the use of a linear model when a quadratic model is appropriate can be found in Atiqullah (1964). Elashoff (1969) provides a somewhat simplified explanation of Atiqullah's paper.

The number of studies in which nonlinearity is a problem does not appear to be great in most areas of the behavioral and social sciences. Even though many studies can be found in which nonlinear models appear to fit the data slightly better than do linear models, the gain is generally insufficient to justify the complexity of interpretation required with nonlinear models.

6.7 NORMALITY OF CONDITIONAL Y SCORES

Dependent variable scores conditional on covariate scores are assumed, in the derivation of the ANCOVA F statistic, to be normally distributed for each treatment group. That is, the Y scores are normally distributed at each level of X as shown in Figure 6.6. It is not, however, assumed that X scores are normally distributed for each value of Y. Rather, as indicated previously, X is assumed to be error free and fixed.

Testing the Assumption of Normality

Many tests of normality are available (Shapiro, et al. 1968, D'Agostino 1971), but at present it is not clear that the normality assumption should be tested before ANCOVA is carried out.

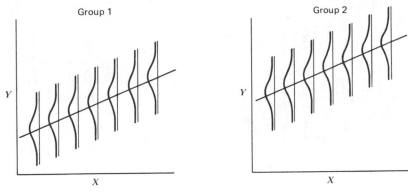

Fig. 6.6 Conditional Y distributions.

Consequences of Violating the Normality Assumption

Analysis of covariance may be more sensitive to departures from normality than is the case with ANOVA. Atiqullah (1964) has shown that the extent to which ANCOVA is biased by nonnormality is largely determined by the distribution of the covariate. Even though normality of the covariate is not an assumption of the ANCOVA model, the bias introduced into the analysis by nonnormality of the dependent variable is greater when X is not normally distributed. However, it has been pointed out by Glass et al. (1972) that Atiqullah's findings were based on distributions with nonnormality reflected in the kurtosis. The effects of skewness were not investigated. Monte Carlo studies of the effects of several combinations of sample size, skewness, and kurtosis are needed to determine the degree of bias introduced by departures from normality. It is likely, however, that nonnormality in the dependent variable has little effect on the ANCOVA F in most behavioral science studies because the covariates employed in these areas are generally at least crude approximations of normally distributed random variables. Atiqullah showed that nonnormality in Y in this type of situation has little effect on the ANCOVA F test in balanced designs. Rank analysis of covariance, described in Chapter 12, is recommended when there is an obviously large deviation of both X and Y distributions from normality.

6.8 HOMOGENEITY OF VARIANCE OF CONDITIONAL Y SCORES

It is assumed that (1) the variance of the conditional Y scores is the same for each treatment group and (2) that the variance of the conditional Y scores does not depend on the value of X. Elashoff (1969) describes two cases in which a violation is present:

Case 1. The variance of the Y scores is constant across treatments for specified X values, but X is related to the variance of Y. Notice in Figure 6.7

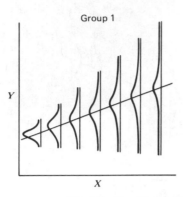
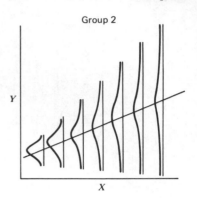

Fig. 6.7 Conditional Y distributions with variances increasing with X.

that the overall variance for the two distributions is the same but that the variance for Y (conditional on individual X values) increases as X increases. (The term often applied to case 1 heterogeneity is *heteroscedasticity*.)

Case 2. The treatment group variances on Y differ, but within each treatment group the variances of the Y scores are constant across different levels of X. An example of Case II violation is illustrated in Figure 6.8.

Tests of Homogeneity of Variance Assumption

The conventional tests of homogeneity of variance can be applied (with appropriate modification of degrees of freedom) to the dependent variable scores conditional on the covariate. Descriptions and evaluations of several tests of homogeneity of variance for the ANOVA can be found in Games, et al. (1972), Gartside (1972), and Layard (1973).

The procedure presented here, which is based on the Bonferroni F_B distribution (Huitema, in preparation), is appropriate for any number of

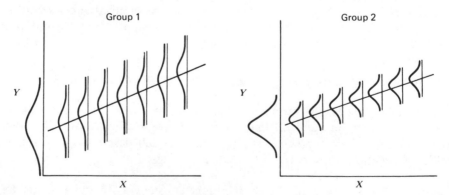

Fig. 6.8 Heterogeneous variances across groups.

groups and covariates and for any combination of sample sizes. The computation procedure consists of the following steps.

1. Compute the residual sum of squares (for each group) around the pooled within-group slope b_w. This can be computed for the jth group by multiplying $(1 - r_w^2)$ by the within-group sum of squares on Y for the jth group:

$$j\text{th group SSres} = (1 - r_w^2)\Sigma y_j^2$$

where

$$r_w = \frac{\Sigma xy_w}{\sqrt{(\Sigma x_w^2)(\Sigma y_w^2)}}$$

2. Compute the conditional variance estimate $(s_{y_j|x}^2)$ for each group by dividing the residual sum of squares by the corresponding degrees of freedom, which are $n_j - 1 - C$, where n_j is the sample size associated with the jth group and C is the number of covariates.

3. Form the F ratio of the largest variance estimate $(s_{y|x_{largest}}^2)$ over the smallest variance estimate $(s_{y|x_{smallest}}^2)$.

4. Compare the result of Step 3 with the critical value of the Bonferroni F_B statistic that is found in Table A.8. For a test at level α experimentwise, the critical value is

$$F_{B(\alpha/2, C', n_{largest} - 1 - C, n_{smallest} - 1 - C)}$$

where $\alpha/2$ is one half the desired experimentwise α level and C' is $[J(J-1)]/2$, which is the "number of comparisons," and J is the number of groups. If the obtained statistic $F_{B_{obt}}$ is equal to or exceeds the critical value of F_B, the hypothesis of equal conditional variances

$$H_0: \sigma_{y_1|x}^2 = \sigma_{y_2|x}^2 = \cdots = \sigma_{y_J|x}^2$$

is rejected.

Computational Example

The data summarized in Table 3.2 are employed in the computational example of the homogeneity of conditional variance test. The hypothesis being tested is

$$H_0: \sigma_{y_1|x}^2 = \sigma_{y_2|x}^2 = \sigma_{y_3|x}^2$$

Step 1.   The residual sum of squares for each of the three groups are as follows:

Sum-of-squares Residual		
Group 1	Group 2	Group 3
$(1 - 0.5247)1064 = 505.72$	$(1 - 0.5247)896 = 425.87$	$(1 - 0.5247)1576 = 749.07$

The value 0.5247 is r_w^2, which was computed by using the sums of squares and cross products shown in Table 3.2.

$$r_w = \frac{\Sigma xy_w}{\sqrt{(\Sigma x_w^2)(\Sigma y_w^2)}} = \frac{3252}{\sqrt{(5700)(3536)}} = 0.72436$$

and $r_w^2 = (0.72436)^2 = 0.5247$.

Step 2. The conditional variance estimates are as follows:

Conditional Variance Estimates

Group 1	Group 2	Group 3
$505.72/10-1-1=63.22$	$425.87/10-1-1=53.23$	$749.07/10-1-1=93.63$

Step 3. The largest : smallest estimates ratio is

$$\frac{93.63}{53.23} = 1.76 = F_{B_{obt}}.$$

Step 4. The critical value of the Bonferroni F_B statistic, using experiment-wise (familywise) $\alpha = .10$, is obtained by entering the table with

$\alpha/2 = .05$

$C' = 3(3-1)/2 = 3 =$ the value for entering the "number of comparisons" column in the F_B table

$df_1 = n_L - 1 - 1 = 10 - 1 - 1 = 8$

$df_2 = n_S - 1 - 1 = 10 - 1 - 1 = 8$, where n_L and n_S are the sample sizes associated with the largest and smallest variances, respectively.

The value found in the table is

$$F_{B_{(.05, 3, 8, 8)}} = 5.10$$

Since the obtained value is less than the critical value (i.e., $1.76 < 5.10$), the hypothesis of equal conditional variances is retained.

Consequences of Violating the Homogeneity of Variance Assumption

The work of Kocher (1974) and Shields (1978) suggests that the violation of the assumption of homogeneity of error variances is not likely to lead to serious discrepancies between actual and nominal Type I error rates unless sample sizes are unequal. However, as in the case of nonnormality, the effects of heterogeneous variances on conditional values of Y are related to the distribution of the covariate. When the covariate is approximately normally distributed, it is unlikely that the ANCOVA F test is affected enough by heterogeneous variances to be of practical concern as long as the design is balanced (i.e., has equal sample sizes). Shields (1978), in her investigation of the effect of heterogeneity of variance in ANCOVA, found essentially the same pattern of results as has been found for ANOVA. When variance sizes and sample sizes differ, the F is conservative if the larger variances are associated with the larger sample sizes and the smaller variances are associated with the smaller sample sizes. When the smaller variances are associated with the larger sample sizes, the bias is liberal (i.e., the true α is greater than the nominal α).

Hence if the differences among sample sizes are large and the conditional variances are not homogeneous, the ANCOVA F may be biased in either a conservative or a liberal direction. With an unbalanced design and a high degree of nonnormality in the distribution of X, heterogeneity of variance may lead to substantial bias in the distribution of F. Rank ANCOVA (see Chapter 12) is recommended in this situation.

6.9 FIXED TREATMENT LEVELS

It is assumed that the treatment levels are fixed by the experimenter. That is, the treatment levels included in the experiment are not selected by randomly sampling the population of possible treatment levels. Rather, the levels selected are the specific levels of interest to the experimenter, and the generalization of the results of the experiment is with respect to these levels.

Test of the Assumption of Fixed Treatment Levels

Obviously, no test of this assumption is possible—the experimenter must decide how treatments are selected.

Consequences of Violating the Assumption of Fixed Treatment Levels

If the experimenter randomly selects treatment levels and then employs a fixed effects ANCOVA, the variance estimates will be correct provided that the design has only one factor. If two or more factors are involved and the levels of one or more factors are randomly selected, an advanced treatment of the analysis of variance (e.g., Kirk 1968, Winer 1971) should be consulted for modifications of computation and interpretation associated with mixed and random effects models. These references describe modifications for ANOVA models, but these modifications generalize in a straightforward manner to ANCOVA. The properties of tests based on these modifications are not, however, well known. Some of the problems associated with ANCOVA in the mixed model situations are described in Federer (1957) and Henderson and Henderson (1979).

6.10 SUMMARY

The assumptions of greatest concern in true experiments are those of random independent sampling, homogeneity of regression, and covariate–treatment independence. If the experimenter has appropriately assigned treatments to subjects and has treated subjects within each group independently, the most important assumption should be satisfied. The homogeneity of regression assumption is easily tested by using the F test designed for this purpose. If the covariate is measured after the treatment is administered, the assumption of covariate–treatment independence should be tested by performing an ANOVA on the covariate. If the covariate is affected by the treatment, the adjusted means must be interpreted very carefully.

The reliability of the covariate is critical in nonrandomized observational studies, but not in most biased assignment and randomized group experiments. The reason for this is that unreliability in nonequivalent group studies can lead to statements concerning adjusted means that are very misleading. With most randomized designs, the problem of unreliability simply results in F tests having lower precision than that in F tests with highly reliable covariates. True-score ANCOVA is recommended as an alternative to conventional ANCOVA as a method of correcting for unreliability in the covariate in nonequivalent groups studies and randomized studies where the treatment affects the covariate.

Linearity of within-group regressions may be considered less important than the previously mentioned assumptions because (1) most behavioral science studies find the linear model adequate, and (2) the effect of moderate nonlinearity is a slight conservative bias on the ANCOVA F test. When nonlinearity is suspected, an alternative method such as nonlinear ANCOVA should be considered. The effects of not meeting the assumptions of normality and homogeneity of variance of the conditional Y scores are similar to the effects of violations of corresponding assumptions with ANOVA. It is known, however, that there is an additional consideration with ANCOVA; the effect of nonnormality and heterogeneity of variances on bias in the ANCOVA F test is related to the distribution of the covariate. It appears unlikely, however, given the types of covariates generally employed in the behavioral sciences, that departure from these assumptions result in serious bias of the F test when balanced designs are employed. Rank ANCOVA is recommended when either or both of these assumptions are obviously violated in randomized experiments.

CHAPTER SEVEN
Design and Interpretation Problems

7.1 INTRODUCTION

The extent to which the assumptions for the analysis of covariance can be met is closely related to the design of the experiment and the data-collection method. Some of the problems associated with the application of this technique are rather subtle and difficult to identify unless the researcher is familiar with a variety of specific situations in which the model may not be appropriate. An overview of several design situations is presented in Table 7.1. These designs are presented in a rough rank order of interpretability.

Table 7.1 Basic Characteristics of Six Designs in Which ANCOVA is Employed

Research Design	Basic Characteristics
I. Completely Randomized Experiment	Subjects are randomly and independently assigned to treatments
II. Randomly Assigned Groups True Experiment	Several intact groups are randomly assigned to each treatment. The mean of each group is used as the unit of analysis
III. Randomly Assigned Groups Quasi Experiment	One group is randomly assigned to each treatment. The individual subject is used as the unit of analysis
IV. Comparative-Developmental Study	Subjects from two or more populations are observed at two points in time
V. Biased Assignment Experiment	Subjects are selected from high and low ranges of a single population on the basis of X and then assigned to treatments
VI. Nonequivalent Groups Study	Subjects are selected from different populations

Interpretation of the first two or three situations is relatively straightforward, but the last design may require many qualifications. Details on these designs and the associated interpretation problems are described in the next section.

7.2 RANDOMIZED DESIGNS

Four varieties of randomized designs are described in this section.

Randomized Experiment—Covariate Measured before Treatment

The completely randomized experiment involves the following steps:

1. Subjects are randomly selected* from some defined population.
2. Covariate measures are obtained.
3. Subjects are randomly assigned to treatments.
4. Treatments are applied.
5. Analysis of covariance is computed.
6. A conditional statement concerning differences among treatment population means is made (e.g., "the difference between adjusted treatment means represents an effect that is independent of the covariate").

If these steps are correctly carried out and the covariate has relatively high correlation with the dependent variable, the power of the experiment will be higher than would be the case if an analysis of variance were employed. The principal reason for employing the covariate in the completely randomized experiment is to reduce the size of the error term rather than to adjust the treatment means. Since the subjects are randomly assigned and the covariate measures are obtained before the treatments, pretreatment differences between group means will generally be slight on both X and Y variables. The major function of ANCOVA in this situation is not to reduce systematic bias since no systematic bias is present with randomization. If a sample difference between group means on X exists, the Y means will be adjusted by an appropriate amount, but this adjustment is for *chance* differences on X—not systematic bias. Completely randomized experiments employing a covariate are interpreted in essentially the same way as an ordinary analysis of variance; the hypotheses associated with both analyses are the same.

ANCOVA versus Randomized-block ANOVA. It may have occurred to the reader that the error-reducing function of the covariate in ANCOVA is

---

*If subjects are not randomly selected from a defined population, inferential tests are still valid if randomization is involved in the assignment of subjects to treatments or treatments to subjects. Generalization of results from the sample to the population from which the subjects were selected must then be based on nonstatistical grounds.

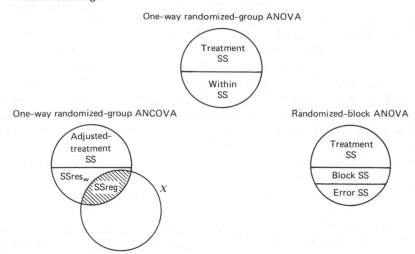

Fig. 7.1 Variability partitions for one-way ANOVA, one-way ANCOVA, and randomized-block ANOVA.

handled by the blocking variable in a randomized block ANOVA. That is, the covariate and the blocking variable are both employed to reduce the size of the error term by systematically accounting for part of the variability that is treated as error in a one-way randomized group ANOVA. This similarity is illustrated in Figure 7.1.

To the extent that the covariate or the blocking variable is correlated with the dependent variable, the error sum of squares is reduced and the power of the test is increased. The choice between the two methods will generally be dictated by design considerations rather than statistical properties. Recall that the randomized-block design requires the experimenter to (1) collect blocking variable measurements on a relatively large number of subjects, (2) select subjects who are homogeneous on the blocking variable for each block, and (3) randomly assign subjects within each block to treatments before the experiment is initiated. If these procedures are feasible, there is little reason to use covariance instead. The advantages of the randomized-block approach over ANCOVA are that the analysis is computationally simple and function free. That is, regardless of the form of the relationship between X and Y, the randomized-block approach will effectively remove between-subject variability. This is not true of ANCOVA. If a great deal of nonlinearity exists and the linear ANCOVA is employed, the effectiveness of the covariate will be reduced or eliminated. Nonlinear ANCOVA can be applied in this situation if nonlinearity is suspected; however, an investigator need not be aware of the correct function when the randomized block design is employed.

There are some problems, however, with the randomized-block approach. First, it is essential that X scores be measured and available *before* the experiment is carried out. It may be inconvenient to measure on X before the treatment groups are formed. Recall that treatments under ANCOVA are simply randomly assigned to subjects; the values of X are not required in the design stage of the experiment. A second problem with the randomized-block design is that it is often difficult to obtain a sufficient number of subjects at the extremes of the X distribution to form homogeneous blocks. If an insufficient number of very high or very low X (blocking) scores is available, the experimenter is forced to either form blocks with large within-group variability or to limit his sample to a narrow range of the X distribution by discarding those subjects having extreme scores. The consequence of both courses of action is a reduction in the power of the test. This problem is avoided with the analysis of covariance because there is no need for an equal number of subjects at various points on the X dimension.

Another problem with blocking is found when the experimenter employs a two-factor factorial design, which has at least two subjects per cell, rather than a randomized block design, which has one subject per cell. In this case the first factor is treatments and the second factor is blocks. The greatest power is obtained with such a design when the optimal number of blocks is employed. The problem is that the experimenter must consider the sample size, the number of treatment levels, and the population within-group correlation ρ_{xy_w} to select the optimal number of blocks (Feldt 1958). Since an estimate of ρ_{xy_w} is frequently not available, the optimal number of blocks is often not employed, and the power of the analysis may be considerably lower than with ANCOVA. If the optimal number of blocks is chosen, studies (Feldt 1958) of the relative precision of ANCOVA and the factorial design employing blocks as one factor reveal that ANCOVA is more powerful when ρ_{xy_w} is $\geqslant .6$; however, a clear advantage for ANCOVA is found only when the correlation is greater than .8. When the correlation is below .4, the precision of ANCOVA is equal to or less than the factorial design. For correlations below .2 and small N, the simple randomized group design may be a reasonable alternative to both ANCOVA and the factorial design because neither of the more complex approaches yields appreciably greater precision. In fact, it can be shown that these more complex approaches can be less powerful than the simple randomized-group ANOVA if the number of treatments is large, the sample size is small and ρ_{xy_w} is very low or negative.

ANCOVA versus Randomized Group ANOVA.   Cochran (1957) has provided a convenient formula for comparing the error variance under randomized group ANOVA and ANCOVA models. If σ_e^2 is the error variance under the ANOVA model, the error variance under the ANCOVA model is approximately

$$\sigma_e^2\left(1 - \rho_{xy_w}^2\right)\frac{N-J}{N-J-1}$$

where

ρ_{xy_w} = correlation between covariate and dependent variable
$N - J$ = error degrees of freedom for ANOVA
$N - J - 1$ = error degrees of freedom for ANCOVA

If, for example, the ANOVA error variance (σ_e^2) is 20, $\rho_{xy_w} = 0.40$, three groups are involved and $N = 30$, the ANCOVA error variance is

$$20(1 - 0.16)\frac{27}{26} = 17.45$$

If $\rho_{xy_w} = 0.90$, the ANCOVA error variance is only 3.95. When $\rho_{xy_w} = 0.19$ or less, the ANCOVA error variance is larger than the ANOVA error variance. Hence in the case of three treatments and 30 subjects, it is possible that ANOVA will be more powerful than ANCOVA, but only in the unlikely case that the correlation between the covariate and the dependent variable is under 0.20.

Randomized Experiment—Pretest–Posttest Design

The pretest–posttest design differs from the situation described in the preceding paragraph in that the covariate is an early measurement on the variable to be used as the dependent variable rather than a separate variable measured in a different way. The premeasure may be planned, or, if the experiment involves the collection of data over a long treatment period as in some learning experiments, the decision to employ early trials as the covariate may be made after the experiment is begun. This procedure consists of the following steps.

1. Subjects are randomly selected from some defined population.
2. Subjects are randomly assigned to treatments.
3. (a) Pretest measures are obtained before treatment, or
   (b) early performance measures are obtained.
4. Analysis of covariance is computed on Y scores observed at the end of the treatment period, pretest or early performance measure is used as the covariate.
5. A conditional statement concerning differences among treatment population means is made (e.g., "the differences between adjusted means measured at the end of the treatment period represents an effect that is independent of the differences among subjects observed during early trials").

This design is frequently analyzed by using another procedure. It is sometimes recommended that a simple analysis of variance on the gain scores rather than ANCOVA be used with this design. A gain score (also called a difference or change score) is defined as the posttest score minus the pretest score. An ANOVA on gain scores may be either more or less powerful than ANCOVA on posttest scores. The appropriate use of the gain-score approach requires that the slope of the pooled within-group regression of the posttest on the pretest equal 1.0. If the slope—which, of course, is not computed as a part of the ANOVA on gains—is in reality far from 1.0, the analysis will be less powerful than ANCOVA. If the slope is very near 1.0, the ANOVA on gain scores can be slightly more powerful than ANCOVA.

Examples of both situations can be found. Edwards (1968, p. 344) computed both analyses on the same data and found:

	ANCOVA	ANOVA on Gains
Error mean square	0.501	1.20
Slope b_w	1.44	

The large departure of the slope from 1.0 resulted in the error term for the gain-score analysis being over twice as large as the ANCOVA error term. Obviously, this difference in the size of the error terms illustrates the fact that the power of ANCOVA can be much greater than ANOVA on gains.

Snedecor and Cochran (1967, p. 424) provide an example of the gain-score approach that yields a smaller error term than does ANCOVA.

	ANCOVA	ANOVA on Gains
Error mean square	17.03	15.45
Slope b_w	0.988	

Although the gain-score error term is smaller, the difference is not large. In general, unless it is known that the slope is close to 1.0, ANCOVA is recommended for the pretest–posttest experimental design.

A third method of analyzing the pretest–posttest experiment is to ignore the pretest scores and carry out an ANOVA on the posttest scores. Generally, this procedure will be less powerful than will ANOVA on gain scores or ANCOVA. Should the correlation between "pre" and "post" measures be low, which is very unlikely, this method can be more powerful than either ANOVA on gains or ANCOVA.

Another method of analyzing this design is ANCOVA on the gain scores. It turns out that this approach is equivalent to ANCOVA on posttest scores; hence it is not pursued here.

The pretest–posttest design can also be analyzed through a repeated-measures ANOVA approach. Under this procedure there are, as shown in the following matrix, J levels on the nonrepeated factor and two levels (pretest and posttest) on the repeated factor.

	Trial	
Treatment	Pretest	Posttest
1		
2		
⋮	⋮	⋮
J		

This analysis turns out to be equivalent to the simple ANOVA on gain scores. That is, the F resulting from the ANOVA on gain scores is the same value and is associated with the same hypothesis as the F on the treatment by trial interaction in the repeated measures ANOVA. The application of these five analysis procedures to one set of data can be seen in Table 7.2.

Notice that, as mentioned earlier, ANOVA on gains and repeated-measures ANOVA are equivalent, ANCOVA on posttest scores and ANCOVA on gains are equivalent, and ANOVA on posttest scores yields a lower F value than does any other procedure. The data for these analyses were contrived. An empirical example is presented next.

A reanalysis of some data collected by Fredericks (1969) is presented in Table 7.3 to illustrate some of the advantages of ANCOVA over ANOVA. Fredericks was interested in evaluating the effectiveness of two methods of improving the motor coordination of mongoloids. The first method was the Domon–Delacato patterning treatment, and the second was a behavior-modification treatment that utilized principles of shaping, reverse chaining, and social reinforcement. Seventy-two mongoloid children were randomly assigned to six groups. Only the first three groups are considered here.

The first group (group A) was treated with the Domon–Delacato patterning procedure four times a day, 15 minutes at a time for a period of 9 weeks. The second group (group B) was treated with the behavior-modification procedure for the same period of time. The third group (group C) was a control group that was not treated but was tested in accordance with the testing schedule employed with the first two groups. Subjects in all three groups were measured on a modified version of the Lincoln–Oseretsky motor development scale and on the Doman–Delacato profile before treatment began, every two weeks during treatment, and at the conclusion of the treatment program. Only the pretest and posttest data are included in our reanalyses.

These data have been chosen to show that (1) annoyingly large differences among pretest means in randomized designs are not atypical, and (2) the analysis of covariance can remove the basic interpretation difficulty associated with this problem. First, observe the pretest differences among the three groups on the two response measures. Fredericks (1969, p. 74), in commenting on his analysis (t tests) of the posttest scores on the Lincoln–Oseretsky scale, expresses concern for the adequacy of this analysis, "The fact that the behavior modification group mean is approximately 13 score points higher than either the Doman–Delacato group or the control group at the start of the experimental program somewhat clouds the interpretability of the obtained trend differences and the much larger posttest difference favoring the behavior modification treatment."

This problem is attended to with the analysis of covariance. Notice in Table 7.3 that the adjusted means associated with ANCOVA differ from the unadjusted values by several points. The adjustments are such that there is less difference among group means. This adjustment effect is even more

Table 7.2. Five Analyses Applied to a Pretest–Posttest Experimental Design

Treatment group	Trial		Gain
	Pretest	Posttest	
1	2	4	2
1	4	9	5
1	3	5	2
2	3	7	4
2	4	8	4
2	4	8	4
3	3	9	6
3	3	8	5
3	2	6	4

Source	SS	df	MS	F	p
ANOVA on Posttest Scores					
Between	5.556	2	2.778	0.86	.47
Within	19.333	6	3.222		
Total	24.889	8			
ANOVA on Gain Scores					
Between	6	2	3.000	2.25	.19
Within	8	6	1.333		
Total	14	8			
ANCOVA on Post Scores					
Adjusted treatment	8.959	2	4.480	7.00	.04
Res_w	3.200	5	0.640		
Res_t	12.159	7			
ANCOVA on Gain Scores					
Adjusted treatment	8.959	2	4.480	7.00	.04
Res_w	3.2	5	0.640		
Res_t	12.159	7			
Repeated-measurement ANOVA					
Between subjects	22.778	8	2.847		
Treatments	4.111	2	2.056	0.66	.55
Subjects within Groups	18.667	6	3.111		
Within subjects	79.000	9	8.778		
Trials	72.000	1	72.000	108.00	.00
Trials × treatments	3.000	2	1.5	2.25	.19
Trials × subjects within groups	4.000	6	0.667		
Total	101.778	17			

Table 7.3 Data and Reanalysis of a Randomized Pretest–Posttest Design

Treatment[a]	Lincoln–Oseretsky Scale						Doman–Delacato Profile					
	A		B		C		A		B		C	
	Pretest	Posttest	Pretest	Posttest	Pretest	Posttest	Pretest	Posttest	Pretest	Posttest	Pretest	Posttest
	31	65	49	62	23	30	35	39.5	52	60	35	39.5
	10	16	0	0	41	68	32	35	12	12	48	54
	9	10	89	117	46	60	15	18	48	56	44	52
	65	76	44	68	0	4	46	54	48	50	18	18
	33	61	5	8	19	38	38	42.5	13	15	33.5	36.5
	0	5	52	105	5	25	6	10.5	39.5	42	23	23
	5	0	0	4	24	45	38	38	17	17	29	33
	0	0	2	5	0	0	16	17	38	39.5	9	9
	6	16	19	85	21	48	29	32	40	42	32	33
	41	52	67	122	26	49	32	35	50	60	37	41
			33	54	0	14			29	33	32	33

Data Analysis

Lincoln–Oseretsky Scale

ANOVA on pretest

$\bar{X}_A = 20.00$
$\bar{X}_B = 32.73$
$\bar{X}_C = 18.73$

Source	SS	df	MS	F	p
Among	1303.51	2	651.8	1.19	.32
Within	15904.37	29	548.4		
Total	17207.88	31			

ANOVA on posttest

$\bar{Y}_A = 30.10$
$\bar{Y}_B = 56.36$
$\bar{Y}_C = 34.64$

Doman–Delacato Profile

ANOVA on pretest

$\bar{X}_A = 28.70$
$\bar{X}_B = 35.14$
$\bar{X}_C = 30.95$

Source	SS	df	MS	F	p
Among	226.00	2	113.0	0.67	.52
Within	4916.37	29	169.5		
Total	5142.37	31			

ANOVA on posttest

$\bar{Y}_A = 32.15$
$\bar{Y}_B = 38.77$
$\bar{Y}_C = 33.82$

131

Table 7.3 *Continued*

Data Analysis

Lincoln–Oseretsky Scale

Source	SS	df	MS	F	p
Among	4226.88	2	2113	1.80	.18
Within	33967.10	29	1171		
Total	38194.88	31			

ANCOVA

$\overline{Y}_{A\,adj} = 35.4$
$\overline{Y}_{B\,adj} = 44.6$
$\overline{Y}_{C\,adj} = 41.6$

Source	SS	df	MS	F	p
Adjusted tr.	441.38	2	220.7	1.15	.33
Res$_w$	5372.17	28	191.9		
Res$_t$	5813.55	30			

Doman–Delacato Profile

Source	SS	df	MS	F	p
Among	253.16	2	126.6	0.56	.58
Within	6536.34	29	225.4		
Total	6789.50	31			

ANCOVA

$\overline{Y}_{A\,adj} = 35.6$
$\overline{Y}_{B\,adj} = 34.8$
$\overline{Y}_{C\,adj} = 34.7$

Source	SS	df	MS	F	p
Adjusted tr.	4.64	2	2.32	0.47	.63
Res$_w$	137.25	28	4.90		
Res$_t$	141.89	30			

Summary of All Analyses

Lincoln–Oseretsky Scale

Treatment	A	B	C	F	p
$\overline{X}$	20.00	32.73	18.73	1.19 (ANOVA)	.32
$\overline{Y}$	30.10	56.36	34.64	1.80 (ANOVA)	.18
$\overline{Y}_{adj}$	35.4	44.6	41.6	1.15 (ANCOVA)	.33

Doman–Delacato Profile

Treatment	A	B	C	F	p
$\overline{X}$	28.70	35.14	30.95	0.67 (ANOVA)	.52
$\overline{Y}$	32.15	38.77	33.82	0.56 (ANOVA)	.58
$\overline{Y}_{adj}$	35.6	34.8	34.7	0.47 (ANCOVA)	.63

[a]Treatments: A, Doman–Delacato; B, behavior modification; C, control

Source: Data from Fredericks (1969).

dramatic on the Doman–Delacato profile. Notice that there is almost a seven-point difference between the highest and the lowest unadjusted means but that the adjusted means are almost identical. Note that all these analyses yield nonsignificant differences. Descriptively, however, the results of ANCOVA are less ambiguous than those of ANOVA because the adjusted differences are the posttest differences that would be expected if the pretest scores for the three groups were equal.

Before concluding the discussion of this design, it should be mentioned that there is a large body of literature devoted to the complexities of the problems associated with the pretest–posttest design. It turns out, however, that most of this work (e.g., Bohrnstedt 1969, Cronbach and Furby 1970, Dyer, et al. 1969, Linn and Slinde 1977, Richards 1975) deals with either the one-group pre-test–posttest design or the nonequivalent group design. Most of the issues associated with these two designs are not relevant to the randomized pre-test–posttest design discussed here.

In summary, ANCOVA is the recommended procedure for the randomized pretest–posttest design. Other procedures (1) do not adjust for chance differences on the pretest and (2) generally have lower power.

Randomized Experiment—Covariate Measured after Treatment

The steps involved in this situation are identical to those of the situation described in the beginning of this section (covariate measured before treatment) except that the covariate measures are obtained after rather than before the treatment. If the treatment cannot affect the subjects' scores on the covariate, this usage is acceptable. To make such an assumption, however, there should be supporting data from previous research or a sound theoretical justification. If the treatments do affect the covariate scores, the experimenter should be aware of three possible consequences:

1. The analysis may be misinterpreted. The completely randomized ANCOVA tests the same hypothesis as the completely randomized ANOVA when the covariate is not affected by the treatment. But when the treatment does affect the covariate, the adjusted and unadjusted means do not have the same expectation. In other words, the unconditional population means and the conditional population means are not the same. Rather than providing an answer to the question of whether there are mean differences due to treatments, ANCOVA answers the question of whether there are treatment effects on Y that are independent of the measured values of X. Since the difference between the adjusted means is affected by the difference between the covariate means even when there is no treatment effect on Y, the distinction between the ANOVA and ANCOVA hypotheses is especially important when the treatment affects the covariate.

2. The adjusted means may not be interpretable in reality. That is, it makes no sense to compare adjusted treatment means if it is not actually feasible to have the same covariate score for different treatment groups. An example of this was provided in Section 6.4.

3. The ANCOVA F may be biased in either a conservative or a liberal direction. Recall that the ANCOVA F and the adjusted means are easily interpreted in most randomized experiments because the expected difference between groups on the X variable is zero. But if the treatments have an effect on the covariate, the expected difference on X is no longer zero, the expected difference between the adjusted means (under a true null hypothesis concerning treatment effects) will no longer be zero, and the F test will be biased.

Suppose that ANCOVA is employed in a situation in which the dependent variable is one form of an achievement test and the covariate is another form of the same test. If both measures are obtained *after* the treatments are applied, it would seem that the covariate should reduce the adjusted treatment effects to zero. That is, if the two forms of the same test measure the same characteristic(s), the mean difference on Y and the mean difference on X should be the same (if the two measures are parallel and equivalent), and it would seem that the difference on Y should be completely explained by the difference on X. But ANCOVA is not likely to yield this intuitively apparent solution. This is because

$$(\mu_{y_1} - \mu_{y_2}) - \beta(\mu_{x_1} - \mu_{x_2}) = 0.0$$

only if the slope is 1.0. That is, the difference between the population means on the dependent variable (i.e., $\mu_{y_1} - \mu_{y_2}$) is equal to the slope times the population mean difference on X (i.e., $\beta(\mu_{x_1} - \mu_{x_2})$, *only* if the slope is 1.0. Keep in mind that we are dealing with the situation in which the X and Y variables have the same means and variances for simplification of the example.

The problem here is that the slope will almost certainly be *less* than 1.0 due to measurement error. Hence even though X and Y measure the same thing and have the same means and variances, the expected difference on Y will not be zero when X is the covariate. Suppose that the population means are as follows:

$$\mu_{y_1} = 70 \qquad \mu_{x_1} = 70$$
$$\mu_{y_2} = 50 \qquad \mu_{x_2} = 50$$

If the correlation between the two variables is less than perfect (which, of course, will be the case), the slope will be less than 1.0. If the slope is 0.90, the expected mean difference on Y is

$$(70 - 50) - 0.90(70 - 50) = 2.0$$

In other words, a two-point difference between the adjusted means is expected even though the treatment has had exactly the same effect on both X and Y. The so-called true-score ANCOVA procedure (see Chapter 14) is recommended in situations where the covariate has been affected by the treatment.

Randomized Experiment—Covariate Measured during Treatment

This situation differs from that described earlier (section on randomized experiment—pretest–posttest design) in that the covariate measurement is obtained during the treatment period on a variable other than the response variable used in the experiment. In some experiments the covariate may be measured during the treatment period because it is most practical or convenient to do so or because a source of bias is discovered after the experiment starts. This procedure consists of the following steps:

1. Subjects are randomly selected from some defined population.
2. Subjects are randomly assigned to treatments.
3. Treatments are initiated.
4. The experimenter discovers a source of bias that accompanies the treatment (e.g., while investigating the effect of three drugs on a vigilance task it is found that the subjects exposed to different drugs obtain different amounts of sleep), and measurements are obtained on the bias-causing variable. Variance on the bias-causing variable (sleep) is not due to treatments.
5. Analysis of covariance is computed by using the bias-causing variable as the covariate (e.g., amount of sleep obtained during the treatment).
6. A conditional statement concerning differences among treatment populations means is made (e.g., "the difference among drug treatment means represents an effect that is independent of bias caused by amount of sleep").

The experimenter must be sure that the covariate employed in this situation is not affected by the treatment. For example, an experimenter interested in the differential effects of three methods of teaching mathematics may discover that students obtaining the highest achievement scores also have the highest state of anxiety. An ANCOVA using state-of-anxiety scores as the covariate could be employed. If, however, the investigator is primarily interested in discovering which treatment yields the highest mean achievement score, it would seem unwise to use anxiety scores as a covariate if it is likely that systematic between-group differences on anxiety are caused by the treatments. On the other hand, anxiety scores would appear to be a reasonable covariate to use if group differences on anxiety are chance differences and the within-group regression coefficient is large. In this situation increased power over a simple analysis of variance can be expected.

7.3   RANDOM ASSIGNMENT OF INTACT GROUPS—TRUE EXPERIMENT

The previously described designs involve the random assignment of individual subjects to treatments. It is not necessary to employ individual subjects as the unit of analysis. If, for example, 10 schools are randomly assigned to two treatment conditions, the mean values on X and Y obtained from each school could be treated as the values to be employed in ANCOVA. In this case the school rather than the individual subject is the unit of analysis. Since the unit is the school, the analysis is based on $N = 10$.

In situations where there is clear evidence that the individual subjects are not likely to respond independently, it is strongly recommended that the individual subjects not to be used as the unit of analysis. Recall that the assumption of independent observations is a critical one and that the investigator should attempt to employ a unit of analysis that is likely to meet this assumption. If individual subjects are not likely to respond independently but a larger unit of analysis is, the latter would be the appropriate unit to analyze.

A major problem with using some molar unit such as a school rather than individual subjects is that the reduction in degrees of freedom for error is drastic. If a reasonable number of molar units (e.g., schools) is available, however, the power of the F test is not as low as one might guess because the error mean square is generally smaller than is the case with individual units. That is, the sampling fluctuation associated with 10 means, for example, can be expected to be much less than with individual subjects. But in the case of only two or three molar units, there will be no degrees of freedom for error, and ANCOVA cannot be computed.

7.4   RANDOM ASSIGNMENT OF INTACT GROUPS—QUASI EXPERIMENT

A quasi-experimental design that has characteristics of both the true experiment, in which subjects are randomly assigned to treatments, and the observational or nonequivalent group study, in which no randomization is employed, can be carried out as follows:

1. Intact groups of subjects are selected.
2. Covariate measurements are obtained.
3. Intact groups are randomly assigned to treatments.
4. Treatments are applied.
5. Analysis of covariance is computed with the subject used as the unit of analysis.
6. A conditional statement concerning differences among treatment population means is made.

It should be recognized that this design differs from the randomly assigned groups true experiment in that (1) there is only one group per treatment, and (2) the individual subject is the unit of analysis. Suppose four treatments are randomly assigned to four schools. Since there is only one school per treatment it is clear that a conventional ANCOVA using the school as the unit of analysis cannot be carried out because within-group variability cannot be estimated in this case. Hence researchers frequently treat the individual subject as the unit of analysis even though the subjects have not been randomly assigned to treatments. Since entire groups rather than individual subjects have been randomly assigned to treatments the design is not a completely randomized design. The extent to which this design should be viewed as an acceptable experimental design appears to depend on two factors pointed out by Evans and Anastasio (1968): (1) the magnitude of the difference between group means on the covariate and (2) the number of groups involved.

If the groups have only slightly different covariate means, the correlation between treatment and covariate can be expected to be small. Any bias produced by such a correlation should also be small. The consequences of having different covariate means are likely to be most severe when only two groups are involved, because, in a sense, randomization has little opportunity to "work" when few groups are available.

An additional and perhaps more basic issue that should be considered in both ANOVA and ANCOVA situations is the effect of using intact groups on the independence of the observation and hence on the independence of the errors ε_{ij}. This problem is discussed in Section 6.2. Problems discussed in Sections 7.2 and 7.3 also apply in the case of intact groups.

7.5   COMPARATIVE–DEVELOPMENTAL STUDY

The analysis of covariance may legitimately be employed in situations where the research problem involves a comparison of the developmental rates of two or more populations. If two groups are to be compared, the following steps are involved:

1. A random sample from population A and a random sample from population B are selected.
2. Response measures from both samples are collected at time 1.
3. Response measures from both samples are collected at time 2.
4. The analysis of variance is computed on gain scores (i.e., time 2 scores minus time 1 scores).
5. An unconditional statement concerning the mean difference in gains for the two populations is made.
6. The analysis of covariance is computed by using time 1 measures as the covariate and gain as the dependent variable.
7. A conditional statement concerning the difference between the population A and B conditional means (i.e., mean gain conditional on time 1 observations) is made.

It is suggested that both ANOVA and ANCOVA be computed because the distinction between conditional and unconditional statements is often unclear in research reports. If both analyses are computed and reported, the results are less likely to be misinterpreted. The fact that the two analyses can lead to different conclusions has been termed "Lord's paradox" (Lord 1967). Analysis of variance leads to an inferential statement concerning the unconditional question, "Is it likely that there is a mean difference between the gains of the populations?" Analysis of covariance allows an inferential statement concerning the question, "Is it likely that there is a difference between populations on mean gains which are conditional on time 1 scores?"

The difference between conditional and unconditional inference can easily be seen in the following example discussed by Bock (1969) and Lord (1967).

Male and female students who eat their meals in university dining halls are weighed at time 1 (beginning of the school year) and at time 2 (end of the school year). Simple gain scores are computed and plotted by sex for all students as shown in Figure 7.2.

The difference between the male and female students in terms of weight gain is then analyzed by two statisticians. The first statistician runs an analysis of variance on the gain scores and concludes that there is not evidence of a differential effect of school diet on the two sexes. The second statistician runs an analysis of covariance on gain scores using initial weight (measured at time 1) as the covariate. His conclusion is that the male students gain more weight than female students when the gain scores are adjusted for difference in initial weight.

Lord's position is that the two conclusions represent a paradox and that confused interpretations are likely in such situations. Bock's position is that there is no paradox in Lord's example. Instead, Bock argues that the two analyses provide appropriate answers to two different inferential problems.

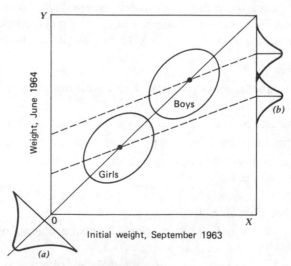

Fig. 7.2 Distribution of gains (*a*) unconditionally and (*b*) conditionally on initial score (adapted from Bock 1969).

Analysis of variance on gain scores supports the unconditional inference that there is no difference in mean weight gain between the male and female populations sampled in the experiment, whereas ANCOVA supports the conditional inference that the means of the male and female conditional distributions given initial weight are not equal.

The answer to the question as to whether the mean gain in weight for males differs from that for females is suggested by distribution *a* in Figure 7.2. Since the distributions for the two sexes are the same, only a single distribution of gains is visible. The lack of a difference in the distributions for the two sexes may provide the information the investigator is seeking. On the other hand, the investigator may have a more specific question in mind that could not be answered by a comparison of the mean gain scores for the two sexes. Suppose the investigator wants to know if 140 lb males are expected to gain the same amount as 140 lb females. This is a conditional question that should be answered with the use of an analysis yielding a conditional answer.

The distributions of weight gain by sex, conditional on initial weight, are shown in the upper right-hand corner of Figure 7.2. The difference in the elevation of the two conditional distributions indicates a difference in the expected gain for males and females when initial weight is considered. The expected weights of the 140 lb (initial weight) male and female can be seen in Figure 7.3.

If the mean female weight is 120 lb and the slope b_1 (obtained by regressing final weight on initial weight) is 0.5, the predicted final weight for a 140 lb female is 130 lb. If the mean male weight is 150 lb and the slope is 0.5, the predicted final weight is 145 lb. Thus females with an initial weight of 140 lb are expected to *lose* 10 lb, whereas males initially weighing 140 lb are

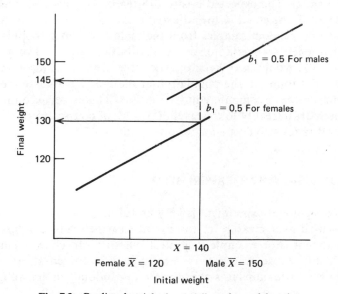

Fig. 7.3 Predicted weight for 140-lb males and females.

expected to *gain* 5 lb. In general, then, females below the female mean are expected to gain and those above the mean are expected to lose. The same is true for males having initial weights above and below the male mean. It can be seen that a subject's weight is expected to regress toward the mean of the population from which he or she has been selected.

Bock's argument clearly shows that the two analyses answer two very different questions with different answers and that the distinction between conditional and unconditional distributions must be kept in mind for correct interpretation of comparative–developemental data analyzed with the use of ANOVA and ANCOVA.

Note that the interpretation of ANCOVA in this comparative–developmental design is not fundamentally different from the interpretation in the case of a randomized groups design. That is, in both cases, the adjusted means are estimates of the mean difference that would be obtained if subjects from different groups had covariate mean scores equal to the grand covariate mean. In the case of the randomized-group design, the expected mean difference under H_0 is zero. With the comparative–developmental design, the expected mean difference on the posttest is *not* zero because the population covariate distributions (i.e., pretest distributions) have different means. In the example of male and female weight scores, the adjusted means are easily interpreted because there are both males and females with weights that are equal to the grand covariate mean. Here it is quite reasonable to consider the possibility of selecting both a sample of males having weights equal to the grand mean and a sample of females having weights equal to the grand mean. Even though the covariate (pretest) means for the two samples would be the same (i.e., matched) in this case, we would *not* expect the posttest sample means to be equal. The expected posttest difference in this contrived matching example would be equal to the adjusted mean difference in the ANCOVA of data based on random samples from the male and female populations.

If there had been no overlap of the X distributions for the two groups, the use of ANCOVA would have been questionable. That is, if male and female weights were not found in the population at the grand covariate mean, there would be no reason to compute adjusted means. This is because there would be no counterpart in reality to the level of X and, of course, no subjects about which generalizations could be made.

7.6 BIASED ASSIGNMENT EXPERIMENT

The biased assignment experiment is very useful in those situations where the covariate is used as a measure of the extent to which subjects "need" some kind of treatment. For example, a pretest measure of achievement may be used to identify those who need some type of compensatory education program. In such situations it is appropriate to randomly assign subjects who have low pretest scores to treatment and control conditions. Such a randomized experiment would be a straightforward method of evaluating the treatment. But practical considerations may yield such a design unusable. It may be impossible to withhold the treatment from any subject who has been identified as having a low pretest score. Even though there may be no data to indicate that a treatment is effective, the political realities surrounding the research project may be such that random assignment simply cannot be employed.

In such cases it may be possible to use some variety of the biased assignment experiment. The biased assignment experiment involves the following steps:

1. A single population of subjects is identified and subjects are randomly selected. (Or, more realistically, a collection of N subjects from a school system, school, or classroom is simply available.)
2. A covariate measure (generally a pretest) is obtained for each subject.
3. Subjects are assigned to treatment and no-treatment conditions *exclusively* on the basis of the covariate score. Suppose that the covariate is a reading pretest measure and that the purpose of the treatment is to improve reading skills. Those subjects with pretest scores below the median (or any other point on the X continuum) will be assigned to the treatment condition; those at or above the median do not receive the treatment. Another variety of the design would involve dividing the N available subjects into three groups, say, thirds, on the basis of the covariate scores. The lowest third would be assigned the treatment and the highest third would be the no treatment group. Subjects in the middle third would be randomly assigned to the treatment and no treatment conditions.
4. The experiment is carried out.
5. Analysis of covariance is computed.
6. A conditional statement concerning the difference between the adjusted sample means is made, such as "The 17-point difference between the adjusted sample posttest means is an unbiased estimate of the difference between the adjusted (expected) population means $\mu_{1\ adj}$ and $\mu_{2\ adj}$; this difference, which is statistically significant at the .01 level, is independent of the pretest difference and represents the increment due to the treatment".

This type of experiment is illustrated in Figure 7.4. Notice that before the treatments are administered, both the covariate and dependent variable means are clearly different. Obviously, a comparison of unadjusted Y means after treatments would be invalid as a measure of evaluating the treatment effects because the Y means were systematically different before treatments were applied. A comparison of adjusted means, however, appears to be

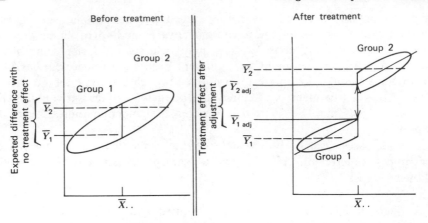

Fig. 7.4 Biased assignment experiment.

reasonable. As long as the assumptions of homogeneous regression slopes and linearity are met, ANCOVA provides the proper adjustment. The issue of measurement error or unreliability is not a concern in the adjustment process with this design. Likewise, the concern that the subjects in the treatment and no-treatment groups may differ on other variables that have not been measured is not relevant. Mathematical arguments for not being concerned about measurement error and other sources of bias with the biased assignment design can be found in Cronbach et al. (1977), and Rubin (1977). It must be kept in mind, however, that the unbiased nature of the treatment effects will not hold if the assignment of the subjects to groups is on the basis of any consideration other than the covariate. If subjects are assigned to treatments on the basis of characteristics other than the covariate, ANCOVA is likely to yield biased results. This problem is described in detail under the nonequivalent-group studies.

7.7 NONEQUIVALENT-GROUP STUDIES

When data are collected from groups that have not been formed by random assignment or on the basis of a covariate, the population means may differ on many variables. These differences are not likely to be predictable from a covariate. Since ANCOVA is not likely to yield unbiased adjusted means and hypothesis tests with nonequivalent-group studies, it is important to be able to identify such studies and to recognize the interpretation problems associated with their analysis.

There are two frequently encountered types of nonequivalent group studies. We refer to the first one as the *nonexperimental observational study* (sometimes called the *contrasted groups design*) and to the second one as the *nonequivalent treatment-group study*.

Nonexperimental Observational Study

In studying the differences between two (or more) existing populations such as chemists and mechanics, blacks and whites, hypertensive and nonhypertensive patients, and so on, attempts to adjust the existing groups for differences explained by some covariate may be undertaken.

Generally, the motivation for employing ANCOVA with this design is to gain an understanding of why the groups differ on the dependent variable. If a covariate is found that yields a nonsignificant ANCOVA F in the case where the ANOVA on the unadjusted means is significant, it is sometimes stated that the covariate has "explained" why there is a difference on Y. For example, an investigator may want to compare the mean intelligence test scores for black and white groups after the means have been adjusted for income level. This procedure consists of the following steps:

1. Subjects are selected from different existing populations (e.g., black and white populations).
2. Covariate measurements (e.g., income scores) are obtained.
3. Measurements on some response variable (e.g., intelligence test scores) are obtained.
4. Analysis of covariance is computed.
5. A conditional statement concerning differences between the adjusted population means is made (e.g., "the mean intelligence test scores for the black and white populations are equal when they are adjusted for differences in income level").

Nonequivalent Treatment Groups Study

The only difference between the nonexperimental observational study and the nonequivalent-treatment-group study is the application of different treatments to the various nonequivalent groups. This procedure consists of the following steps:

1. Subjects are selected from different existing populations (e.g., schools from different areas).
2. Covariate measurements (e.g., aptitude) are obtained.
3. Treatments (e.g., methods of teaching) are administered.
4. Measurements on some response variable (e.g., achievement) are obtained.
5. Analysis of covariance is computed.
6. A conditional statement concerning differences between treatments means adjusted for covariate differences is made (e.g., "the difference between teaching methods is statistically significant when aptitude differences are held constant").

These usages of ANCOVA are frequently encountered in the literature; unfortunately, the results of such analyses are often misinterpreted. Many investigators state that the difference between the adjusted means is what would be expected if the covariate had been controlled experimentally. This is not true. Suppose we carry out the type of investigation previously mentioned in which intelligence-test scores for black and white subjects are adjusted for income level. We *cannot* legitimately say that the difference between the adjusted means is what would be expected if the mean income level for the two groups had been the same. The problem here is the elementary "correlation-equals-cause fallacy." Unfortunately, even though this fallacy is well understood by virtually all researchers who deal with simple correlation problems, the reasoning does not seem to readily transfer when the interpretation of the analysis of covariance is involved.

Let us say that we regress intelligence-test scores on income scores and that b_w, the pooled within-group regression coefficient, is 0.50. The hypothetical X and Y means for the black and white groups follow:

Black		White	
Income	Intelligence	Income	Intelligence
20	90	60	110

The adjusted means are:

$$\overline{Y}_{b\ adj} = 90 - 0.5(20 - 40) = 100$$

$$\overline{Y}_{w\ adj} = 110 - 0.5(60 - 40) = 100$$

The results of the adjustments *appear* to be clear—there is no difference between the mean intelligence test scores for the black and white groups when intelligence is adjusted for differences on income. What does this statement mean? Before attempting to interpret the meaning of adjusted scores, a logical analysis of the correlation between X and Y should be undertaken. In the present example we should ask *why* are intelligence and income level correlated? It may be that (1) income level causes differences in intelligence test scores, (2) intelligence differences cause differences in the income levels people reach, or (3) some other variable(s) cause intelligence test scores and income level to covary. Since all these explanations (and others) are plausible, it should be obvious that a simple correlation coefficient or regression coefficient tells us nothing about the cause of the relationship. If explanation (1) is correct the adjusted means can be interpreted as the mean values we would expect if all subjects had been assigned to the common income level of 40. If explanation (2) or (3) is correct, there is no justification for stating that the adjusted means are the means one would expect to obtain if income level had been controlled experimentally.

Suppose (unrealistically) that the factor actually responsible for the mean difference on intelligence is the amount of a specific substance contained in the diet. Most whites eat the substance, but most blacks do not. If the critical dietary substance is found in the diet of the white group regardless of income, but not in the diet of the black group regardless of income, we would expect a correlation between income and intelligence scores because whites just happen to include the critical substance in their diet for cultural reasons unrelated to income. We *would not* expect intelligence to change simply because income is changed. That is, if we randomly assign all black and white subjects to a common income level, the blacks will not necessarily be expected to add the critical substance to their diets, and intelligence-test scores should not be expected to increase since the critical dietary factor has not been added. Hence it can be seen that the correct interpretation of ANCOVA and adjusted means depends on the *reason* for the correlation between the X and Y scores. Unfortunately, the causal factors are frequently unknown when the analysis of covariance is employed with naturally occurring groups. The investigator must be aware that if the analysis of covariance is employed when subjects have been assigned to groups neither by random assignment nor on the basis of the covariate, causal inference is valid only if all variables not included in the analysis are uncorrelated with the treatment or the covariate, or if the system is closed. If, for example, there were evidence that either (1) all variables possibly related to intelligence are uncorrelated with race or income level or (2) the assumption of a closed system is realistic (e.g., race, income, and intelligence comprise a system of variables not possibly affected by other variables), causal statements would be possible. If the statements are to be taken seriously, the reasons for believing that outside variables are uncorrelated with the treatment and covariate or that the system is closed would have to be very powerful indeed. It should be obvious in the race example that the system is not closed and that many variables not included in the analysis are very likely to influence intelligence and to also be related to race and income levels.

If the purpose of the analysis is *not* to estimate what the mean difference would be if all subjects were assigned to a common income level, ANCOVA may be useful even when the reasons for the correlation between X and Y are not known. If the covariate does not cause variance on the dependent variable but is correlated with some variable that does, the ANCOVA can be computed, but the interpretation of the F and the adjusted means is lowered from a causal level to a correlational level. That is, inferences concerning differences between adjusted means are mere statements that differences do (or do not) exist for intact groups after covariate adjustments.

The interpretation problem with ANCOVA in this situation is identical to the problem of interpreting partial correlation coefficients when the variable statistically held constant is not causally related to the dependent variable. Recall that the partial correlation $r_{yx_1 \cdot x_2}$ provides an estimate of the correlation between Y and X_1, when X_2 is held constant experimentally—*if X_2 is* causally related to Y. If X_2 is not causally related to Y, the interpretation of the partial correlation becomes somewhat ambiguous. The same problem occurs in ANCOVA applied to existing groups. The adjusted means are not necessarily the means expected if subjects are randomly assigned to a common covariate value. In general, if the analysis of covariance is computed on scores obtained from existing rather than randomly formed groups or groups formed on the basis of the covariate, the design should be considered a correlational or observational one, and the results should be interpreted with the same qualifications associated with any other correlation study.

The preceding discussion should not be interpreted as a criticism of correlational research. Indeed, in many situations there are no alternatives. It is quite reasonable to select subjects from different populations, such as obese and nonobese, and then attempt to identify variables that are related to or discriminate between such groups. This is an excellent method of generating hypotheses concerning causal variables. Ideally, those variables that appear to "explain" group differences will be employed in future *experiments* to determine whether the explanation is causal or simply correlational. The use of ANCOVA is appropriate for attempting to identify explanatory variables; it is clearly not, however, a replacement for experimentation. As can be seen in the remainder of this section, there are several problems with ANCOVA that should be of concern, even if the cause–correlation fallacy is well understood.

The Extrapolation Problem.   Researchers should be aware of three problems in addition to the "causal–correlational" issue that are associated with the use of covariance analysis on nonequivalent group data. These are labeled problems of *extrapolation*, *precision*, and *regression artifacts*. The extrapolation problem is illustrated in Figure 7.5. Here we have a typical educational research problem in which the difference in the effectiveness of two different schools is to be evaluated. This is an example of a nonequivalent group design because students have not been randomly assigned to schools. Suppose the covariate is some measure of academic aptitude and the dependent variable is a test of academic achievement. As expected, it can be seen that students attending school 1 are superior in terms of both aptitude and achievement. It appears, however, that the difference in achievement is not explained by the difference in aptitude. That is, when we examine Figure 7.5*b*, we see that there is a difference between the adjusted means. This difference is not as large as the unadjusted difference, but it appears that there is a definite "school" effect. The parallel regression lines and the dotted extensions of these lines suggest that the school effect can be expected regardless of the aptitude level of the student. It is tempting to state that school 1 is superior to school 2 for students of all aptitude levels. In fact, the additional contention that there is no aptitude by school interaction by computing the homogeneity of regression test might seem reasonable. A close look at the data, however, reveals problems with both the statement that school 1 is superior to school 2 regardless of student aptitude, and with the

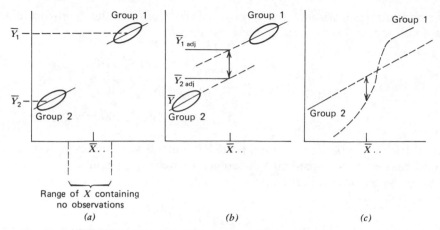

Fig. 7.5 Extrapolation problem: (*a*) obtained data; (*b*) extrapolated or assumed relationships; (*c*) actual relationships.

homogeneity of regression test. The regression lines of Figure 7.5*a* are based on data actually collected, whereas the dotted lines of Figure 7.5*b* are extrapolated lines based on the *assumption* that the slope is the same regardless of the level of X. If this assumption (and all other ANCOVA assumptions) were met, the results of ANCOVA would be valid. It can be seen in Figure 7.5*c*, however, that the assumption is false. Unfortunately, the researcher has no way of knowing that the school 1 slope changes drastically at lower levels of X unless lower levels of X are, in fact, included in the sample of school 1 students. As Werts and Linn (1971, p. 97) point out, in this situation "the interpretation of the ANCOVA results is speculative and should be so labeled." This is true whenever there is a large no-man's land or region in which there are no observations on X separating the observational groups.

The homogeneity of regression test is only useful for testing the assumption that the slopes are parallel for the range of values actually included in the sample. It is not sensitive to the problem of extrapolation since the value of the test statistic is a function of neither the covariate mean difference nor the amount of extrapolation associated with the adjusted means.

The Precision Problem. When ANCOVA is applied in the analysis of a randomized experiment, the researcher can expect to detect smaller treatment effects than would be the case if ANOVA were employed. That is, the ANCOVA has greater precision or power (when the assumptions are met). With a nonequivalent group study, however, the power may be very low indeed. A close look at the formulas in Chapter 5 will reveal that as the difference between covariate means increase, the power of the analysis on pairwise differences, for example, decreases because the error terms used in the multiple comparison tests become larger. Consider the following data based on a hypothetical *randomized* experiment.

$$\begin{aligned}
&\bar{Y}_{1\,adj}=9 \quad \text{MSres}_w=5\\
n_1=n_2=n_3=10 \quad &\bar{Y}_{2\,adj}=12 \quad \text{MS}_{b_x}=100\\
&\bar{Y}_{3\,adj}=20 \quad \text{SS}_{w_x}=1400
\end{aligned}$$

All pairwise comparisons are evaluated by the Bryant–Paulson test. It is found that even the three-point difference between $\bar{Y}_{1\,adj}$ and $\bar{Y}_{2\,adj}$ is significant at the 5% level if the following error term is used for each comparison.

$$\sqrt{\frac{5[1+100/1400]}{10}}=0.732$$

Adjusted Mean Difference	Absolute Obtained Q_p	Critical Value of Q_p
Groups 1,2 = 9−12 = 3	4.10	3.59
Groups 1,3 = 9−20 = 11	15.03	3.59
Groups 2,3 = 12−20 = 8	10.93	3.59

Now suppose that the same adjusted mean differences and MSres$_w$ are found in an observational study, but, as would be expected in this type of study, large differences are present among the covariate means. The effects of differences among the covariate means on the error term in the Bryant–Paulson tests (or any other multiple comparison procedure) can be seen as follows.

$$\begin{aligned}
n_1=10 \quad &\bar{X}_1=20 \quad \bar{Y}_{1\,adj}=9 \quad \text{MSres}_w=5\\
n_2=10 \quad &\bar{X}_2=40 \quad \bar{Y}_{2\,adj}=12\\
n_3=10 \quad &\bar{X}_3=95 \quad \bar{Y}_{3\,adj}=20 \quad \text{SS}_{w_x}=1400
\end{aligned}$$

Adjusted Mean Difference	Error Term for the Adjusted Difference	Absolute Obtained Q_p	Critical Value of Q_p
3	$\sqrt{\dfrac{5\left[\frac{2}{10}+(20-40)^2/1400\right]}{2}}=1.10$	2.72	3.59
11	$\sqrt{\dfrac{5\left[\frac{2}{10}+(20-95)^2/1400\right]}{2}}=3.25$	3.39	3.59
8	$\sqrt{\dfrac{5\left[\frac{2}{10}+(40-95)^2/1400\right]}{2}}=2.43$	3.29	3.59

Even the largest difference is not significant in this analysis. Clearly, when large differences on X exist, the power of ANCOVA is lower than when small X differences exist. This problem can lead to the conclusion that differences on Y do not exist after the relationship with X is taken into account. Snedecor and Cochran (1967, p. 430) point out that "a sounder interpretation is that the adjusted differences are so imprecise that only very large effects could have been detected."

Regression Artifact Problem. Another problem to be considered in the interpretation of ANCOVA applied to observational or nonequivalent-group studies is that of the regression artifact. A special case of this problem was described in Section 6.5, where it was pointed out that the difference between adjusted means in the nonrandomized study is partly a function of the reliability of the covariate if the covariate means are not equal. Figures 6.4 and 6.5 may help the reader to recall that the difference between adjusted means may be too large or too small when the covariate is unreliable and the groups are not randomly equivalent. The meaning of the terms "too large" or "too small" should be clarified. Actually, the use of ANCOVA with non-equivalent-group studies should be viewed as yielding biased results (adjusted mean differences too large or too small) only if the purpose of the analysis is conceptualized in certain ways. If the purpose of the analysis is to (1) explain variability on the dependent variable in terms of other basic (true-score) dimensions or (2) estimate treatment effects when different treatments are applied to nonequivalent groups, ANCOVA will not yield appropriate results. On the other hand, if there is interest in estimating the expected mean difference on Y for subjects who have the same measured value on the covariate, ANCOVA is appropriate. Since the latter purpose is not consistent with many behavioral science studies, ANCOVA is frequently inappropriately employed with nonequivalent-group studies.

Since the issue of regression artifacts (and measurement error) is of great concern for nonequivalent groups, it may be worthwhile to present an example of how measurement error leads to the regression artifact by using either a matched-group design or ANCOVA.

Suppose that we are interested in evaluating the effects of a new reading program on the reading skills of third-grade children. Suppose also that the researcher has decided that it is appropriate and practical to match subjects from two different schools on the basis of a reading pretest (Distributions of contrived pretest scores are presented in Figure 7.6.) Let us say, to keep the example simple, that the population means for schools A and B are the same as the sample means $\overline{X}_A$ and $\overline{X}_B$. That is, $\mu_{x_A} = 11$ and $\mu_{x_B} = 21$; clearly, this is a nonequivalent group situation because the population means are different. Intuitively, it seems reasonable to (1) match the subjects from the two schools, (2) apply the conventional program to one group (say, A) and the experimental program to group B, and (3) administer a posttest after the programs are completed. Before the data from the subjects who have been

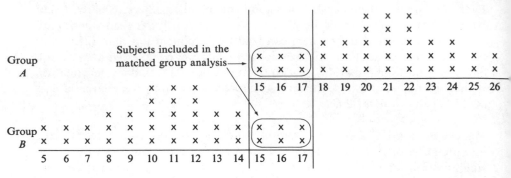

Fig. 7.6 Pretest reading score distributions for two nonequivalent groups.

matched are analyzed, the researcher should ask what the expected mean difference on the posttest would be had there been no treatment effect. That is, if there is no difference in the effectiveness of the two methods of teaching reading, what is the predicted value of the mean difference on the posttest? It would seem that the difference should be zero since subjects from the two groups were matched. Unfortunately, common sense fails in this situation; the predicted difference is *not* zero unless the reliability of the pretest is perfect. The problem here is that the subjects have been matched on observed scores, not true scores. Observed scores contain measurement error. In classic psychometric theory, an observed score is viewed as being the sum of a true-score component and a measurement error component. If there is no measurement error (i.e., perfect reliability), the observed score is equal to the true score; and, in the absence of differential treatment effects, the predicted value of the posttest difference is equal to the pretest difference. This situation is, unfortunately, quite unrealistic—measurement error virtually always exists.

It has been pointed out that, for the matched subjects, the pretest means are expected to equal the posttest means when the pretest is perfectly reliable. We now consider the other extreme: zero reliability. If the pretest scores in distribution A, for example, are completely unreliable, we have no reason to expect a subject who gets a pretest score of, say, 16 to get a posttest score of 16. Likewise, there is no reason to expect the mean of the six lowest scores in distribution A, which is 16, to be equal to the posttest mean for these same subjects. It turns out, however, that a good guess can be made of the posttest mean score for these six subjects. The best guess is the mean of the population from which the subjects were selected. Hence even though the pretest mean was 16 for the six subjects at the lower end of distribution A, the predicted posttest mean (in the absence of treatment effects, maturation effects, etc.) is 21. Similarly, the predicted mean score for the six subjects at the upper end of distribution B is not 16; it is 11. It can be seen, then, that when the pretest is completely unreliable, the predicted posttest value for a subject or group of subjects from any point on the pretest distribution is the mean of the population from which the subjects have been selected. The scores are said to "regress" to the mean of the population because, regardless of the pretest score, the expected posttest score is the population mean. In the case of perfect reliability of the pretest, no regression toward the mean will take place. In the case of zero reliability, complete regression is expected. That is, in the former case the expected posttest score is equal to the pretest score. In the latter case the expected posttest score is equal to the mean of the population from which the pretest score was selected. A simple formula can be employed to formalize this relationship between the correlation of the pretest with the posttest and the expected value of the posttest:

$$\hat{Y}_{ij} = \mu_{x_j} + \rho_{xy}(X_{ij} - \mu_{x_j})$$

where

$\hat{Y}_{ij}$ = predicted value of posttest for the ith subject in jth group

ρ_{xy} = within-group correlation between pretest and posttest

X_{ij} = pretest score for ith subject in jth group

μ_{x_j} = mean of the entire pretest distribution for the jth group

Note that if the correlation is 1.0 and this formula is applied to the matched data in Figure 7.6, the predicted posttest scores are as follows:

	Matched Subjects Predicted Posttest Score	Matched Subjects Pretest Score
$\hat{\bar{Y}}_A = 16$	15	$= 21 + 1(15 - 21)$
	15	$= 21 + 1(15 - 21)$
	16	$= 21 + 1(16 - 21)$
	16	$= 21 + 1(16 - 21)$
	17	$= 21 + 1(17 - 21)$
	17	$= 21 + 1(17 - 21)$
$\hat{\bar{Y}}_B = 16$	15	$= 11 + 1(15 - 11)$
	15	$= 11 + 1(15 - 11)$
	16	$= 11 + 1(16 - 11)$
	16	$= 11 + 1(16 - 11)$
	17	$= 11 + 1(17 - 11)$
	17	$= 11 + 1(17 - 11)$

If the reliability of the pretest is zero, the predicted posttest scores are as follows:

	Matched Subjects Predicted Posttest Score	Matched Subjects Pretest Score
$\hat{\bar{Y}}_A = 21$	$\begin{cases} 21 \\ 21 \\ 21 \\ 21 \\ 21 \\ 21 \end{cases}$	$= 21 + 0(15 - 21)$ $= 21 + 0(15 - 21)$ $= 21 + 0(16 - 21)$ $= 21 + 0(16 - 21)$ $= 21 + 0(17 - 21)$ $= 21 + 0(17 - 21)$
$\hat{\bar{Y}}_B = 11$	$\begin{cases} 11 \\ 11 \\ 11 \\ 11 \\ 11 \\ 11 \end{cases}$	$= 11 + 0(15 - 11)$ $= 11 + 0(15 - 11)$ $= 11 + 0(16 - 11)$ $= 11 + 0(16 - 11)$ $= 11 + 0(17 - 11)$ $= 11 + 0(17 - 11)$

Reliability coefficients of 1.0 and 0.0 are not generally encountered in practice. A value of 0.70 is fairly typical. If we apply the prediction formula to the matched data using 0.70 as the reliability coefficient, we find:

	Matched Subjects Predicted Posttest Mean		Matched Subjects Pretest Mean
A	17.5	=	$21 + 0.70(16 - 21)$
B	14.5	=	$11 + 0.70(16 - 11)$

When the expected mean difference on the posttest is computed for each of the three reliability coefficients 1.0, 0.70, and 0.0 respectively, we find

$$16 - 16 = 0.0$$
$$17.5 - 14.5 = 3.0$$
$$21 - 11 = 10.0$$

Since the expected posttest difference is not zero for the matched subjects when $\rho_{xy} \neq 1.0$ and there is no treatment, such a difference is said to be a *regression artifact*. It should be clear, then, that the expected mean difference for the matched subjects on the posttest is a function of the reliability of the pretest and the difference between the pretest population means. If the pretest population means are equal, however, the expected posttest difference between means is zero regardless of the reliability of the measure employed. This is one reason why randomized designs are preferable to many other designs.

The purpose of the preceding discussion is to point out the regression problem associated with the matching design that was employed. It can be seen that a t test on the posttest means is *not* a test of the hypothesis of zero treatment effects with this design because the expected mean difference is not generally zero in the absence of treatment effects. What is the relevance of this discussion to the analysis of covariance? It turns out that "statistical matching" through the application of ANCOVA runs into exactly the same problem!

If ANCOVA is applied to data obtained from two nonequivalent groups, the adjusted mean difference is an estimate of the expected mean difference on Y for subjects who fall at the grand mean on X. In the nonequivalent group pretest–posttest design the pretest is the covariate and the posttest is the dependent variable. If the pretest data of Figure 7.6 are employed as the covariate, the difference between adjusted means would be an estimate of the expected mean difference for subjects who have covariate scores of 16, which is the grand covariate mean. Suppose that the standard deviations for both groups are the same on the posttest as on the pretest. In this case, the slope b_w will be equal to the within-group correlation between the pre- and posttests. Hence we can compare the results of the matched-groups study with results expected using ANCOVA. The expected mean posttest difference in the absence of treatment effects was shown above to be three points when $r_{xy} = 0.70$. This difference was based on 12 matched observations. The application of ANCOVA will allow all observations in the two sample distributions to be employed. The adjusted means associated with this analysis follow:

Group	$\bar{Y}_{adj}$	=	$\bar{Y}_j - b_w(\bar{X}_j - \bar{X}..)$
A	17.5	=	$21 - 0.70(21 - 16)$
B	14.5	=	$11 - 0.70(11 - 16)$

The mean difference is three points—exactly the same expected difference associated with the matched-group design. Clearly, the analysis of covariance does not yield an unbiased estimate of the treatment effect when the non-equivalent group design is employed.

The likely consequences of employing ANCOVA with the two varieties of nonequivalent group studies can be summarized as follows. First, with the nonexperimental observational study, the extent to which the differences among the groups on Y are explained by true X scores will be underestimated. That is, if the difference between two groups on Y is completely explained by the true scores on X, the adjusted mean difference that is based on *observed* X scores will not be zero. In other words, conventional ANCOVA adjusts for X *as measured*, not for the true scores on X. The adjustment made on the basis of observed X scores is an *underestimate* of what the adjustment would be if true scores were available. Second, in the case of nonequivalent treatment group study, the underadjustment of the treatment means can result in the inappropriate conclusion that: (1) treatment effects exist (when they don't), (2) no treatment effects exist (when they do), (3) a particular treatment is helpful (when it is actually harmful), and (4) a particular treatment is harmful (when it is actually helpful).

A method of attempting to correct for the problem of measurement error is described in Chapter 14, where it is pointed out that measurement error in X is not the only reason for regression artifacts and that problems not mentioned in this section should be considered in the analysis of nonequivalent-treatment-group-studies.

Summary of Interpretation Problems Associated with ANCOVA Applied to Nonequivalent-Group Studies.   There are four basic problems associated with the application of ANCOVA to nonequivalent group studies: (1) the cause–correlation fallacy, (2) extrapolation, (3) low precision, and (4) regression artifacts. If it is not known that variability on the dependent variable is caused by variability on the covariate, there is no reason to believe that the results of ANCOVA are a valid estimate of the results that would be expected with the use of experimental methods to control for covariate differences. It cannot be stated that statistical control through ANCOVA is equivalent to experimental control.

Large differences between covariate means sometimes occur with nonequivalent group studies. If the difference between the two (or more) X distributions is extreme, there may be no observations in a part of the region between the two means. Since ANCOVA tests the hypothesis that the adjusted Y means are equal at the grand mean on X, an extrapolation problem occurs. The problem is that the ANCOVA F test may be applied to test the difference between means adjusted to a covariate value for which there are no observations. If there are no observations at the grand X mean, little confidence can be placed in the outcome of the test unless it is known that the relationship between X and Y can be appropriately extrapolated to the region having no observations.

The precision or power of tests on adjusted means is related to the difference between the covariate means. As the covariate difference increases, the precision of the tests decreases. Since covariate mean differences can be large in nonequivalent group studies, the precision of the tests can be very low.

If the covariate contains measurement error (which is always the case) or if the covariate and the dependent variable do not measure the same dimension (which is frequently the case), regression artifacts will occur. The effect of the regression artifact is to bias the estimated difference between adjusted means and, correspondingly, the ANCOVA F test.

In general, ANCOVA is not an appropriate procedure for the analysis of nonequivalent group studies—especially if the purpose is to evaluate treatment effects. Alternative methods of analysis are described in Chapters 14 and 15.

7.8 SUMMARY

Six varieties of research design relevant to ANCOVA include: the completely randomized experiment, the randomly assigned group true experiment, the randomly assigned group quasi experiment, the comparative–developmental study, the biased assignment experiment, and the nonequivalent group study.

The completely randomized experiment can be of several forms, including the case where the covariate is measured before the treatments are administered, the pretest–posttest design, the case where the covariate is measured after the treatments are administered, and the case where the covariate is measured during the treatment administration. The ANCOVA is easily interpreted when the covariate is measured before treatments are administered or during the treatment administration where the treatment does not affect the covariate. Analysis of covariance is generally the preferred method of analyzing the randomized pretest–posttest design. Designs in which the covariate is affected by the treatment yield results that may not be consistent with the experimenter's question; care must be taken in deciding whether the adjusted means are providing information of experimental interest. True-score ANCOVA should be considered for the analysis of designs where the covariate has been affected by the treatments.

Intact groups rather than individual subjects should be considered as the unit of analysis in situations where the assumption of independence is likely to be violated with individual subjects.

The randomly assigned group quasi experiment involves elements of the completely randomized experiment and the randomly assigned group experiment; individual subjects are the unit of analysis, but treatments are randomly assigned to existing intact groups.

The comparative–developmental study involves the observation of subjects from two or more populations at two points in time. The purpose of such a design may be to decide whether the gains are the same for the different populations, or it may be to answer the question of whether subjects from the different populations with the same pretest or X score are likely to gain the same amount. These two different questions require different analyses. One is answered with an ANOVA of the gains, whereas the other is answered with ANCOVA.

The biased assignment experiment is a very useful solution to the problem of not being able to randomly assign subjects to treatments. If subjects can be assigned to treatments on the basis of the covariate score alone, causal statements can be made. The problem of measurement error is not relevant to this design as a source of treatment effect bias.

Two varieties of the nonequivalent-group study include the nonexperimental observational study and the nonequivalent-treatment-group study. In the case of the former, ANCOVA is employed to attempt to evaluate whether a covariate "explains" the differences among groups. In the case of the latter study, interest is focused on the use of a covariate to "equate" groups to evaluate treatment effects when different treatments have been applied to nonequivalent groups. Analysis of covariance is not generally an adequate method of analysis for either variety. Logical, statistical, and measurement problems are associated with the analysis of these designs. Analysis of covariance can sometimes provide useful information in these cases, but great care must be exercised in the interpretation of results. Alternative methods of analysis should be considered with these designs.

PART TWO
COMPLEX ANALYSIS OF COVARIANCE

CHAPTER EIGHT
Multiple Covariance Analysis

8.1 INTRODUCTION

It is not unusual for a researcher to be interested in controlling more than one source of unwanted variability. Suppose that a state department of education is interested in comparing the effectiveness of a sex education instructional package that includes a large television component with a similar package that does not include the television aspect. If classrooms from a large geographic area are randomly assigned to the two treatments, there is likely to be sizable sampling variability both between and within treatment groups with respect to variables such as race, parent's income, previous educational experiences, and religion. If measures are available on these variables, multiple covariance analysis can be employed as a control procedure by using all the measures as covariates. The issues discussed in previous chapters for the case of a single covariate generalize directly to the case of multiple covariates.

There are other procedures for controlling several sources of unwanted (nuisance) variability, such as direct experimental control, selection, and factorial designs. An investigator could control for race, parent's income, previous educational experiences, and religion by selecting subjects who belong to the same race and religion, have parents at one income level, and have the same educational experiences. This approach would require a considerable amount of effort to obtain a reasonable sample size; more importantly, it would require that the selected subjects in some way be isolated from other students for treatment purposes. It is not realistic to think that such subjects could be (or should be) isolated for this purpose. The use of each of these variables as a factor (in addition to the treatment factor) in a factorial ANOVA design is a more practical strategy. But there are weaknesses with this approach also. When a control variable is a continuous variable (e.g., income or amount of experience), information is lost by forming just a few levels of such factors; thus if income is broken down into low, medium, and high levels, there will be information lost concerning differences in income within any one of these crude categories. On the other hand, if a large number of levels is included (say, 10 levels of income) there will be a correspondingly large loss of error degrees of freedom that will tend to decrease the power of the analysis.

Multiple ANCOVA circumvents the problems mentioned above. All available subjects can be included in the analysis, there is no problem of coarse grouping, and each covariate (regardless of the number of different values of X) has one degree of freedom. All covariates can be included simultaneously in one analysis of covariance. The adjusted treatment mean differences are then interpreted as being independent of all variables included as covariates. This interpretation is straightforward when the data are based on a randomized design.

Multiple ANCOVA is often employed with nonrandomized studies to eliminate bias in the comparison of several groups. But the problems here are essentially the same as were mentioned with one covariate. Multiple ANCOVA will *not* equate groups that differ on the covariates. It may *reduce* bias on Y that is predictable from the covariates, but it will generally not *eliminate* bias.

Suppose an evaluation of the effect of an advertising program is undertaken by comparing the sales in one city where the program was used with another city where the program was not used. It is reasonable in this case, to ask whether the cities are equivalent on all variables relevant to sales. Let us say that income level, average age, and political conservatism are believed to be the major dimensions that distinguish the two cities. If these variables are employed as multiple covariates and the adjusted mean difference on Y is statistically significant, this does *not* mean that the advertising program was effective. Two problems must be kept in mind in the interpretation of such an analysis. First, it is unlikely that differences between the groups can be *completely* explained by a few variables. If all variables that are required to explain differences between groups on a particular dependent variable are not included as covariates, the groups will not be equated. Second, even if the appropriate variables are included in the analysis, measurement error associated with the measurement of these variables will generally lead to an underadjustment of the means. Multiple ANCOVA will yield means on Y adjusted for the variability accounted for by the covariates *as measured*. It would be appropriate to state that the groups were equated on the measured or obtained scores but not on the true scores.

Fortunately, these problems do not yield ANCOVA and multiple ANCOVA useless with observational studies. As long as it is recognized that these analyses generally *reduce* rather than remove bias in the comparison of groups, there is no problem. Very often the best that can be done with these designs is to employ as many covariates as can reasonably be expected to explain group differences.

When covariates are chosen, an attempt should be made to select variables that are not highly redundant. That is, the correlations among the covariates should be considered. If the correlations are too high, it should be recognized that (1) the estimates of the parameters may be unstable and (2) the computer programs employed to estimate the multiple regression equations required in multiple ANCOVA may not perform properly. The latter problem is easy to identify since most computer programs will generally abort the computation and provide a message stating that the "inverse of the matrix can't be computed," that the matrix is "singular," or that "a linear dependence among predictor variables exists." All these messages refer to the problem of extremely high correlation among the predictor or covariate variables. These problems are referred to in the regression literature under the topic of multicollinearity.

It is possible to use up to $N - (J + 1)$ covariates, where N is the total number of subjects and J is the number of groups. But it is good practice to limit the number of covariates to the extent that the ratio

$$\frac{C + (J - 1)}{N}$$

(where C is the number of covariates) does not exceed 0.10. If this ratio is greater than 0.10, the ANCOVA F test is valid but the estimates of the adjusted means are likely to be unstable. That is, if a study with a high $C + (J - 1)/N$ ratio is cross-validated, it can be expected that the equation that is used to estimate the adjusted means in the original study will yield very different estimates for another sample from the same population.

The hypothesis tested with multiple ANCOVA is the same as with simple ANCOVA, except that the adjusted means are adjusted with respect to multiple covariates rather than a single covariate. A graphic representation of simple and multiple ANCOVA is presented in Figure 8.1.

It can be see that the adjusted population means in simple ANCOVA are conditional on the single covariate X. Under multiple ANCOVA the adjusted population means are conditional on both X_1 and X_2. In general, if C covariates are employed, the adjusted means are conditional on C covariates and the multiple ANCOVA F test is a test of the equality of the elevations of J hyperplanes.

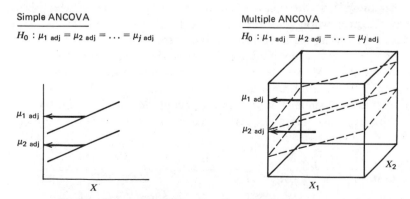

Fig. 8.1 Comparison of simple and multiple covariance analysis.

The interpretation problems described in Chapters 6 and 7 for simple ANCOVA also apply to the multiple ANCOVA model, which is

$$Y_{ij} = \mu + \alpha_j + \beta_1\left(X_{1_{ij}} - \bar{X}_{1..}\right) + \beta_2\left(X_{2_{ij}} - \bar{X}_{2..}\right) + \cdots + \beta_C\left(X_{C_{ij}} - \bar{X}_{C..}\right) + \varepsilon_{ij}$$

where

μ = grand population mean of all Y scores

α_j = effect of jth treatment

$\beta_1, \beta_2, \cdots, \beta_C$ = partial regression coefficients associated with covariates 1 through C

$X_{1_{ij}}, X_{2_{ij}} \cdots X_{C_{ij}}$ = scores on covariates X_1 through X for the ith subject in jth group

$\bar{X}_{1..}, \bar{X}_{2..} \cdots \bar{X}_{C..}$ = grand sample means of covariates 1 through C

ε_{ij} = deviation of the observation from adjusted mean of group of which is it a member (i.e., $Y_{ij} - \mu_{j \text{ adj}}$).

The assumptions associated with this model are also straightforward extensions of those associated with simple ANCOVA.

8.2 MULTIPLE COVARIANCE ANALYSIS THROUGH MULTIPLE REGRESSION

The amount of computation involved in multiple covariance analysis becomes extreme if the analysis is carried out by hand for more than two covariates. Fortunately, if a multiple regression computer program is available, multiple ANCOVA can easily be carried out for any number of covariates. The rationale and computation are straightforward extensions of simple ANCOVA. Suppose that two covariates had been involved in the previously described (Chapter 3) behavioral objectives study rather than one. In addition to the aptitude test scores previously employed as the covariate, we are going to use scores from an academic motivation test. The design matrix for the complete multiple covariance analysis is presented in Table 8.1. The academic motivation covariate scores are in the X_2 column.

By regressing Y on the first four columns, we obtain the R^2 which provides us with the proportion of the variability in Y accounted for by both covariates and the dummy variables. We then regress Y on X_1 and X_2 to obtain the proportion (R^2) accounted for by only the covariates. The difference between the two R^2 values represents the proportion of the variability accounted for by the dummy variables (i.e., group membership or treatment effect) that is independent of variability accounted for by X_1 and X_2. The F test of the adjusted treatment effects takes the general form tabulated as follows:

Source	SS	df	MS	F
Adjusted treatment	$(R^2_{yD,X} - R^2_{yX})$SST	$J-1$	SSAT$/J-1$	MSAT/MSres$_w$
Within residual	$(1 - R^2_{yD,X})$SST	$N-J-C$	SSres$_w/N-J-C$	
Total residual	$(1 - R^2_{yX})$SST	$N-C-1$		

Table 8.1   Design Matrix for Multiple ANCOVA and Associated Tests

(1) D_1	(2) D_2	(3) X_1	(4) X_2	(5) D_1X_1	(6) D_2X_1	(7) D_1X_2	(8) D_2X_2	Y
1	0	29	3	29	0	3	0	15
1	0	49	3	49	0	3	0	19
1	0	48	2	48	0	2	0	21
1	0	35	5	35	0	5	0	27
1	0	53	5	53	0	5	0	35
1	0	47	9	47	0	9	0	39
1	0	46	3	46	0	3	0	23
1	0	74	7	74	0	7	0	38
1	0	72	6	72	0	6	0	33
1	0	67	8	67	0	8	0	50
0	1	22	3	0	22	0	3	20
0	1	24	2	0	24	0	2	34
0	1	49	4	0	49	0	4	28
0	1	46	4	0	46	0	4	35
0	1	52	5	0	52	0	5	42
0	1	43	4	0	43	0	4	44
0	1	64	8	0	64	0	8	46
0	1	61	7	0	61	0	7	47
0	1	55	6	0	55	0	6	40
0	1	54	5	0	54	0	5	54
0	0	33	2	0	0	0	0	14
0	0	45	1	0	0	0	0	20
0	0	35	5	0	0	0	0	30
0	0	39	4	0	0	0	0	32
0	0	36	3	0	0	0	0	34
0	0	48	8	0	0	0	0	42
0	0	63	8	0	0	0	0	40
0	0	57	4	0	0	0	0	38
0	0	56	9	0	0	0	0	54
0	0	78	7	0	0	0	0	56

or

$$\frac{\left(R^2_{yD,X} - R^2_{yX}\right)/J - 1}{\left(1 - R^2_{yD,X}\right)/N - J - C} = F$$

where

$R^2_{yD,X}$ = coefficient of multiple determination obtained by regressing dependent variable on group membership dummy variables and covariates

R^2_{yX} = coefficient of multiple determination obtained by regressing dependent variable on covariates

N = total number of subjects

J = number of groups

C = number of covariates

The obtained F is evaluated with $F_{(\alpha, J-1, N-J-C)}$

For the data in Table 8.1, we can see that columns 1 and $2 = D$ and columns 3 and $4 = X$. The R^2 values are

$$R_{yD,X}^2 = R_{y1234}^2 = 0.754186$$

$$R_{yX}^2 = R_{y34}^2 = 0.596264$$

The difference or unique contribution of $D = 0.157922$.

Source	SS	df	MS	F
Adjusted treatments	(0.157922)3956 = 624.74	2	312.37	8.03
Multiple within residual	(0.245814)3956 = 972.44	25	38.90	
Multiple total residual	(0.403736)3956 = 1597.18	27		

or

$$\frac{0.157922/2}{0.245814/25} = \frac{0.78961}{0.009833} = 8.03$$

The critical values of F for the .05 and .01 significance levels are 3.39 and 5.57, respectively. Multiple covariance has resulted in a higher F value in this case than either ANOVA or simple ANCOVA. The ANOVA F is 1.60 (not significant), and the simple ANCOVA F is 5.48 ($p < .05$).

8.3   TESTING HOMOGENEITY OF REGRESSION PLANES

Just as we assume homogeneous regression slopes in simple ANCOVA, we assume homogeneous regression planes or hyperplanes in multiple covariance analysis. A test of homogeneous regression planes should be carried out before proceeding with the interpretation of multiple ANCOVA. The general linear regression approach to this test is described in this section. The advantage of the approach presented here is that the conceptual framework and the computation can be seen to be just a slight extension of the main ANCOVA. The test is as follows:

$$\frac{\left(R_{yD,X,DX}^2 - R_{yD,X}^2\right)/C(J-1)}{\left(1 - R_{yD,X,DX}^2\right)/N - \left[J(C+1)\right]} = F$$

where

$R_{yD,X,DX}^2 =$ coefficient of multiple determination obtained by regressing dependent variable on group membership dummy variables, covariates, and products of dummy variables and covariates

$R_{yD,X}^2 =$ coefficient of multiple determination obtained by regressing dependent variable on group membership dummy variables and covariates

$N =$ total number of subjects

$J =$ number of groups

$C =$ number of covariates

The obtained F is evaluated with $F_{[\alpha, C(J-1), N-[J(C+1)]]}$. The right-hand R^2 in the numerator is obtained as a step in the computation of the main ANCOVA analysis; $R^2_{yD,X,DX}$, the left-hand R^2 in the formula, will exceed $R^2_{yD,X}$ if multiple regressions based on separate groups have smaller residuals than is obtained with a pooled within-group multiple regression equation.

For the example data of Table 8.1, we see that columns 1 and 2 = D, columns 3 and 4 = X, and columns 5 through 8 = DX. Hence

$$R^2_{yD,X,DX} = R^2_{y12345678} = 0.78573$$

$$R^2_{yD,X} = R^2_{y1234} = 0.754186$$

The difference or unique contribution of $DX = 0.031547$. The test is

$$\frac{0.03154/4}{0.214267/21} = 0.77 = F$$

The critical value of F using $\alpha = .05$ is 2.84. Equivalently, the form found in the following table can be used:

Source	SS	df	MS	F
Heterogeneity of planes	$(R^2_{yD,X,DX} - R^2_{yD,X})$SST	$C(J-1)$	MShet	MShet/MSres$_i$
Multiple residual$_i$	$(1 - R^2_{yD,X,DX})$SST	$N-[J(C+1)]$	MSres$_i$	
Multiple residual$_w$	$(1 - R^2_{yD,X})$SST	$N-J-C$		

With the example data, the summary is:

Source	SS	df	MS	F
Heterogeneity of Planes	124.80	4	31.20	0.77
Multiple residual$_i$	847.64	21	40.36	
Multiple residual$_w$	972.44	25		

Since the obtained F is less than the critical value of F we conclude that the regression planes are homogeneous for the three treatment groups. Hence we do not have to be concerned with the problems of treatment–covariate interaction.

8.4 COMPUTATION OF ADJUSTED MEANS

Adjusted means in multiple ANCOVA may be computed in a manner similar to the procedure suggested for simple ANCOVA. After the regression equation associated with $R^2_{yD,X}$ and the grand covariate means are obtained, each adjusted mean is computed from the equation

$$\overline{Y}_{j\,\text{adj}} = b_0 + b(d_1) + \cdots + b(d_{J-1}) + b(\overline{X}_1..) + \cdots + b(\overline{X}_C..)$$

The regression equation associated with $R^2_{yD,X}$ for the example of this section (i.e., R^2_{y1234}) includes the following terms:

$$b_0 = 7.9890045$$
$$b_1 = -6.82978$$
$$b_2 = 4.40365$$
$$b_3 = 0.276592$$
$$b_4 = 2.83490$$

The grand means for the two covariates are

$$\bar{X}_{1\cdot\cdot} = 49.333$$
$$\bar{X}_{2\cdot\cdot} = 5.000$$

Adjusted means are then obtained from the equation

$$\bar{Y}_{J\,adj} = 7.9890045 - 6.82978(d_1) + 4.40365(d_2) + 0.276592(\bar{X}_{1\cdot\cdot}) + 2.83490(\bar{X}_{2\cdot\cdot})$$

The adjusted means for the three groups are

$$\bar{Y}_{1\,adj} = 7.9890045 - 6.82978(1) + 4.40365(0) + 0.276592(49.333) + 2.8349(5.0)$$
$$= 7.9890045 - 6.82978 + 0 + 13.645113 + 14.1745$$
$$= \underline{28.98}$$

$$\bar{Y}_{2\,adj} = 7.9890045 - 6.82978(0) + 4.40365(1) + 0.276592(49.333) + 2.8349(5.0)$$
$$= 7.9890045 - 0 + 4.40365 + 13.645113 + 14.1745$$
$$= \underline{40.21}$$

$$\bar{Y}_{3\,adj} = 7.9890045 - 6.82978(0) + 4.40365(0) + 0.276592(49.333) + 2.8349(5.0)$$
$$= 7.9890045 - 0 + 0 + 13.645113 + 14.1745$$
$$= \underline{35.81}$$

8.5 MULTIPLE COMPARISON PROCEDURES FOR MULTIPLE COVARIANCE ANALYSIS

The formulas presented in Chapter 5 for multiple comparison tests must be modified for multiple covariance analysis. If a randomized experiment is involved, the formulas presented in Table 8.2 are suitable for any number of covariates. The formulas appropriate for nonrandomized studies with any number of covariates are presented in Table 8.3. Many readers will not be familiar with the matrix notation employed in these formulas. Most textbooks

Table 8.2 Formulas for Multiple Comparisons with Multiple Covariates in Randomized Experiments

Procedure	Formula	Critical Value
Protected LSD	$$\dfrac{\overline{Y}_{i\,\text{adj}} - \overline{Y}_{j\,\text{adj}}}{\sqrt{\text{MS multiple res}_w\left[(1/n_i)+(1/n_j)\right]\left[1+\text{tr}(\mathbf{W}_x^{-1}\mathbf{S}_{b_x})\right]}}$$	Student's $t_{(\alpha, N-J-C)}$
Bryant–Paulson	$$\dfrac{\overline{Y}_{i\,\text{adj}} - \overline{Y}_{j\,\text{adj}}}{\sqrt{\text{MS multiple res}_w\left[1+\text{tr}(\mathbf{W}_x^{-1}\mathbf{S}_{b_x})\right]/n}}$$	Bryant–Paulson generalized Studentized range $Q_{p(\alpha:C,J,N-J-C)}$
Dunn–Bonferroni	$$\dfrac{c_1\overline{Y}_{1\,\text{adj}} + c_2\overline{Y}_{2\,\text{adj}} + \cdots + c_J\overline{Y}_{J\,\text{adj}}}{\sqrt{\text{MS multiple res}_w\left[(c_1^2/n_1)+(c_2^2/n_2)+\cdots+(c_J^2/n_J)\right]\left[1+\text{tr}(\mathbf{W}_x^{-1}\mathbf{S}_{b_x})\right]}}$$	Dunn–Bonferroni $t_{DB(\alpha,C',N-J-C)}$
Scheffé	(Same formula as Dunn–Bonferroni)	$F' = \sqrt{(J-1)F_{\alpha,J-1,N-J-C}}$

on applied multivariate analysis (e.g., Harris 1975) contain chapters on matrix algebra.

The terms in the formulas presented in Tables 8.2 and 8.3 are defined as follows: MS multiple res_w is the error mean square employed in overall multiple ANCOVA; $\mathbf{W}_x^{-1}$ is the inverse of the corrected *within*-group sum of products matrix for the covariates, that is,

$$\mathbf{W}_x = \begin{bmatrix} \Sigma x_1^2 & \Sigma x_1 x_2 & \cdots & \Sigma x_1 x_C \\ \Sigma x_2 x_1 & \Sigma x_2^2 & \cdots & \Sigma x_2 x_C \\ \cdot & \cdot & \cdots & \cdot \\ \cdot & \cdot & \cdots & \cdot \\ \cdot & \cdot & \cdots & \cdot \\ \Sigma x_C x_1 & \Sigma x_C x_2 & \cdots & \Sigma x_C^2 \end{bmatrix}$$

$\mathbf{S}_{b_x}$ is the between-group variance–covariance matrix for the covariates; that is, $\mathbf{S}_{b_x} = \dfrac{1}{J-1}\mathbf{B}_x$ where $\mathbf{B}_x = \mathbf{T}_x - \mathbf{W}_x$ and $\mathbf{T}_x$ is the corrected total sum of products matrix for the covariates; and $\mathrm{tr}(\mathbf{W}_x^{-1}\mathbf{S}_{b_x})$ is the trace of the product $\mathbf{W}_x^{-1}\mathbf{S}_{b_x}$.

The vector $\mathbf{d}$ in the formulas in Table 8.3 is the column vector of differences between the ith and jth group means on the covariates. For example, to test the difference between adjusted means 1 and 3 in a three-group experiment with C covariates, we have

$$
\begin{array}{c}
\\
\text{Covariate 1} \\
\text{Covariate 2} \\
\vdots \\
\text{Covariate } C
\end{array}
\begin{array}{cc}
\text{Group 1} & \text{Group 3} \\
\end{array}
\qquad \mathbf{d}
$$

$$
\begin{array}{c}
\text{Covariate 1} \\
\text{Covariate 2} \\
\vdots \\
\text{Covariate } C
\end{array}
\begin{bmatrix}
\bar{X}_{1,1} & - & \bar{X}_{1,3} \\
\bar{X}_{2,1} & - & \bar{X}_{2,3} \\
\vdots & \vdots & \vdots \\
\bar{X}_{C,1} & - & \bar{X}_{C,3}
\end{bmatrix}
=
\begin{bmatrix}
d_1 \\
d_2 \\
\vdots \\
d_C
\end{bmatrix}
$$

If the Dunn–Bonferroni or Scheffé procedures are employed to evaluate complex contrasts, the vector $\mathbf{d}$ is

$$
\begin{array}{c}
\\
\text{Covariate 1} \\
\text{Covariate 2} \\
\cdot \\
\cdot \\
\cdot \\
\text{Covariate } C
\end{array}
\begin{array}{cccc}
\multicolumn{4}{c}{\text{Group}} \\
1 & 2 & \cdots & J
\end{array}
\qquad \mathbf{d}
$$

$$
\begin{bmatrix}
c_1\bar{X}_{1,1} & + c_2\bar{X}_{1,2} & + \cdots + & c_J\bar{X}_{1,J} \\
c_1\bar{X}_{2,1} & + c_2\bar{X}_{2,2} & + \cdots + & c_J\bar{X}_{2,J} \\
\cdot & \cdot & \cdots & \cdot \\
\cdot & \cdot & \cdots & \cdot \\
\cdot & \cdot & \cdots & \cdot \\
c_1\bar{X}_{C,1} & + c_2\bar{X}_{C,2} & + \cdots + & c_J\bar{X}_{C,J}
\end{bmatrix}
=
\begin{bmatrix}
d_1 \\
d_2 \\
\cdot \\
\cdot \\
\cdot \\
d_C
\end{bmatrix}
$$

The transpose of $\mathbf{d}$ is the row vector $\mathbf{d}'$.

Table 8.3 Formulas for Multiple Comparisons with Multiple Covariates in Nonrandomized Studies

Procedure	Formula	Critical Value
Protected LSD	$$\dfrac{\overline{Y}_{i\,\text{adj}} - \overline{Y}_{j\,\text{adj}}}{\sqrt{\text{MS multiple res}_{\text{w}}\left[(1/n_i) + (1/n_j) + \mathbf{d}'\mathbf{W}_x^{-1}\mathbf{d}\right]}}$$	Student's $t_{(\alpha,\,N-J-C)}$
Bryant–Paulson	$$\dfrac{\overline{Y}_{i\,\text{adj}} - \overline{Y}_{j\,\text{adj}}}{\sqrt{\dfrac{\text{MS multiple res}_{\text{w}}\left[(2/n) + \mathbf{d}'\mathbf{W}_x^{-1}\mathbf{d}\right]}{2}}}$$	Bryant–Paulson generalized Studentized range $Q_{p_{(\alpha:C,\,N-J-C)}}$
Dunn–Bonferroni	$$\dfrac{c_1\overline{Y}_{1\,\text{adj}} + c_2\overline{Y}_{2\,\text{adj}} + \cdots + c_J\overline{Y}_{J\,\text{adj}}}{\sqrt{\text{MS multiple res}_{\text{w}}\left[(c_1^2/n_1) + (c_2^2/n_2) + \cdots + (c_J^2/n_J) + \mathbf{d}'\mathbf{W}_x^{-1}\mathbf{d}\right]}}$$	Dunn–Bonferroni $t_{\text{DB}_{(\alpha,\,C,\,N-J-C)}}$
Scheffé	(Same formula as Dunn–Bonferroni)	$F' = \sqrt{(J-1)F_{(\alpha,\,J-1,\,N-J-C)}}$

Example: Pairwise Comparisons Using Protected LSD Tests

Multiple covariate analysis on the data presented in Table 8.1 yields the following adjusted means, covariate means and error mean square:

$$\overline{Y}_{1\,adj} = 28.98 \qquad \overline{X}_{1,1} = 52.00 \qquad \overline{X}_{2,1} = 5.1$$

$$\overline{Y}_{2\,adj} = 40.21 \qquad \overline{X}_{1,2} = 47.00 \qquad \overline{X}_{2,2} = 4.8$$

$$\overline{Y}_{3\,adj} = 35.81 \qquad \overline{X}_{1,3} = 49.00 \qquad \overline{X}_{2,3} = 5.1$$

$$\text{MS Multiple Res}_w = 38.90$$

Since the experiment was based on a randomized design the differences among covariate means are small for both covariates (ANOVA F on $X_1 = 0.30$ and ANOVA F on $X_2 = 0.05$). The formula for the protected LSD test found in Table 8.2 is appropriate for all pairwise contrasts. This formula requires the computation of $tr(W_x^{-1}S_{b_x})$, this computation involves the following steps:

1. Compute W_x.

$$W_x = \begin{bmatrix} 5700.0 & 591.0 \\ 591.0 & 149.4 \end{bmatrix}$$

2. Compute the inverse of W_x.

$$W_x^{-1} = \begin{bmatrix} 0.000297 & -0.001177 \\ -0.001177 & 0.011348 \end{bmatrix}$$

3. Compute the corrected between-group sum of products matrix B_x.

$$B_x = T_x - W_x = \begin{bmatrix} 5826.67 & 598.00 \\ 598.00 & 150 \end{bmatrix} - \begin{bmatrix} 5700.00 & 591.00 \\ 591.00 & 149.40 \end{bmatrix}$$

$$= \begin{bmatrix} 126.67 & 7.00 \\ 7.00 & 0.60 \end{bmatrix}$$

4. Multiply B_x by the scalar $(1/J-1)$ to obtain the between-group variance–covarince matrix S_{b_x}.

$$S_{b_x} = \frac{1}{2}\begin{bmatrix} 126.67 & 7.00 \\ 7.00 & 0.60 \end{bmatrix} = \begin{bmatrix} 63.335 & 3.50 \\ 3.50 & .030 \end{bmatrix}$$

5. Compute the product $W_x^{-1}S_{b_x}$.

$$W_x^{-1}S_{b_x} = \begin{bmatrix} 0.000297 & -0.001177 \\ -0.001177 & 0.011348 \end{bmatrix}\begin{bmatrix} 63.335 & 3.50 \\ 3.50 & 0.30 \end{bmatrix}$$

$$= \begin{bmatrix} 0.014691 & 0.000686 \\ -0.034827 & -0.000715 \end{bmatrix}$$

6. Compute the trace $\mathbf{W}_x^{-1}\mathbf{S}_{b_x}$.

$$\text{tr}\left(\mathbf{W}_x^{-1}\mathbf{S}_{b_x}\right) = \begin{bmatrix} 0.014691 & 0.000686 \\ -0.034827 & -0.000715 \end{bmatrix} = 0.0139759$$

The standard error of the difference between adjusted means is

$$\sqrt{\text{MS multiple res}_w\left[(1/n_i)+(1/n_j)\right]\left[1+\text{tr}\left(\mathbf{W}^{-1}\mathbf{S}_{b_x}\right)\right]} =$$
$$\sqrt{38.90\left(\tfrac{1}{10}+\tfrac{1}{10}\right)\left[1+0.0139759\right]} = 2.81$$

All pairwise contrasts are divided by this term to obtain t.

Groups	Adjusted Mean Difference	Obtained t
1 vs. 2	$28.98 - 40.21 = -11.23$	-4.00
1 vs. 3	$28.98 - 35.81 = -6.83$	-2.43
2 vs. 3	$40.21 - 35.81 = 4.40$	1.57

The absolute value of t_{obt} is evaluated with $t_{(\alpha, N-J-C)}$ or, in this case, $t_{(.05, 30-3-2)} = t_{(.05, 25)} = 2.06$. Two of the three pairwise contrasts are significant.

If the study had not involved randomized groups, the differences among means on the covariates could have been large. In this case the LSD formula in Table 8.3 would be appropriate. The data presented in Table 8.1 are once again employed to illustrate the application of the nonrandomized or non-equivalent groups formula for the protected LSD tests. The formula is

$$\frac{\overline{Y}_{i\,\text{adj}} - \overline{Y}_{j\,\text{adj}}}{\sqrt{\text{MS multiple res}_w\left[(1/n_i)+(1/n_j)+\mathbf{d}'\mathbf{W}_x^{-1}\mathbf{d}\right]}} = t$$

The adjusted means, MS multiple res$_w$, and $\mathbf{W}^{-1}$ have already been computed. The column vector $\mathbf{d}$ is simply the vector of differences between the ith and jth group means on the covariate. Since there are three groups and three pairwise contrasts in this example, the $\mathbf{d}$ vectors associated with the three contrasts are

Groups	Covariate Mean Differences	$= \mathbf{d}$
1 vs. 2	$\overline{X}_{1,1} - \overline{X}_{1,2} = 52 - 47$	$= \begin{bmatrix} 5.0 \\ 0.3 \end{bmatrix}$
	$\overline{X}_{2,1} - \overline{X}_{2,2} = 5.1 - 4.8$	
1 vs. 3	$\overline{X}_{1,1} - \overline{X}_{1,3} = 52 - 49$	$= \begin{bmatrix} 3.0 \\ 0.0 \end{bmatrix}$
	$\overline{X}_{2,1} - \overline{X}_{2,3} = 5.1 - 5.1$	
2 vs. 3	$\overline{X}_{1,2} - \overline{X}_{1,3} = 47 - 49$	$= \begin{bmatrix} -2.0 \\ -0.3 \end{bmatrix}$
	$\overline{X}_{2,2} - \overline{X}_{2,3} = 4.8 - 5.1$	

If we carry out the computations for the three pairwise contrasts by using the nonrandomized study formula, we find that the three obtained t values are

1 vs. 2 $\qquad \dfrac{-11.23}{\sqrt{38.90\left[\frac{1}{10}+\frac{1}{10}+0.004915\right]}} = \dfrac{-11.23}{2.823} = -3.98 = t_{obt}$

where

$$0.004915 = \begin{bmatrix} 5.0 & 0.3 \end{bmatrix} \begin{bmatrix} 0.000297 & -0.001177 \\ -0.001177 & 0.011348 \end{bmatrix} \begin{bmatrix} 5.0 \\ 0.3 \end{bmatrix}$$

1 vs. 3 $\qquad \dfrac{-6.83}{\sqrt{38.90[0.2+0.002673]}} = \dfrac{-6.83}{2.808} = -2.43 = t_{obt}$

where

$$0.002673 = \begin{bmatrix} 3.0 & 0.0 \end{bmatrix} \begin{bmatrix} 0.000297 & -0.001177 \\ -0.001177 & 0.011348 \end{bmatrix} \begin{bmatrix} 3.0 \\ 0.0 \end{bmatrix}$$

2 vs. 3 $\qquad \dfrac{4.40}{\sqrt{38.90[0.2+0.000797]}} = \dfrac{4.40}{2.795} = 1.57 = t_{obt}$

where

$$0.000797 = \begin{bmatrix} -2.0 & -0.3 \end{bmatrix} \begin{bmatrix} 0.000297 & -0.001177 \\ -0.001177 & 0.011348 \end{bmatrix} \begin{bmatrix} -2.0 \\ -0.3 \end{bmatrix}$$

The critical value of t is, as before, 2.06.

A comparison of the t values based on the two formulas reveals almost no difference.

Contrast	t_{obt} with Randomized Experiment Formula	t_{obt} with Nonrandomized Study Formula
-11.23	-4.00	-3.98
-6.83	-2.43	-2.43
4.40	1.57	1.57

The similarity of the results using the two formulas is expected when the differences on the covariates are small. If the square of the error terms employed in the computation of the nonrandomized study t values are averaged (i.e., $[(2.823)^2+(2.808)^2+(2.795)^2]/3=7.89$), we obtain the square of the value of the error term of the randomized experiment formula. If large differences among covariate means exist, the t values will not be similar if the two different formulas are used because the error estimate based on the randomized experiment formula is an average value.

Simultaneous Confidence Intervals

The radical terms associated with the Bryant–Paulson, Dunn–Bonferroni, and Scheffé formulas in Tables 8.2 and 8.3 can be employed to construct simultaneous confidence intervals. A specified adjusted mean difference plus and minus the product of the radical term times the associated critical value yields the confidence interval.

Example of Simultaneous Confidence Intervals

The 95% Bryant–Paulson simultaneous confidence intervals, for the data of Table 8.1, require the same matrix manipulations as have been carried out earlier for the protected LSD tests. The interval for the contrast of the ith and jth adjusted means is

$$\left(\overline{Y}_{i\,\text{adj}} - \overline{Y}_{j\,\text{adj}}\right) \pm \left(\sqrt{\frac{\text{MS multiple res}_w\left[1 + \text{tr}\left(W_x^{-1}S_{b_x}\right)\right]}{n}}\right)\left(Q_{p(\alpha:\,C,J,\,N-J-C)}\right)$$

or

$$\left(\overline{Y}_{i\,\text{adj}} - \overline{Y}_{j\,\text{adj}}\right) \pm \left(\sqrt{\frac{38.90\left[1 + 0.0139759\right]}{10}}\right)\left(Q_{p(.05,2,3,25)}\right) =$$

$$\left(\overline{Y}_{i\,\text{adj}} - \overline{Y}_{j\,\text{adj}}\right) \pm (1.986)(3.68)$$

The 95% simultaneous confidence intervals for the three pairwise contrasts are:

Groups	Adjusted Mean Difference	Interval
1 vs. 2	-11.23	$-11.23 \pm 7.308 = (-18.54, -3.92)$
1 vs. 3	-6.83	$-6.83 \pm 7.308 = (-14.14, 0.48)$
2 vs. 3	4.40	$4.40 \pm 7.308 = (-2.91, 11.71)$

8.6 SUMMARY

Multiple covariance analysis is a straightforward extension of ANCOVA with one covariate. Several covariates can be expected to increase the power of an experiment more than a single covariate if the covariates are well chosen. Also, the use of more than one covariate will generally remove more bias when means are compared. Still, even the use of many covariates will not generally *eliminate* bias in the comparison of groups from different populations.

The computation required for multiple ANCOVA is a slight extension of the general linear regression approach employed with simple ANCOVA. The dependent variable is first regressed on all group-membership dummy variables and all covariates. Next, the dependent variable is regressed on all covariates. The difference between the R^2 values associated with these two regressions represents the proportion of the total variability in Y that is accounted for by group differences that are independent of the covariates.

CHAPTER NINE
Nonlinear ANCOVA

9.1 INTRODUCTION

The relationship between the covariate and the dependent variable scores is not always linear. Since an assumption underlying the ANCOVA model is that the within-group relationship between X and Y is linear, researchers should be aware of the problem of nonlinearlity. If ANCOVA is employed when the data are nonlinear, the power of the F test is decreased and the adjusted means may be poor representations of the treatment effects.

Two reasons for nonlinear relationships between X and Y are inherent nonlinearity of characteristics and scaling error. It is quite possible that the basic characteristics being measured are not linearly related. For example, the relationship between extroversion (X) and industrial sales performance (Y) could be predicted to be nonlinear. Those salespeople with very low extroversion scores may have poor sales performance because they have difficulty interacting with clients. Those with very high extroversion scores may be viewed as overly social and not serious about their work. Hence, very low or very high extroversion scores may be associated with low sales performance, whereas intermediate extroversion scores may be associated with high sales performance.

Another example of expected nonlinearity might be found between certain measures of motivation (X) and performance (Y). Psychologists working in the area of motivation sometimes hypothesize that there is an optimal level of motivation or arousal for an individual working on a specific task. At very low or very high levels of arousal, performance is lower than at the optimal level of arousal. In both examples the relationship between X and Y scores is expected to be nonlinear because the relationship between the basic characteristic underlying the observed (measured) scores is expected to be nonlinear. This distinction between the measured and underlying or basic scores is important. It is quite possible that the relationship between *observed* X and Y scores is nonlinear when the relationship between the basic X and Y characteristics is linear. When this occurs, the problems of scaling error are involved.

There are several types of scaling error that can produce nonlinearity, but probably the most frequently encountered type results in either "ceiling" or "floor" effects. In either case the problem is that the instrumentation or scale used in the measurement of either the X or the Y variable (or both) may not be adequate to reflect real differences in the characteristics being measured.

For example, if most of the subjects employed in a study obtain nearly the highest possible score on a measure, there are likely to be unmeasured differences among those who get the same high score. The measurement procedures simply does not have sufficient "ceiling" to reflect differences among the subjects on the characteristics being measured. Suppose most subjects get a score of 50 on a 50-point pretest that is employed as a covariate; the test is much too easy for the subjects included in the experi-

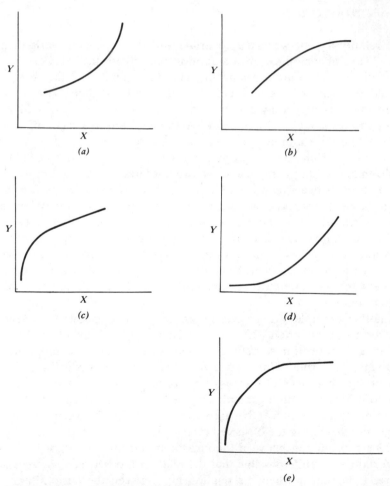

Fig. 9.1 Representations of ceiling and floor effects: (*a*) ceiling effect in X; (*b*) ceiling effect in Y; (*c*) floor effect in X; (*d*) floor effect in Y; (*e*) floor effect in X and ceiling effect in Y.

ment. If the scores on this measure are plotted against scores on a posttest that is of the appropriate difficulty level, nonlinearity will be observable. This situation is presented in Figure 9.1*a*. (Other types of nonlinearity are illustrated in Figures 9.1*b-e*.) Here the inherent relationship between the X and Y characteristics is linear, but the obtained relationship between the *observed* measures is not linear. Hence one reason for nonlinearity in the XY relationship is scaling error or inappropriate measurement. Regardless of the reason for nonlinearity, the linear ANCOVA model is inappropriate if the degree of nonlinearity is severe.

9.2   METHODS OF DEALING WITH NONLINEARITY

A routine aspect of any data analysis should be to plot the data. This preliminary step involves plotting the Y scores against the X scores for each group. Severe nonlinearity will generally be obvious in both the trend observed in the scatter plot and in the shape of the marginal distributions. Most computer statistical packages include options that will provide the required plots. Other approaches to the problem of identifying nonlinearity include plotting the residuals of the regression of Y on X and fitting various models to the data. Additional details on these procedures can be found in Draper and Smith (1966).

Once it has been decided that nonlinearity is problematic, the next step is to (1) seek a transformation of the original X and/or Y scores that will result in a linear relationship for the transformed data or (2) fit a polynomial ANCOVA model to the original data.

Data Transformations

If the relationship between X and Y is nonlinear but monotonic (i.e., Y increases when X increases but the function is not linear), a transformation of X should be attempted. Logarithmic, square root, and reciprocal transformations are most commonly used because they usually yield the desired linearity. Advanced treatments of regression analysis should be consulted for details on these and other types of transformation (e.g., Ezekiel and Fox 1959, Draper and Smith 1966).

Once a transformation has been selected, ANCOVA is carried out in the usual way on the transformed data. For example, if there is reason to believe that the relationship between $\log_e X$ and Y is linear, ANCOVA is carried out using $\log_e X$ as the covariate. It must be pointed out in the interpretation of the analysis, however, that $\log_e X$ rather than X was the covariate.

A method of determining whether a transformation has improved the fit of the model to the data is to plot the scores and compute ANCOVA for both untransformed and transformed data. A comparison of the plots and ANCOVAs will reveal the effect of the transformation.

Polynomial ANCOVA Models

If the relationship between X and Y is not monotonic, a simple transformation will not result in linearity. The reason for this may be understood by examining Figure 9.2. In the nonlinear–monotonic situation the values of Y increase as value of X increase. In the nonlinear–nonmonotonic situation, Y increases as X increases only up to a point, and then Y decreases as X increases. If we transform X to $\log_e X$ for the nonmonotonic situation, the $\log_e X$ values increase as X increases and nonlinearity is still present when $\log_e X$ and Y are plotted. The simplest alternative in this case is to fit a second-degree polynomial (quadratic) ANCOVA model. This model is written

$$\overline{Y}_{ij} = \mu + \alpha_j + \beta_1\left(X_{ij} - \overline{X}..\right) + \beta_2\left(X_{ij}^2 - \left(\overline{X^2}..\right)\right) + \varepsilon_{ij}$$

where

$Y_{ij} =$ dependent variable score of ith individual in jth group
$\mu =$ population mean common to all observations on Y
$\alpha_j =$ effect of treatment j
$\beta_1 =$ linear effect regression coefficient
$X_{ij} =$ covariate score for ith individual in jth group
$\overline{X}.. =$ mean of all observations on covariate
$\beta_2 =$ curvature effect coefficient
$X_{ij}^2 =$ squared covariate score for ith individual in jth group
$(\overline{X^2}..) =$ mean of squared observations on covariate (i.e., $\sum_{i=1}^{N} X_{ij}^2 / N$)
$\varepsilon_{ij} =$ error component associated with ith individual in jth group

This model differs from the linear model in that it contains the curvature effect term $\beta_2(X_{ij}^2 - (\overline{X^2}..))$. If the dependent variable scores are a quadratic rather than a linear function of the covariate, this model will provide a better fit and will generally yield greater power with respect to tests on adjusted means.

The quadratic ANCOVA is computed by using X and X^2 as if they were two covariates in a multiple covariate analysis. The main ANCOVA test, the homogeneity of regression test, the computation of adjusted means, and multiple comparison tests are all carried out as with an ordinary two-covariate ANCOVA.

If the relationship between X and Y is more complex than a quadratic function, a higher-degree polynomial may be useful. The third-degree polynomial (cubic) ANCOVA model is written

$$Y_{ij} = \mu + \alpha_j + \beta_1\left(X_{ij} - \overline{X}..\right) + \beta_2\left(X_{ij}^2 - \left(\overline{X^2}..\right)\right) + \beta_3\left(X_{ij}^3 - \left(\overline{X^3}..\right)\right) + \varepsilon_{ij}$$

This model will provide a good fit if the relationship between the covariate

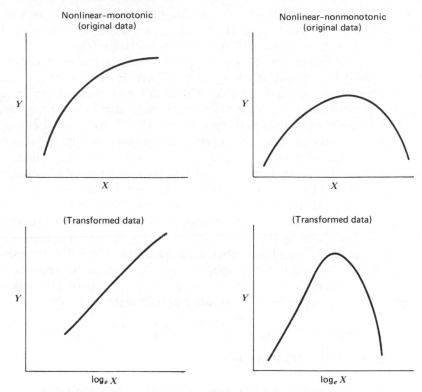

Fig. 9.2   Effect of log transformation on two types of nonlinearity.

and the dependent variable is a cubic function. Cubic ANCOVA is carried out by employing X, X^2, and X^3 as covariates in a multiple covariance analysis. Higher-degree polynomials can be employed for more complex functions, but it is very unusual to encounter such situations.

Higher-degree polynomial models virtually always fit sample data better than do simpler polynomial models, but this does not mean that the more complex models are preferable to the simpler ones. Care must be taken not to employ a more complex model than is required for two reasons. First, a degree of freedom is lost from the ANCOVA error mean square (i.e., $MSres_w$) for each additional term in the ANCOVA model. If the number of subjects is not large, the loss of degrees of freedom can easily offset the sum-of-squares advantage of a better fit afforded by the more complex model. Hence even though the sum of squares error (residual$_w$) is smaller with more complex models, the mean-square error can be considerably larger with complex models. The consequences of the larger error term are less precise estimates of the adjusted means, and, correspondingly, less precise tests on the difference between adjusted means. This problem is illustrated in Section 9.3. The second reason for not employing a more complex model than is required is the law of parsimony. If a linear model fits the data almost as well as a quadratic model, the simpler model should usually be chosen because the interpretation and generalization of results is more straightforward.

Two additional points on the use of polynomial regression models are relevant to the polynomial ANCOVA described here. First, it is not necessary that the covariate be a fixed variable. This point was made earlier in the discussion of assumptions for ANCOVA but is reiterated here for nonlinear ANCOVA because, as Cramer and Appelbaum (1979) have pointed out, it is sometimes mistakenly believed that polynomial regression is appropriate only with X fixed. Second, the parameters of the polynomial regression are sometimes difficult to estimate with certain multiple regression computer programs because these programs will not, with certain data sets, yield the inverse of the required matrix. This problem develops because X, X^2, X^3, and so on are all highly correlated. These computational difficulties can generally be solved by transforming the raw X scores to deviation scores before the regression analysis is carried out. That is, in quadratic ANCOVA for example, $(X - \overline{X})$ and $(X - \overline{X})^2$ rather than X and X^2 should be used as the covariates. Additional details on this problem in the context of conventional regression analysis can be found in Bradley and Srivastava (1979).

9.3 COMPUTATION AND EXAMPLE

The comuputation and rationale for quadratic ANCOVA are essentially the same as for multiple ANCOVA. Consider the following data:

(1) Experimental Group		(2) Control Group	
X	Y	X	Y
13	18	11	13
7	14	2	1
17	7	19	2
14	14	15	9
3	8	8	10
12	19	11	15

Previous research or theoretical consideration may suggest that the relationship between X and Y is best described as a quadratic function. A scatter plot of these data appears to support a quadratic model. Hence the experimenter has a reasonable basis for deciding to employ the quadratic ANCOVA model. The computation of the complete quadratic ANCOVA through the general linear regression procedure is based on the following design matrix:

(1) D	(2) X	(3) X^2	(4) DX	(5) DX^2	(6) Y
1	13	169	13	169	18
1	7	49	7	49	14
1	17	289	17	289	7
1	14	196	14	196	14
1	3	9	3	9	8
1	12	144	12	144	19
0	11	121	0	0	13
0	2	4	0	0	1
0	19	361	0	0	2
0	15	225	0	0	9
0	8	64	0	0	10
0	11	121	0	0	15

We now proceed as if we were performing a multiple covariance analysis using X and X^2 as the covariates. As before, the main test on adjusted treatment effects is based on the coefficients of multiple determination R_{yX}^2 and $R_{yD,X}^2$.

The term R_{yX}^2 represents the proportion of the total variability explained by the quadratic regression (i.e., the regression of Y on X and X^2); whereas $R_{yD,X}^2$ represents the proportion of the total variability explained by the quadratic regression and the treatments. Hence the difference between the two coefficients represents the proportion of the variability accounted for by the treatments that is independent of that accounted for by quadratic regression. The proportion of unexplained variability is, of course, $1 - R_{yD,X}^2$.

Column 1 in this example is the only dummy variable (since there are only $J-1$ dummy variables), columns 2 and 3 are the covariate columns, columns 4 and 5 are the interaction columns (not used in the main analysis), and column 6 contains the dependent variable scores. The regression analyses yield the following tabulated data:

$$R_{yD,X}^2 = R_{y123}^2 \qquad\qquad = 0.918903$$

$$R_{yX}^2 = R_{y23}^2 \qquad\qquad = 0.799091$$

Difference or unique contribution of dummy
variable beyond quadratic regression $\qquad = 0.119812$

Total sum of squares $\qquad\qquad\qquad = 361.67$

The general form of the quadratic ANCOVA summary is as follows:

Source	SS	df	MS	F
Adjusted treatment	$(R^2_{yD.x} - R^2_{yx})$SST	$J-1$	SSAT$/J-1$	MSAT$/$MSres$_w$
Quadratic residual$_w$	$(1-R^2_{yD.x})$SST	$N-J-C$	SSres$_w/N-J-C$	
Quadratic residual$_t$	$(1-R^2_{yx})$SST	$N-1-C$		

The quadratic ANCOVA summary for the example data is as follows:

Source	SS	df	MS	F
Adjusted treatment	$(0.119812)361.67 = 43.33$	1	43.33	11.81
Quadratic residual$_w$	$(1-0.918903)361.67 = 29.33$	8	3.67	
Quadratic residual$_t$	$(1-0.799091)361.67 = 72.66$	9		

The obtained F is significant beyond the 1% level.

Adjusted means and multiple comparison procedures are also dealt with as they are under the multiple ANCOVA model. The adjusted means for the example data are obtained through the regression equation associated with R^2_{y123}. The intercept and regression weights are:

$$b_0 = -5.847359$$
$$b_1 = 3.83111$$
$$b_2 = 3.66943$$
$$b_3 = -0.17533$$

The group 1 dummy score, the grand mean covariate score, and the grand mean of the squared covariate scores are 1, 11, and 146 respectively. Hence $\bar{Y}_{1\,adj} = -5.847539 + 3.83111(1) + 3.66943(11) - 0.17533(146) = 12.75$. The group 2 dummy score, the grand mean covariate score, and the grand mean of the squared covariate scores are 0, 11, and 146 respectively. Hence $\bar{Y}_{2\,adj} = -5.847359 + 3.83111(0) + 3.66943(11) - 0.175333(146) = 8.92$.

Just as the test of the homogeneity of regression planes is an important adjunct to the main F test in multiple ANCOVA, the test of the homogeneity of the quadratic regressions for the separate groups should be carried out in quadratic ANCOVA. This test is computed in the same manner as the test of the homogeneity of regression planes.

The form of the summary is as follows:

Source	SS	df	MS	F
Heterogeneity of quadratic regression	$(R^2_{yD.x,DX} - R^2_{yD.x})$SST	$2(J-1)$	MShet	$\dfrac{\text{MShet}}{\text{MSres}_i}$
Quadratic residual$_i$	$(1-R^2_{yD.x,DX})$SST	$N-(J3)$	MSres$_i$	
Quadratic residual$_w$	$(1-R^2_{yD.x})$SST	$N-J-2$		

A more general form, appropriate for testing the homogeneity of any degree polynomial regression, is as follows:

Source	SS	df	MS	F
Heterogeneity of polynomial regression	$(R^2_{yD,X,DX} - R^2_{yD,X})\text{SST}$	$C(J-1)$	MShet	$\dfrac{\text{MShet}}{\text{MSres}_i}$
Polynomial residual$_i$	$(1 - R^2_{yD,X,DX})\text{SST}$	$N - J(C+1)$	MSres$_i$	
Polynomial residual$_w$	$(1 - R^2_{yD,X})\text{SST}$	$N - J - C$		

For the example data, the necessary quantities are

$$R^2_{yD,X,DX} = R^2_{y12345} \qquad = 0.944817$$

$$R^2_{yD,X} = R^2_{y123} \qquad = 0.918903$$

Difference or heterogeneity of regression $= 0.025914$

Total sum of squares $\qquad\qquad = 361.67$

and the summary is as follows:

Source	SS	df	MS	F
Heterogeneity of quadratic regression	$(0.025914)361.67 = 9.37$	2	4.69	1.41
Quadratic residual$_i$	$(1 - 0.944817)361.67 = 19.96$	6	3.33	
Quadratic residual$_w$	$(1 - 0.918903)361.67 = 29.33$	8		

The obtained F value is clearly not significant, and we conclude that the quadratic regressions for the experimental and control groups are homogeneous.

Comparison of Quadratic ANCOVA with Other Models

It was mentioned earlier that the complexity of the model employed should be sufficient to adequately describe the data but that it should not be more complex than is required. The results of applying four different models to the data of the example problem are tabulated as follows:

Model	Obtained F	Degrees of Freedom	Probability
ANOVA	2.62	1,10	$>.10$
Linear ANCOVA	2.38	1,9	$>.10$
Quadratic ANCOVA	11.81	1,8	$<.01$
Cubic ANCOVA	9.95	1,7	$<.025$

The F of the simplest model, ANOVA, when compared with the linear ANCOVA F, illustrates the fact that ANOVA can be more powerful than ANCOVA when the correlation between the covariate and the dependent variable is low. The F of the most complex of the four models, cubic ANCOVA, when compared with the quadratic F, illustrates the fact that more complex models do not necessarily lead to greater precision. The greatest precision is obtained with the model that is neither too simple nor more complex than is necessary for an adequate fit.

9.4 SUMMARY

The assumption of the conventional ANCOVA model that the covariate and the dependent variable are linearly related will not always be met. Severe nonlinearity generally can be easily identified by inspecting the XY scatter plot within groups. If the relationship is nonlinear but monotonic, it is likely that a simple transformation (generally of the X variable) can be found that will yield a linear relationship between transformed X and Y. Analysis of covariance is then applied by using the transformed variable as the covariate. If the relationship is not monotonic, the simple transformation approach will not be satisfactory, and more complex approach of employing some polynomial of X should be attempted. Generally, a quadratic or cubic ANCOVA model will fit the data. Complex polynomial models should be employed only if simpler ones are obviously inadequate. Simpler models are preferred because results based on complex models are (1) more difficult to interpret and generalize and (2) less stable. When polynomial ANCOVA models are clearly called for, the computation involves a straightforward extension of multiple ANCOVA.

CHAPTER TEN
Complex Experimental Designs

10.1 INTRODUCTION

Designs other than randomized group designs have been briefly mentioned in several previous chapters. Randomized-block ANOVA, repeated-measurement ANOVA, and two-factor ANOVA were referred to as alternatives to the randomized-group ANCOVA. In this chapter we present an overview of the essentials of these alternatives and then describe the application of covariance analysis to independent sample and repeated measurement two-factor designs. It is shown that the advantages of blocking, repeated measurement, and two-factor ANOVA designs are included among the strengths of two-factor ANCOVA designs.

10.2 ALTERNATIVES TO ONE-FACTOR RANDOMIZED-GROUP ANALYSIS OF COVARIANCE

The randomized-group one-factor or single-classification design with covariance analysis is preferable to the analysis of variance of this design because the effects of the covariate can be removed from both the error and the treatment estimates. This generally results in a more powerful and perhaps more meaningful test of treatment effects because the estimated error variance and the adjusted means are free of nuisance variability (i.e., variability due to any source not wanted in the experiment) that is measured by the covariate. There are, however, other ways of attempting to cope with the problem of nuisance variables.

One alternative to ANCOVA is an ANOVA design that includes one or more nuisance variables. We describe two alternatives to the one-factor randomized-group design that employ such nuisance variables. These alternatives are the one-factor randomized-block design and the one-factor repeated-measurement design. Each design is utilized to test the same null hypothesis associated with one-factor randomized-group designs (i.e., H_0: $\mu_1 = \mu_2 = \cdots = \mu_J$), but the design layout, randomization procedures, and analysis approaches differ. Like the analysis of covariance, these designs are generally employed as power increasing approaches. Unlike covariance analysis, there is no adjustment of dependent variable means for chance differences between treatment groups that exist before the experiment is carried out. The details of the design and the analysis of these approaches and the relationships between them are presented in Sections 10.3 and 10.4.

10.3 ONE-FACTOR RANDOMIZED-BLOCK DESIGN

The one-factor randomized-block design, like the one-factor randomized-group design, is employed to test the hypothesis that J population means are equal. The basic difference between these two designs can be found in the method of assigning subjects to treatments. With the randomized group design the available subjects are randomly assigned to J treatment conditions. With the randomized-block design there are three steps involved in the assignment of subjects to treatments. First, information on subjects that can be employed to form blocks is collected. A block is simply a group of subjects that is relatively homogeneous with respect to some variable (called the *blocking variable*) that is believed to be related to the dependent variable. Blocking variables may be continuous variables such as age, weight, intelligence, experience, and pretest score, as well as nominal variables such as sex, occupation type, classroom, litter, and observer.

Once the blocking-variable information is available for all subjects, the second step, the formation of the blocks, can be carried out. If the blocking data are in the form of continuous scores, the blocks are formed by simply (1) ordering the subjects according to their score on the blocking variable and (2) assigning the highest J subjects to block 1, the next highest group of J subjects to block 2, and so on. For example, an experimenter may believe that anxiety as measured by the galvanic skin response (GSR) is related to the performance of subjects on a complex motor skills task. Suppose that 25 subjects are available and that the purpose of the experiment is to compare the effectiveness of five different methods of training employees to perform a complex task. First, the 25 subjects are measured on the blocking variable, the GSR. Then the subjects are ordered from highest to lowest according to their GSR scores. Since $J = 5$ (i.e., there are five treatments) the subjects associated with the top five GSR scores constitute the first block. The subjects associated with the next highest five scores constitute the second block and so on, through the subjects associated with the lowest five scores who constitute the last block (block 5).

The third step in setting up the randomized block design is randomization. After the blocks have been formed, the subjects in each block are randomly assigned to treatments. This randomization is carried out independently for each block. For the training example, the five subjects in the first block are randomly assigned to the five treatments. Then the five subjects in block 2 are randomly assigned to the five treatments, and so on. A comparison of the randomization procedures for the randomized group and randomized block designs can be seen in Figure 10.1. In this example there are five treatments and five blocks. It is not necessary, however, that the number of treatments and the number of blocks be the same. If 15 subjects are available and five treatments are to be studied, there would be three blocks. Notice there is one subject for every combination of block and treatment with this design.

It may be helpful to compare the one-factor randomized group ANCOVA with the randomized-block ANOVA. Recall that information on some variable believed to be related to the dependent variable is employed as the covariate in the one-factor randomized-group ANCOVA and as the blocking variable in the one-factor randomized-block design. In both cases the power of the test on treatment effects is increased over what it would be with a one-factor randomized-group ANOVA because the concomitant information, whether employed in the form of a covariate or blocks, reduces the size of the error term in the F test.

The randomized-block design is often chosen over the randomized-group ANCOVA because the assumption of linearity in the XY relationship can be ignored and the computation is simpler. It should be kept in mind, however, that the randomized-block design involves a design constraint that is not associated with the randomized-group ANCOVA. Information on the concomitant variable must be available *before* the experiment is carried out if the

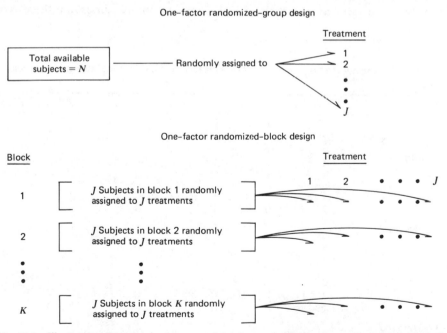

Fig. 10.1 Illustration of randomization procedures for randomized-group and randomized-block designs.

randomized-block design is employed. That is, the blocks must be formed before the subjects are assigned to treatments. Since the concomitant information is employed only in the analysis (not in the randomization procedure) with ANCOVA, it is possible to employ covariance analysis in situations where the randomized-block design is not practical. If the concomitant information is available before the experiment is carried out, it generally makes little difference in results (given that the assumptions are met) whether ANCOVA with the randomized-group design or the randomized-block ANOVA design is utilized.

Analysis Procedure

The computation procedure suggested for the analysis of the randomized-block design is based on a regression procedure much like the one described in Chapter 4 for the randomized-group design. The required design matrix is presented in Table 10.1. As before, there are J (the number of treatments) minus one dummy variables. If a subject falls in the first treatment, a "one" is entered in column D_1; otherwise, a "zero." If a subject falls in the second treatment, a "one" is entered in column D_2; otherwise, a "zero." If a subject falls in the $J-1$th treatment a "one" is entered in column D_{J-1}; otherwise, a zero. The next column is column T. The T stands for the block total; hence the first entry in column T is the sum of *all* raw scores in the first block. That is, the summation is across all treatments for a given block. This means that it will be necessary to enter the block sum as many times as there are treatments.

Table 10.1 Design matrix for Analysis of Variance of One-factor Randomized-block Design

	Block	D_1	$D_2 \cdots$	D_{J-1}	T	Y
	1	1	0	0		
	2					
Treatment 1	⋮	⋮	⋮	⋮		
	K	1	0	0		
	1	0	1	0		
	2	0	1	0		
Treatment 2	⋮	⋮	⋮	⋮		
	K	0	1	0		
⋮						
	1	0	0	0		
	2	0	0	0		
Treatment J	⋮	⋮	⋮	⋮		
	K	0	0	0		

Table 10.2 Contrived Data from a One-factor Randomized-block Experiment

	Treatment			
	1	2	3	Total
Block	Y	Y	Y	Y
1	7	5	8	20
2	10	11	13	34
3	14	13	17	44
4	20	21	24	65
	$Y_1 = 12.75$	$Y_2 = 12.5$	$Y_3 = 15.5$	

[It is more conventional to employ $K-1$ dummy variables to indicate block membership, but it has been shown by Pedhazur (1977) that the use of what we call the T variable in place of these dummy variables is simpler.] Column Y contains the dependent variable scores. An example may be helpful at this point.

Consider the data in Table 10.2. There are three treatments ($J=3$) and four blocks ($K=4$). Suppose the treatments are three methods of treating snake phobia and the blocks are formed on the basis of a measure of the severity of the phobia before the treatments are applied. The dependent variable is another measure of the severity of the phobia taken after the treatment administration. The design matrix is presented in Table 10.3. It can be seen that this matrix is exactly the same as it would be for a randomized-group design except for the third column, which contains the total block scores. Notice that the entries in this column are the same for each treatment group. That is, 20 is the sum of all raw Y scores in block 1. It is obtained by adding all Y scores in block 1: $7+5+8=20$. The three remaining block totals of 34, 44, and 65 are likewise the sums of the Y scores associated with blocks 2, 3,

Table 10.3 Design Matrix for Contrived Snake-phobia Study

	Block	(1) D_1	(2) D_2	(3) T	(4) Y
Treatment 1	1	1	0	20	7
	2	1	0	34	10
	3	1	0	44	14
	4	1	0	65	20
Treatment 2	1	0	1	20	5
	2	0	1	34	11
	3	0	1	44	13
	4	0	1	65	21
Treatment 3	1	0	0	20	8
	2	0	0	34	13
	3	0	0	44	17
	4	0	0	65	24

and 4, respectively. It is important to remember that the same block totals must be entered for each treatment group.

After the design matrix is completed, ANOVA can be carried out by performing the following steps:

1. Regress the dependent variable (Y) on all dummy variables (D) and variable T. The resulting value $R^2_{yD,T}$ yields the proportion of the total variability that is accounted for by treatments and blocks.
2. Regress the dependent variable (Y) on variable T. The resulting value R^2_{yT} yields the proportion of the total variability that is accounted for by blocks.
3. Subtract R^2_{yT} from $R^2_{yD,T}$ to obtain the proportion of the total variability that is accounted for by treatments.
4. Form the F ratio by using

$$F = \frac{(R^2_{yD,T} - R^2_{yT})/J - 1}{(1 - R^2_{yD,T})/(J-1)(K-1)}$$

or complete the summary shown in Table 10.4.
5. Compare the obtained F statistic with the critical value

$$F_{[\alpha, J-1, (J-1)(K-1)]}.$$

If the obtained value exceeds the critical value for the level of α selected, reject H_0.

The application of these steps to the data of Tables 10.2 and 10.3 follows.

Step 1. $R^2_{yD,T} = 0.98831$
Step 2. $R^2_{yT} = 0.93072$
Step 3. $R^2_{yD,T} - R^2_{yT} = 0.05759$

Table 10.4.   Analysis of Variance Summary Table for One-factor Randomized-block Design

Source	SS	df	MS	F
Treatments	$(R^2_{yD,T} - R^2_{yT})\text{SST} = \text{SSTR}$	$J-1$	$\text{SSTR}/J-1$	MSTR/MSE
Blocks	$(R^2_{yT})\text{SST} = \text{SSBK}$	$K-1$	$\text{SSBK}/K-1$	
Error	$(1 - R^2_{yD,T})\text{SST} = \text{SSE}$	$(J-1)(K-1)$	$\text{SSE}/(J-1)(K-1)$	
Total	SST	$N-1$		

Step 4. $F = \dfrac{(R_{yD,T}^2 - R_{yT}^2)/J - 1}{(1 - R_{yD,T}^2)/(J-1)(K-1)} = \dfrac{0.028795}{0.001948} = 14.78$

or

Source	SS	df	MS	F
Treatments	22.17	2	11.08	14.78
Blocks	358.25	3	119.42	
Error	4.50	6	.75	
Total	384.92	11		

Step 5. $F_{obt} = 14.78$

$F_{(.05,2,6)} = 5.14$, $\therefore$ Reject H_0

Multiple Comparison Procedures

Formulas for multiple comparison tests for the randomized-block design are presented in Table 10.5. The denominators of these formulas (except the protected LSD) can, of course, be employed in establishing simultaneous confidence intervals.

Numerical Example. The Tukey multiple comparison tests and simultaneous confidence intervals for the data of Table 10.3 are shown in the tables that

Table 10.5 Multiple Comparison Formulas for Randomized-block Analysis of Variance Designs

Procedure	Formula[a]	Critical Value
Protected LSD	$\dfrac{\bar{Y}_i - \bar{Y}_j}{\sqrt{MSE[(1/n_i) + (1/n_j)]}}$	Student's $t_{(\alpha,(J-1)(K-1))}$
Tukey	$\dfrac{\bar{Y}_i - \bar{Y}_j}{\sqrt{MSE/n}}$	Studentized range $q_{(\alpha,J,(J-1)(K-1))}$
Dunn–Bonferroni	$\dfrac{c_1\bar{Y}_1 + c_2\bar{Y}_2 + \cdots + c_J\bar{Y}_J}{\sqrt{MSE[(c_1^2/n_1) + (c_2^2/n_2) + \cdots + (c_J^2/n_J)]}}$	$t_{DB_{(\alpha,C',(J-1)(K-1))}}$
Scheffé	(Same formula as Dunn–Bonferroni)	$\sqrt{(J-1)F_{(\alpha,J-1,(J-1)(K-1))}}$

[a]Notation:
    MSE = SSE/$(J-1)(K-1)$
   n_i and n_j = number of subjects associated with ith and jth treatments, respectively
       n = common sample size when equal sample sizes are employed

follow. The critical value for the Tukey procedure (which is based on the Studentized range statistic) is found in Table A. 5.

Treatments	Mean Difference	Error Term	Absolute q_{obt}	$q_{(.05,3,6)}$
1,2	$12.75 - 12.5 = 0.25$	$\sqrt{0.75/4} = 0.433$	0.58	4.34
1,3	$12.75 - 15.5 = -2.75$	" "	6.35	4.34
2,3	$12.5 - 15.5 = -3.00$	" "	6.92	4.34

Population Contrast	95% Simultaneous Confidence Intervals
$\mu_1 - \mu_2$	$(0.25) \pm 0.433(4.34) = (-1.63, 2.13)$
$\mu_1 - \mu_3$	$(-2.75) \pm 0.433(4.34) = (-4.63, -0.87)$
$\mu_2 - \mu_3$	$(-3.00) \pm 0.433(4.34) = (-4.88, -1.12)$

10.4 ONE-FACTOR REPEATED-MEASUREMENT DESIGN

An alternative to the one-factor randomized-group and randomized-block ANOVA designs is the one-factor repeated-measurement design. This design is sometimes described as a variety of the randomized-block design because the analysis for both designs is the same. But there is a clear difference between the two designs, and they probably should be treated separately even though the analysis procedures are identical.

The repeated-measurement design is a within-subject design in which each subject is exposed to all treatments, whereas the randomized-block design is a between-subject design where each subject is exposed to only one treatment. The general form of the repeated-measurement design can be seen in Figure 10.2. The first version of this design, version *a* in the table, involves random assignment of the order of the treatment administration for each subject. The second version, version *b*, does not involve the random assignment of treatment order to each subject. Instead, all subjects are measured on the response variable in the same sequence. Each measurement occasion is sometimes called a *trial* because the subjects are observed performing some behavior on each of several different sessions (or trials) across time. Version *a* is often referred to as a *subjects-by-treatments* design, and version *b* is sometimes called a *subjects-by-trials* design. It can be seen that both forms are within-subject designs because each subject is observed under all treatments or trials. Version *a* is frequently employed instead of a randomized-group or randomized-block design because fewer subjects can be employed. This is an important consideration in many research projects, but it should be remembered that this design does not answer the same question that is answered by the randomized-group and randomized-block designs. Results of between-subject designs (i.e., randomized-group and randomized-block designs) generalize to a population of subjects who have been exposed to only one treatment. Results of within-subject designs generalize to a population of

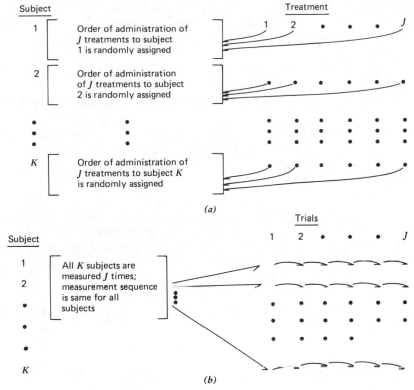

Fig. 10.2   Illustration of two varieties of one-factor repeated-measurement design.

subjects who have been exposed to all the treatments. For example, suppose
an experimenter has decided to investigate the effects of four different
incentive programs on employee output of an industrial product. If this
problem is investigated with a randomized-block or randomized-group design, each group will be exposed to only one type of incentive.

With the repeated-measurement design, each subject will be exposed to all
types of incentive. The results of the randomized-group and randomized-
block designs generalize to subjects who have been exposed to only one of the
incentive programs. But with the repeated measurement design, since every
subject has been exposed to all the different incentive conditions, the results
generalize to a population of subjects who have been exposed to all incentive
types. In using the repeated-measurement design, if incentive number 2 yields
a significantly higher mean that the other three, it can be concluded that this
incentive type is superior for a population of subjects who have been exposed
to all four types of incentive. As noted previously, repeated-measurement
designs are analyzed in exactly the same way as is the randomized-block
design. The independence assumption is sometimes violated with these designs, but there are straightforward procedures for both identifying and

compensating for this problem. Most experimental design texts such as Winer (1971) describe these procedures. A recent discussion of statistical issues associated with repeated measurement designs can be found in Huynk and Feldt (1979).

10.5   TWO-FACTOR ANCOVA FOR INDEPENDENT SAMPLE DESIGNS

The analyses associated with the randomized-block and repeated-measurement ANOVA designs described earlier in this chapter are employed to test the same hypothesis as is tested with completely randomized ANOVA and ANCOVA designs. That is, $H_0 : \mu_1 = \mu_2 = \cdots = \mu_J$. With all these designs, then, there is just one factor that is of experimental interest. A comparison of the randomized-group ANOVA with the randomized-group ANCOVA, the randomized-block ANOVA and the repeated-measurement ANOVA designs, reveals a common difference in the way the total sum of squares is partitioned. A summary of the partitioning for each of these designs is shown in Figure 10.3. It can be seen that there is one source of variability other than that associated with treatments in the randomized-group ANOVA; all other analyses involve breaking down the within treatment sum of squares into two separate components.

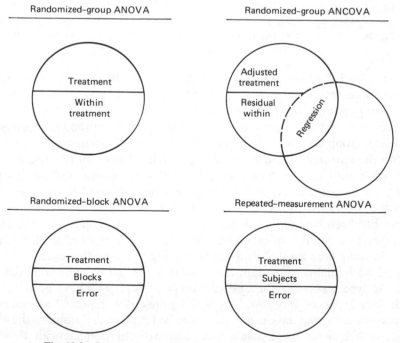

Fig. 10.3   Summary of partitioning for four one-factor analyses.

Variability accounted for by X in ANCOVA, by blocks in the randomized-block ANOVA, and by subjects in the repeated-measurement design has the same effect for all these analyses. These sources of variability are removed from the within-treatment variability; consequently, the error sum of squares is reduced. The effect of this is to increase the power of the analysis. These strategies of reducing the size of the error variability by employing covariates, blocks, and subjects can be generalized to designs in which the experimenter is interested in evaluating the effects of two independent variables.

The application of covariance analysis to the two-factor independent sample design is described in this section. The two-factor repeated measurement design is described in Section 10.6. If the reader has not been previously exposed to two-factor ANOVA designs, we recommend that a sound introductory text such as Hopkins and Glass (1978) or Hays (1973) be consulted before completing this section.

Review of Purpose of Independent Sample Two-factor ANOVA

It is typical for experimenters to be interested in evaluating the effects of more than one independent variable in a single experiment. For example, an experimenter might be interested in studying the effects of two different types of reinforcement (monetary and social) and two different creative thinking programs (Purdue and Khatena) on the frequency of novel responses produced under controlled conditions. The two factors in this design are (1) type of reinforcement (factor A) and (2) type of program (factor B). Each factor has two levels. An advantage of setting the experiment up as one two-factor design rather than as two one-factor designs is that possible interaction effects can be studied. Separate studies of the effects of type of reinforcement and type of program will provide no information on the effects of both factors applied simultaneously. The three major hypotheses that are tested with a two-factor design are

$$H_0: \mu_{A_1} = \mu_{A_2} = \cdots = \mu_{A_I}$$

$$H_0: \mu_{B_1} = \mu_{B_2} = \cdots = \mu_{B_J}$$

H_0: Simple effects of factor A are consistent
across all levels of factor B

where μ_{A_1} through μ_{A_I} are the population main effects means associated with treatment levels A_1 through A_I and μ_{B_1} through μ_{B_J} are the population main effects means associated with treatment levels B_1 through B_J. These three hypotheses are equivalent to stating that the A, B, and $A \times B$ effects in the two-factor ANOVA model are equal to zero. The model is

$$Y_{ijk} = \mu + \alpha_i + \gamma_j + \alpha\gamma_{ij} + \varepsilon_{ijk}$$

where

μ = population mean common to all observations

α_i = effect of ith level of factor A

γ_j = effect of jth level of factor B

$\alpha\gamma_{ij}$ = interaction effect of ith level of factor A and jth level of factor B

ε_{ijk} = error component associated with kth observation in ith level of factor A and jth level of factor B

Hence if population data from a two-factor experiment that has two levels of factor A (i.e., $I=2$) and two levels of factor B (i.e., $J=2$) are available, the *marginal means* reflect the overall (main) effects of the two factors. It can be seen in Table 10.6 that each factor A marginal mean is averaged across both levels of factor B and that each factor B marginal mean is averaged across both levels of factor A. The *cell means* are employed in describing possible interaction between the two factors. It can be seen in the interaction plot in the lower part of Table 10.6 that the difference between cell means $\mu_{A_1B_1}$ and $\mu_{A_2B_1}$ (the simple effect of A at B_1) is not the same as the difference between $\mu_{A_1B_2}$ and $\mu_{A_2B_2}$ (the simple effect of A at B_2). Since these two differences are

Table 10.6   Example of Population Hypothetical Means Associated with a Two-factor Design

		Factor B		Marginal Means for
		Level B_1	Level B_2	Factor A
Factor A	Level A_1	$\mu_{A_1B_1}=10$	$\mu_{A_1B_2}=30$	$\mu_{A_1}=20$
	Level A_2	$\mu_{A_2B_1}=5$	$\mu_{A_2B_2}=15$	$\mu_{A_2}=10$
Marginal Means→ for factor B		$\mu_{B_1}=7.5$	$\mu_{B_2}=22.5$	

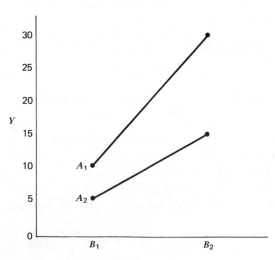

not equal, it is said that the simple effects of factor A are not consistent across the two levels of factor B and, therefore, that factor A and B interact.

When a two-factor experiment is carried out, the analysis of variance is employed to test the hypotheses that (1) the I factor A marginal means are equal, (2) the J factor B marginal means are equal and, (3) the difference between the factor A cell means is the same at all levels of factor B. The summary table for the independent sample two-factor analysis of variance is shown in Table 10.7, along with a diagrammatic representation of the partitioning of the total sum of squares. It can be seen in the summary table that three F ratios are obtained, one for each of the null hypotheses described earlier. Each obtained F value is compared with the corresponding critical value of F. In other words

Obtained F	Compared with	Critical F
F_A	"	$F_{(\alpha, I-1, N-IJ)}$
F_B	"	$F_{(\alpha, J-1, N-IJ)}$
$F_{A \times B}$	"	$F_{[\alpha,(I-1)(J-1), N-IJ]}$

If the obtained value equals or exceeds the critical value, the associated null hypothesis is rejected.

It is generally pointed out in most introductions to the two-factor ANOVA that the sample sizes associated with the various cells of the design should be either equal or proportional. An example of unequal but proportional sample

Table 10.7 ANOVA Summary and Partitioning

Source	SS	df	MS	F
Factor A	SS_A	$I-1$	$SS_A/I-1$	MS_A/MS_{wc}
Factor B	SS_B	$J-1$	$SS_B/J-1$	MS_B/MS_{wc}
$A \times B$ Interaction	$SS_{A \times B}$	$(I-1)(J-1)$	$SS_{A \times B}/(I-1)(J-1)$	$MS_{A \times B}/MS_{wc}$
Within cell	SS_{wc}	$N-IJ$	$SS_{wc}/N-IJ$	
Total		$N-1$		

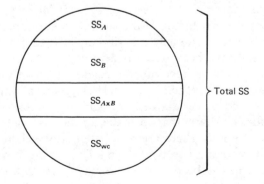

sizes is as follows:

	B_1	B_2
A_1	$n = 15$	$n = 30$
A_2	$n = 10$	$n = 20$
A_3	$n = 20$	$n = 40$

Notice that the number of subjects in each B_2 cell is twice the number in the B_1 cell of the same row. There are, of course, many ways of achieving proportionality of the cell frequencies. The reasons for designing the experiment so that equality or portionality of cell sizes exists are computational and conceptual. If the sample sizes are disproportional, the design is said to be nonorthogonal. Essentially, this means that the various sources of variability in the design are not independent of each other. When the sample sizes are proportional, the design is said to be orthogonal because various sums of squares (i.e., SS_A, SS_B, $SS_{A \times B}$, and SS_{wc}) are all independent of each other. That is, the sum of squares for any one source of variability in the experiment has nothing to do with the sum of squares for any other source. Suppose that the following sums of squares have been computed for a given set of data:

$$SS_A = 400$$
$$SS_B = 200$$
$$SS_{A \times B} = 225$$
$$SS_{wc} = 500$$

If a constant value of 50, for example, is added to the score of each subject who falls in one of the levels of factor A, the sum of squares for factor A will change. But the other sums of squares will *not* change if the design is proportional. If it is disproportional, the B sum of squares can be affected by the change in SS_A. Hence the sources of variability are not independent if the design involves disproportionality. This problem occurs if the conventional procedure for the computation of the sums of squares is followed. An interpretation issue develops if the sum of squares for one factor affects the sum of squares for another factor. If the sum of squares are not independent, it is not possible to unambiguously describe the effects of a particular factor.

Many procedures have been developed to cope with this problem. There are some statisticians who recommend randomly discarding observations from cells to achieve proportionality; others suggest that values be made up (estimated) and added to certain cells to achieve proportionality. An argument can be made, however, that to make up data is unethical and that to throw it away is immoral! There are other alternatives. Unfortunately, there are too many alternatives. The controversy concerning how to analyze the nonorthogonal design has been running for many years, and, unfortunately, the confusion has had an impact on the computer programs that have been written to analyze this design. It is quite possible to run a given set of data through a half-dozen different widely used computer programs that are supposed to handle nonorthogonal designs and to get a half-dozen different results. This is absurd. There is one approach, however, that appears (to the author at least) to be the method that should almost always be employed simply *because it tests the same hypotheses as are tested with orthogonal designs*. This makes sense. (It does not make sense to us to change one's research question just because the design is disproportional.) There are those who will disagree with our recommendation, but it is suggested that papers by Carlson and Timm (1974), Overall et al. (1975), and Speed et al. (1978) be studied very carefully by those who favor other approaches. The latter paper is especially important; it cuts through most of the extant confusion. The recommended approach for two-factor ANOVA is presented along with the two-factor analysis of covariance in the remainder of this section.

Purpose of ANCOVA for Two-factor Orthogonal and Nonorthogonal Designs

The application of covariance analysis to the two-factor independent sample design serves the same purpose as in the case of one-factor designs. That is, power is increased and means are adjusted for chance pretreatment differences measured by the covariate. The hypotheses associated with the two-factor ANCOVA are

$$H_0: \mu_{A_1\,\text{adj}} = \mu_{A_2\,\text{adj}} = \cdots \mu_{A_I\,\text{adj}}$$

$$H_0: \mu_{B_1\,\text{adj}} = \mu_{B_2\,\text{adj}} = \cdots = \mu_{B_J\,\text{adj}}$$

H_0: Adjusted simple effects of factor A are
      consistent across all levels of factor B

As was the case with two-factor ANOVA, these hypotheses are equivalent to stating that A, B, and $A \times B$ effects in the two-factor ANCOVA model are equal to zero. The model is

$$Y_{ijk} = \mu + \alpha_i + \gamma_j + \alpha\gamma_{ij} + \beta_1(X_{ijk} - \overline{X}..) + \varepsilon_{ijk}$$

where

μ = overall population mean
α_i = effect of ith level of factor A
γ_j = effect of jth level of factor B
$\alpha\gamma_{ij}$ = interaction effect of the ith level of factor A and jth
      level of factor B
$\beta_1(X_{ijk} - \overline{X}..)$ = effect of covariate
ε_{ijk} = error component associated with kth observation in ith
      level of factor A and jth level of factor B

In the case of a completely randomized two-factor design, the adjusted population means and the unadjusted population means are equal when there are no treatment effects, and the null hypotheses associated with ANCOVA and ANOVA are equivalent.

Suppose, in the previously described two-factor experiment, where the two factors were type of reinforcement and type of creativity training program, that the experimenter decides that it would be useful to control for verbal fluency. If the dependent variable is the number of novel responses produced in a specific period of time, it could be argued that much of the variability in the production of such responses should be predictable from a measure of verbal fluency. Differences among subjects in verbal fluency will contribute to the within-cell error variance. If a measure of verbal fluency can be obtained before the treatments are applied, the power of the tests on all effects will be increased because within-group variability on the number of novel responses produced (Y) will be decreased by information contained in the verbal fluency measure (X).

Construction of Design Matrices for Two-factor ANOVA and ANCOVA. The computation procedure described in this section is appropriate for both orthogonal and nonorthogonal designs. The same hypotheses are tested and the same interpretations are made in either case. Table 10.8 contains data from an independent sample experiment that has two levels of factor A and two levels of factor B. This is described as a 2×2 independent sample ANCOVA design. The design matrix required for the appropriate analysis of these data is provided in Table 10.9. It can be seen that the first three columns contain dummy variables. For two-factor designs, we employ the procedure described in the following paragraphs to construct the dummy variables.

First, the total number of dummy variables required in a two-factor design is equal to the number of cells minus one. The number of cells is equal to the

Table 10.8 Data for a 2×2 Independent Sample ANCOVA

		Factor B			
		B_1		B_2	
		X	Y	X	Y
		2	1	1	4
		5	8	7	7
	A_1	3	2	3	5
		9	9	6	8
Factor A		X	Y	X	Y
		3	2	9	6
	A_2	4	2	1	3
		7	6	3	4
		8	5	5	4

Table 10.9 Design Matrix for Computation of ANCOVA

	(1) d_{a_1}	(2) d_{b_1}	(3) $d_{a_1}d_{b_1}$	(4) X	(5) $d_{a_1}X$	(6) $d_{b_1}X$	(7) $d_{a_1}d_{b_1}X$	Y
Cell A_1B_1	1	1	1	2	2	2	2	1
	1	1	1	5	5	5	5	8
	1	1	1	3	3	3	3	2
	1	1	1	9	9	9	9	9
Cell A_1B_2	1	-1	-1	1	1	-1	-1	4
	1	-1	-1	7	7	-7	-7	7
	1	-1	-1	3	3	-3	-3	5
	1	-1	-1	6	6	-6	-6	8
Cell A_2B_1	-1	1	-1	3	-3	3	-3	2
	-1	1	-1	4	-4	4	-4	2
	-1	1	-1	7	-7	7	-7	6
	-1	1	-1	8	-8	8	-8	5
Cell A_2B_2	-1	-1	1	9	-9	-9	9	6
	-1	-1	1	1	-1	-1	1	3
	-1	-1	1	3	-3	-3	3	4
	-1	-1	1	5	-5	-5	5	4

number of levels on factor A times the number of levels on factor B. Hence a 2×2 design contains four cells and requires three dummy variables; a 2×3 design requires five dummy variables, and so on. The sum of the number of dummy variables associated with factor A, factor B, and the $A \times B$ interaction is equal to the total number required. If there are I levels of factor A, there must be $I - 1$ dummy variables to identify the levels of factor A. If there are J levels of factor B, there must be $J - 1$ dummy variables to identify the levels of factor B. The $A \times B$ interaction requires $(I - 1)(J - 1)$ dummy variables. It can be seen, then, that $I - 1$, $J - 1$, and $(I - 1)(J - 1)$ add up to $(IJ) - 1$ dummy variables. The form of the dummy variables required for the analysis is not the same as was employed in Chapter 4 for the one-factor design. The dummy variables differ in that the possible values are 1, -1, and 0 rather than 1 and 0. The rules for assigning these values are quite straightforward.

In each dummy-variable column each member of the cell being identified is assigned the value 1; those subjects who fall in the last cell are assigned the value -1 and all others, the value 0. If there are only two levels of a factor, the subjects in the first level are assigned the value 1 and those in the second level, the value -1; there is no "zero" in the dummy variable column because in the case of two levels the subjects are members of either the first or the last level. A few examples should clarify the procedure.

Consider the data in Table 10.8. Since this is a 2×2 design, the total number of dummy variables required is $(2 \times 2) - 1$, or 3. The first dummy variable is labeled d_{a_1} because this variable will identify which subjects are members of level A_1 and which subjects are members of A_2. The subjects in level A_1 are assigned 1s, and those in level A_2 are assigned -1s. This dummy variable can be seen in the first column of Table 10.9. Notice that the first eight subjects are members of level A_1 and the last eight subjects are members of level A_2. The second dummy variable column (d_{b_1}) is used to identify members of levels B_1 and B_2 using 1s and -1s, respectively. The third dummy variable column ($d_{a_1}d_{b_1}$) is the $A \times B$ interaction column. The values in this column are obtained by computing the products of the values in columns d_{a_1} and d_{b_1}.

Covariate values are entered in the fourth column (X), and the next three columns are simply the products of the dummy variables and the covariate. That is, the values in d_{a_1} times the values in X yield the values in $d_{a_1}X$. The corresponding situation holds for $d_{b_1}X$ and $d_{a_1}d_{b_1}X$. The last column contains scores on the dependent variable.

Examples of dummy variables for 2×3 and 3×4 designs can be seen in Table 10.10. (The columns for the products of the dummy variables and covariate have been omitted from the 3×4 matrix to save space.)

The notation employed with the dummy variables is as follows:

$D_a = all\ I-1$ dummy variables required to identify group membership of levels of factor A

$D_b = all\ J-1$ dummy variables required to identify group membership in levels of factor B

$D_{a\times b} = all\ (I-1)(J-1)$ product dummy variables

Table 10.10   Design Matrices for 2×3 and 3×4 ANCOVA Designs

2×3 Design Matrix

Cell	d_{a_1}	d_{b_1}	d_{b_2}	$d_{a_1}d_{b_1}$	$d_{a_1}d_{b_2}$	X	$d_{a_1}X$	$d_{b_1}X$	$d_{b_2}X$	$d_{a_1}d_{b_1}X$	$d_{a_1}d_{b_2}X$	Y
A_1B_1	1	1	0	1	0							
A_1B_2	1	0	1	0	1							
A_1B_3	1	-1	-1	-1	-1							
A_2B_1	-1	1	0	-1	0							
A_2B_2	-1	0	1	0	-1							
A_2B_3	-1	-1	-1	1	1							

3×4 Design Matrix

Cell	d_{a_1}	d_{a_2}	d_{b_1}	d_{b_2}	d_{b_3}	$d_{a_1}d_{b_1}$	$d_{a_1}d_{b_2}$	$d_{a_1}d_{b_3}$	$d_{a_2}d_{b_1}$	$d_{a_2}d_{b_2}$	$d_{a_2}d_{b_3}$	X	Y
A_1B_1	1	0	1	0	0	1	0	0	0	0	0		
A_1B_2	1	0	0	1	0	0	1	0	0	0	0		
A_1B_3	1	0	0	0	1	0	0	1	0	0	0		
A_1B_4	1	0	-1	-1	-1	-1	-1	-1	0	0	0		
A_2B_1	0	1	1	0	0	0	0	0	1	0	0		
A_2B_2	0	1	0	1	0	0	0	0	0	1	0		
A_2B_3	0	1	0	0	1	0	0	0	0	0	1		
A_2B_4	0	1	-1	-1	-1	0	0	0	-1	-1	-1		
A_3B_1	-1	-1	1	0	0	-1	0	0	-1	0	0		
A_3B_2	-1	-1	0	1	0	0	-1	0	0	-1	0		
A_3B_3	-1	-1	0	0	1	0	0	-1	0	0	-1		
A_3B_4	-1	-1	-1	-1	-1	1	1	1	1	1	1		

d_{a_1} = label attached to first dummy variable associated with factor A (if there are more than two levels of factor A, the labels attached to the $I-1$ dummy variables are $d_{a_1}, d_{a_2}, d_{a_3}, \ldots, d_{a_{I-1}}$)

d_{b_1} = label attached to first dummy variable associated with factor B (if there are more than two levels of factor B, labels attached to remaining dummy variables are $d_{b_2}, d_{b_3}, \ldots, d_{b_{J-1}}$)

$d_{a_1} d_{b_1}$ = label attached to product of d_{a_1} and d_{b_1}, similar labels are attached to the additional product dummy variables required when there are more than two levels of either factor

Once the design matrix has been constructed, the two-factor ANOVA, ANCOVA, and homogeneity of regression tests can easily be carried out with the aid of a multiple linear regression computer program. The procedures for carrying out each of these analyses are described in the following three subsections.

Two-factor ANOVA Computation. For *two-factor ANOVA, the required R^2 values are:*

1. $R^2_{yD_a, D_b, D_{a \times b}}$. This is the squared multiple correlation coefficient associated with the regression of the dependent variable (Y) on all dummy variables used to identify the levels of factor A (D_a), all dummy variables used to identify the levels of factor B (D_b), and all interaction dummy variables ($D_{a \times b}$). This coefficient yields the proportion of the total sum of squares that is accounted for by factor A, factor B, and the $A \times B$ interaction.

2. $R^2_{yD_b, D_{a \times b}}$. This is the squared multiple correlation coefficient associated with the regression of the dependent variable (Y) on all dummy variables used to identify the levels of factor B (D_b) and all interaction dummy variables ($D_{a \times b}$). This coefficient yields the proportion of the total sum of squares that is accounted for by factor B and the $A \times B$ interaction.

3. $R^2_{yD_a, D_{a \times b}}$. This is the squared multiple correlation coefficient associated with the regression of the dependent variable (Y) on all dummy variables used to identify the levels of factor A (D_a) and all interaction dummy variables ($D_{a \times b}$). This coefficient yields the proportion of the total sum of squares that is accounted for by factor A and the $A \times B$ interaction.

4. $R^2_{yD_a, D_b}$. This is the squared multiple correlation coefficient associated with the regression of the dependent variable (Y) on all dummy variables used to identify the levels of factor A (D_a) and all dummy variables used to identify the levels of factor B (D_b). This coefficient yields the proportion of the total sum of squares that is accounted for by factors A and B.

These R^2 values are then employed to compute the required sums of squares. Table 10.11 provides the format for the summary of the analysis.

The first line of the summary is for factor A. Notice that the R^2 based on all A, B, and $A \times B$ dummy variables is the first coefficient and that the R^2 based on B and $A \times B$ dummy variables is subtracted from the former

Table 10.11 Summary Table for Two-factor Independent Sample ANOVA for Orthogonal or Nonorthogonal Designs

Source	SS	df	MS	F
Factor A	SST$(R^2_{y.D_a,D_b,D_{a\times b}} - R^2_{y.D_b,D_{a\times b}})$	$I-1$	SSA$/I-1$	MS$A/$MSwc
Factor B	SST$(R^2_{y.D_a,D_b,D_{a\times b}} - R^2_{y.D_a,D_{a\times b}})$	$J-1$	SSB$/J-1$	MS$B/$MSwc
$A \times B$				
Interaction	SST$(R^2_{y.D_a,D_b,D_{a\times b}} - R^2_{y.D_a,D_b})$	$(I-1)(J-1)$	SSA$\times B/(I-1)(J-1)$	MS$A \times B/$MSwc
Within cell	SST$(1 - R^2_{y.D_a,D_b,D_{a\times b}})$	$N-IJ$	SSwc$/N-IJ$	
Total	(SST)	$N-1$		

Cell means are based on the following regression equation:

$$\bar{Y}_{ij} = b_0 + b_1(d_1) + b_2(d_2) + \cdots + b_{IJ-1}(d_{IJ-1})$$

coefficient. The only reason for a difference between these two R^2 values can be seen in the subscripts. If the R^2 based on A, B, and $A \times B$ is larger than the R^2 based on B and $A \times B$, the difference is explained by factor A. Hence the difference between the two R^2 values yields the proportion of Y that is accounted for by factor A independent of factor B and the $A \times B$ interaction. If this difference is multiplied by the total sum of squares, the sum of squares for factor A is obtained.

The second line of the summary table follows the same pattern as the first. The left-hand R^2 is based on A, B, and $A \times B$ dummy variables. The right-hand R^2 is based on A and $A \times B$ dummy variables. The difference between the two coefficients describes the proportion of Y that is accounted for by factor B independent of factor A and the $A \times B$ interaction. This difference times the total sum of squares yields the sum of squares for factor B. The third line of the summary table follows the same pattern as the first row. Once again, the left-hand R^2 is based on A, B, and $A \times B$ dummy variables. The right-hand R^2 is based on A and B dummy variables. The difference between the two R^2 values can be explained by the difference between the sets of dummy variables associated with each value. Since the right-hand R^2 was based on A and B whereas the left-hand R^2 was based on A, B, and $A \times B$, the $A \times B$ interaction must account for the difference. This difference is the proportion of the total sum of squares that is accounted for by the $A \times B$ interaction independent of the A and B main effects. The product of the total sum of squares times the difference between these two R^2 values yields the $A \times B$ interaction sum of squares.

The next line involves the subtraction of the R^2 based on A, B, and $A \times B$ dummy variables from one. Since the R^2 value describes the proportion of the total sum of squares that is accounted for by all the systematic sources in this design (i.e., factors A, B, and $A \times B$ interaction), the difference between this value and one yields the proportion that is not systematic (i.e., error). The product of this difference times the total sum of squares yields the within cell sum of squares.

If the design is orthogonal, the sum of the first four lines will be equal to the total sum of squares. If the design is nonorthogonal (i.e., disproportionality of the cell sample size frequencies), the first four lines will not add up to the total sum of squares. In either case the F tests are appropriate for the purpose of testing the hypotheses

$$H_0 : \mu_{A_1} = \mu_{A_2} = \cdots = \mu_{A_I}$$

$$H_0 : \mu_{B_1} = \mu_{B_2} = \cdots = \mu_{B_J}$$

H_0 : Simple effects of A are consistent at all levels of B

The regression equation associated with $R^2_{yD_a, D_b, D_{a \times b}}$ is the regression of Y on all dummy variables associated with A, B, and $A \times B$; this equation is shown at the bottom of Table 10.11. Notice that it can be used to compute the cell means. When the dummy values for cell i,j are entered into the equation, the mean of cell i,j is the value predicted by the equation.

It can be shown that the intercept b_0 is the grand (unweighted) mean of the cells and that each partial regression coefficient b_i is the effect (deviation of the cell mean from the grand unweighted mean) of membership in the cell associated with the dummy variable. For example, the effect of membership in the first level of factor A is equal to the first partial regression coefficient. If there are only two levels of factor A, the effect of membership in the second level is *minus* the value of the first partial regression coefficient. If there are three levels of A, the first two partial regression coefficients are equal to the effects of levels A_1 and A_2, respectively. The effect of membership in A_3 (in the case of three levels of A) is *minus* the sum of the first two partial regression coefficients. This method of interpretation of these coefficients holds for all levels and factors.

Example

The design matrix provided in Table 10.9 was employed to compute ANOVA on the Y variable. The required R^2 values* are

$$R^2_{yD_a, D_b, D_{a \times b}} = R^2_{y1,2,3} = 0.12921$$

$$R^2_{yD_b, D_{a \times b}} = R^2_{y2,3} = 0.02809$$

$$R^2_{yD_a, D_{a \times b}} = R^2_{y1,3} = 0.10393$$

$$R^2_{yD_a, D_b} = R^2_{y1,2} = 0.12640$$

The total sum of squares $= 89.00$. The analysis is summarized in Table 10.12. The regression equation associated with $R^2_{y1,2,3} = R^2_{yD_a, D_b, D_{a \times b}}$ is

$$\hat{Y} = 4.75 + 0.75(d_1) - 0.375(d_2) - 0.125(d_3)$$

The cell means are

$$\overline{Y}_{A_1B_1} = 4.75 + 0.75(1) - 0.375(1) - 0.125(1) = 5$$

$$\overline{Y}_{A_1B_2} = 4.75 + 0.75(1) - 0.375(-1) - 0.125(-1) = 6$$

$$\overline{Y}_{A_2B_1} = 4.75 + 0.75(-1) - 0.375(1) - 0.125(-1) = 3.75$$

$$\overline{Y}_{A_2B_2} = 4.75 + 0.75(-1) - 0.375(-1) - 0.125(1) = 4.25$$

---

*The numerical subscripts attached to the second column of R^2 values refer to the specific column numbers in Table 10.9 that identify the various dummy variables.

Table 10.12   ANOVA (on Y) Summary for Data of Tables 10.8 and 10.9

Source	SS	df	MS	F
A	$89(0.12921 - 0.02809) = 9.00$	1	9.00	1.39
B	$89(0.12921 - 0.10393) = 2.25$	1	2.25	.35
$A \times B$	$89(0.12921 - 0.12640) = 0.25$	1	0.25	.05
Within cell	$89(1 - 0.12921) = 77.50$	12	6.46	
Total	89	15		

The procedure used in this ANOVA is appropriate for both orthogonal and nonorthogonal designs and for any number of levels of the two factors. It should be mentioned, however, that in the case of two levels of each factor (i.e., a 2×2 design), it is not necessary to compute all the regressions employed in this general procedure. If the computer program provides a test of the significance of each partial regression coefficient, the analysis is complete in one pass. That is, if Y is regressed on d_{a_1}, d_{b_1}, and $d_{a_1 \times b_1}$, the tests of significance of the three associated partial regression coefficients are equivalent to the three F tests in the ANOVA summary table. Most multiple regression computer programs provide these tests; they may be either t or F values. If they are F values, they will equal the values in the ANOVA summary table. If the output provides t tests for the coefficients, the squares of the t values are equal to the ANOVA F values.

Intuitively, it is reasonable that these tests should be equivalent. Partial regression coefficients are measures of the independent effects of the variables in the equation. Likewise, the two-factor ANOVA tests the independent effects of factors A, B, and $A \times B$ interaction. Since the dummy variables are independent of each other across factors, and since each one is an indicator of group membership, the equivalence should, at least, not seem bizarre.

When more than two levels are associated with factors A or B, there will probably be interest in multiple comparison tests or simultaneous confidence intervals for the factor or factors that have three or more marginal means. The formulas in Table 10.13 provide the appropriate error terms for these tests or intervals.

Computation Procedure for Two-factor Independent Sample ANCOVA with Orthogonal or Nonorthogonal Designs.   The two-factor ANCOVA hypotheses of equal adjusted marginal means for factors A and B, no interaction of $A \times B$, and the homogeneity of regression can be tested using the computation described in the following paragraphs. As before with other designs, this procedure relies on the use of a multiple regression computer program.

The first step involves the construction of the design matrix using $1, 0, -1$ dummy variables. This step has been described earlier in this section. The second step involves the computation of the following squared multiple

Table 10.13 Multiple Comparison Formulas for Comparisons of Marginal Means in Two-factor Independent Sample ANOVA[a]

Procedure	Error Term	Critical Value
Protected LSD	$\sqrt{MS_{wc}\left[\dfrac{1}{n_{m_i}} + \dfrac{1}{n_{m_j}}\right]}$	Student's $t_{(\alpha, N-IJ)}$
Tukey HSD	$\sqrt{MS_{wc}/n_m}$	Studentized range $q_{(\alpha, L, N-IJ)}$
Dunn–Bonferroni	$\sqrt{MS_{wc}\left[(c_1^2/n_{m_1}) + (c_2^2/n_{m_2}) + \cdots + (c_L^2/n_{m_L})\right]}$	Dunn–Bonferroni $t_{DB(\alpha, C', N-IJ)}$
Scheffe'	(Same formula as Dunn–Bonferroni)	$\sqrt{(L-1)F_{(\alpha, L-1, N-IJ)}}$

[a]*Notation*:

n_m = Common number of subjects associated with the two marginal means involved in comparison (with unequal marginal sample sizes, use the harmonic means of the two for the Tukey procedure)

n_{m_i}, n_{m_j} = Number of subjects associated with ith and jth marginal means involved in comparison

L = Number of levels associated with factor involved in comparison; it will be equal to either I (number of levels of factor A) or J (number of levels of factor B)

All α values refer to nondirectional tests.

correlation coefficients:

1. $R^2_{yD_a, D_b, D_{a \times b}, X, D_a X, D_b X, D_{a \times b} X}$. This is the coefficient associated with the re-gression of Y on all dummy variables used to identify levels of factor A, all dummy variables used to identify levels of factor B, all dummy variables used to identify the $A \times B$ interaction, the covariate(s), and the product of each of these dummy variables with the covariate. This R^2 value is interpreted as the proportion of the total sum of squares that is accounted for by factor A, factor B, the $A \times B$ interaction, the covariate(s), and the interaction of the covariate and the treatments.

2. $R^2_{yD_a, D_b, D_{a \times b}, X}$. This is the coefficient associated with the regression of Y on all dummy variables used to identify levels of factor A, all dummy variables used to identify levels of factor B, the $A \times B$ interaction dummy variables, and the covariate(s). This R^2 value is interpreted as the propor-tion of the total sum of squares that is accounted for by factor A, factor B, the $A \times B$ interaction, and the covariate(s).

3. $R^2_{yD_b, D_{a \times b}, X}$. This is the coefficient associated with the regression of Y on all dummy variables used to identify levels of factor B, the $A \times B$ interac-tion dummy variables, and the covariate(s). This R^2 value is interpreted as the proportion of the total sum of squares that is accounted for by factor B, the $A \times B$ interaction, and the covariate(s).

4. $R^2_{yD_a, D_{a \times b}, X}$. This is the coefficient associated with the regression of Y on all dummy variables used to identify levels of factor A, the $A \times B$ interac-tion dummy variables, and the covariate. This R^2 value is interpreted as the proportion of the total sum of squares that is accounted for by factor A, the $A \times B$ interaction, and the covariate(s).

5. $R^2_{yD_a, D_b, X}$. This is the coefficient associated with the regression of Y on all dummy variables used to identify levels of factor A, levels of factor B, and

the covariate(s). This R^2 value is interpreted as the proportion of the total sum of squares that is accounted for by factor A, factor B, and the covariate(s).

These five R^2 values are employed in the summary tables for the two-factor ANCOVA and the homogeneity of regression slopes test. Both summaries can be seen in Table 10.14.

The rationale for employing the R^2 values shown in Table 10.14 for the ANCOVA and homogeneity of regression tests is as follows. The adjusted effects of factor A are the effects of membership in the different levels of factor A that are independent of factor B, the $A \times B$ interaction, and the covariate(s). The difference between the R^2 value that is based on all variables that represent A, B, $A \times B$, and X and the R^2 value based on all these variables except those representing levels of factor A, is explained by the elimination of factor A variables from the equation. The value $R^2_{yD_a, D_b, D_{a \times b}, X}$ will exceed $R^2_{yD_b, D_{a \times b}, X}$ only if there are some differences among levels of A that are independent of the effects of B, $A \times B$, and X. Hence the first line of the ANCOVA summary includes the difference between these two R^2 values. Since this difference is the proportion of the total sum of squares that is independently accounted for by factor A, the product of the total sum of squares times this difference yields the adjusted factor A sum of squares.

The same approach is taken in the second line of the summary. The left-hand R^2 is based on A, B, $A \times B$, and X, but the right-hand R^2 is based on A, $A \times B$, and X. The total sum of squares times the difference between these two R^2 values yields the sum of squares accounted for by factor B that is independent of the effects of A, $A \times B$, and X.

The adjusted $A \times B$ sum of squares is based on the same approach. Notice in the third line of the summary that, once again, the left-hand R^2 is based on A, B, $A \times B$, and X. The right-hand R^2 is based on A, B, and X. The difference between the two R^2 values is the proportion of the total sum of squares that is accounted for by the $A \times B$ interaction independently of the effects of factor A, factor B, and the covariate(s). The product of the total sum of squares times this difference between R^2 values yields the adjusted $A \times B$ interaction sum of squares.

Under the two-factor ANCOVA model all variability not explained by A, B, $A \times B$, and X is termed *error* or *residual variability*. Since $R^2_{yD_a, D_b, D_{a \times b}, X}$ is the proportion of the total variability accounted for by factor A, factor B, the $A \times B$ interaction and X, one minus this value must be the proportion of the total sum of squares that is not explained by these sources of variability. The product of the total sum of squares times $(1 - R^2_{yD_a, D_b, D_{a \times b}, X})$ yields the residual within cell sum of squares that is entered on the fourth line of the summary. The last line of the summary contains the total residual sum of squares that is all variability not accounted for by X. The total sum of squares times $(1 - R^2_{yX})$ yields this sum of squares.

Table 10.14 Summary Tables for Independent Sample Two-factor ANCOVA and Homogeneity of Within-cell Regression[a]

ANCOVA

Source	SS	df	MS	F
Adjusted A	SST$(R^2_{y,D_a,D_b,D_{a\times b},X} - R^2_{y,D_b,D_{a\times b},X})$	$I-1$	SS$A_{\mathrm{adj}}/I-1$	MSA_{adj}/MSres$_{\mathrm{wc}}$
Adjusted B	SST$(R^2_{y,D_a,D_b,D_{a\times b},X} - R^2_{y,D_a,D_{a\times b},X})$	$J-1$	SS$B_{\mathrm{adj}}/J-1$	MSB_{adj}/MSres$_{\mathrm{wc}}$
Adjusted $A\times B$	SST$(R^2_{y,D_a,D_b,D_{a\times b},X} - R^2_{y,D_a,D_b,X})$	$(I-1)(J-1)$	SS$A\times B_{\mathrm{adj}}/(I-1)(J-1)$	MS$A\times B_{\mathrm{adj}}$/MSres$_{\mathrm{wc}}$
Res$_{\mathrm{wc}}$	SST$(1 - R^2_{y,D_a,D_b,D_{a\times b},X})$	$N-IJ-C$	SSres$_{\mathrm{wc}}/N-IJ-C$	
Res$_t$	SST$(1 - R^2_{y,X})$	$N-C-1$		

Homogeneity of Within-cell Regression test

Source	SS	df	MS	F
Hetero. reg.	SST$(R^2_{y,D_a,D_b,D_{a\times b},X,D_aX,D_bX,D_{a\times b}X} - R^2_{y,D_a,D_b,D_{a\times b},X})$	$C(IJ-1)$	SSHR$/C(IJ-1)$	MSHR/MSres$_i$
Res$_i$	SST$(1 - R^2_{y,D_a,D_b,D_{a\times b},X,D_aX,D_bX,D_{a\times b}X})$	$N-[IJ(C+1)]$	SSres$_i/N-[IJ(C+1)]$	
Res$_{\mathrm{wc}}$	SST$(1 - R^2_{y,D_a,D_b,D_{a\times b},X})$	$N-IJ-C$		

[a]Adjusted cell means are based on the regression equation associated with $R^2_{y,D_a,D_b,D_{a\times b},X}$.

$$\bar{Y}_{ij\,\mathrm{adj}} = b_0 + b_1(d_1) + b_2(d_2) + \cdots + b_{IJ-1}(d_{IJ-1}) + b_{X_1}(\bar{X}_{1\cdots})$$
$$+ b_{X_2}(\bar{X}_{2\cdots}) + \cdots + b_{X_C}(\bar{X}_{C\cdots})$$

where

$d_1-d_{IJ-1} = IJ-1$ dummy variables

b_1-b_{IJ-1} = partial regression coefficients associated with the $IJ-1$ dummy variables

$b_{X_1}-b_{X_C}$ = partial regression coefficients associated with covariates 1 through C

A diagrammatic representation of the partitioning just described can be seen in the upper part of Figure 10.4. The lower part of this figure illustrates the manner in which the sum of squares residual within cells is further partitioned into heterogeneity of regression sum of squares and individual residual sum of squares. This additional partitioning is required for the homogeneity of within-cell regression test that is shown in the lower part of Table 10.14. The rationale for the partitioning is the same as in the case of a one-factor design.

It is assumed under the two-factor ANCOVA model that the regression slope (or plane or hyperplane in the case of multiple covariates) of Y on X is the same within each level of A for tests on A, within each level of B for tests on B, and within each cell for tests on $A \times B$ interaction. If these assumptions are not met, the most likely consequences are that the F tests for adjusted A, B, and $A \times B$ effects will be conservatively biased and the adjusted means will be difficult to interpret. Typically, the approach to evaluating these assumptions is to test the homogeneity of the within-cell regressions. The method of

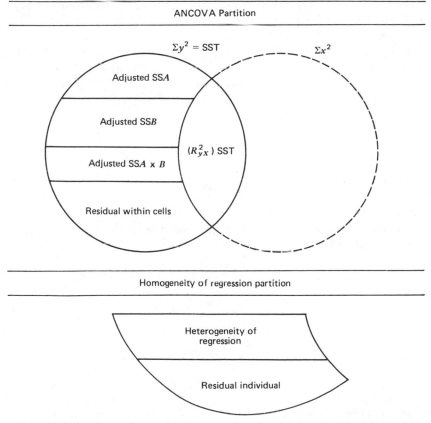

Fig. 10.4 Analysis of covariance and homogeneity of regression partitions for orthogonal two-factor designs.

performing this test involves first computing the within-cell residual sum of squares using the pooled within cell slope. As can be seen in Table 10.14, this sum of squares is the error or within-cell residual sum of squares used in the two-factor ANCOVA. The second step is to compute the residual sum of squares within cells based on individual within-cell regressions rather than on the pooled within-cell regression. This can be accomplished by multiplying the total sum of squares times $(1 - R^2_{yD_a, D_b, D_{a\times b}, X, D_aX, D_bX, D_{a\times b}X})$. The difference between the pooled within-cell residuals and the individual within-cell residuals reflects heterogeneity of the regression slopes. A somewhat more direct method of computing the heterogeneity of regression sum of squares is to compute the product of the total sum of squares times the difference between the two R^2 values shown in last two lines of Table 10.14. The mean-square heterogeneity of regression over the mean-square individual within-cell residual provides the F test for the homogeneity of within-cell regression. If this F test yields a nonsignificant F value, the assumption of homogeneous within-cell regression slopes is retained, and the adjusted-cell and marginal means and the ANCOVA F tests can be interpreted in a straightforward manner. If the homogeneity of regression slopes test yields a significant F value, the methods of Chapter 13 should be considered as substitutes for (or additions to) the ANCOVA.

The computational formula for the adjusted-cell means is shown below the homogeneity of within-cell regression test in Table 10.14. Notice that the equation is the one associated with the regression of Y on all dummy variables and all covariates, (i.e., the equation associated with $R^2_{yD_a, D_b, D_{a\times b}, X}$). An example follows.

Example

The design matrix provided in Table 10.9 was employed to compute ANCOVA, and the homogeneity of regression tests that are summarized in Table 10.15. The R^2 values that are required for this analysis are as follows (both general notation and specific design matrix column labels used with this data set are shown):

$$R^2_{yD_a, D_b, D_{a\times b}, X, D_aX, D_bX, D_{a\times b}X} = R^2_{y1,2,3,4,5,6,7} = 0.84842$$

$$R^2_{yD_a, D_b, D_{a\times b}, X} = R^2_{y1,2,3,4} = 0.72369$$

$$R^2_{yD_b, D_{a\times b}, X} = R^2_{y2,3,4} = 0.57010$$

$$R^2_{yD_a, D_{a\times b}, X} = R^2_{y1,3,4} = 0.65092$$

$$R^2_{yD_a, D_b, X} = R^2_{y1,2,4} = 0.72346$$

The total sum of squares (SST) = 89.00 and R^2_{yX}, which is not essential in the analysis, is 0.49995.

Table 10.15 Two-factor Independent Sample ANCOVA and Homogeneity of Within-group Regression Test for Data of Table 10.8

Source	SS	df	MS	F
	ANCOVA			
Adjusted A	$89(0.72369-0.57010)=13.67$	1	13.67	6.11
Adjusted B	$89(0.72369-0.65092)=6.48$	1	6.48	2.90
Adjusted $A \times B$	$89(0.72369-0.72346)=0.02$	1	0.02	0.01
Res_{wc}	$89(1-0.72369)=24.59$	11	2.235	
Res_t	$89(1-0.49995)=44.50^a$			
Critical value $= F_{(.05,1,11)}=4.84$				
	Homogeneity of Within-cell Regression Test			
Hetero. reg.	$89(0.84842-0.72369)=11.10$	3	3.70	2.19
Res_i	$89(1-0.84842)=13.49$	8	1.69	
Res_{wc}	$89(1-0.72369)=24.59$	11		
Critical value $= F_{(.05,3,8)}=4.07$				

^{a}This value deviates slightly from the sum of the four values above it as a result of rounding error.

Regression equation associated with $R^2_{yD_a,D_b,D_{a \times b},X}$:

$$\hat{Y}=1.3539+0.9287(d_{a_1})-0.6431(d_{b_1})-0.0356(d_{a_1}d_{b_1})+0.7150(X)$$

Grand unweighted mean (i.e., mean of cell means) on $X=4.75$
Adjusted *Cell means*:

$$\bar{Y}_{A_1B_1}=1.3539+0.9287(1)-0.6431(1)-0.0356(1)+0.7150(4.75)=5.00$$

$$\bar{Y}_{A_1B_2}=1.3539+0.9287(1)-0.6431(-1)-0.0356(-1)+0.7150(4.75)=6.36$$

$$\bar{Y}_{A_2B_1}=1.3539+0.9287(-1)-0.6431(1)-0.0356(-1)+0.7150(4.75)=3.21$$

$$\bar{Y}_{A_2B_2}=1.3539+0.9287(-1)-0.6431(-1)-0.0356(1)+0.7150(4.75)=4.43$$

		Factor B		Adjusted Marginal A Means
		B_1	B_2	
Factor A	A_1	5.00	6.36	$\bar{Y}_{A_{1adj}}=5.68$
	A_2	3.21	4.43	$\bar{Y}_{A_{2adj}}=3.82$

Adjusted Marginal B Means $\bar{Y}_{B_{1adj}}=4.11$ $\bar{Y}_{B_{2adj}}=5.39$

213

The adjusted marginal means in Table 10.15 were computed by averaging the adjusted-cell means associated with each factor level. Alternatively, they can be computed directly from the regression equation used to compute the cell means. When this approach is used, all terms in the equation are ignored except the intercept, the partial regression coefficient associated with the dummy variable(s) associated with the factor for which the adjusted marginal mean is being computed, and the partial regression coefficient associated with the covariate(s). The grand unweighted mean on Y is equal to the intercept plus the product of partial regression coefficient for X times the grand unweighted mean on X. The deviation from the Y unweighted mean due to the effect of level A_1 is equal to the partial regression coefficient associated with level A_1. The effect of level A_2 is minus the partial regression coefficient associated with A_1. The same approach applies to the main effects on factor B.

For the example data, the regression equation has the following coefficients:

$$b_0 = 1.3539$$
$$b_1 = 0.9287$$
$$b_2 = -0.6431$$
$$b_3 = -0.0356$$
$$b_4 = 0.7150$$

The grand unweighted mean on $X = 4.75$. The grand unweighted mean on Y is $b_0 + b_4(4.75) = 4.75$. The main effect of level A_1 is b_1. The main effect of level A_2 is $-b_1$; hence the adjusted marginal means for factor A are

$$\overline{Y}_{A_{1\,\text{adj}}} = 4.75 + 0.93 = 5.68$$

$$\overline{Y}_{A_{2\,\text{adj}}} = 4.75 - 0.93 = 3.82$$

Likewise, the adjusted marginal means for factor B are

$$\overline{Y}_{B_{1\,\text{adj}}} = 4.75 + (-0.64) = 4.11$$

$$\overline{Y}_{B_{2\,\text{adj}}} = 4.75 - (-0.64) = 5.39$$

At this point it may be helpful to compare the results of ANOVA and ANCOVA on the example data. Both analyses are summarized in Table 10.16. Notice that type of reinforcement, type of program, and the interaction of the two are nonsignificant in the case of ANOVA. With ANCOVA, the type of reinforcement has a statistically significant effect. The type of program and the interaction of reinforcement type and program type are nonsignificant. The reasons ANOVA and ANCOVA reach different conclusions

Table 10.16   Comparison of Two-factor Independent Sample ANOVA and ANCOVA for Data in Table 10.8

ANOVA		ANCOVA	
Source	F	Source	F
Factor A (type of reinforcement)	1.39	Adjusted A	6.11
Factor B (type of program)	0.35	Adjusted B	2.90
$A \times B$ Interaction	0.05	Adjusted $A \times B$	0.01
(error MS = 6.46)		(error MS = 2.24)	
Critical value of F (for $\alpha = .05) = 4.75$		Critical value of F (for $\alpha = .05) = 4.84$	

$$\begin{array}{c c c} & B_1 & B_2 \\ A_1 & \overline{Y} = 5.00 & \overline{Y} = 6.00 \\ A_2 & \overline{Y} = 3.75 & \overline{Y} = 4.25 \\ & \overline{Y}_{B_1} = 4.38 & \overline{Y}_{B_2} = 5.12 \end{array}$$

$\overline{Y}_{A_1} = 5.50$
$\overline{Y}_{A_2} = 4.00$ } ANOVA

$$\begin{array}{c c c} & B_1 & B_2 \\ A_1 & \overline{Y}_{adj} = 5.00 & \overline{Y}_{adj} = 6.36 \\ A_2 & \overline{Y}_{adj} = 3.21 & \overline{Y}_{adj} = 4.43 \\ & \overline{Y}_{B_{1\,adj}} = 4.11 & \overline{Y}_{B_{2\,adj}} = 5.39 \end{array}$$

$\overline{Y}_{A_{1\,adj}} = 5.68$
$\overline{Y}_{A_{2\,adj}} = 3.82$

} ANCOVA

$$\begin{array}{c c c} & B_1 & B_2 \\ A_1 & \overline{X} = 4.75 & \overline{X} = 4.25 \\ A_2 & \overline{X} = 5.50 & \overline{X} = 4.50 \\ & \overline{X}_{B_1} = 5.12 & \overline{X}_{B_2} = 4.38 \end{array}$$

$\overline{X}_{A_1} = 4.50$
$\overline{X}_{A_2} = 5.00$

on the effect of type of reinforcement (factor A) can be seen by comparing the size of the error terms associated with the two models and by comparing the size of the effects being tested. The error mean square for ANOVA (the within-cell MS) is 6.46, whereas the error MS for ANCOVA (the within-cell MS residual) is only 2.24. Also, the difference between the marginal means is smaller than the difference between adjusted marginal means. The larger difference between adjusted marginal means is explained by the fact that the level with the lower covariate mean (A_1) had the higher dependent variable mean. Recall that the adjustment process works in such a way that when there is a positive relationship between the covariate and the dependent variable, the Y mean is adjusted upward for groups that fall below the X mean and downward for groups that fall above the X mean. In the example, the unadjusted mean difference is 1.5 points and the adjusted difference is

1.86 points. The overall result of the much smaller error term and the larger mean difference associated with ANCOVA is a much larger F value.

COMPUTATIONAL SIMPLIFICATION FOR TWO-FACTOR INDEPENDENT SAMPLE DESIGNS WITH TWO LEVELS OF EACH FACTOR: ANY NUMBER OF COVARIATES. When a two-factor design contains only two levels of each factor, the general computation procedure described in Table 10.14 is appropriate, but there is a very efficient shortcut that can be employed. This shortcut is applicable to orthogonal and nonorthogonal designs with any number of covariates. The typical multiple linear regression computer program provides, as a part of the standard output, a test of significance (either t or F) of each partial regression coefficient in the equation. If the design matrix is set up using the same approach that is illustrated in Table 10.9, the regression of Y on the dummy variables associated with A, B, $A \times B$, and the covariate(s), is all that is required for the tests on the adjusted A, B, and $A \times B$ effects. The test on the first partial regression coefficient is equivalent to the test for adjusted A effects described in Table 10.14. The tests of significance of the second and third partial regression coefficients are equivalent to the tests of adjusted B and $A \times B$ effects, respectively.

CONFIDENCE INTERVALS FOR TWO-FACTOR DESIGNS WITH TWO LEVELS ON EITHER FACTOR. In previous chapters we expressed a preference for confidence intervals over significance tests. That preference also holds for two-factor designs. When a factor consists of only two levels, the following formula is appropriate:

$$\left(\bar{Y}_{m_{1adj}} - \bar{Y}_{m_{2adj}} \right) \pm \sqrt{ \text{MSres}_{wc} \left[\frac{1}{n_{m_1}} + \frac{1}{n_{m_2}} + \frac{\left(\bar{X}_{m_1} - \bar{X}_{m_2} \right)^2}{\text{SS}_{wc_x}} \right] } \quad \left(t_{(\alpha, N-IJ-1)} \right)$$

where

$\bar{Y}_{m_{1adj}}, \bar{Y}_{m_{2adj}}$ = adjusted marginal means

$\bar{X}_{m_1}, \bar{X}_{m_2}$ = marginal means on covariate associated with same factor levels as $\bar{Y}_{m_{1adj}}$ and $\bar{Y}_{m_{2adj}}$

n_{m_1}, n_{m_2} = number of subjects associated with marginal means 1 and 2

MSres_{wc} = mean square residual within cell associated with two-factor ANCOVA

SS_{wc_x} = sum of squares within cell associated with a two-factor ANOVA on X [which can be computed by using SST_x $(1 - R^2_{xD_a, D_b, D_{a \times b}})$, where SST_x is total sum of squares on covariate and $R^2_{xD_a, D_b, D_{a \times b}}$ is squared multiple correlation coefficient associated with regression of X on dummy variables indicating levels of A, B, and $A \times B$]

$t_{(\alpha, N-IJ-1)}$ = critical value of Student's t based on N (total number of subjects) minus IJ (total number of cells) minus one degree of freedom

This formula can be applied to the example data described previously in this section. The required terms can be found in Tables 10.15 and 10.16 except for SS_{wc_x}. This value is equal to

$$SST_x\left(1 - R^2_{xD_a, D_b, D_{a \times b}}\right) = 107(1 - 0.03271) = 103.50$$

The 95% confidence interval associated with the difference between the two levels of the factor A adjusted marginal means is

$$\left(\bar{Y}_{A_{1adj}} - \bar{Y}_{A_{2adj}}\right) \pm \sqrt{2.235\left[\frac{1}{8} + \frac{1}{8} + \frac{(4.50 - 5.00)^2}{103.50}\right]} (2.201) =$$
$$(5.68 - 3.82) \pm .751(2.201) = (.21, 3.51).$$

The 95% confidence interval associated with the difference between the two levels of the factor B adjusted marginal means is

$$\left(\bar{Y}_{B_{1adj}} - \bar{Y}_{B_{2adj}}\right) \pm \sqrt{2.235\left[\frac{1}{8} + \frac{1}{8} + \frac{(5.12 - 4.38)^2}{103.50}\right]} (2.201) =$$
$$(4.11 - 5.39) \pm 0.755(2.201) = (-2.94, .38)$$

MULTIPLE COMPARISON TESTS FOR MARGINAL ADJUSTED MEANS. If more than two levels exist for a factor, multiple comparison procedures should be used for the analysis of specific contrasts among the marginal means. The formulas presented in Table 10.17 provide the appropriate error terms for multiple comparison hypothesis tests and simultaneous confidence intervals for the case of one covariate. If more than one covariate is involved, the formulas in Table 10.18 are appropriate.

The terms in the formulas contained in Tables 10.17 and 10.18 are defined as follows:

$MSres_{wc}$ = within-cell residual MS that is the error term in two-factor ANCOVA

n_{m_i}, n_{m_j} = sample sizes associated with ith and jth marginal means that are being compared

$n_{m_i} - n_{m_L}$ = sample sizes associated with each level of L
L = number of levels of factor being analyzed (equal to either I or J)

$\bar{X}_{m_i}, \bar{X}_{m_j}$ = marginal means on covariate for ith and jth levels of factor being analyzed

SS_{wc_x} = within-cell sum of squares from two-factor ANOVA on covariate

Table 10.17 Multiple Comparison Formulas for Two-factor Independent Sample Designs with One Covariate

Procedure	Error Term	Critical Value
Protected LSD	$\sqrt{\mathrm{MSres}_{wc}\left[\dfrac{1}{n_{m_i}} + \dfrac{1}{n_{m_j}} + \dfrac{(\overline{X}_{m_i} - \overline{X}_{m_j})^2}{\mathrm{SS}_{wc_x}}\right]}$	Student's $t_{(\alpha,\, N - IJ - 1)}$
Bryant–Paulson	$\sqrt{\mathrm{MSres}_{wc}\left[\dfrac{2}{n_m} + (\overline{X}_{m_i} - \overline{X}_{m_j})^2 / \mathrm{SS}_{wc_x}\right] / 2}$	Bryant–Paulson generalized Studentized range $Q_{P_{(\alpha:\,C,\,N - IJ - 1)}}$
Dunn–Bonferroni	$\sqrt{\mathrm{MSres}_{wc}\left[\dfrac{c_1^2}{n_{m_1}} + \dfrac{c_2^2}{n_{m_2}} + \cdots + \dfrac{c_L^2}{n_{m_L}} + \dfrac{(c_1\overline{X}_{m_1} + c_2\overline{X}_{m_2} + \cdots + c_L\overline{X}_{m_L})^2}{\mathrm{SS}_{wc_x}}\right]}$	$t_{\mathrm{DB}_{(\alpha,\,C',\,N - IJ - 1)}}$
Scheffé	(Same formula as Dunn–Bonferroni)	$\sqrt{(L-1)F_{(\alpha,\,L-1,\,N-IJ-1)}}$

Table 10.18 Multiple Comparison Formulas for Two-factor Independent Sample Designs with Any Number of Covariates

Procedure	Error Term	Critical Value
Protected LSD	$\sqrt{MSres_{wc}\left[\dfrac{1}{n_{m_i}} + \dfrac{1}{n_{m_j}} + \mathbf{d}'\mathbf{W}_{wc_x}^{-1}\mathbf{d}\right]}$	Student's $t_{(\alpha, N-IJ-C)}$
Bryant–Paulson	$\sqrt{MSres_{wc}\left[\dfrac{2}{n_m} + \mathbf{d}'\mathbf{W}_{wc_x}^{-1}\mathbf{d}\right]\bigg/2}$	Bryant–Paulson generalized Studentized range $Q_{p_{(\alpha:C,L,N-IJ-C)}}$
Dunn–Bonferroni	$\sqrt{MSres_{wc}\left[\dfrac{c_1^2}{n_{m_1}} + \dfrac{c_1^2}{n_{m_2}} + \cdots + \dfrac{c_L^2}{n_{m_L}} + \mathbf{d}'\mathbf{W}_{wc_x}^{-1}\mathbf{d}\right]}$	Dunn–Bonferroni $t_{DB_{(\alpha,C,N-IJ-C)}}$
Scheffé	(Same formula as Dunn–Bonferroni)	$\sqrt{(L-1)F_{(\alpha,L-1,N-IJ-C)}}$

$c_1 - c_L$ = contrast coefficients associated with levels 1 through L of factor being analyzed

N = total number of subjects

I, J = number of levels of factors A and B, respectively

C' = number of planned comparisons

C = number of covariates

$\mathbf{d}$ = column vector of differences between ith and jth marginal means on covariates

$\mathbf{d}'$ = transpose of $\mathbf{d}$

$\mathbf{W}_{wc_x}^{-1}$ = inverse of within-cell sum of products matrix for covariates

An example of the term $\mathbf{d}$ (column vector) is as follows. If an investigator is interested in comparing marginal means 1 and 2 on factor A, the $\mathbf{d}$ vector would be

$$
\begin{array}{c}
\\
\text{Covariate 1} \\
\text{Covariate 2} \\
\cdot \\
\cdot \\
\cdot \\
\text{Covariate } C
\end{array}
\begin{array}{c}
\text{Level } A_1 \\
\left[\begin{array}{c} \bar{X}_{m_{1,1}} \\ \bar{X}_{m_{2,1}} \\ \cdot \\ \cdot \\ \cdot \\ \bar{X}_{m_{C,1}} \end{array} \right.
\end{array}
\begin{array}{c}
\\ - \\ - \\ \cdot \\ \cdot \\ \cdot \\ -
\end{array}
\begin{array}{c}
\text{Level } A_2 \\
\left. \begin{array}{c} \bar{X}_{m_{1,2}} \\ \bar{X}_{m_{2,2}} \\ \cdot \\ \cdot \\ \cdot \\ \bar{X}_{m_{C,2}} \end{array} \right]
\end{array}
=
\begin{array}{c}
\mathbf{d} \\
\left[\begin{array}{c} d_1 \\ d_2 \\ \cdot \\ \cdot \\ \cdot \\ d_C \end{array} \right]
\end{array}
$$

where

$\bar{X}_{m_{1,1}} = A_1$ marginal mean on covariate 1

$\bar{X}_{m_{1,2}} = A_2$ marginal mean on covariate 1

$d_1 = (\bar{X}_{m_{1,1}} - \bar{X}_{m_{1,2}})$, and so on

SIMPLE MAIN EFFECTS. The main effects associated with the two-factor design are no more difficult to interpret than are the adjusted means in a one-factor design as long as the $A \times B$ interaction is small or nonsignificant. If the interaction is significant, the interpretation of the main effects becomes ambiguous because the effects of A are not consistent across the different levels of B. In this case the main effects of A (or B) are still representative of the overall differences due to factor A (or B) but "overall" differences, which are reflected in the marginal means, are generally of little interest if the interaction is large. Figure 10.5 contains the interaction plot from two studies. The first plot is based on the data summary presented in Table 10.15; the interaction is not significant with these data. The second interaction plot is based on data (not presented here) that yield a clear interaction. The values plotted are the adjusted cell means in both cases. Notice in the left-hand plot that the difference between adjusted cell means A_1 and A_2 at level B_1 is about the same as the difference between adjusted cell means A_1 and A_2 at level B_2. Since the simple effect of A is consistent at B_1 and B_2, the difference between the adjusted marginal means A_1 and A_2 is a meaningful difference to analyze.

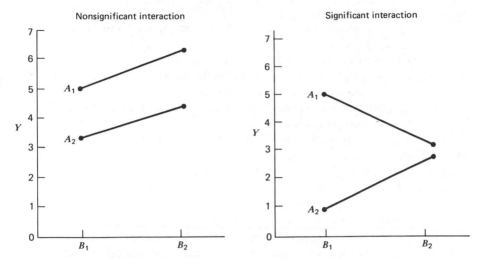

Fig. 10.5   Adjusted cell mean plots for nonsignificant and significant interactions.

But this is not true for the right-hand plot. Notice that in this case the difference between A_1 and A_2 is clearly not the same at B_1 as it is at B_2. The adjusted marginal means A_1 and A_2 are not representative of the A effects at B_1 or at B_2. This difference in the simple effects of A at the two levels of B will be reflected in a large interaction F. We recommend that the conventional two-factor ANCOVA be supplemented with "simple main effects" tests (or simple main effects confidence intervals) when the interaction is significant. "Simple main effects" refer to cell means, whereas main effects refer to marginal means.

The simple main effects tests associated with the 2×2 ANCOVA design refer to tests on the following:

1. Differences between adjusted cell means A_1 and A_2 at level B_1
2. Differences between adjusted cell means A_1 and A_2 at level B_2
3. Differences between adjusted cell means B_1 and B_2 at level A_1
4. Differences between adjusted cell means B_1 and B_2 at level A_2

A convenient way of carrying out these tests is to employ the following formula for simple main effects comparisons associated with a given factor:

$$\frac{\bar{Y}_{cm_{i\,adj}} - \bar{Y}_{cm_{j\,adj}}}{\sqrt{MSres_{wc}\left[\dfrac{1}{n_{cm_i}} + \dfrac{1}{n_{cm_j}} + \dfrac{\left(\bar{X}_{cm_i} - \bar{X}_{cm_j}\right)^2}{SS_{wc_x}}\right]}} = t$$

where

$\overline{Y}_{cm_{i\,adj}}, \overline{Y}_{cm_{j\,adj}}$ = adjusted cell means involved in simple main effects comparison

$MSres_{wc}$ = within-cell residual mean square

n_{cm_i}, n_{cm_j} = sample sizes associated with ith and jth cells involved in simple main effects comparison

$\overline{X}_{cm_i}, \overline{X}_{cm_j}$ = covariate means associated with cells involved in simple main effects comparison

SS_{wc_x} = within-cell sum of squares on covariate

The obtained t value is compared with the critical value of the Dunn–Bonferroni statistic $t_{DB_{(\alpha, C', N - IJ - 1)}}$, where α is the same level as is employed in the overall main effects tests and C' is the number of simple main effects tests associated with the factor being analyzed.

The denominator of this t formula can be employed in the construction of simultaneous confidence intervals for the simple main effects comparisons. That is, if we denote the error term of the t described previously as $s_{\overline{Y}_{cm_{i\,adj}} - \overline{Y}_{cm_{j\,adj}}}$, the simultaneous confidence intervals are constructed by using

$$\left(\overline{Y}_{cm_{i\,adj}} - \overline{Y}_{cm_{j\,adj}} \right) \pm s_{\overline{Y}_{cm_{i\,adj}} - \overline{Y}_{cm_{j\,adj}}} \left[t_{DB_{(\alpha, C', N - IJ - 1)}} \right]$$

This procedure is employed in the paragraphs that follow with the example data originally presented in Table 10.8 (and plotted in the left-hand plot of Figure 10.5) to illustrate the computations. Note, however, that simple main effects procedures are not required for these data because the interaction is not significant. Only the simple main effects comparisons associated with factor A are carried out in the paragraphs that follow.

The simple main effects comparisons for factor A involve the adjusted cell means difference between A_1 and A_2 at B_1 and at B_2. It can be seen in Table 10.16 that the difference at B_1 is $(5.00 - 3.21) = 1.79$ and at B_2, $(6.36 - 4.43) = 1.93$.

The confidence intervals associated with these two simple main effects comparisons are

$$(1.79) \pm \sqrt{2.235 \left[\frac{1}{4} + \frac{1}{4} + \frac{(4.75 - 5.50)^2}{103.50} \right]} \left[t_{DB_{(.05, 2, 11)}} \right]$$

$$= (1.79) \pm 1.063(2.593) = (-.97, 4.55)$$

$$(1.93) \pm \sqrt{2.235 \left[\frac{1}{4} + \frac{1}{4} + \frac{(4.25 - 4.50)^2}{103.50} \right]} \left[t_{DB_{(.05, 2, 11)}} \right]$$

$$= (1.93) \pm 1.058(2.593) = (-.81, 4.67)$$

The probability is at least .95 that the difference $(\mu_{A_{1\,adj}} - \mu_{A_{2\,adj}})$ at B_1 and $(\mu_{A_{1\,adj}} - \mu_{A_{2\,adj}})$ at B_2 are included in these intervals. You may notice that both of these intervals are wider than the interval that was constructed for the adjusted marginal mean difference between A_1 and A_2. There are three reasons for the width of the simple effects intervals to differ from the width of the overall (marginal) interval. First, there is more sampling error associated with the adjusted cell means than with the adjusted marginal means because the cell values are based on fewer observations. Second, the differences between the covariate cell means (which are involved in the error term for simple main effects) are not the same as the difference between the marginal covariate means. Third, two statements are made in the case of the simple effects; only one is made in the case of the overall main effects. The Bonferroni approach controls α for the whole collection (two in this case) of simple main effects comparisons. The more comparisons there are, the wider the confidence intervals will be. The probability is at least $1 - \alpha$ that the whole collection of intervals will span the true differences between the adjusted cell means included in the simple main effects comparisons.

If multiple covariates are employed, the following formula for simple main effects comparisons is appropriate:

$$\frac{\overline{Y}_{\text{cm}_{i\,\text{adj}}} - \overline{Y}_{\text{cm}_{j\,\text{adj}}}}{\sqrt{\text{MSres}_{\text{wc}}\left[\dfrac{1}{n_{\text{cm}_i}} + \dfrac{1}{n_{\text{cm}_j}} + \mathbf{d}'\mathbf{W}_{\text{wc}_x}^{-1}\mathbf{d}\right]}} = t$$

where the terms are defined as before and the obtained t value is compared with $t_{\text{DB}(\alpha,C',\,N-IJ-C)}$.

10.6 TWO-FACTOR ANOVA AND ANCOVA FOR REPEATED-MEASUREMENT DESIGNS

The two-factor design of the previous section involves the use of independent subjects in each cell. The design of this section involves the use of independent subjects in the various levels of factor A, but the same subjects are used in the various levels of factor B. This design is sometimes called a *split-plot* design or a *two-factor design with one repeated-measurement factor*. The layout of the design is presented in Figure 10.6. It can be seen that this design involves aspects of both the one-factor randomized-group design and the repeated-measurement one-factor design. The randomized-group aspect is involved in the random assignment of S subjects to the different treatment levels of factor A. In some varieties of this design, however, the assignment to the levels of A may not be random. Rather, subjects may be selected from different existing populations. For example, A_1 may be male subjects and A_2 female subjects. Hence the levels of factor A may be under experimental control, or they may simply be classification levels. Factor B, the repeated-measurement factor, may involve levels of several types. If the levels of factor

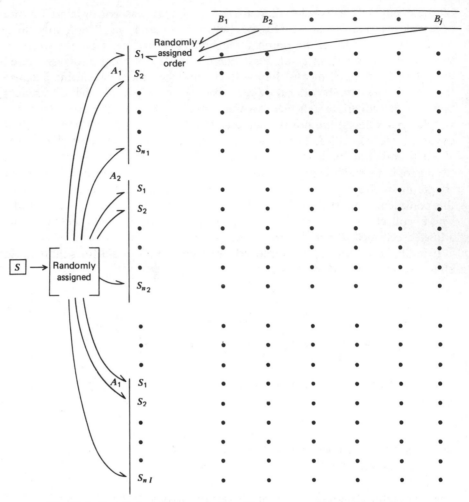

Fig. 10.6 Layout for two-factor design with one repeated-measurement factor.

B are different treatments, the order in which a subject is exposed to the various treatments is randomly assigned for each subject independently. If the levels of factor B are different trials or observations of the same behavior at J different time points, there is no randomization for this factor.

An advantage of this design over the independent sample two-factor design is that the number of subjects required is smaller. Also, since each subject is observed under all levels of factor B, the size of the error term for testing the effects of this factor is generally smaller than with independent sample designs. It must be kept in mind, however, that the results with respect to factor B are generalized to a population of subjects who have been exposed to repeated-measurement conditions. Additional details on advantages and disadvantages of repeated measurement designs can be found in Greenwald (1976), Kirk (1968), and Winer (1971).

Suppose that a researcher has interest in evaluating the effects of two drugs and three situations in which the drugs are administered, on anxiety. The two drugs are LSD and psilocybin and the three conditions of administration are alone, with a group of friends, and with a group of strangers. Hence the two factors are A, drug type; and B, administration situation. The dependent variable is a scale of situational anxiety.

Ten subjects are randomly assigned to the two drug levels A_1 (LSD) and A_2 (psilocybin). Then the order of administration of the three levels of factor B is randomly assigned for each subject independently. Before the experiment is initiated, the investigator notices apparently large differences in general (trait) anxiety among the subjects. His or her concern is that individual differences in anxiety level (that can be measured by using a general anxiety scale) will have an effect on the way subjects respond on the dependent variable. Hence a measure of general anxiety is obtained from each subject before the experiment is started. This measure is then used as a covariate in the analysis after the experiment is completed. Contrived data appear in Table 10.19. Notice that there are five subjects under each drug type. The five subjects associated with A_1 are not the same subjects associated with level A_2. The covariate score associated with each subject is shown in the X column, the three scores on the dependent variable (Y) under the three levels of B are in the next three columns, and the sum of the three Y scores for each subject is entered in the T column.

The design matrix required for the analysis is shown in Table 10.20. The $\{1, 0, -1\}$ dummy variables are constructed in exactly the same way as was

Table 10.19 Example Data for a Two-factor Design with One Repeated-measurement Factor and One Covariate

Factor A Drug Type		Factor B, Administration Situation			
		B_1 = Alone	B_2 = Friends	B_3 = Strangers	
		X	Y	Y	Y	T
A_1 LSD	S_1	6	9	6	7	22
	S_2	5	6	7	7	20
	S_3	4	8	5	6	19
	S_4	7	9	8	8	25
	S_5	4	7	5	6	18
A_2 Psilocybin	S_1	9	8	7	6	21
	S_2	6	4	3	4	11
	S_3	3	3	3	2	8
	S_4	5	4	4	3	11
	S_5	6	3	4	4	11

Table 10.20 Design Matrix for 2×3 Two-factor Design with One Repeated-Measurement Factor and One Covariate

Cell	(1) d_{a_1}	(2) d_{b_1}	(3) d_{b_2}	(4) $d_{a_1}d_{b_1}$	(5) $d_{a_1}d_{b_2}$	(6) X	(7) $d_{a_1}X$	(8) T	Y
	1	1	0	1	0	6	6	22	9
	1	1	0	1	0	5	5	20	6
A_1B_1	1	1	0	1	0	4	4	19	8
	1	1	0	1	0	7	7	25	9
	1	1	0	1	0	4	4	18	7
	1	0	1	0	1	6	6	22	6
	1	0	1	0	1	5	5	20	7
A_2B_2	1	0	1	0	1	4	4	19	5
	1	0	1	0	1	7	7	25	8
	1	0	1	0	1	4	4	18	5
	1	-1	-1	-1	-1	6	6	22	7
	1	-1	-1	-1	-1	5	5	20	7
A_1B_3	1	-1	-1	-1	-1	4	4	19	6
	1	-1	-1	-1	-1	7	7	25	8
	1	-1	-1	-1	-1	4	4	18	6
	-1	1	0	-1	0	9	-9	21	8
	-1	1	0	-1	0	6	-6	11	4
A_2B_1	-1	1	0	-1	0	3	-3	8	3
	-1	1	0	-1	0	5	-5	11	4
	-1	1	0	-1	0	6	-6	11	3
	-1	0	1	0	-1	9	-9	21	7
	-1	0	1	0	-1	6	-6	11	3
A_2B_2	-1	0	1	0	-1	3	-3	8	3
	-1	0	1	0	-1	5	-5	11	4
	-1	0	1	0	-1	6	-6	11	4
	-1	-1	-1	1	1	9	-9	21	6
	-1	-1	-1	1	1	6	-6	11	4
A_2B_3	-1	-1	-1	1	1	3	-3	8	2
	-1	-1	-1	1	1	5	-5	11	3
	-1	-1	-1	1	1	6	-6	11	4

described in the previous section for the two-factor independent sample design. The "ones" in the first column (d_{a_1}) identify observations associated with level A_1; the "minus ones" in this column identify observations associated with level A_2. Since there are two levels of factor A only one dummy variable is required for this factor (i.e., the number of dummy variables required for factor $A = I - 1 = 2 - 1 = 1$). Factor B has three levels; two dummy variables are required for this factor because the required number is the number of levels of B minus one (i.e., $J - 1 = 3 - 1 = 2$). A "one" is entered in column 2 (d_{b_1}) for each observation that is associated with level B_1. A "minus one" is entered for each observation associated with the last level (i.e., B_3). "Zero" is entered for observations that are associated with neither B_1 nor B_3 (i.e., B_2). The third column (d_{b_2}) contains "ones" to identify observations associated with level B_2, "minus ones" to identify those associated with B_3, and "zeros" to identify observations associated with neither B_2 nor B_3. The next two columns $(d_{a_1}d_{b_1}$ and $d_{a_1}d_{b_2})$ are product columns. Column 6 is the covariate column, and column 7 is the product of the d_{a_1} and X columns. Column 8 is the T column; each entry in this column is the total of the three Y scores across the three levels of B. Notice that each covariate score and each T score appears three times for each subject. This occurs because each subject contributes three scores to the Y column; one score is obtained under each level of factor B. Since X and T scores for a given subject are the same regardless of the level of B, each X and T score must appear as many times as there are levels of B. Dependent variable scores are entered in the last column.

Unlike the design matrix associated with the two-factor independent sample design, this design matrix does not contain the products of the dummy variables used to identify levels of B and X. Notice that there are no columns labeled $d_{b_1}X$ or $d_{b_2}X$. This is because only factor A is adjusted by the covariate with this design. The covariate accounts for neither between-level nor within-level variability on factor B because the covariate score is constant for a given subject at all levels of this factor. The results of this are that (1)

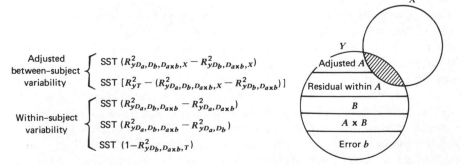

Fig. 10.7 Diagrammatic representations of partitioning associated with two-factor designs.

the covariate does not adjust the differences among the marginal factor B means, (2) the $A \times B$ interaction is not related to the covariate, and (3) the error term used to test the B and $A \times B$ effects is not related to the covariate.

A comparison of the design matrix of this section with the design matrix of the previous section reveals two basic differences. Whereas the design matrix of this section contains no product of B and X columns, it does contain the T column that is not found in the other matrix. This difference in matrices is necessary because the partitioning of the total sum of squares is different with these two varieties of two-factor design. These differences can be seen in Figure 10.7.

Notice that the variability in the independent sample design is all between-subject variability. The repeated-measurement design contains variability that is partly attributed to differences between subjects and partly attributed to differences within subjects. The computation of ANOVA, ANCOVA, and homogeneity of regression tests for the repeated-measurement design requires the following R^2 values for the example data summarized in Tables 10.19 and 10.20:

$$R^2_{yD_a, D_b, D_{a \times b}, X} = R^2_{y1,2,3,4,5,6} = 0.87961$$

$$R^2_{yD_b, D_{a \times b}, X} = R^2_{y2,3,4,5,6} = 0.24823$$

$$R^2_{yT} = R^2_{y8} = 0.85491$$

$$R^2_{yD_a, D_b, D_{a \times b}} = R^2_{y1,2,3,4,5} = 0.55469$$

$$R^2_{yD_a, D_{a \times b}} = R^2_{y1,4,5} = 0.51395$$

$$R^2_{yD_a, D_b} = R^2_{y1,2,3} = 0.53292$$

$$R^2_{yD_b, D_{a \times b}, T} = R^2_{y2,3,4,5,8} = 0.91741$$

$$R^2_{yD_a, X, D_a X} = R^2_{y1,6,7} = 0.81716$$

$$R^2_{yD_b, D_{a \times b}} = R^2_{y2,3,4,5} = 0.06250$$

$$R^2_{yD_a, X} = R^2_{y1,6} = 0.81711$$

The appropriate R^2 values are entered in the ANCOVA summary provided in Table 10.21. The difference between the two R^2 values in the first line of the table is the proportion of the total sum of squares accounted for by factor A independently of B, $A \times B$, and X. The first R^2 in the second line provides the proportion of the total sum of squares that is accounted for by between subject differences (i.e., 0.85491). The value subtracted from 0.85491 is 0.81711, which is the proportion of the total sum of squares accounted for by A and X independently of B and $A \times B$ (i.e., 0.87691 − 0.06250). Hence the first two lines of the analysis (i.e., adjusted A effects and the residual within levels of A) contain between-subject sums of squares that are independent of B, $A \times B$, and X.

Table 10.21 General Form of ANCOVA Summary for Two-factor Design with One Repeated-Measurement Factor[a]

Source	SS	df	MS	F
Adjusted A	$SST(R^2_{y,D_a,D_b,D_{a\times b},X} - R^2_{y,D_b,D_{a\times b},X})$	$I-1$	$SSA_{adj}/I-1$	$MSA_{adj}/MSres_{wa}$
Res_{wa}	$SST[R^2_{yT} - (R^2_{y,D_a,D_b,D_{a\times b},X} - R^2_{y,D_b,D_{a\times b}})]$	$S-I-C$	$SSRes_{wa}/S-I-C$	
B	$SST(R^2_{y,D_a,D_b,D_{a\times b}} - R^2_{y,D_a,D_{a\times b}})$	$J-1$	$SSB/J-1$	MSB/MSE_b
$A\times B$	$SST(R^2_{y,D_a,D_b,D_{a\times b}} - R^2_{y,D_a,D_b})$	$(I-1)(J-1)$	$SSA\times B/(I-1)(J-1)$	$MSA\times B/MSE_b$
Error b	$SST(1-R^2_{y,D_b,D_{a\times b},T})$	$\sum_{i=1}^{I}(J-1)(n_i-1)$	$SSE_b/\sum_{i=1}^{I}(J-1)(n_i-1)$	
Res_t	$SST(1-R^2_{yX})$	$N-1-C$		

Homogeneity of Regression Summary

Heterogeneity	$SSres_{wa} - SSres_i$	$C(I-1)$	$SShet/C(I-1)$	$MShet/MSres_i$
Res_i	$SST(R^2_{yT} - R^2_{y,D_a,X,D_aX})$	$S-[I(C+1)]$	$SSres_i/S-[I(C+1)]$	

Computation of Cell Means with Any Number of Levels of Factors A and B and Any Number of Covariates

$$\overline{Y}_{ij} = b_0 + b_1(d_1) + b_2(d_2) + \cdots + b_{IJ-1}(d_{IJ-1}) + b_{X_1}(\overline{X}_{\cdot\cdot\cdot 1}) + b_{X_2}(\overline{X}_{\cdot\cdot\cdot 2}) + \cdots + b_{X_C}(\overline{X}_{\cdot\cdot\cdot C})$$

[a]*Notation:*

Res_{wa} = residual within levels of A

Res_t = total residual

I = number of levels of factor A

J = number of levels of factor B

S = total number of subjects in experiment

n_i = number of subjects in ith level of factor A

N = total number of observations in experiment

C = number of covariates

The difference between the two R^2 values in line 3 is the proportion of the total sum of squares accounted for by B independently of A and $A \times B$; line 4 provides the R^2 values required to obtain the sum of squares for the $A \times B$ interaction.

The partitioning associated with B and $A \times B$ does not involve the use of the covariate in the computation of the R^2 values, but these sources of variability are still independent of all other sources of variability in the design including the covariate. The error term for the B and $A \times B$ effects (error b) is based on the sum of squares that is independent of B, $A \times B$, and all between-subject variability.

The ANOVA of this design differs from the ANCOVA only in the first two lines. The unadjusted A sum of squares is computed by using

$$\mathrm{SST}\left(R^2_{yD_a, D_b, D_{a \times b}} - R^2_{yD_b, D_{a \times b}}\right)$$

The factor A error sum of squares (error a) is computed by using $\mathrm{SST}(R^2_{yT})$ minus the unadjusted A sum of squares.

Comparison of ANOVA and ANCOVA

The difference between the general forms of ANOVA and ANCOVA for the two-factor repeated-measurement design can be seen by comparing Tables 10.21 and 10.23a. The computational examples of these two analyses can be compared in Tables 10.22 and 10.23b. Notice that the marginal means for factor A are different under the two analyses but the factor B marginal means are identical. As was mentioned previously, this outcome is to be expected because the covariate adjusts the factor A means but not the factor B means.

Confidence Intervals for Adjusted Factor A Marginal Means

If there are only two levels of factor A (the nonrepeated factor), the confidence interval for the adjusted marginal means is computed by using the error term of the protected LSD formula that is presented in Table 10.24. This formula is not appropriate if there are three or more levels of factor A and simultaneous confidence intervals are desired.

The application of this formula to the example data yields the following 95% confidence level:

$$\left(\bar{Y}_{m_1\,\mathrm{adj}} - \bar{Y}_{m_2\,\mathrm{adj}}\right) \pm \sqrt{\mathrm{MSres}_{\mathrm{wa}}\left[\frac{1}{Jn_1} + \frac{1}{Jn_2} + \frac{\left(\bar{X}_{m_1} - \bar{X}_{m_2}\right)^2}{\mathrm{SS}_{\mathrm{wa}_x}}\right]}\;\left(t_{(\alpha,\,S-I-1)}\right)$$

$$(6.95 - 3.72)) \pm \sqrt{0.646\left[\frac{1}{15} + \frac{1}{15} + \frac{(5.2 - 5.8)^2}{25.6}\right]}\;(2.365)$$

$$(3.23) \pm 0.73 = (2.50, 3.96)$$

Table 10.22   ANCOVA Summary for Data of Table 10.19

Source	SS	df	MS	F
Adjusted A	$119.47(0.87961-0.24823)=75.43$	1	75.43	117
Res_{wa}	$119.47(0.85491-0.81711)=4.52$	7	0.646	
B	$119.47(0.55469-0.51395)=4.87$	2	2.44	3.95
$A\times B$	$119.47(0.55469-0.53292)=2.63$	2	1.32	2.14
Error b	$119.47(1-0.91741)=9.87$	16	0.617	
Res_t	$119.47(1-0.18573)=97.3$	28		
	Homogeneity of Regression Test			
Hetero. regr.	$(\text{res}_{wa}-\text{res}_i)=0.01$	1	0.01	0.01
Res_i	$119.47(0.85491-0.81716)=4.51$	6	0.752	

Cell Means

Regression equation associated with $R^2_{yD_a,\,D_b,\,D_{a\times b},X}=R^2_{y1,2,3,4,5,6}$:

$$\hat{Y}=1.623+1.613(d_1)+0.567(d_2)-0.333(d_3)+0.300(d_4)-0.400(d_5)+0.711(X)$$

$$\bar{Y}_{1,1\,adj}=1.623+1.613(1)+0.567(1)-0.333(0)+0.300(1)-0.400(0)+0.711(5.5)=8.01$$

$$\bar{Y}_{1,2\,adj}=1.623+1.613(1)+0.567(0)-0.333(1)+0.300(0)-0.400(1)+0.711(5.5)=6.41$$

$$\bar{Y}_{1,3\,adj}=1.623+1.613(1)+0.567(-1)-0.333(-1)+0.300(-1)-0.400(-1)+0.711(5.5)$$
$$=7.01$$

$$\bar{Y}_{2,1\,adj}=1.623+1.613(-1)+0.567(1)-0.333(0)+0.300(-1)-0.400(0)+0.711(5.5)$$
$$=4.19$$

$$\bar{Y}_{2,2\,adj}=1.623+1.613(-1)+0.567(0)-0.333(1)+0.300(0)-0.400(-1)+0.711(5.5)$$
$$=3.99$$

$$\bar{Y}_{2,3\,adj}=1.623+1.613(-1)+0.567(-1)-0.333(-1)+0.300(1)-0.400(1)+0.711(5.5)$$
$$=3.59$$

		B_1	B_2	B_3	Factor A Marginal Means
	A_1	8.01	6.41	7.01	7.14
Factor B					
Marginal	A_2	4.19	3.99	3.59	3.92
Means		6.10	5.20	5.30	

When three or more levels of factor A are involved, one of the other error term formulas in Table 10.24 should be used. The considerations in selecting among the different formulas are the same as those discussed in Chapter 5 for one-factor designs.

Confidence Intervals for Factor B Marginal Means

If two levels of factor B are involved, the error term for the protected LSD procedure described in the bottom of Table 10.24 is appropriate for the confidence interval. When three or more levels of factor B are included in the

Table 10.23 General Form of ANOVA on Factor A and Summary for Data of Table 10.19[a]

Source	SS	df	MS	F
	a. General Form—ANOVA			
A	$\mathrm{SST}(R^2_{yD_a,D_b,D_{a\times b}} - R^2_{yD_b,D_{a\times b}})$	$I-1$	$\mathrm{SS}A/I-1$	$\mathrm{MS}A/\mathrm{MSE}_{wa}$
Error_{wa}	$\mathrm{SST}[R^2_{yT} - (R^2_{yD_a,D_b,D_{a\times b}} - R^2_{yD_b,D_{a\times b}})]$	$S-I$	$\mathrm{SSE}_a/S-I$	
B $A\times B$ $\mathrm{Error}\ b$	} (Same as shown in Table 10.21 for ANCOVA)			
	b. ANOVA Summary—Data of Table 10.19			
A	$119.47(0.55496-0.06250)=58.80$	1	58.80	10.85
$\mathrm{Error}\ A$	$119.47(0.85491-0.49219)=43.33$	8	5.42	
B $A\times B$ $\mathrm{Error}\ b$	} (Same as shown in Table 10.22 for ANCOVA)			
Total	119.47	29		

[a] Computation of cell means is based on the regression equation associated with $R^2_{yD_a,D_b,D_{a\times b}} = R^2_{y1,2,3,4,5}$:

$$\bar{Y}_{ij}=b_0+b_1(d_1)+b_2(d_2)+\cdots+b_{IJ-1}(d_{IJ-1})=5.533+1.400+0.567-0.333+0.300-0.400$$

$$\bar{Y}_{1,1}=5.533+1.400(1)+0.567(1)-0.333(0)+0.300(1)-0.400(0)=7.80$$

$$\bar{Y}_{1,2}=5.533+1.400(1)+0.567(0)-0.333(1)+0.300(0)-0.400(1)=6.20$$

$$\bar{Y}_{1,3}=5.533+1.400(1)+0.567(-1)-0.333(-1)+0.300(-1)-0.400(-1)=6.80$$

$$\bar{Y}_{2,1}=5.533+1.400(-1)+0.567(1)-0.333(0)+0.300(-1)-0.400(0)=4.40$$

$$\bar{Y}_{2,2}=5.533+1.400(-1)+0.567(0)-0.333(1)+0.300(0)-0.400(-1)=4.20$$

$$\bar{Y}_{2,3}=5.533+1.400(-1)+0.567(-1)-0.333(-1)+0.300(1)-0.400(1)=3.80$$

	B_1	B_2	B_3	Factor A Marginal Means
A_1	7.80	6.2	6.8	6.93
A_2	4.4	4.2	3.8	4.13
Factor B Marginal Means $\longrightarrow$	6.1	5.2	5.3	

design, one of the other formulas should be used. The simultaneous 95% confidence intervals for pairwise comparisons using the Tukey procedure are shown as follows for the example data:

$$\left(\bar{Y}_{m_i}-\bar{Y}_{m_j}\right)\pm\sqrt{\mathrm{MSE}_b/S}\ \left(q_{(\alpha,J,\Sigma(J-1)(n_i-1))}\right)$$

$$(6.1-5.2)\pm\sqrt{0.617/10}\ \left(q_{(.05,3,16)}\right)=(0.9)\pm0.91=(-.01,1.81)$$

$$(6.1-5.3)\pm\sqrt{0.617/10}\ \left(q_{(.05,3,16)}\right)=(0.8)\pm0.91=(-0.11,1.71)$$

$$(5.2-5.3)\pm\sqrt{0.617/10}\ \left(q_{(.05,3,16)}\right)=(-0.1)\pm0.91=(-1.01,0.81)$$

Table 10.24 Multiple Comparison Formulas for Two-factor Designs with One Repeated-measurement Factor and One Covariate

Procedure	Error Term	Critical Value
	Formulas for Factor A—Nonrepeated (Between Subject) Factor	
Protected LSD	$\sqrt{\text{MSres}_{wa}\left[\dfrac{1}{Jn_i}+\dfrac{1}{Jn_j}+\dfrac{(\bar{X}_{m_i}-\bar{X}_{m_j})^2}{SS_{wa_x}}\right]}$	Student's $t_{(\alpha,\,S-I-1)}$
Bryant–Paulson	$\sqrt{\text{MSres}_{wa}\left[\dfrac{2}{Jn_m}+(\bar{X}_{m_i}-\bar{X}_{m_j})^2/SS_{wa_x}\right]/2}$	Bryant–Paulson generalized Studentized range $Q_{P(\alpha:\,1,I,S-I-1)}$
Dunn–Bonferroni	$\sqrt{\text{MSres}_{wa}\left[\dfrac{c_1^2}{Jn_1}+\dfrac{c_2^2}{Jn_2}+\cdots+\dfrac{c_I^2}{Jn_I}+\dfrac{(c_1\bar{X}_{m_1}+c_2\bar{X}_{m_2}+\cdots+c_I\bar{X}_{m_I})^2}{SS_{wa_x}}\right]}$	$t_{DB(\alpha,\,C',\,S-I-1)}$
Scheffé	(Same formula as Dunn–Bonferroni)	$F'=(I-1)\sqrt{F_{(\alpha,I-1,S-I-1)}}$
	Formulas for Factor B—Repeated-measurement (Within-subject) Factor	
Protected LSD	$\sqrt{\text{MSE}_b\left[\dfrac{2}{S}\right]}$	Student's $t_{(\alpha,\,\Sigma(J-1)(n_i-1))}$
Tukey	$\sqrt{\text{MSE}_b}/S$	Studentized range $q_{(\alpha,J,\Sigma(J-1)(n_i-1))}$
Dunn–Bonferroni	$\sqrt{\text{MSE}_b\left[\dfrac{c_1^2}{S}+\dfrac{c_2^2}{S}+\cdots+\dfrac{c_J^2}{S}\right]}$	Dunn–Bonferroni $t_{DB(\alpha,\,C',\,\Sigma(J-1)(n_i-1))}$
Scheffé	(Same as Dunn–Bonferroni)	$F'=(J-1)\sqrt{F_{(\alpha,J-1,\Sigma(J-1)(n_i-1))}}$

Simple Main Effects—Factor A. If the $A \times B$ interaction is significant, comparisons between cell means may provide more relevant information than comparisons of marginal means. Simple main effects comparisons between levels of A at each level of B are based on the following term and critical value:

Error Term	Critical Value

$$\sqrt{\left[\frac{\text{SSres}_{\text{wa}} + \text{SSE}_b}{(S - I - 1) + [\Sigma(J - 1)(n_i - 1)]}\right]\left[\frac{1}{n_i} + \frac{1}{n_j} + \frac{(\bar{X}_{A_i} - \bar{X}_{A_j})^2}{\text{SS}_{\text{wa}_X}}\right]} \qquad t_{\text{DB}_{[\alpha, C', S - I - 1 + \Sigma(J - 1)(n_i - 1)]}}$$

The number of comparisons (C') associated with all simple main effects tests on factor A is computed by using $J[I(I - 1)/2]$.

If multiple covariates are used, the error-term formula and the critical value are

$$\sqrt{\left[\frac{\text{SSres}_{\text{wa}} + \text{SSE}_b}{(S - I - C) + [\Sigma(J - 1)(n_i - 1)]}\right]\left[\frac{1}{n_i} + \frac{1}{n_j} + \mathbf{d'W}_{a_x}^{-1}\mathbf{d}\right]}$$

and $t_{\text{DB}_{[\alpha, C', S - I - C + (\Sigma(J - 1)(n_i - 1))]}}$.

Simple Main Effects—Factor B. Simple main effects comparisons between the ith and jth levels of factor B at each level of A involve the following formulas for the error term and the critical value.

Error Term	Critical Value

$$\sqrt{\text{MSE}_b\left[\frac{1}{n_{\text{cm}_i}} + \frac{1}{n_{\text{cm}_j}}\right]} \qquad t_{\text{DB}_{[\alpha, C', \Sigma(J - 1)(n_i - 1)]}}$$

The number of comparisons $C' = I[J(J - 1)/2]$ and n_{cm_i} and n_{cm_j} are the sample sizes associated with the ith and jth cell means being compared.

Independence of Errors Assumption. It was pointed out in the section on the one-factor randomized-block and repeated-measurement design that the independence of errors assumption should be questioned. The same issue is of concern for the repeated-measurement factor in two-factor designs. The methods of testing factor B and $A \times B$ effects described in this section will lead to positively biased F values if the independence of errors assumption is violated. The test of compound symmetry (Box 1950) can be employed to test this assumption. This test, which is described in most texts on experimental design (e.g., Winer 1971), can be avoided by (1) computing the F test on B using degrees of freedom equal to 1 and $\Sigma_{i=1}^{I}(n_i - 1)$. The latter approach yields the "conservative" F (Geisser and Greenhouse 1958). If the conservative test is significant, it is known that the conventional F must also be significant because the conservative test is based on a higher critical value. If the conventional F is not significant, there is no need to compute the conservative test because it will also be nonsignificant. If the conventional test is significant but the conservative test is not, there are three acceptable courses of action: (1) accept the results of the conservative test, (2) compute the more exact test described by Box (1954), or (3) compute the exact multivariate test known as Hotelling's T^2. Options (2) and (3) involve matrix manipulations that we do not describe here. Most experimental design and multivariate analysis texts (e.g., Bock 1975), describe the multivariate approach [see also Ceurvorst and Stock (1978)].

10.7   OTHER COMPLEX DESIGNS

Covariance analysis can be applied to any of the more complex ANOVA designs frequently described in the experimental design literature. Three-factor independent sample designs, for example, require no new principles. Additional dummy variables are required for the third factor and its interaction with the other factors, but the basic procedure involved in the construction of dummy variables and testing effects is unchanged. Federer and Henderson (1978a, 1978b) have evaluated several statistical computer packages that have been developed for complex ANCOVA designs.

10.8   SUMMARY

Analysis of covariance is not the only procedure for attempting to remove the effects of unwanted variability from an experiment. Other approaches include blocking and the use of subjects as their own control in repeated-measurement designs. The basic advantages of ANCOVA relative to these other approaches are that (1) the response measure is adjusted for chance differences that exist before treatments are applied, and (2) fewer steps are sometimes required in the design of the experiment because homogeneous blocks do not have to be formed.

Two two-factor ANCOVA designs include the independent sample and repeated measurement varieties. Both designs provide information on the overall (main) effects of two different independent variables or factors as well as information on the interaction of the two factors. The application of covariance analysis to the independent sample two-factor design generally results in greater power for tests on main effects and interaction. Effects are also adjusted for chance differences on the covariate that are related to Y.

The two-factor ANOVA design with one repeated-measurement factor involves between-subject comparisons on factor A and within-subject comparisons on factor B. This design generally has very high power on comparisons among levels of factor B but relatively low power on comparisons among levels of factor A. When covariance analysis is applied to this design, the power of factor A comparisons can be markedly increased if the covariate is well chosen. Hence the covariance analysis of this design can result in very high power on factor A (if the covariate is highly correlated with the dependent variable); tests on B and the $A \times B$ interaction also have high power, but this is a result of the repeated measurement aspect of the design. Covariance analysis can be applied to virtually any other complex experimental design.

CHAPTER ELEVEN
Multivariate Analysis of Covariance

11.1 INTRODUCTION

It is not unusual for more than one dependent variable to be employed in a single study in many types of experimental and nonexperimental research. For example, in a study of the effects of viewing a film that was expected to produce stress, Oken (1967) obtained measurements on six dependent variables: heart rate, systolic blood pressure, diastolic blood pressure, respiration rate, skin conductance, and finger temperature. Daniel and Hall (cited in Bock 1966) compared the effectiveness of a programmed and a conventional text covering similar content (elementary psychology) on a final examination that consisted of six subscales. Specific content was measured by subscales 1 and 2, knowledge of concepts and principles was measured by subscales 3 and 4, applications were measured by subscale 5, and problem solving was measured by subscale 6. Studies of the effects of drugs frequently use many response variables such as pupillary dilation, psychomotor proficiency, time perception, responsiveness to suggestions, and visual perception.

A common procedure in the analysis of studies such as these with multiple dependent variables is to compute a conventional (univariate) analysis on each dependent variable. Two issues associated with the use of separate univariate analyses on multiple dependent variables are generally overlooked and should be considered: error rate and loss of information. These issues are discussed in the remainder of this chapter, generally within the ANCOVA context. But it should be understood that the same problems apply to ANOVA and, more generally, to any design (regardless of the type of analysis) in which multiple dependent variables are analyzed.

Error Rate

If more than one dependent variable is employed, the probability of making a Type I error for each separate ANCOVA F test is α. The probability that one or more of the ANCOVA F tests in the set of F values will result in Type I error is not, in general, α. If the researcher is interested in keeping the family error rate (i.e., the probability that one or more of the *set* of tests will result in Type I error) at α, the use of conventional univariate ANCOVA procedures is not appropriate. When several dependent variables are employed, the only situation in which the use of conventional ANCOVA procedures on each dependent variable will yield a family error rate of α is with perfectly correlated dependent variables. Since we do not expect to encounter a set of perfectly correlated dependent variables in practice, this extreme situation is of little practical concern. Obviously, if several dependent variables are perfectly correlated, there is no reason to employ more than one of them since the others would contribute no unique information. The reason for mentioning this extreme situation is simply to point out that when perfectly correlated dependent variables are analyzed, the results of ANCOVA on any one variable will be the same on any other variable. All tests beyond the first are redundant. Likewise, the error rate for a single dependent variable must be the same as the error rate for the whole set, that is, $\alpha_{individual} = \alpha_{family}$.

If the dependent variables are all uncorrelated (another unlikely situation), the probability of making a Type I error in the set or family is quite different from the error rate for a single dependent variable. That is, $\alpha_{family} \neq \alpha_{individual}$. If there is no correlation among the dependent variables, the error rate for the family of F tests is $1 - (1 - \alpha_{individual})^p$ where p is the number of dependent variables. Suppose that a study employing 10 dependent variables is analyzed by computing a univariate ANCOVA F test on each dependent variable. If each test is run at the .05 level (i.e., $\alpha_{individual} = .05$), the family error rate is $1 - (1 - .05)^{10}$ or .40. This is the probability that one or more of the 10 F tests will be significant when the null hypothesis is true for each variable. Since this family error rate is quite high the researcher may want to employ a procedure that will keep the family error rate at .05 rather than to employ the conventional univariate procedure, which keeps the error rate for each individual F at .05.

Loss of Information

The use of more than one dependent variable in an experiment provides a source of information that will be lost with separate univariate analyses on the set of dependent variables. If two or more dependent variables are employed, the researcher may gain insights concerning the nature of the data by analyzing the treatment effects with the use of procedures that utilize the relationships among the dependent variables. Procedures that deal with the issues of error rate and/or loss of information are described in the two following sections.

11.2 BONFERRONI F PROCEDURE

It has been pointed out that the family error rate for an experiment depends on the degree of relationship among the dependent variables. When the correlation among them is perfect, the individual error rate is also the family error rate. If there is no correlation among the dependent variables, the family error rate is $1-(1-\alpha_{\text{individual}})^p$. Hence the probability that one or more of a set of F tests will be declared significant when the null hypothesis is true for each variable lies between $\alpha_{\text{individual}}$ and $1-(1-\alpha_{\text{individual}})^p$. If we are interested in keeping the probability of making a Type I error at less than α_{family}, it is reasonable to simply divide the desired α_{family} by the number of dependent variables to obtain the $\alpha_{\text{individual}}$ that should be employed for each test. That is, if a family α of .05 is desired and 10 dependent variables are employed, we find $.05/10=.005=\alpha_{\text{individual}}$. If each test is run at the .005 level, the sum of the 10 individual alpha values can be no more than .05. Hence

$$\sum_{i=1}^{p} \alpha_{\text{individual}} \leqslant \alpha_{\text{family}}$$

You may have noticed that $p(\alpha_{\text{individual}})$ or $10(.005)$ was used here as the maximum possible family error rate rather than the previously defined expression $1-(1-\alpha_{\text{individual}})^p$. It turns out that $p(\alpha_{\text{individual}})$, which is an approximate formula, yields values very close to those obtained with $1-(1-\alpha_{\text{individual}})^p$ (the exact formula) for conventional α values. For the present example, we find $1-(1-.005)^{10}=.04889$, whereas $10(.005)=.05$. The construction of the table of critical values of the Bonferroni F statistic has been simplified by using $p(\alpha_{\text{individual}})$ as the family error rate.

Computational Steps for Bonferroni F Analytic Procedure

1. Compute F_{obt} for each univariate ANCOVA.
2. Evaluate each F_{obt} by consulting the Bonferroni F table (Table A.8) for the critical value. Suppose that a three-group experiment is carried out employing five dependent variables and one covariate. If a family error rate of .05 is employed and $N=30$, the critical value for each F test is found (in Table A.8) to be 5.53. (This is the value associated with five comparisons, $df_1=J-1=2$ and $df_2=N-J-C=26$.) Notice that the only difference between this procedure and the ordinary univariate F procedure is that the obtained F values are evaluated with the Bonferroni F statistic based on p dependent variables and $J-1$ and $N-J-C$ degrees of freedom rather than F with $J-1$ and $N-J-C$ degrees of freedom.
3. Conclude that the F_{obt} values that equal or exceed the Bonferroni F critical value are significant.
4. Compute the appropriate multiple comparison tests on the dependent variables (if any) associated with significant Bonferroni F values.

11.3   MULTIVARIATE ANALYSIS OF COVARIANCE

A procedure known as *multivariate analysis of covariance* (MANCOVA) is an alternative to the Bonferroni F procedure when multiple dependent variables are employed. Like the Bonferroni F procedure, MANCOVA is employed to control the family error rate. Unlike the Bonferroni F procedure, MANCOVA utilizes the relationships among the dependent variables and provides an overall test on treatment differences. If the overall test is significant, a univariate test on each dependent variable or some other more complex multivariate procedure is generally employed.

Only the basic notions of MANCOVA are described in this section. A thorough understanding of the more complex aspects of the procedure is based on a prerequisite familiarity with multivariate analysis of variance, canonical correlation, and discriminant analysis. Texts on multivariate analysis (e.g., Bock 1975, Cooley and Lohnes 1971, Harris 1975, Tatsuoka 1971) cover these topics in detail. An excellent one chapter introduction can be found in Jones (1966).

The similarity of univariate and multivariate analysis of covariance can be seen in Table 11.1 for the reader familiar with matrix algebra. Notice that the null hypothesis associated with MANCOVA is similar to the hypothesis associated with univariate ANCOVA, but adjusted population means ($\mu_{j\,\text{adj}}$) are replaced with adjusted population centroids or multivariate means ($\boldsymbol{\mu}_{j\,\text{adj}}$). The adjusted population centroid is a vector of adjusted population means. For example, if p dependent variable adjusted means are obtained for population j, the centroid is

$$
\boldsymbol{\mu}_{j\,\text{adj}} = \begin{bmatrix} \mu_j^{(1)}{}_{\text{adj}} \\ \mu_j^{(2)}{}_{\text{adj}} \\ \vdots \\ \mu_j^{(p)}{}_{\text{adj}} \end{bmatrix}
$$

If several populations are involved, $p\mu$ values exist for each population and it can be seen that each population has a centroid. Thus the test of the hypothesis $\boldsymbol{\mu}_{1\,\text{adj}} = \boldsymbol{\mu}_{2\,\text{adj}} = \cdots = \boldsymbol{\mu}_{J\,\text{adj}}$ is a test that all populations have equal adjusted means on each dependent variable:

$$
\begin{bmatrix} \mu_{1\,\text{adj}}^{(1)} \\ \mu_{1\,\text{adj}}^{(2)} \\ \vdots \\ \mu_{1\,\text{adj}}^{(p)} \end{bmatrix} = \begin{bmatrix} \mu_{2\,\text{adj}}^{(1)} \\ \mu_{2\,\text{adj}}^{(2)} \\ \vdots \\ \mu_{2\,\text{adj}}^{(p)} \end{bmatrix} = \cdots = \begin{bmatrix} \mu_{J\,\text{adj}}^{(1)} \\ \mu_{J\,\text{adj}}^{(2)} \\ \vdots \\ \mu_{J\,\text{adj}}^{(p)} \end{bmatrix}
$$

The purpose of MANCOVA is to test this hypothesis. If the multivariate F test is not significant, the analysis is terminated because there is insufficient data to conclude that the adjusted population centroids are not equal. If the multivariate F is significant, several procedures are available to determine which variable(s) is (are) responsible for the overall hypothesis being rejected.

Table 11.1 Comparison of Basic Steps in ANCOVA and MANCOVA

ANCOVA	MANCOVA
$H_0:\ \mu_{1\ \text{adj}}=\mu_{2\ \text{adj}}=\cdots=\mu_{J\ \text{adj}}$	$H_0:\ \mu_{1\ \text{adj}}=\mu_{2\ \text{adj}}=\cdots=\mu_{J\ \text{adj}}$
Step 1. Compute total sum of squares $$\Sigma y_t^2 = SS_t$$	*Step 1.* Compute total sum of products matrix $\mathbf{T}_{yy}$
Step 2. Compute total residual sum of squares $$\Sigma y_t^2 - (\Sigma xy_t)^2/\Sigma x_t^2 = SSres_t$$	*Step 2.* Compute total residual sum of products matrix $$\mathbf{T}_{yy} - \mathbf{T}_{yx}\mathbf{T}_{xx}^{-1}\mathbf{T}_{xy} = \mathbf{T}_{y\cdot x}$$
Step 3. Compute within-group sum of squares $$\Sigma y_w^2 = SS_w$$	*Step 3.* Compute within-group sum of products matrix $\mathbf{W}_{yy}$
Step 4. Compute within-group residual sum of squares $$\Sigma y_w^2 - (\Sigma xy_w)^2/\Sigma x_w^2 = SSres_w$$	*Step 4.* Compute within-group residual sum of products matrix $$\mathbf{W}_{yy} - \mathbf{W}_{yx}\mathbf{W}_{xx}^{-1}\mathbf{W}_{xy} = \mathbf{W}_{y\cdot x}$$
Step 5. Compute adjusted treatment effect sum of squares $$SSres_t - SSres_w = SSAT$$	*Step 5.* Compute Wilks Lambda for adjusted treatment effects $$\frac{|\mathbf{W}_{y\cdot x}|}{|\mathbf{T}_{y\cdot x}|} = \Lambda$$
Step 6. Compute F ratio for adjusted treatment effects where $$F = \frac{SSAT/J-1}{SSres_w/N-J-C}$$	*Step 6.* Compute multivariate F using Rao's approximation, which is $$F = \left(\frac{1-\Lambda^{1/s}}{\Lambda^{1/s}}\right)\left(\frac{ms-v}{p(J-1)}\right)$$ where $$s = \sqrt{\frac{p^2(J-1)^2-4}{p^2+(J-1)^2-5}}$$ $$m = N-C-1-\frac{p+J}{2}$$ $$v = \frac{p(J-1)-2}{2}$$
Step 7. Evaluate F_{obt} with $F_{(\alpha, J-1, N-J-C)}$	*Step 7.* Evaluate multivariate F with $F_{(\alpha, p(J-1),\ ms-v)}$
Step 8. Carry out multiple comparisons if appropriate	*Step 8.* If multivariate F does not exceed the critical value, retain H_0: $\mu_{1\,\text{adj}}=\mu_{2\,\text{adj}}=\cdots=\mu_{J\,\text{adj}}$ and terminate the analysis. If multivariate F is significant, reject H_0 and proceed to Step 9.

241

Table 11.1 *Continued*

ANCOVA	MANCOVA
	Step 9. Carry out univariate ANCOVA on each dependent variable to identify variable(s) responsible for significant overall multivariate F (if multivariate F is significant but no univariate F is significant, conclude that some linear combination(s) of dependent variables differentiates groups even though no single dependent variable does)
	Step 10. Carry out multiple ANCOVAs by employing X and first dependent variable as covariates; then employ X and second dependent variable as covariates, and so on, to examine unique contribution (if any) of each dependent variable in separating treatment groups (if relevant).
	Step 11. Carry out multiple comparison tests on each dependent variable if appropriate

The simplest procedure is to compute a univariate F test on each dependent variable.

After these univariate tests have been computed, it may be of interest to compute multiple ANCOVA using the original covariate X *and* one or more of the dependent variables as multiple covariates to evaluate the treatment effects that are independent of both X and certain dependent variables. This procedure is described in the example that follows.

11.4 COMPUTATIONAL EXAMPLE

Suppose that two dependent variables were employed in the behavioral-objectives study described in Chapter 3 rather than one (the hypothetical data are

Table 11.2 Raw Data and Basic Matrices for MANCOVA Examples

Group 1			Group 2			Group 3		
Y_1	Y_2	X	Y_1	Y_2	X	Y_1	Y_2	X
15	1	29	20	2	22	14	1	33
19	3	49	34	3	24	20	5	45
21	5	48	28	2	49	30	6	35
27	4	35	35	4	46	32	6	39
35	6	53	42	5	52	34	5	36
39	6	47	44	4	43	42	7	48
23	5	46	46	5	64	40	7	63
38	6	74	47	5	61	38	6	57
33	6	72	40	4	55	54	7	56
50	7	67	54	6	54	56	7	78

presented in Table 11.2). The Y_1 values are scores on a biology achievement test, the Y_2 values are scores measuring interest in the biological sciences, and the X values are scores on an aptitude test. The problem is to test the hypothesis that the three adjusted population centroids are equal. That is, H_0: $\mu_{1\ adj} = \mu_{2\ adj} = \mu_{3\ adj}$. The preliminary starting point for MANCOVA is the computation of the total and within-group deviation sum of squares and sum of products supermatrices T and W. The total supermatrix is of the following form:

$$\mathbf{T} = \left[\begin{array}{c|c} \mathbf{T}_{yy} & \mathbf{T}_{yx} \\ \hline \mathbf{T}_{xy} & \mathbf{T}_{xx} \end{array}\right]$$

T is a symmetric matrix of order $(p+C) \times (p+C)$. That is, the number of rows (or columns) is equal to the number of dependent variables plus the number of covariates C. The submatrices $\mathbf{T}_{xx}$, $\mathbf{T}_{yx}$, $\mathbf{T}_{xy}$, and $\mathbf{T}_{yy}$ are total-deviation sum of squares and sum of products matrices (generally simply called *sum of products matrices*) associated with the variables indicated by the subscripts.

The supermatrix W is also symmetric and of the order $(p+C) \times (p+C)$. It differs from T only in that it is based on pooled within-group rather than total-deviation sums of squares and products:

$$\mathbf{W} = \left[\begin{array}{c|c} \mathbf{W}_{yy} & \mathbf{W}_{yx} \\ \hline \mathbf{W}_{xy} & \mathbf{W}_{xx} \end{array}\right]$$

and it is understood that $\mathbf{W} = \mathbf{W}_1 + \mathbf{W}_2 + \cdots + \mathbf{W}_J$, where the subscripts refer to groups 1 through J.

The $\mathbf{T}$ and $\mathbf{W}$ supermatrices for the example data are presented in the matrices that follow.

$$
\mathbf{T} = \quad
\begin{array}{c}
y_1 \\ y_2 \\ x
\end{array}
\begin{array}{ccc}
y_1 & y_2 & x \\
\left[\begin{array}{cc|c}
3956 & 411 & 3022 \\
411 & 89.47 & 466.33 \\
\hline
3022 & 466.33 & 5826.67
\end{array}\right]
\end{array}
$$

$$
\mathbf{W_1} = \quad
\begin{array}{c}
y_1 \\ y_2 \\ x
\end{array}
\begin{array}{ccc}
y_1 & y_2 & x \\
\left[\begin{array}{cc|c}
1064 & 150 & 1003 \\
150 & 28.9 & 183 \\
\hline
1003 & 183 & 2014
\end{array}\right]
\end{array}
$$

$$
\mathbf{W_2} = \quad
\begin{array}{c}
y_1 \\ y_2 \\ x
\end{array}
\begin{array}{ccc}
y_1 & y_2 & x \\
\left[\begin{array}{cc|c}
896 & 113 & 911 \\
113 & 16 & 119 \\
\hline
911 & 119 & 1798
\end{array}\right]
\end{array}
$$

$$
\mathbf{W_3} = \quad
\begin{array}{c}
y_1 \\ y_2 \\ x
\end{array}
\begin{array}{ccc}
y_1 & y_2 & x \\
\left[\begin{array}{cc|c}
1576 & 176 & 1338 \\
176 & 30.01 & 146 \\
\hline
1338 & 146 & 1888
\end{array}\right]
\end{array}
$$

$$
\mathbf{W} = \mathbf{W_1} + \mathbf{W_2} + \mathbf{W_3} = \quad
\begin{array}{c}
y_1 \\ y_2 \\ x
\end{array}
\begin{array}{ccc}
y_1 & y_2 & x \\
\left[\begin{array}{cc|c}
3536 & 439 & 3252 \\
439 & 75 & 448 \\
\hline
3252 & 448 & 5700
\end{array}\right]
\end{array}
$$

Since the basic $\mathbf{T}$ and $\mathbf{W}$ matrices are now available, the nine steps described in Table 11.1 can be carried out.

1. The total sum of products matrix $\mathbf{T}_{yy}$ is found in the upper left quadrant of the supermatrix $\mathbf{T}$.

$$
\mathbf{T}_{yy} = \quad
\begin{array}{c}
y_1 \\ \\ y_2
\end{array}
\begin{array}{cc}
y_1 & y_2 \\
\left[\begin{array}{cc}
3956 & 411 \\
\\
411 & 89.47
\end{array}\right]
\end{array}
$$

2. The total residual sum of products matrix is

$$
\mathbf{T}_{y \cdot x} = \mathbf{T}_{yy} - \mathbf{T}_{yx}\mathbf{T}_{xy}^{-1}\mathbf{T}_{xy}
$$

$$
= \begin{bmatrix} 3956 & 411 \\ 411 & 89.47 \end{bmatrix} - \begin{bmatrix} 3022 \\ 466.33 \end{bmatrix} [0.000172] \begin{bmatrix} 3022 & 466.33 \end{bmatrix}
$$

$$
= \begin{bmatrix} 2388.64 & 169.14 \\ 169.14 & 52.15 \end{bmatrix}
$$

3. The within-group sum of products matrix is found in the upper left quadrant of the supermatrix $\mathbf{W}$:

$$\mathbf{W}_{yy} = \begin{array}{cc} & \begin{array}{cc} y_1 & y_2 \end{array} \\ \begin{array}{c} y_1 \\ \\ y_2 \end{array} & \left[\begin{array}{cc} 3536 & 439 \\ \\ 439 & 75 \end{array} \right] \end{array}$$

4. The within-group residual sum of products matrix is

$$\mathbf{W}_{y \cdot x} = \mathbf{W}_{yy} - \mathbf{W}_{yx} \mathbf{W}_{xx}^{-1} \mathbf{W}_{xy}$$

$$= \left[\begin{array}{cc} 3536 & 439 \\ 439 & 75 \end{array} \right] - \left[\begin{array}{c} 3252 \\ 448 \end{array} \right] [0.000175] \quad [3252 \quad 448]$$

$$= \left[\begin{array}{cc} 1680.65 & 183.40 \\ 183.40 & 39.79 \end{array} \right]$$

5. Wilks Λ is computed.

$$\frac{|\mathbf{W}_{y \cdot x}|}{|\mathbf{T}_{y \cdot x}|} = \frac{33233.83}{95954.13} = 0.34635 = \Lambda$$

6. The F_{obt} is

$$\left(\frac{1 - (0.34635)^{1/2}}{(0.34635)} \right) \left(\frac{[(25.5)2] - 1}{2(2)} \right) = 8.74$$

where

$$s = \sqrt{\frac{2^2(3-1)^2 - 4}{2^2 + (3-1)^2 - 5}} = 2$$

$$m = 30 - 1 - 1 - \frac{5}{2} = 25.5$$

$$v = \frac{2(2) - 2}{2} = 1$$

7. Using $\alpha = .05$, the critical value of F is

$$F_{(.05, p(J-1), ms-v)} = F_{(.05, 4, 50)} = 2.57$$

8. Since the obtained F exceeds the critical value of F, we conclude that the three adjusted population centroids are not equal. That is, we reject

$$H_0 \colon \mu_{1 \text{ adj}} = \mu_{2 \text{ adj}} = \mu_{3 \text{ adj}}$$

9. Separate univariate ANCOVA F tests are then carried out on the two dependent variables. These tests can be carried out with very little additional computation if the $\mathbf{T}_{y \cdot x}$ and $\mathbf{W}_{y \cdot x}$ matrices, which were employed in the overall MANCOVA F test, are available. These matrices are

$$\mathbf{T}_{y \cdot x} = \begin{bmatrix} 2388.64 & 169.14 \\ 169.14 & 52.15 \end{bmatrix}$$

$$\mathbf{W}_{y \cdot x} = \begin{bmatrix} 1680.65 & 183.40 \\ 183.40 & 39.79 \end{bmatrix}$$

The element in the first row and first column of the $\mathbf{T}_{y \cdot x}$ matrix is the SSres$_t$ for the first dependent variable. The corresponding element in $\mathbf{W}_{y \cdot x}$ is SSres$_w$ for the first dependent variable. Hence the ANCOVA F for the first dependent variable is easily computed because SSres$_t$ $-$ SSres$_w$ = SS adjusted treatment effects. In this case $2388.64 - 1680.65 = 707.99$, and the F test is as follows:

Source	SS	df	MS	F
Adjusted treatments	707.99	2	354	5.48
Res$_w$	1680.65	26	64.64	
Res$_t$	2388.64	28		

The ANCOVA F test for the second dependent variable is based on the values found in the second row and second column of the $\mathbf{T}_{y \cdot x}$ and $\mathbf{W}_{y \cdot x}$ matrices. These values are the SSres$_t$ and SSres$_w$ for Y_2. The difference $52.15 - 39.79 = 12.36$ is the SS adjusted treatment effects for the second dependent variable. The univariate F test for this variable is as follows:

Source	SS	df	MS	F
Adjusted treatments	12.36	2	6.18	4.04
Res$_w$	39.79	26	1.53	
Res$_t$	52.15	28		

The critical value of F is 3.37 for $\alpha = .05$.

10. Additional analyses of covariance may be of interest for the example data. Recall that two dependent variables, achievement and interest, were employed along with the covariate aptitude. The overall multivariate analysis of covariance (MANCOVA) and the two univariate tests support the notion that the three treatments are not equally effective in terms of achievement or interest in the biological sciences. (Pairwise comparisons on both dependent variables have not been computed here, but they generally should be reported.) The researcher may now question whether treatment differences on achievement exist that are independent of both

aptitude and interest. If both X (aptitude) and Y_2 (interest) are conceptualized as covariates and achievement is viewed as the only dependent variable, we have a multiple covariance problem. That is, achievement is the dependent variable, and aptitude and interest are held constant statistically by employing them as multiple covariates. The results of this analysis are as follows:

Source	SS	df	MS	F
Adjusted treatment	1004.8	2	502.4	15.04
Multiple res$_w$	835.26	25	33.41	
Multiple res$_t$	1840.06	27		

The critical value of F for $\alpha = .05$ is 3.39.

It appears that there are treatment effects on achievement that are independent of interest and aptitude.

The researcher may also be interested in evaluating whether the treatment effects on interest are independent of both aptitude and achievement. The ANCOVA previously carried out on interest with aptitude as the covariate indicated that there are treatment effects on interest that are independent of aptitude. It is not unreasonable, however, to hypothesize that differences in interest can be explained by a combination of aptitude and achievement scores. This hypothesis can be tested by a multiple ANCOVA in which interest is the dependent variable and aptitude and achievement are the multiple covariates. The summary of this analysis is as follows:

Source	SS	df	MS	F
Adjusted treatment effect	20.40	2	10.20	12.90
Multiple res$_w$	19.77	25	0.79	
Multiple res$_t$	401.7	27		

Since the critical value of F is 3.39 for $\alpha = .05$ we conclude that treatment effects on interest exist that are independent of both aptitude and achievement. There is an issue of interpretation, however, that should be kept in mind when ANCOVA is employed as described in this section. The conditional statements that are made concerning the treatment effects (e.g., effects on interest independent of aptitude and achievement) hold only for the obtained or fallible covariate measures. Although this procedure is frequently recommended (e.g., Bock and Haggard 1968), measurement-error problems (discussed in Chapters 6, 7, and 14) should be considered because the differences between adjusted means are *not* generally the differences that would be expected if true (completely reliable) covariate measures were available. The problem here is similar to the problem of applying ANCOVA to nonequivalent groups.

If the treatments affect the variables employed as covariates, the expected difference between the adjusted means is not zero under a true null hypothesis unless the covariates are perfectly reliable. In the preceding example the use of aptitude and achievement as covariates and interest as the dependent variable has resulted in a significant F value. Only when there is no effect on aptitude and achievement can this F be interpreted as testing the hypothesis of no treatment effects on interest. If the covariate is affected by the treatment, the use of true-score ANCOVA (see Chapter 14) is recommended if a less biased test of adjusted treatments effects is desired. The procedures of the present section are acceptable if it is understood that the conditional statements refer to the covariate scores as measured, not true scores.

11.5 MANCOVA THROUGH MULTIPLE REGRESSION ANALYSIS—TWO GROUPS ONLY

The computation of MANCOVA for the two-group situation can be carried out by using a multiple regression analysis computer program. Unlike the univariate case in which the regression approach can easily be applied to any number of groups, the multivariate case is conveniently analyzed through regression analysis only when two groups are involved. The similarity of the two-group MANCOVA procedure and the univariate ANCOVA procedure (described in Chapter 4) is described in Table 11.3.

Table 11.3 Comparison of Steps Involved in Computing ANCOVA and MANCOVA through Multiple Linear Regression Procedures

One-factor ANCOVA—J Groups	One-factor MANCOVA—Two Groups
Step 1. Construct $J-1$ dummy variables indicating group membership	*Step 1.* Construct one dummy variable indicating group membership
Step 2. Regress Y on $J-1$ dummy variables and covariate(s) to obtain $R^2_{yD,X}$	*Step 2.* Regress the dummy variable on all p dependent variables and covariate(s) to obtain $R^2_{dY,X}$
Step 3. Regress Y on covariate(s) to obtain R^2_{yX}	*Step 3.* Regress dummy variable on covariate(s) to obtain R^2_{dX}
Step 4. Compute F by using	*Step 4.* Compute multivariate F by using
$$\frac{\left(R^2_{yD,X}-R^2_{yX}\right)/J-1}{\left(1-R^2_{yD,X}\right)/N-J-C}=F$$	$$\frac{\left(R^2_{dY,X}-R^2_{dX}\right)/p}{\left(1-R^2_{dY,X}\right)/N-p-C-1}=F$$
Step 5. Evaluate F_{obt} with $F_{(\alpha,J-1,N-J-C)}$	*Step 5.* Evaluate multivariate F_{obt} with $F_{(\alpha,p,N-p-C-1)}$

Computational Example

The MANCOVA steps described in Table 11.3 have been applied to the following data:

	D	Y_1	Y_2	Y_3	Y_4	Y_5	X
	1	1	7	10	10	17	22
	1	1	6	7	6	14	17
Group 1	1	5	3	13	12	20	14
	1	7	3	5	6	18	14
	1	1	1	7	6	18	9
	0	10	13	15	17	10	24
	0	12	13	18	12	13	18
Group 2	0	9	13	18	14	10	14
	0	10	10	18	13	8	15
	0	6	6	11	13	16	6

1. The dummy variable is constructed by assigning a "one" to all subjects in the first group and a "zero" to all subjects in the second group. Column 1 is the dummy variable, columns 2 through 6 are dependent variables, and column 7 is the covariate.
2. Column 1, the dummy variable, is regressed on columns 2 through 7 to obtain $R_{dY,X}^2 = 0.96943$.
3. Column 1 is regressed on column 7, the covariate, to obtain $R_{dX}^2 = 0.00038$.
4. The multivariate F is

$$\frac{(0.96943 - 0.00038)/5}{(1 - 0.96943)/3} = \frac{0.19381}{0.0191} = 19.02$$

5. Evaluate F_{obt} with $F_{(\alpha, 5, 10 - 5 - 1 - 1)}$.

If $\alpha = .05$, then $F_{(5,3)} = 9.01$ and the null hypothesis is rejected. Univariate ANCOVA on each dependent variable would be appropriate at this stage.

11.6   ISSUES ASSOCIATED WITH USE OF BONFERRONI F AND MANCOVA PROCEDURES

There is little agreement among statisticians concerning the most appropriate method of dealing with experiments having multiple dependent variables. Some prefer to disregard the problem of high family error rate and analyze each dependent variable as if the others do not exist. Others employ MANCOVA and associated procedures such as canonical analysis and discriminant analysis, even with small samples. The problems of making a choice among the various multivariate test strategies currently available can sometimes overwhelm the most sophisticated researchers. The present author considers the Bonferroni F procedure presented in this chapter to be generally more appropriate for typical experimental research than are the more complex multivariate methods. This recommendation is based on a consideration of three factors: (1) the relevant dependent variables, (2) statistical power, and (3) computational simplicity.

Relevant Dependent Variables

Most experimenters select dependent variables on the basis of information that is available before the experiment is carried out. A review of previous research and a logical analysis of available response measures will generally yield a set of variables of interest to the experimenter a priori. If the measures have been carefully chosen on the basis of the information that each should provide concerning the treatment effects, there is probably little to be gained by employing a complex multivariate technique. The research question is "On which dependent variables (if any) are treatment effects significant?" The Bonferroni procedure answers this question. The MANCOVA multivariate F test answers the somewhat different question, "Are treatment effects significant on an optimum linear combination of the dependent variables?" The experimenter may have no particular interest in the test on the linear combination if the dependent variables were carefully chosen to yield the information of experimental interest. A problem associated with the difference in the hypotheses tested by the MANCOVA and Bonferroni procedures may arise when MANCOVA F tests are followed up with univariate F tests on each dependent variable.

Suppose that two dependent variables are employed in a two-group experiment and the multivariate F is significant. It is quite possible for both univariate F tests to be nonsignificant. In this case it is concluded that treatment effects have been detected, but the statement concerning the existence of effects does not apply to either of the dependent variables! Rather, it is concluded that an optimum linear combination of the two dependent variables results in a composite variable that *does* reflect the treatment effect. The linear combination responsible for the significant multivariate F can be specified (it is known as the discriminant function), but the experimenter may not be particularly interested in it.

Undoubtedly, there are situations in which preliminary studies are run when little is known about the characteristics of the response measures. Much applied research must be carried out by using variables chosen for convenience rather than for maximum information yield. These situations appear to be appropriate for the application of MANCOVA and discriminant analysis because the nature of the variables can be clarified through such procedures. In general, however, it seems preferable to investigate the characteristics of the response measures before carrying out an experiment.

Statistical Power

The sample sizes employed in much experimental research are often relatively small. If many dependent variables are employed, the MANCOVA F test may have very low power. When small sample sizes (say, under 30) are employed, it is not unusual to encounter a nonsignificant multivariate F but significant univariate F values. This is no problem if, for example, an experiment with $\alpha = .05$ and 20 dependent variables yields a nonsignificant MANCOVA F and one significant univariate F. In this case we expect to obtain one significant univariate F by chance, and the nonsignificant MANCOVA F suggests that there is insufficient information to conclude otherwise. That is, the multivariate F protects us from falsely concluding that treatment effects on one dependent variable are present in the population. Unfortunately, however, there are situations in which the use of the MANCOVA F as a preliminary test can result in a serious logic problem. Suppose, in an experiment with 20 dependent variables, that 15 or 20 univariate F values are significant at $\alpha = .05$ but that the multivariate F is not. This disturbing situation can occur when many dependent variables are employed because the power of multivariate tests tends to decrease more rapidly with reductions in sample size than is the case with univariate tests. The Bonferroni F procedure circumvents this whole problem.

As an example of this problem in the case of ANOVA, both multivariate and univariate analyses were applied to the data in Table 11.4. The multivariate test is nonsignificant if $\alpha = .05$ is used, but each univariate test is

Table 11.4   Fictitious Data for a Two-group Experiment with Five Dependent Variables

	Subject	Y_1	Y_2	Y_3	Y_4	Y_5
	1	1	7	10	10	17
	2	1	6	7	6	14
Group 1	3	5	3	13	12	20
	4	7	3	5	6	18
	5	1	1	7	6	18
	6	10	13	15	17	10
	7	12	13	18	12	13
Group 2	8	9	13	18	14	10
	9	10	10	18	13	8
	10	6	6	11	13	16

Multivariate $F_{obt} = 4.38$	Multivariate critical value $= 6.26$
Univariate F_{obt} on $Y_1 = 16.00$	Univariate critical value $= 5.32$
Univariate F_{obt} on $Y_2 = 15.81$	Univariate critical value $= 5.32$
Univariate F_{obt} on $Y_3 = 14.96$	Univariate critical value $= 5.32$
Univariate F_{obt} on $Y_4 = 14.38$	Univariate critical value $= 5.32$
Univariate F_{obt} on $Y_5 = 12.33$	Univariate critical value $= 5.32$
	Bonferroni F critical value $= 11.26$

significant. The univariate tests are also significant with the use of the Bonferroni procedure. It is suggested that this example of an inconsistency in the multivariate and Bonferroni conclusions be resolved by accepting the result of the Bonferroni procedure.

Even though the power of the multivariate test is generally low with small sample sizes, it should be pointed out that the joint distribution of the dependent variables *may* be such that the multivariate test is very sensitive. It can be shown that it is possible, with certain distributions, for small samples to yield a very large multivariate F value and very small univariate F values (Huitema 1974).

Computational Simplicity

Obviously, the MANCOVA procedure is more complex than the Bonferroni F. The Bonferroni procedure can be carried out given a multiple regression computer program and the Bonferroni F tables. If more than two groups are involved, the MANCOVA procedure involves somewhat more computation, but computational complexity should be a minor consideration in the choice among competing procedures. Multivariate analysis computer programs are widely available, but the documentation associated with almost all of them is very poor.

11.7 SUMMARY

Studies that contain more than one dependent variable can be analyzed in several different ways. One approach is to carry out conventional F tests on each dependent variable. The basic problem with this approach is that the probability of making a Type I error in the whole family of F tests is much greater than the α associated with each F test if there are many dependent variables. A very simple way to control the error rate for the whole family of F tests is to employ the Bonferroni F statistic as the critical value for each separate F test. Another method of controlling the error rate is to employ a multivariate analysis test. This method considers the number of dependent variables and the relationship among all the variables in the overall multivariate F test. Two disadvantages of the multivariate approach are interpretation difficulties and, in many cases, low power with small sample sizes.

If ANCOVA is used to clarify the nature of a treatment effect by utilizing some dependent variables as covariates, caution should be exercised in the interpretation of the adjusted effects. This usage of ANCOVA is one in which the covariate is very likely to be affected by the treatment. When the population covariate means are not equal, the expected adjusted mean difference is not zero when there are no treatment effects if the covariate is imperfectly measured. True-score ANCOVA should be considered in this case.

PART THREE
ALTERNATIVES TO ANALYSIS OF COVARIANCE

CHAPTER TWELVE
Rank Analysis of Covariance

12.1 INTRODUCTION

There are two situations in which the method described in this chapter, rank analysis of covariance (Quade 1967), should be considered. The first situation is one in which the data are originally measured in the form of ranks. Suppose that 30 subjects are rank ordered on a submissiveness dimension in a study of the effects of assertiveness training on the behavior of participants in such a program. They are then randomly assigned to either an assertiveness training group or a control group. After the training is completed, the submissiveness behavior of all subjects is again ranked. Both the pretest (X) and the posttest (Y) measures are in the form of ranks; rank ANCOVA is an appropriate method for analysis of these data.

The second situation in which rank analysis of covariance should be considered is when the original data, which are not in the form of ranks, are distributed in such a way that the assumptions of normality, homogeneity of conditional variances, and/or linearity of regression slopes are seriously violated. Extreme violations of these assumptions, especially in the case of unequal sample sizes, can lead to biased F ratios. Rank analysis of covariance is appropriate if X and/or $Y|X$ are not normally distributed, if the conditional variances are heterogeneous, and/or if the relationship between X and Y is nonlinear but monotonic. This procedure is also recommended as an alternative to the well-known Friedman test when the blocking variable is not qualitative.

The choice between rank ANCOVA and conventional parametric ANCOVA in the situation described in the preceding paragraph is not always clear-cut. If the sample sizes are equal, violations of the assumptions of normality and homogeneity of conditional variances are not likely to have drastic effects on parametric ANCOVA, but there are additional considerations. As is the case with other nonparametric procedures, rank ANCOVA does not test the same hypothesis as parametric ANCOVA when the assumptions mentioned previously are seriously violated. Rather than testing the equality of adjusted means, rank ANCOVA tests the hypothesis that the conditional population distributions, which are of unspecified form, are identical. Estimates of adjusted population means are not provided.

It is reasonable to compute both rank ANCOVA and parametric ANCOVA when the normality and homogeneity of conditional variance assumptions are seriously violated. Even though parametric ANCOVA may yield a biased F test (and biased confidence intervals), the estimates of the adjusted means will not, in general, be biased. The descriptive information provided by these estimates will often be more relevant in interpreting the outcome of an experiment than will the hypothesis tests. If rank ANCOVA is also computed, valid inferential statements can be made about whether the distributions are identical. It is appropriate to report the information gleaned from both analyses.

12.2   COMPUTATION: ANY NUMBER OF GROUPS, ONE COVARIATE

This computation consists of the following steps:

Step 1. If the original data are not in rank form, transform scores on the covariate and the dependent variable to ranks. Rank all N scores from the smallest first. That is, the lowest score is assigned the rank 1, the second lowest score is assigned rank 2, and so on. If ties occur, assign average ranks.

Step 2. Transform ranks on X and Y to deviation ranks. A deviation rank is simply the rank minus the mean rank; in other words:

$$X_{\text{rank}} - \overline{X}_{\text{rank}} = x_{\text{rank}} \text{ and } Y_{\text{rank}} - \overline{Y}_{\text{rank}} = y_{\text{rank}}$$

Step 3. Regress all y_{rank} values on x_{rank} values by using an ordinary linear regression program or run these x and y values through a program that provides Pearson correlation coefficients. The output described as the slope with the regression program or the correlation coefficient with the correlation program will yield the Spearman rank-order correlation coefficient r_S. (The Spearman is equivalent to the Pearson correlation applied to ranks.)

Step 4. Multiply each x_{rank} by the Spearman correlation of step 3 to obtain the estimated deviation rank on Y (i.e., $\hat{y}_{\text{rank}}$):

$$r_S(x_{\text{rank}}) = \hat{y}_{\text{rank}}$$

Step 5. Subtract each estimated deviation rank on Y from the corresponding observed deviation rank on Y to obtain the residual Z for each subject;

$$Z_{ij} = y_{\text{rank } ij} - \hat{y}_{\text{rank } ij}$$

Many multiple regression computer programs yield the residual $Y - \hat{Y}$, where

$\hat{Y}$ is based on the multiple regression equation

$$b_0 + b_1 X_1 + b_2 X_2 + \cdots + b_m X_m.$$

As long as all X and Y variables are transformed to deviation ranks before entering the data, the residuals labeled $Y - \hat{Y}$ on the typical computer output are the Z residuals required for rank analysis of covariance.

Step 6.   Sum the residuals associated with each J group separately (i.e., obtain $\Sigma_1^n Z_i$ for each group).

Step 7.   Square each sum of residuals.

Step 8.   Divided each squared sum of residuals by n_j, which is the sample size associated with the corresponding sum. That is, compute $(\Sigma_1^n Z_{ij})^2 / n_j$ for each group.

Step 9.   Sum all J values obtained in step 8. This sum is the numerator or treatment sum of squares. This sum is written as follows:

$$\Sigma_1^J \frac{(\Sigma_1^n Z_{ij})^2}{n_j}$$

Step 10.   Square each Z value obtained in step 5. This quantity, $\Sigma_1^J \Sigma_1^n Z_{ij}^2$, is the total residual sum of squares. This sum can also be computed using the formula $(1 - r_S^2)\frac{1}{12} N(N^2 - 1)$ when there are no tied scores.

Step 11.   The result of step 10 minus the result of step 9 yields the denominator or error sum of squares.

Step 12.   The F statistic is computed by using

$$\frac{\text{Treatment sum of squares}/J - 1}{\text{Error sum of squares}/N - J} = F$$

The obtained F is evaluated with $F_{(\alpha, J-1, N-J)}$. If the obtained F equals or exceeds the critical value, it is concluded that the conditional distribution of Y given X is not the same for all treatment populations. In summary, the rank analysis of covariance is based on the variance ratio indicated in the following table:

Source	SS	df	MS	F
Treatment	$\Sigma_1^J [(\Sigma_1^n Z_{ij})^2 / n_j]$	$J - 1$	$\text{SStr}/J - 1$	MStr/MSE
Error	$\Sigma_1^J \Sigma_1^n Z_{ij}^2 - \Sigma_1^J [(\Sigma_1^n Z_{ij})^2 / n_j]$	$N - J$	$\text{SSE}/N - J$	
Total residual	$\Sigma_1^J \Sigma_1^n Z_{ij}^2$	$N - 1$		

A close look at the summary table will reveal that Quade's procedure is simply a parametric analysis of variance on the residuals Z_{ij}.

Computational Example: Three Groups, One Covariate

The data employed in Chapters 3 and 4 are reanalyzed in this section with the use of rank analysis of covariance. Raw-score data and the computational steps are as follows:

Group 1		Group 2		Group 3	
Y	X	Y	X	Y	X
15	29	20	22	14	33
19	49	34	24	20	45
21	48	28	49	30	35
27	35	35	46	32	39
35	53	42	52	34	36
39	47	44	43	42	48
23	46	46	64	40	63
38	74	47	61	38	57
33	72	40	55	54	56
50	67	54	54	56	78

Step 1. Transform all X scores (combined) to ranks and transform all Y scores (combined) to ranks.

Group 1		Group 2		Group 3	
Y_{rank}	X_{rank}	Y_{rank}	X_{rank}	Y_{rank}	X_{rank}
2	3	4.5	1	1	4
3	16.5	13.5	2	4.5	10
6	14.5	9	16.5	10	5.5
8	5.5	15.5	11.5	11	8
15.5	19	22.5	18	13.5	7
19	13	24	9	22.5	14.5
7	11.5	25	26	20.5	25
17.5	29	26	24	17.5	23
12	28	20.5	21	28.5	22
27	27	28.5	20	30	30

Step 2. Transform ranks on X and Y to deviation ranks by subtracting the mean rank (15.5) from each rank.

Group 1		Group 2		Group 3	
y_{rank}	x_{rank}	y_{rank}	x_{rank}	y_{rank}	x_{rank}
-13.5	-12.5	-11.0	-14.5	-14.5	-11.5
-12.5	1.0	-2.0	-13.5	-11.0	-5.5
-9.5	-1.0	-6.5	1.0	-5.5	-10.0
-7.5	-10.0	0.0	-4.0	-4.5	-7.5
0.0	3.5	7.0	2.5	-2.0	-8.5
3.5	-2.5	8.5	-6.5	7.0	-1.0
-8.5	-4.0	9.5	10.5	5.0	9.5
2.0	13.5	10.5	8.5	2.0	7.5
-3.5	12.5	5.0	5.5	13.0	6.5
11.5	11.5	13.0	4.5	14.5	14.5

Step 3. Regress all 30 y_{rank} values on x_{rank} values ($r = b_1 = r_S = 0.66$).

Step 4. Compute estimated deviation ranks on Y by using $0.66(x_{rank}) = \hat{y}_{rank}$.

Step 5. Compute residuals Z_{ij} by using $Z_{ij} = y_{rank\ ij} - 0.66(x_{rank\ ij})$.

	Subject	Observed y_{rank}	(Step 4) Estimated y_{rank} $= 0.66(x_{rank})$	(Step 5) Residual Z	(Step 6) Sum of Residuals for Each Group
	1	−13.5	$0.66(−12.5) = −8.25$	−5.25	
	2	−12.5	$0.66(1.0) = 0.66$	−13.16	
	3	−9.5	$0.66(−1.0) = −0.66$	−8.84	
	4	−7.5	$0.66(−10.0) = −6.66$	−0.90	
Group	5	0.0	$0.66(3.5) = 2.31$	−2.31	−45.92
1	6	3.5	$0.66(−2.5) = −1.65$	5.15	
	7	−8.5	$0.66(−4.0) = −2.64$	−5.86	
	8	2.0	$0.66(13.5) = 8.91$	−6.91	
	9	−3.5	$0.66(12.5) = 8.25$	−11.75	
	10	11.5	$0.66(11.5) = 7.59$	3.91	
	11	−11.0	$0.66(−14.5) = −9.57$	−1.43	
	12	−2.0	$0.66(−13.5) = −8.91$	6.91	
	13	−6.5	$0.66(1.0) = 0.66$	−7.16	
	14	0.0	$0.66(−4.0) = −2.64$	2.64	
Group	15	7.0	$0.66(2.5) = 1.65$	5.35	37.96
2	16	8.5	$0.66(−6.5) = −4.29$	12.79	
	17	9.5	$0.66(10.5) = 6.93$	2.57	
	18	10.5	$0.66(8.5) = 5.61$	4.89	
	19	5.0	$0.66(5.5) = 3.63$	1.37	
	20	13.0	$0.66(4.5) = 2.97$	10.03	
	21	−14.5	$0.66(−11.5) = −7.59$	−6.91	
	22	−11.0	$0.66(−5.5) = −3.63$	−7.37	
	23	−5.5	$0.66(−10.0) = −6.60$	1.10	
	24	−4.5	$0.66(−7.5) = −4.95$	0.45	
Group	25	−2.0	$0.66(−8.5) = −5.61$	3.61	7.96
3	26	7.0	$0.66(−1.0) = −0.66$	7.66	
	27	5.0	$0.66(9.5) = 6.27$	−1.27	
	28	2.0	$0.66(7.5) = 4.95$	−2.95	
	29	13.0	$0.66(6.5) = 4.29$	8.71	
	30	14.5	$0.66(14.5) = 9.57$	4.93	

Step 6. Sum the residuals associated with each of the J groups.

Step 7. Square each sum obtained in step 6.

$$(\Sigma_1^n Z_{ij})^2$$

Group 1	$(−45.92)^2 = 2108.65$
Group 2	$(37.96)^2 = 1440.96$
Group 3	$(7.96)^2 = 63.36$

Step 8. Divide each squared sum obtained in step 7 by n_j.

$$\frac{\left(\Sigma_1^n Z_{ij}\right)^2}{n_j}$$

Group 1	$2108.65/10 = 210.865$
Group 2	$1440.96/10 = 144.096$
Group 3	$63.36/10 = 6.336$

Step 9. Sum the values obtained in step 8 to obtain the treatment sum of squares:

$$\sum_1^J \frac{\left(\sum_1^n Z_{ij}\right)^2}{n_j} = 210.865 + 144.096 + 6.336 = 361.3$$

Step 10. Compute the total residual sum of squares:

$$\sum_1^J \sum_1^n Z_{ij}^2 = 1265.9$$

Step 11.. Subtract the result of step 9 from the result of step 10 to obtain the error sum of squares:

$$1265.9 - 361.3 = 904.6$$

Step 12. Compute F.

Source	SS	df	MS	F
Treatment	361.3	2	180.7	5.39
Error	904.6	27	33.5	
Total residual	1265.9	29		

The critical value of F using $\alpha = .05$ is 3.35.

It is concluded that the conditional distribution of Y given X is not the same for the three treatment populations. That is, the $Y|X$ values in some treatment populations tend to be greater than in others.

12.3 MULTIPLE COMPARISON TESTS

Procedures for testing individual comparisons are not described in Quade's 1967 paper. The procedure suggested here is an analogue to the protected LSD procedure for parametric ANCOVA. The characteristics of this procedure have not been investigated, but the probability of making a Type I error (experimentwise) is not greater than the experimentwise error associated with the overall F test. If the overall rank ANCOVA F test is significant, it is appropriate to compute pairwise tests by the following steps:

Step 1.   Compute the mean deviation rank $\bar{y}$ for each group.
Step 2.   Compute pairwise differences among all $\bar{y}$.
Step 3.   Divide each difference $\bar{y}_i - \bar{y}_j$ by

$$\sqrt{\left[\frac{1}{n_i} + \frac{1}{n_j}\right]\text{MSE}}$$

where MSE is the mean square error of step twelve in Section 12.2.
Step 4.   Evaluate the result of step 3 with $t_{(\alpha, N-J)}$.

Computational Example

The overall F obtained in the example of Secfion 12.2 exceeds the critical value, and the hypothesis of identical $Y|X$ distributions for the three populations is rejected. Since this test is significant, all pairwise tests are justified. These tests are carried out by the following steps:

Step 1.

$$\bar{y}_1 = -45.92/10 = -4.592$$
$$\bar{y}_2 = 37.96/10 = 3.796$$
$$\bar{y}_3 = 7.96/10 = 0.796$$

Step 2.

$$\bar{y}_1 - \bar{y}_2 = -8.388$$
$$\bar{y}_1 - \bar{y}_3 = -5.388$$
$$\bar{y}_2 - \bar{y}_3 = 3.000$$

Step 3.   Group 1 versus group 2

$$\frac{-8.388}{\sqrt{\left[\frac{1}{10} + \frac{1}{10}\right]33.5}} = \frac{-8.388}{2.592} = -3.24 = t$$

Group 1 versus group 3

$$\frac{-5.388}{\sqrt{\left[\frac{1}{10} + \frac{1}{10}\right]33.5}} = \frac{-5.388}{2.592} = -2.08 = t$$

Group 2 versus group 3

$$\frac{3.000}{\sqrt{\left[\dfrac{1}{10} + \dfrac{1}{10}\right]33.5}} = \frac{3.000}{2.592} = 1.16 = t$$

Step 4.   The absolute value of each t is compared with the critical value of t, which is 2.052 for $\alpha = .05$ and 27 degrees of freedom. It is concluded that the conditional distribution of Y given X for population 1 differs from the conditional distribution for populations 2 and 3. There are insufficient data to conclude that the conditional distribution for population 2 is not identical to the conditional distribution for population 3.

12.4 MULTIPLE RANK ANALYSIS OF COVARIANCE: ANY NUMBER OF GROUPS, ANY NUMBER OF COVARIATES

Just as parametric ANCOVA can be extended to handle multiple covariates, rank ANCOVA can also be extended to the multiple covariate case. An example was described earlier in which pretest submissiveness rank was the covariate and posttest submissiveness rank was the dependent variable. Suppose that additional pretreatment data in the form of ranks on physical attractiveness are also available. In this case both pretreatment rank and physical attractiveness rank are suitable covariates for multiple rank ANCOVA.

The computational steps for multiple rank ANCOVA are described in the steps that follow. These steps are applied to the three-group, two-covariate data set shown in Table 12.1. The computational procedure differs little from the one-covariate case.

Table 12.1 Data for Three-group, Two-covariate Rank ANCOVA

Sample 1			Sample 2			Sample 3		
Y	X_1	X_2	Y	X_1	X_2	Y	X_1	X_2
16	26	12	44	21	17	17	1	8
60	10	21	67	28	2	28	19	1
82	42	24	87	5	40	105	41	9
126	49	29	100	12	38	149	48	28
137	55	34	142	58	36	160	35	16

Source: Data from Quade (1967).

Step 1. Transform all raw scores to ranks.

	Sample 1			Sample 2			Sample 3	
Y_{rank}	$X_{1\ rank}$	$X_{2\ rank}$	Y_{rank}	$X_{1\ rank}$	$X_{2\ rank}$	Y_{rank}	$X_{1\ rank}$	$X_{2\ rank}$
1	7	5	4	6	7	2	1	3
5	3	8	6	8	2	3	5	1
7	11	9	8	2	15	10	10	4
11	13	11	9	4	14	14	12	10
12	14	12	13	15	13	15	9	6

Step 2. Transform all ranks to deviation ranks.

	Sample 1			Sample 2			Sample 3	
y_{rank}	$x_{1\ rank}$	$x_{2\ rank}$	y_{rank}	$x_{1\ rank}$	$x_{2\ rank}$	y_{rank}	$x_{1\ rank}$	$x_{2\ rank}$
-7	-1	-3	-4	-2	-1	-6	-7	-5
-3	-5	0	-2	0	-6	-5	-3	-7
-1	3	1	0	-6	7	2	2	-4
3	5	3	1	-4	6	6	4	2
4	6	4	5	7	5	7	1	-2

Step 3. Regress all y_{rank} values on $x_{1\ rank}$ and $x_{2\ rank}$ values by using an ordinary multiple linear regression program. The multiple Spearman correlation coefficient is $=0.76877$, and the regression equation is

$$0.0+0.58(x_{1\ rank})+0.38(x_{2\ rank})=\hat{y}_{rank}$$

(*Note*: The regression intercept will always equal zero when the data are deviation ranks.)

Step 4. Multiply each x_{rank} by the corresponding regression coefficient to obtain the estimated deviation rank on Y:

$$b_1x_{1\ rank}+b_2x_{2\ rank}+\cdots+b_Cx_{C\ rank}=\hat{y}_{rank}$$

Step 5. Subtract each estimated deviation rank on Y from the corresponding observed deviation rank on Y to obtain the residual Z for each subject. For example, for subject 1, $y_{rank}=-7$, $\hat{y}_{rank}=0.58(-1)+0.38(-3)=-1.72$, and $Z=-5.28$.

	Subject	Observed y_{rank}	(Step 4) Estimated y_{rank}	(Step 5) Residual Z	(Step 6) Sum of Residuals for Each Group
	1	-7	-1.72	-5.28	
	2	-3	-2.90	-0.10	
Sample 1	3	-1	2.12	-3.12	-10.54
	4	3	4.04	-1.04	
	5	4	5.00	-1.00	

Subject	Observed y_{rank}	(Step 4) Estimated y_{rank}	(Step 5) Residual Z	(Step 6) Sum of Residuals for Each Group

	6	−4	−1.54	−2.46	
	7	−2	−2.28	0.28	
Sample 2	8	0	−0.82	0.82	−1.28
	9	1	−0.04	1.04	
	10	5	5.96	−0.96	
	11	−6	−5.96	−0.04	
	12	−5	−4.40	−0.60	
Sample 3	13	2	−0.36	2.36	11.82
	14	6	3.08	2.92	
	15	7	−0.18	7.18	

Step 6. Sum the residuals associated with each J group separately.

$$\sum_{1}^{n} Z_{ij}$$

Sample 1	−10.54
Sample 2	−1.28
Sample 3	11.82

Step 7. Square each sum of residuals.

$$\left(\sum_{1}^{n} Z_{ij} \right)^2$$

Sample 1	$(-10.54)^2 = 111.092$
Sample 2	$(-1.28)^2 = 1.638$
Sample 3	$(11.82)^2 = 139.971$

Step 8. Divide each squared sum of residuals by the corresponding sample size.

$$\frac{\left(\sum_{1}^{n} Z_{ij} \right)^2}{n_j}$$

Sample 1	$111.092/5 = 22.22$
Sample 2	$1.638/5 = 0.33$
Sample 3	$139.971/5 = 27.94$

Step 9. Sum the values obtained in step 8 [i.e., $\sum_{1}^{J}(\sum_{1}^{n}Z_{ij})^2/n_j$]:

$$22.22 + 0.33 + 27.94 = 50.49$$

This is the treatment sum of squares.

Step 10. Compute the total residual sum of squares by squaring each residual Z or by employing the formula $(1-R_S^2)(\frac{1}{12})N(N^2-1)$, where R_S is the multiple Spearman correlation coefficient. For the example data, we find

$$(1-0.591)(1/12)15(225-1)=114.52$$

Step 11. The result of step 10 minus the result of step 9 yields the error sum of squares:

$$114.52-50.49=64.03$$

Step 12. The F statistic is computed by forming the ratio of treatment sum of squares$/(J-1)$ over error sum of squares$/(N-J)$:

$$\frac{50.49/2}{64.03/12}=\frac{23.245}{5.3358}=4.73$$

The obtained F value is compared with $F_{(\alpha,J-1,N-J)}$. If the obtained F equals or exceeds the critical value, it is concluded that the conditional distribution of Y given $X_1,X_2,\ldots,X_C$ is not the same for all treatment populations.

Multiple Comparison Tests

The multiple comparison procedure described in Section 12.3 also applies in the multiple covariate case. The computation of the three pairwise tests on Quade's data proceeds as follows:

Step 1.

$$\bar{y}_1=-10.54/5=-2.11$$
$$\bar{y}_2=-1.28/5=-0.26$$
$$\bar{y}_3=11.82/5=2.36$$

Step 2.

$$\bar{y}_1-\bar{y}_2=-1.85$$
$$\bar{y}_1-\bar{y}_3=-4.47$$
$$\bar{y}_2-\bar{y}_3=-2.62$$

Step 3. Group 1 versus group 2:

$$\frac{-1.85}{\sqrt{[\frac{1}{5}+\frac{1}{5}]5.3358}}=\frac{-1.85}{1.46}=-1.27=t$$

Group 1 versus group 3:

$$\frac{-4.47}{\sqrt{[\frac{1}{5}+\frac{1}{5}]5.3358}} = \frac{-4.47}{1.46} = -3.06 = t$$

Group 2 versus group 3:

$$\frac{2.62}{\sqrt{[\frac{1}{5}+\frac{1}{5}]5.3358}} = \frac{2.62}{1.46} = 1.79 = t$$

Step 4. Since an α of .05 was employed in evaluating the overall F, the same α is employed for the multiple comparison tests. The critical value for each t is

$$t_{(.05, N-J)} = t_{(.05, 12)} = 2.179$$

It is concluded that the conditional distributions of populations 1 and 3 are not identical because the absolute value of the t statistic exceeds the critical value. The hypothesis of identical population distributions is retained for the other comparisons.

12.5 ASSUMPTIONS

The major advantage of Quade's rank ANCOVA procedure over parametric ANCOVA is that the assumptions of the normality of the conditional distributions of Y, the linearity of the regression of Y on X, and the homogeneity of the conditional Y distributions can be relaxed. This does not mean, however, that rank ANCOVA is assumption free. The basic assumptions for rank ANCOVA are:

1. Subjects are randomly and independently assigned to treatment groups.
2. The covariate(s) and the dependent variable have been measured on at least on an ordinal scale; a dichotomous scale is acceptable.
3. The marginal distributions of the covariate(s) are identical. This assumption can be tested by using the Kruskal–Wallis test on the covariate(s); however, if random assignment has been employed, there is little reason to run this test unless the treatment is applied before the covariate is measured.
4. The relationship between the dependent variable and the covariate(s) is monotonic. This assumption can be checked by plotting the data.
5. The degree of monotonicity is the same for each treatment population. An approximate test of the hypothesis

$$\rho_{s1} = \rho_{s2} = \cdots = \rho_{sJ}$$

where ρ_s is the population Spearman correlation coefficient, can be computed by applying the homogeneity of regression slopes test (described in Chapter 3) to ranks X and Y.

12.6 COMPARISON OF RANK ANCOVA WITH OTHER PROCEDURES

The power of Quade's rank ANCOVA has been compared with several competing procedures by McSweeney and Porter (1971), Porter and McSweeney (1971), and Hamilton (1974, 1975 and 1976). An approximate rank order of the power of parametric ANCOVA, Quade's rank ANCOVA, parametric ANCOVA applied to ranks, Puri and Sen's nonparametric ANCOVA (Puri and Sen 1969), two-way ANOVA, one-way ANOVA, Friedman's ANOVA, and the Kruskal–Wallis ANOVA for each of three XY correlations is presented in Table 12.2.

Table 12.2 provides only a rough idea of the relative power of these procedures because factors such as sample size, number of treatment levels, and α level sometimes slightly change the rank order. Also, it must be kept in mind that the comparisons are based on the situation in which the assumptions for the one-way parametric ANCOVA are met. If these assumptions are not met, the parametric and nonparametric procedures are not easily compared because they should no longer be viewed as testing the same hypothesis. That is, the parametric ANOVA and ANCOVA procedures test hypotheses associated with population means and adjusted means (parameters). The nonparametric procedures test hypotheses associated with entire population distributions of Y or $Y|X$. Only when the parametric assumptions are met does the statement that $\mu_1 = \mu_2$ also mean that the two population distributions of Y are identical. Likewise, only when the parametric assumptions are met does the statement that $\mu_{1\,adj} = \mu_{2\,adj}$ also mean that the two population distributions of $Y|X$ are identical. The reason for including the

Table 12.2 Rank Order of Power of Eight Procedures When Parametric ANCOVA Assumptions Are Met (Rank 1 = Most Powerful)

	$\rho_{yx} = 0.0$		$\rho_{yx} = 0.40$		$\rho_{yx} \geqslant 0.60$
Rank	Procedure	Rank	Procedure	Rank	Procedure
1	One-way ANOVA	1.5	ANCOVA	1	ANCOVA
2	Two-way ANOVA	1.5	Two-way ANOVA	2	Two-way ANOVA
3	ANCOVA	4	Quade's ANCOVA	4	Quade's ANCOVA
5.5	Quade's ANCOVA	4	Puri–Sen ANCOVA	4	Puri–Sen ANCOVA
5.5	Puri–Sen ANCOVA	4	ANCOVA on ranks	4	ANCOVA on ranks
5.5	ANCOVA on ranks	6	One-way ANOVA	6	Friedman
5.5	Kruskal–Wallis	7	Friedman	7	One-way ANOVA
8	Friedman	8	Kruskal–Wallis	8	Kruskal–Wallis

two-factor procedures (i.e., the paramteric two-way ANOVA and the non-parametric Friedman two-way ANOVA) in Table 12.2 is to cover the situation in which one of the two factors is a collection of levels on the X variable formed before the treatments are administered.

A close inspection of the table reveals two important generalizations: (1) the parametric ANCOVA is always more powerful than the competing nonparametric procedures, and (2) Quade's rank ANCOVA procedure is always at least as powerful as the Kruskal–Wallis and Friedman procedures. It appears to be difficult to err in always selecting the rank ANCOVA over the Friedman and Kruskal–Wallis approaches. Porter and McSweeney suggest, however, that the Friedman procedure should be considered if the covariate is a nominal (qualitative) blocking variable or classification factor (e.g., sex, litter, race, or state of residence). The Quade procedure is not designed to handle covariates that are not ordered.

The power of Quade's procedure differs little from that of the Puri–Sen (1969) procedure or parametric ANCOVA applied to ranks. When the three nonparametric ANCOVA procedures are compared with parametric ANCOVA, the discrepancy between the power of the parametric and non-parametric tests increases as the correlation between the covariate and the dependent variable increases. The asymptotic relative efficiency for the Quade and Puri–Sen procedures is between 0.866 and 0.955. Porter and McSweeney's Monte Carlo comparison of the Quade and parametric ANCOVA on ranks procedures, and Hamilton's Monte Carlo comparison of the Quade, Puri–Sen, and ANCOVA on ranks procedures are in essential agreement that the three procedures differ very little in terms of power. A slight preference for the Quade and Puri–Sen procedures over parametric ANCOVA on ranks has been expressed by Porter and McSweeney (1971, p. 38) on the basis of theoretical considerations.

An advantage of all nonparametric alternatives over the parametric approach is insensitivity to "outliers." That is, if the data contain, say, one or two scores that are extreme deviations from what they should be as a result of recording error or instrumentation failure, the nonparametric procedures will be affected less by these invalid outlying scores. When it is impossible to evaluate the adequacy of the measurement of the responses, the use of nonparametric procedures is one useful alternative to parametric methods. It is preferable, of course, to carefully investigate the adequacy of the data-collection procedure and to eliminate invalid scores before the analysis is carried out. Another approach is to run one analysis with all data and then run a second analysis with suspicious observations omitted from the data set. If the results are in essential agreement, there is no interpretation problem. If the results are quite different, both should be reported along with a warning that it is impossible to know which analysis is appropriate. The experiment should, of course, be replicated if possible, with more careful monitoring of the data-collection procedure.

12.7 SUMMARY

Rank analysis of covariance is an alternative to conventional parametric ANCOVA that should be considered when (1) the original data are in the form of ranks or (2) when it appears that there is extreme departure from the ANCOVA assumptions of normality and homogeneity of conditional variances in the unequal sample size situation. Also, if the relationship between X and Y is nonlinear but monotonic, rank ANCOVA should be considered. The hypothesis tested with rank ANCOVA is not the same as the hypothesis tested with parametric ANCOVA when the assumptions for the latter are violated. In this case rank ANCOVA provides a test that the J conditional population distributions are identical. It does not yield estimates of adjusted mean differences.

If the assumptions for parametric ANCOVA are met, the rank ANCOVA tests the same null hypothesis, but with lower power. The power advantage of parametric ANCOVA over rank ANCOVA increases as the sample size increases and as the correlation between the covariate and the dependent variable increases. The rank ANCOVA is less sensitive to outliers than is parametric ANCOVA.

CHAPTER THIRTEEN
Analysis of Heterogeneous Regression Case: Johnson–Neyman Technique

13.1 INTRODUCTION

One of the most important assumptions underlying the appropriate use of ANCOVA is that the regression slopes associated with the different groups in the design are homogeneous. This assumption is not always met. It is not unusual for the homogeneity of regression slopes test, which is a routine part of a complete ANCOVA, to reveal heterogeneity of slopes. In this case an alternative to ANCOVA is required. Several approaches have been suggested for this case (e.g., Robson and Atkinson 1960; Searle 1971, Section 8.2b; Steel and Federer 1955); however, the emphasis in this chapter is on an approach known as the Johnson–Neyman technique (Johnson and Neyman 1936). Before describing this technique, we review the basic problems associated with the failure to meet the homogeneous regression assumption.

Heterogeneous regression slopes associated with ANCOVA present interpretation problems because the magnitude of the treatment effect is not the same at different levels of X. Three varieties of this problem are illustrated in Figure 13.1. Figure 13.1a illustrates a situation (situation A) in which the ANCOVA F is nonsignificant even though group 1 is superior to group 2 at low values of X and inferior to group 2 at high levels of X. If the data were not plotted or if the homogeneity of regression slopes test were not carried out, it might be concluded that there were no treatment effects because the adjusted means are equal. Figure 13.1b (situation B) illustrates a case in which the adjusted means are different, but the true state of affairs is that there are no treatment effects at low levels of X and large effects at high levels of X. Figure 13.1c (situation C) illustrates the case where adjusted means are different and one group is superior to the other group at all levels of X included in the experiment. The only problem with ANCOVA and the associated adjusted means in this case is that the extent to which one group is superior to the other changes with the value of X. The difference between

270

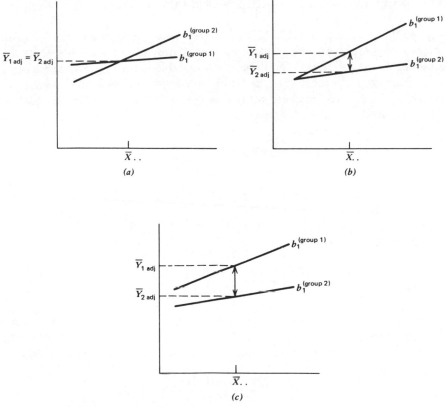

Fig. 13.1 Three examples of heterogeneous regression slopes.

adjusted means is not an adequate estimate of the treatment effects since it overestimates the effects at some levels of X and underestimates the effects at other levels of X.

The homogeneity of regression slopes test in the one-factor ANCOVA is analogous to the $A \times B$ interaction test in a two-factor ANOVA. Likewise, the Johnson–Neyman technique in one-factor ANCOVA designs with heterogeneous slopes is analogous to simple main effects tests in two-factor ANOVA designs in which the interaction is significant. That is, in two-factor ANOVA with interaction, there is interest in identifying the levels of B at which simple A effects are present (and/or levels of A at which B effects are present). The purpose of the Johnson–Neyman procedure is to identify the values of X that are associated with significant group differences on Y. That is, it allows a researcher to make statements about the regions of nonsignificance and significance. The computation procedure required to establish these regions is presented in Section 13.2 for the simple case of two independent groups and one covariate. The procedures for more than two groups, multiple covariates, correlated samples, and two-factor designs are described in other sections.

13.2 COMPUTATION: TWO GROUPS, ONE COVARIATE

Suppose that the data in Table 13.1 are based on an experiment in which two methods of therapy are the treatments, and scores on a sociability scale are employed as the covariate. The dependent variable is the aggressiveness score on a behavioral checklist. The analysis of covariance and homogeneity of regression slopes tests are based on the design matrix shown in Table 13.2.
The required R^2 values for these tests are

$$R^2_{yD,X,DX} = 0.93829$$

$$R^2_{yD,X} = 0.60153$$

$$R^2_{yX} = 0.56784$$

and the total sum of squares on Y is 176.967.
The ANCOVA F is

$$\frac{(0.60153 - 0.56784)/1}{(1 - 0.60153)/27} = 2.28$$

which is not significant if $\alpha = .05$ is used. If the ANCOVA assumptions are met, the investigator can conclude that there are no treatment effects. The test on the assumption of the homogeneity of the regression slopes yields

$$F_{obt} = \frac{(0.93829 - 0.60153)/1}{(1 - 0.93829)/26} = 141.89$$

Table 13.1   Data from a Two- Group One- Covariate Experiment

Therapy 1		Therapy 2	
X	Y	X	Y
1	10	1	5
2	10	1.5	6
2	11	2.5	6
3	10	3.5	7
4	11	4.5	8
5	11	4.5	9
5	10	5	9
6	11	6	9
6	11.5	6	10.5
7	12	7	11
8	12	7	12.5
8	11	7.5	12.5
9	11	8	14
10	12.5	9	14.5
11	12	10	16

Table 13.2 Design Matrix for Data of Table 13.1

Subject	Y	X	D	DX
1	10	1	1	1
2	10	2	1	2
3	11	2	1	2
4	10	3	1	3
5	11	4	1	4
6	11	5	1	5
7	10	5	1	5
8	11	6	1	6
9	11.5	6	1	6
10	12	7	1	7
11	12	8	1	8
12	11	8	1	8
13	11	9	1	9
14	12.5	10	1	10
15	12	11	1	11
16	5	1	0	0
17	6	1.5	0	0
18	6	2.5	0	0
19	7	3.5	0	0
20	8	4.5	0	0
21	9	4.5	0	0
22	9	5	0	0
23	9	6	0	0
24	10.5	6	0	0
25	11	7	0	0
26	12.5	7	0	0
27	12.5	7.5	0	0
28	14	8	0	0
29	14.5	9	0	0
30	16	10	0	0

which, when compared with the critical $F_{(.05, 1, 26)} = 4.23$, is obviously significant. A plot of the data clearly reveals the nature of the heterogeneous
regression slopes. It can be seen in Figure 13.2 that the ANCOVA is quite
inappropriate because the test on the difference between adjusted means does
not attend to the apparent treatment effects at the high and low levels of X.
The relevant questions at this stage appear to be;

1. "What are the values on X associated with nonsignificant treatment
   effects?"
2. "What are the values on X associated with significant treatment effects?"

These questions are answered with the Johnson–Neyman technique as
follows.

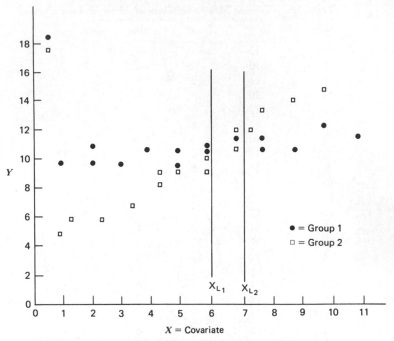

Fig. 13.2 Plot of example data illustrating heterogeneous slopes.

Calculation of Nonsignificance Region

The limits of the region of nonsignificance on X are computed by using

$$X_{L1} = \frac{-B - \sqrt{B^2 - AC}}{A}$$

$$X_{L2} = \frac{-B + \sqrt{B^2 - AC}}{A}$$

where
  X_{L1} and X_{L2} = limits of nonsignificance region

$$A = \frac{-F_{(\alpha,1,N-4)}}{N-4}(SSres_i)\left(\frac{1}{\Sigma x_1^2} + \frac{1}{\Sigma x_2^2}\right) + (b_1^{(\text{group 1})} - b_1^{(\text{group 2})})^2,$$

$$B = \frac{F_{(\alpha,1,N-4)}}{N-4}(SSres_i)\left(\frac{\overline{X}_1}{\Sigma x_1^2} + \frac{\overline{X}_2}{\Sigma x_2^2}\right)$$
$$+ (b_0^{(\text{group 1})} - b_0^{(\text{group 2})})(b_1^{(\text{group 1})} - b_1^{(\text{group 2})})$$

$$C = \frac{-F_{(\alpha,1,N-4)}}{N-4}(SSres_i)\left(\frac{N}{n_1 n_2} + \frac{\overline{X}_1^2}{\Sigma x_1^2} + \frac{\overline{X}_2^2}{\Sigma x_2^2}\right)$$
$$+ (b_0^{(\text{group 1})} - b_0^{(\text{group 2})})^2$$

$F_{(\alpha, 1, N-4)}$ = critical value of F statistic for desired level of α and 1 and $N-4$ degrees of freedom (N = total number of subjects; i.e., $n_1 + n_2$

$SSres_i$ = individual residual sum of squares described in Sections 3.5 and 4.6

$\bar{X}_1, \bar{X}_2$ = covariate means for samples 1 and 2 respectively

$\Sigma x_1^2, \Sigma x_2^2$ = covariate sums of squares for samples 1 and 2, respectively

$b_0^{(group\ 1)}$ = regression intercept for sample 1

$b_0^{(group\ 2)}$ = regression intercept for sample 2

$b_1^{(group\ 1)}$ = regression slope for sample 1

$b_1^{(group\ 2)}$ = regression slope for sample 2.

The summary statistics for the example data are:

Group 1	Group 2
$n_1 = 15$	$n_2 = 15$
$\bar{X}_1 = 5.8$	$\bar{X}_2 = 5.533$
$\Sigma x_1^2 = 130.4$	$\Sigma x_2^2 = 99.23$
$b_0^{(group\ 1)} = 9.857$	$b_0^{(group\ 2)} = 3.155$
$b_1^{(group\ 1)} = 0.209$	$b_1^{(group\ 2)} = 1.237$
$F_{(.05, 1, 26)} = 4.23$	

The values for A, B, and C are

$$A = 1.0253304$$
$$B = -6.7155607$$
$$C = 43.71518698$$
$$X_{L1} = \frac{6.7155607 - 0.525590}{1.0253304} = 6.04$$
$$X_{L2} = \frac{6.7155607 + 0.525590}{1.0253304} = 7.06$$

Hence the region of nonsignificance is 6.04 through 7.06 using $\alpha = .05$. If we select a specific point on X that falls in this region, we conclude that the two methods of therapy do not differ on Y for subjects falling at the specified point on the sociability scale. If we select instead a point below 6.04, we conclude that method 1 is superior for subjects falling at that point on X. If we select a point above 7.06, we conclude that method 2 is superior.

A Weak Alternative Analysis

Suppose that the data of Table 13.1 are viewed as a two-factor ANOVA problem. Factor A is type of therapy (group 1 vs. group 2), and factor B is sociability (classified as high or low). Subjects with sociability scores equal to or less than 5 are classified as "low"; scores above 5 are classified as "high". The data arrangement (Y scores) is as follows:

	Factor B: Sociability	
	Low	High
Group 1 Factor A: Therapy	10, 10, 11, 10, 11, 11, 10	11, 11.5, 12, 12, 11, 11, 12.5, 12
Group 2	5, 6, 6, 7, 8, 9, 9	9, 10.5, 11, 12.5, 12.5, 14, 14.5, 16

The cell means are:

	B_1	B_2
A_1	10.43	11.63
A_2	7.14	12.50

The ANOVA summary is as follows:

Source	SS	df	MS	F
A: Therapy	8.53	1	8.53	3.97
B: Sociability	80.17	1	80.17	37.26
$A \times B$	32.31	1	32.31	15.02
Within cell	55.95	26	2.15	
Total	176.97	29		

The $A \times B$ interaction is significant, which tells us that there is an inconsistency in the therapy efffects across the two levels of sociability. Simple main-effects tests are carried out to identify levels associated with treatment effects.

Source	SS	df	MS	F
A_1 vs. A_2 at B_1	37.79	1	37.79	17.57
A_1 vs. A_2 at B_2	3.06	1	3.06	1.42
Within cell	55.95	26	2.15	

These tests lead us to conclude that the difference between therapies is significant for subjects classified as "low" on sociability, and not for subjects classified as "high".

A comparison of these results with the outcome of the Johnson–Neyman analysis reveals two advantages of the latter approach. First, the $J - N$ technique allows a statement about treatment differences for *any* value of X. The classification of X into "low" and "high" levels for ANOVA has discarded some information on this dimension. The two simple main effects tests test the hypotheses that (1) there are no treatment effects for subjects with an average sociability score of 3.18 (i.e., the average X score for all subjects classified as "low" on X) and (2) there are no treatment effects for subjects with an average sociability score of 7.84 (i.e., the average X score for all subjects classified as "high" on X). Differences on X within the "low" and "high" categories are ignored when these tests are used.

Second, the precision of the tests is much lower with the ANOVA approach. The difference between the two types of therapy is clearly significant for an X score of 7.84 according to the outcome of the $J - N$ analysis. The ANOVA simple main-effects test on this same difference (i.e., A_1, vs. A_2 at B_2) is clearly *not* significant. Why is the same mean difference significant with $J - N$ but not with the ANOVA simple main-effects test? Notice the size of the error mean square (MS) for the $J - N$ and the ANOVA tests:

$$\text{MSres}_i = 0.42(\text{used in } J - N)$$

$$\text{MS within cell} = 2.15(\text{used in ANOVA})$$

In this example the pooled within-treatment residuals (res_i) are *much* smaller than the within-cell variability because a high linear relationship exists between X and Y. A lower relationship between X and Y would yield a smaller power advantage for the $J - N$ technique.

In summary, the Johnson–Neyman technique is preferable to a two-factor ANOVA and simple main effects tests because it (1) is more powerful and (2) allows tests of treatment effects for any and all levels of the covariate. If all levels of the covariate are included as one of the factors (rather than just "high" and "low" levels), each level beyond the first level will consume one degree of freedom. Hence, the use of two-factor ANOVA with many levels of X will not yield the power of the Johnson–Neyman approach because the error degrees of freedom are larger with the latter. It should be pointed out, however, that there is a special type of ANOVA technique that has been developed by Cronbach and Snow (1976, pp. 78, 314) that is a very useful alternative to the Johnson–Neyman technique. This approach can be employed in the case where two extreme groups have been selected on the X dimension. The Cronbach–Snow reference is by far the most comprehensive work available on the interaction between treatments and aptitudes (covariates).

Assumptions

The assumptions underlying the appropriate use of the Johnson–Neyman technique are essentially the same as those associated with ANCOVA—with one exception. It is not, of course, assumed that the regressions are homogeneous. The general equivalence of the assumptions of these two procedures is not surprising because the purpose and computations are similar for both.

The basic difference between the two procedures is that treatment effects are estimated at the grand covariate mean with ANCOVA, but with the Johnson–Neyman technique the treatment effects are estimated as a function of the covariate score. A complete description of all the assumptions are not presented here since the equivalent assumptions are described in Chapter 6. We do, however, summarize the findings of studies on the consequences of violations of several of the assumptions. The major assumptions are that:

1. The residuals of the individual within-group regressions of Y on X are independent, and subjects have been randomly assigned to treatments.
2. The residuals are normally distributed.
3. The residuals have homogeneous variance for each value of X (sometimes called the *homoscedasticity assumption*).
4. The residuals have homogeneous variance across treatment groups (i.e., the conditional variances for all J groups are equal).
5. The regression of Y on X is linear.
6. The levels of the covariate are fixed.
7. The covariate is measured without error.

It is known that the independence assumption is critical and that violations of this assumption will generally lead to an increase in the probability of Type I error. The random assignment of subjects to treatments is also critical. The consequences of violating the normality assumption appear to be of little importance (Mendro 1975). Violation of the assumption of homogeneity of conditional variances across treatments has relatively little effect. Shields (1978) found that when this assumption was violated, the effects were essentially the same as is found with ANOVA and ANCOVA. That is, if sample sizes are equal, the probability of Type I error is little affected. But this is not the case with unequal sample sizes. When the larger variance is combined with the larger group, the probability of Type I error is less than the nominal α. When large variances are combined with small groups, the probability of Type I error is greater than the nominal α. These results are consistent with those of three other studies. Erlander and Gustavsson (1965) evaluated the effects of heterogeneous conditional variances where one population variance was twice the size of the other. This degree of heterogeneity had little effect on either the confidence band for the treatment difference or on the region of significance. Borich and Wunderlich (1973) came to a similar conclusion. Mendro (1975) investigated the effects of severe heterogeneity of conditional variances. When the larger: smaller conditional variance ratio was 4, the probability of Type I error was found to follow the same pattern reported by Shields, that is, positive bias in F when the smaller samples are associated with the larger variance and negative bias when the smaller samples are associated with the smaller variance.

With respect to homoscedasticity, Shields found the Johnson–Neyman technique to be sensitive to violations with both equal and unequal sample sizes. She found that as the variance for a fixed value of the covariate increased, the probability of including that X value in the region of significance increased.

The effects of violating the assumption of fixed error-free covariates have been investigated by Rogosa (1977). If the covariates are random error-free variables rather than fixed, the Type I error rate of the Johnson–Neyman technique is unchanged. The power of the test, however, is lower in this case of random variables. Rogosa also investigated the effects of measurement error.

Three major consequences of measurement error in the covariate are:

1. An inevitable shrinking of the Johnson–Neyman region of significance
2. Reduced probability that the homogeneity of regression slopes test will identify heterogeneous slopes (this means that the Johnson–Neyman technique will be used less frequently when measurement error is present because ANCOVA is used when the slopes are evaluated as being homogeneous)
3. Possible increase in Type I error

Although it may seem that points 1 and 3 are contradictory, they are not. When observed scores (scores that contain measurement error) are used, values of X that lie in the region of significance may lie in the region of nonsignificance if true scores are used. Details on how this can occur can be found in the excellent work of Rogosa (1977).

13.3 MULTIPLE GROUPS, ONE COVARIATE

The original presentation of the Johnson–Neyman technique (Johnson and Neyman 1936) considered the case of two groups. If three or more groups are involved, several alternatives, two of which are described by Potthoff (1964), are available. The simplest procedure is to employ the approach described previously for two groups to each pair of groups in the experiment with one slight modification. The Bonferroni F statistic should be employed in place of the conventional F statistic where F is called for in the computational formulas.

If four groups are involved one might be interested in as many as

$$\frac{J(J-1)}{2} = \frac{4(3)}{2} = 6$$

comparisons of pairs of groups using the Johnson–Neyman technique. The Bonferroni F table is entered with $df_1 = 1$, $df_2 = N - J(C+1)$, and $C' = 6$ for the desired level of α. If the investigator has interest in and plans (before data are collected) to analyze only four of the six possible comparisons, the Bonferroni F table should be entered using $df_1 = 1$, $df_2 = N - J(C + 1)$, and $C' = 4$.

Alternative Multiple-group Technique

An alternative to the multiple-group procedure just described involves the following procedure. First, run the homogeneity of regression test. If this test yields a significant F value, the overall homogeneity of regression hypothesis is rejected, but it does not lead to the conclusion that *all* slopes differ from each other. Follow up the overall test with additional homogeneity of regression tests on all combinations of pairs of slopes. Frequently, the slopes are homogeneous for most of the groups. Apply ANCOVA to the collection of groups that have homogeneous slopes; apply the Johnson–Neyman procedure for comparisons involving groups with heterogeneous slopes. For example, if four groups are employed in an experiment and the overall homogeneity of regression test is significant, run follow-up homogeneity of regression tests for the $[J(J-1)/2] = [4(3)/2] = 6$ comparisons. Suppose the following pattern of homogeneity of regression test results is obtained:

Groups Involved in Slope Comparison	Significant
1,2	No
1,3	No
1,4	Yes
2,3	No
2,4	Yes
3,4	Yes

Since comparisons of all groups suggest that the slopes of groups 1, 2, and 3 are homogeneous, ANCOVA is appropriate for *these* groups. The slope of group 4 differs from the slopes of the others; Johnson–Neyman comparisons between group 4 and each of the other groups would then be carried out. Use Bonferroni F values based on $C' = 4$ for the ANCOVA F and three Johnson–Neyman analyses if there is interest in controlling the error rate experimentwise.

13.4 TWO GROUPS, ANY NUMBER OF COVARIATES

If two covariates are involved, the Johnson–Neyman technique is employed to identify the region of significant differences in the plane of the covariates. An illustration of heterogeneous regression planes can be seen in Figure 13.3. If three or more covariates are employed, a region of significant differences in

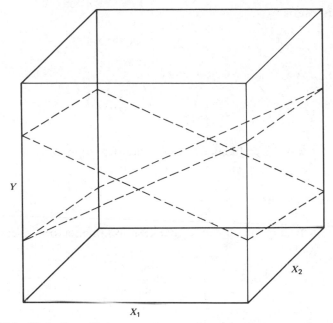

Fig. 13.3 Illustration of heterogeneous regression planes in two-covariate case.

the hyperspace of the covariates can be obtained. Since four-dimensional representations are beyond the author's psychomotor skills, an illustration of heterogeneous regression hyperplanes is not provided.

If multiple covariates are employed, it may be wise to abandon attempts to plot regions of significance. A simpler approach, which can be employed with any number of covariates, is to compute an F test on the difference between treatments for subjects with specified covariate scores. Formally, the null hypothesis associated with this test is

$$H_0 : x_s' \Delta = 0.0$$

where

$x_s = $ a $(C+1) \times 1$ column vector of a specified set of covariate scores augmented by scalar 1 as first element

$\Delta = $ difference vector $\beta^{(\text{group } 2)} - \beta^{(\text{group } 1)}$, where $\beta^{(\text{group } 1)}$ is the vector of least-squares regression parameters obtained by regressing Y on covariates 1 through C for population 1; in other words,

$$\beta^{(\text{group } 1)} = \begin{bmatrix} \beta_0^{(\text{group } 1)} \\ \beta_1^{(\text{group } 1)} \\ \vdots \\ \beta_C^{(\text{group } 1)} \end{bmatrix}$$

correspondingly

$$\boldsymbol{\beta}^{(\text{group 2})} = \begin{bmatrix} \beta_0^{(\text{group 2})} \\ \beta_1^{(\text{group 2})} \\ \vdots \\ \beta_C^{(\text{group 2})} \end{bmatrix}$$

The product $\mathbf{x}_s'\boldsymbol{\Delta}$ is the difference between the two population means on Y at point $\mathbf{x}_s$.

Computation Procedure

The F ratio for testing the difference between treatments for a specified set of covariate scores is

$$F = \frac{\mathbf{x}_s'\mathbf{D}\mathbf{D}'\mathbf{x}_s}{\mathbf{x}_s'\mathbf{V}\mathbf{x}_s}$$

where

$\mathbf{x}_s =$ a $(C+1) \times 1$ column vector of a specified set of covariate scores augmented by scalar 1 as first element; that is,

$$\mathbf{x}_s = \begin{bmatrix} 1 \\ X_1 \\ X_2 \\ \vdots \\ X_C \end{bmatrix}$$

$X_1, X_2, \cdots X_C =$ covariate scores associated with a specific subject on covariates 1 through C

$\mathbf{D} =$ a $(C+1) \times 1$ column vector of differences in least-squares estimates $\mathbf{b}^{(\text{group 1})}$ and $\mathbf{b}^{(\text{group 2})}$; that is,

$$\mathbf{D} = \begin{bmatrix} b_0^{(\text{group 2})} - b_0^{(\text{group 1})} \\ b_1^{(\text{group 2})} - b_1^{(\text{group 1})} \\ b_2^{(\text{group 2})} - b_2^{(\text{group 1})} \\ \vdots \quad \vdots \quad \vdots \\ b_C^{(\text{group 2})} - b_C^{(\text{group 1})} \end{bmatrix} = \left[\mathbf{b}^{(\text{group 2})} - \mathbf{b}^{(\text{group 1})} \right]$$

The variance of the difference in estimated regression planes or hyperplanes is

$$\mathbf{V} = \text{MSres}_i[(\mathbf{X}_1'\mathbf{X}_1)^{-1} + (\mathbf{X}_2'\mathbf{X}_2)^{-1}]$$

where

$\text{MSres}_i = $ error MS employed in homogeneity of regression test described in Section 8.3; that is,

$$\frac{(1 - R^2_{yD,X,DX})\text{SST}}{N - J(C+1)}$$

$\mathbf{X}_1 = n \times (C+1)$ matrix of raw covariate scores for group 1 augmented by a column vector of "ones" as the first column of the matrix; that is,

$$\mathbf{X}_1 = \begin{bmatrix} 1 & X_{11} & X_{21} & \cdot & \cdot & \cdot & X_{C1} \\ 1 & X_{12} & X_{22} & \cdot & \cdot & \cdot & X_{C2} \\ \vdots & \vdots & \vdots & \vdots & \vdots & \vdots & \vdots \\ 1 & X_{1n_1} & X_{2n_1} & \cdot & \cdot & \cdot & X_{Cn_1} \end{bmatrix}$$

$\mathbf{X}_2 = $ corresponding matrix for group 2

The obtained F is evaluated with $F_{[\alpha, 1, N - J(C+1)]}$.

Computational Example: Two Groups, One Covariate

The data presented at the beginning of Section 13.2 are employed to illustrate the equivalence of the procedure of that section for establishing regions of significance (and nonsignificance) and the F test procedure described in this section.

Suppose that we are interested in testing for treatment effects for subjects who have X scores of 8. More specifically, "Is there a difference between the effectiveness of therapy 1 and therapy 2 for a population of subjects having sociability scores of 8?"

The required vectors and matrices are

$$\mathbf{x}_s = \begin{bmatrix} 1 \\ 8 \end{bmatrix}$$

$$\mathbf{D} = \begin{bmatrix} 9.86 - 3.16 \\ 0.208 - 1.236 \end{bmatrix} = \begin{bmatrix} 6.70 \\ -1.028 \end{bmatrix}$$

$$\text{MSres}_i = \frac{(1 - .93829)176.967}{26} = 0.42$$

$$
X_1 = \begin{bmatrix}
1 & 1 \\
1 & 2 \\
1 & 2 \\
1 & 3 \\
1 & 4 \\
1 & 5 \\
1 & 5 \\
1 & 6 \\
1 & 6 \\
1 & 7 \\
1 & 8 \\
1 & 8 \\
1 & 9 \\
1 & 10 \\
1 & 11
\end{bmatrix}
\qquad
X_2 = \begin{bmatrix}
1 & 1 \\
1 & 1.5 \\
1 & 2.5 \\
1 & 3.5 \\
1 & 4.5 \\
1 & 4.5 \\
1 & 5 \\
1 & 6 \\
1 & 6 \\
1 & 7 \\
1 & 7 \\
1 & 7.5 \\
1 & 8 \\
1 & 9 \\
1 & 10
\end{bmatrix}
$$

$$
(X_1'X_1)^{-1} = \begin{bmatrix} 0.324642 & -0.044479 \\ -0.044479 & 0.007669 \end{bmatrix}
$$

$$
(X_2'X_2)^{-1} = \begin{bmatrix} 0.375210 & -0.055761 \\ -0.055761 & 0.010077 \end{bmatrix}
$$

$$
V = \begin{bmatrix} .42 & 0 \\ 0 & 0.42 \end{bmatrix} \begin{bmatrix} 0.699852 & -0.100240 \\ -0.100240 & 0.017746 \end{bmatrix}
$$

$$
= \begin{bmatrix} 0.293938 & -0.042101 \\ -0.042101 & 0.007453 \end{bmatrix}.
$$

$$
F_{obt} = \frac{\begin{bmatrix} 1 & 8 \end{bmatrix} \begin{bmatrix} 6.705 \\ -1.028 \end{bmatrix} \begin{bmatrix} 6.705 & -1.028 \end{bmatrix} \begin{bmatrix} 1 \\ 8 \end{bmatrix}}{\begin{bmatrix} 1 & 8 \end{bmatrix} \begin{bmatrix} 0.29398 & -0.042101 \\ -0.042101 & 0.007453 \end{bmatrix} \begin{bmatrix} 1 \\ 8 \end{bmatrix}}
$$

$$
= 2.30736/0.09731
$$

$$
= 23.71
$$

The obtained F far exceeds the critical value of F based on 1 and $N - J(C + 1)$ degrees of freedom. It is concluded that therapy 2 is superior to therapy 1 for a population of subjects whose sociability scores fall at 8. This result is consistent with the previously described approach of computing the region of nonsignificance that is 6.04 through 7.06.

It turns out that all values of x_s that satisfy the inequality

$$
x_s'DD'x_s - F_{[\alpha, 1, N - J(C + 1)]}x_s'Vx_s \geq 0
$$

define the region of significance. If only one covariate is employed, the procedure described earlier for establishing the region of significance and nonsignificance involves less computation than does the procedure described here. If more than one covariate is employed, however, the simpler procedure does not apply, and various x_s values must be employed to determine the combinations of covariate scores associated with the regions of significance and nonsignificance.

Confidence Intervals

When confidence intervals rather than significance tests are desired, compute

$$x_s'D \pm \sqrt{F_{[\alpha, 1, N - J(C+1)]}} \ \sqrt{x_s'Vx_s}$$

For the example data, the 95% confidence interval (using a covariate score of 8) is

$$\begin{bmatrix} 1 & 8 \end{bmatrix} \begin{bmatrix} 6.705 \\ -1.028 \end{bmatrix} \pm \sqrt{4.23} \ \sqrt{0.09731} = -1.519 \pm (2.057)(0.3119)$$

$$= (-2.161, -0.877)$$

For the specified individual point x_s the probability is .95 that the obtained confidence interval will cover the true population mean difference. The negative sign indicates that group 2 is superior to group 1 for the specified point; all values of $x_s'D$ are negative for covariate scores above the point of intersection of the two slopes and positive for covariate values below the point of intersection. The point of intersection (on the X dimension) for the two-group one-covariate case can be computed using the following formula:

$$\frac{b_0^{(group\ 1)} - b_0^{(group\ 2)}}{b_1^{(group\ 2)} - b_1^{(group\ 1)}}$$

For multiple covariates, there will generally be many combinations of covariate scores that are associated with zero group differences on Y.

Computational Example: Two Groups, Three Covariates

Suppose that the following data were collected in a two-group three-covariate experiment.

	Group 1				Group 2		
X_1	X_2	X_3	Y	X_1	X_2	X_3	Y
1	9	3	10	1	8	2	5
2	7	5	10	1.5	9	4	6
2	9	4	11	2.5	9	4	6
3	9	3	10	3.5	8	5	7
4	9	3	11	4.5	8	4	8
5	8	4	11	4.5	8	4	9
5	7	4	10	5	7	3	9

6	9	4	11	6	8	3	9
6	6	3	11.5	6	6	4	10.5
7	8	5	12	7	7	3	11
8	7	4	12	7	5	5	12.5
8	7	5	11	7.5	8	5	12.5
9	6	6	11	8	6	6	14
10	6	6	12.5	9	7	5	14.5
11	6	6	12	10	6	6	16

The design matrix for ANCOVA and the homogeneity of regression hyperplanes tests is as follows:

D	X_1	X_2	X_3	DX_1	DX_2	DX_3	Y
1	1	9	3	1	9	3	10
1	2	7	5	2	7	5	10
1	2	9	4	2	9	4	11
1	3	9	3	3	9	3	10
1	4	9	3	4	9	3	11
1	5	8	4	5	8	4	11
1	5	7	4	5	7	4	10
1	6	9	4	6	9	4	11
1	6	6	3	6	6	3	11.5
1	7	8	5	7	8	5	12
1	8	7	4	8	7	4	12
1	8	7	5	8	7	5	11
1	9	6	6	9	6	6	11
1	10	6	6	10	6	6	12.5
1	11	6	6	11	6	6	12
0	1	8	2	0	0	0	5
0	1.5	9	4	0	0	0	6
0	2.5	9	4	0	0	0	6
0	3.5	8	5	0	0	0	7
0	4.5	8	4	0	0	0	8
0	4.5	8	4	0	0	0	9
0	5	7	3	0	0	0	9
0	6	8	3	0	0	0	9
0	6	6	4	0	0	0	10.5
0	7	7	3	0	0	0	11
0	7	5	5	0	0	0	12.5
0	7.5	8	5	0	0	0	12.5
0	8	6	6	0	0	0	14
0	9	7	5	0	0	0	14.5
0	10	6	6	0	0	0	16

The R^2 values required for these tests are

$$R^2_{yD,X,DX} = 0.95698$$

$$R^2_{yD,X} = 0.60763$$

$$R^2_{yX} = 0.57399$$

and the total sum of squares on $Y = 176.967$. The ANCOVA F is

$$\frac{(0.60763 - 0.57399)/1}{(1 - 0.60763)/25} = 2.14$$

which is not significant if $\alpha = .05$ is used. The homogeneity of regression hyperplanes F is

$$\frac{(0.95698 - 0.60763)/3}{(1 - 0.95698)/22} = 59.55$$

which far exceeds 4.3, which is the critical value of F based on 1 and 22 degrees of freedom for $\alpha = .05$.

Since the regression hyperplanes are clearly heterogeneous, we ignore the outcome of ANCOVA and employ the Johnson–Neyman technique. Suppose that we want to determine whether the two treatments differ for a population of subjects having the following covariate scores:

$$X_1 = 4$$
$$X_2 = 9$$
$$X_3 = 3$$

$$\mathbf{x}_s = \begin{bmatrix} 1 \\ 4 \\ 9 \\ 3 \end{bmatrix}$$

The vector $\mathbf{b}^{(\text{group 1})}$ is the set of least-squares estimates obtained by regressing Y on X_1, X_2, and X_3 for subjects in the first group.

$$\mathbf{b}^{(\text{group 1})} = \begin{bmatrix} 8.9129 \\ 0.2634 \\ 0.1237 \\ -0.0705 \end{bmatrix}$$

The vector $\mathbf{b}^{(\text{group 2})}$ is the set of least-squares estimates obtained by regressing Y on X_1, X_2, and X_3 for subjects in the second group.

$$\mathbf{b}^{(\text{group2})} = \begin{bmatrix} 4.8574 \\ 1.0243 \\ -0.3180 \\ 0.4302 \end{bmatrix}$$

$$\mathbf{D} = \mathbf{b}^{(\text{group 2})} - \mathbf{b}^{(\text{group 1})}$$

$$= \begin{bmatrix} 4.8574 \\ 1.0243 \\ -0.3180 \\ 0.4302 \end{bmatrix} - \begin{bmatrix} 8.9129 \\ 0.2634 \\ 0.1237 \\ -0.0705 \end{bmatrix} = \begin{bmatrix} -4.0555 \\ 0.7609 \\ -0.4417 \\ 0.5007 \end{bmatrix}$$

$$\mathbf{X}_1 = \begin{bmatrix} 1 & 1 & 9 & 3 \\ 1 & 2 & 7 & 5 \\ 1 & 2 & 9 & 4 \\ 1 & 3 & 9 & 3 \\ 1 & 4 & 9 & 3 \\ 1 & 5 & 8 & 4 \\ 1 & 5 & 7 & 4 \\ 1 & 6 & 9 & 4 \\ 1 & 6 & 6 & 3 \\ 1 & 7 & 8 & 5 \\ 1 & 8 & 7 & 4 \\ 1 & 8 & 7 & 5 \\ 1 & 9 & 6 & 6 \\ 1 & 10 & 6 & 6 \\ 1 & 11 & 6 & 6 \end{bmatrix} \qquad \mathbf{X}_2 = \begin{bmatrix} 1 & 1 & 8 & 2 \\ 1 & 1.5 & 9 & 4 \\ 1 & 2.5 & 9 & 4 \\ 1 & 3.5 & 8 & 5 \\ 1 & 4.5 & 8 & 4 \\ 1 & 4.5 & 8 & 4 \\ 1 & 5 & 7 & 3 \\ 1 & 6 & 8 & 3 \\ 1 & 6 & 6 & 4 \\ 1 & 7 & 7 & 3 \\ 1 & 7 & 5 & 5 \\ 1 & 7.5 & 8 & 5 \\ 1 & 8 & 6 & 6 \\ 1 & 9 & 7 & 5 \\ 1 & 10 & 6 & 6 \end{bmatrix}$$

$$(\mathbf{X}_1'\mathbf{X}_1)^{-1} = \begin{bmatrix} 11.9518 & -0.1822 & -1.0555 & -0.6640 \\ -0.1822 & 0.0204 & 0.0222 & -0.0238 \\ -1.055 & 0.0222 & 0.1042 & 0.0327 \\ -0.6640 & -0.0238 & 0.0327 & 0.1283 \end{bmatrix}$$

$$(\mathbf{X}_2'\mathbf{X}_2)^{-1} = \begin{bmatrix} 9.2590 & -0.2778 & -0.9139 & -0.2270 \\ -0.2778 & 0.0262 & 0.0317 & -0.0238 \\ -0.9139 & 0.0317 & 0.1015 & -0.0014 \\ -0.2270 & -0.0238 & -0.0014 & 0.0879 \end{bmatrix}$$

$$\text{MSres}_i = \frac{(1 - 0.95698)176.967}{22} = 0.346$$

$$V = \begin{bmatrix} 0.346 & 0 & 0 & 0 \\ 0 & 0.346 & 0 & 0 \\ 0 & 0 & 0.346 & 0 \\ 0 & 0 & 0 & 0.346 \end{bmatrix}$$

$$\times \begin{bmatrix} 21.2108 & -0.4600 & -1.9694 & -0.8910 \\ -0.4600 & 0.0466 & 0.0539 & -0.0477 \\ -1.9694 & 0.0539 & 0.2057 & 0.0313 \\ -0.8910 & -0.0477 & 0.0313 & 0.2163 \end{bmatrix}$$

$$= \begin{bmatrix} 7.339 & -0.159 & -0.681 & -0.308 \\ -0.159 & 0.016 & 0.019 & -0.017 \\ -0.681 & 0.019 & 0.071 & 0.011 \\ -0.308 & -0.017 & 0.011 & 0.075 \end{bmatrix}$$

$$F_{obt} = \frac{\begin{bmatrix} 1 & 4 & 9 & 3 \end{bmatrix} \begin{bmatrix} -4.0555 \\ 0.7609 \\ -0.4417 \\ 0.5007 \end{bmatrix} \begin{bmatrix} -4.0555, 0.7609, -0.4417, 0.5007 \end{bmatrix} \begin{bmatrix} 1 \\ 4 \\ 9 \\ 3 \end{bmatrix}}{\begin{bmatrix} 1 & 4 & 9 & 3 \end{bmatrix} \begin{bmatrix} 7.339 & -0.159 & -0.681 & -0.308 \\ -0.159 & 0.016 & 0.019 & -0.017 \\ -0.681 & 0.019 & 0.071 & 0.011 \\ -0.308 & -0.017 & 0.011 & 0.075 \end{bmatrix} \begin{bmatrix} 1 \\ 4 \\ 9 \\ 3 \end{bmatrix}}$$

$$= \frac{12.146}{0.179384}$$

$$= 67.71$$

The value F_{obt} is compared with $F_{(.05, 1, 22)}$, which is 4.3, and we conclude that there are treatment effects for a population of subjects with covariate scores of 4, 9, and 3 on covariates X_1, X_2 and X_3, respectively. The 95% confidence interval is

$$\begin{bmatrix} 1 & 4 & 9 & 3 \end{bmatrix} \begin{bmatrix} -4.0555 \\ 0.7609 \\ -0.4417 \\ 0.5007 \end{bmatrix} \pm \sqrt{4.3} \sqrt{0.179384} = -3.4851 \pm (2.0736)(.4235375)$$

$$= (-4.363, -2.607)$$

The probability is .95 that the obtained confidence interval will cover the population mean difference for subjects who have covariate scores of 4, 9, and 3 on covariates X_1, X_2, and X_3, respectively. Since the sign associated with $x_s'D$ (i.e., -3.4851) is negative, we know that the second treatment group is superior to the first. As the confidence interval does not contain zero, we conclude that the superiority of the second treatment over the first is significant.

13.5 ANY NUMBER OF GROUPS, ANY NUMBER OF COVARIATES

Situations in which more than two groups are involved can be analyzed pairwise regardless of the number of covariates. The Bonferroni F statistic is substituted for the conventional F statistic in all computations if the experimenter desires to maintain the error rate at α for the whole collection of pairwise contrasts.

Suppose that three groups had been involved in the example described in Section 13.4 rather than two. If the experimenter selects a point x_s and computes $x_s'D$ for each pair of groups, he or she will have three $x_s'D$ values. Each value can then be tested by using the Johnson–Neyman technique. Since three tests are being carried out the critical value is based on the Bonferroni F table, which is entered by using

$$df_1 = 1$$

$$df_2 = 22$$

$$C' = 3$$

rather than the conventional F table. The probability of making a Type I error in the collection of three tests is equal to or less than the α associated with the table employed. (A Type I error is committed when the difference $x_s'D$ is declared significant when the true difference $x_s'\Delta$ is zero.)

If simultaneous confidence intervals rather than tests are of interest, the Bonferroni F critical value is employed in the formula for confidence intervals described in Section 13.4.

13.6 TWO-FACTOR DESIGNS

The homogeneity of regression slopes assumption applies to the case of two-factor designs as well as to one-factor designs. The conventional homogeneity of regression test associated with the independent sample two-factor design is a test of the equality of the cell slopes (see Chapter 10 for computational details of this test). This test is directly relevant to the interpretation of the $A \times B$ interaction and any simple main-effects tests that may be of interest. If the cell slopes are homogeneous, main-effects tests, the interaction test, and simple main-effects tests are all appropriate. Whereas heterogeneity of cell slopes invalidates the $A \times B$ interaction test, the main-effects tests may be appropriate in this case. Additional testing is appropriate to help decide whether the main effects should be interpreted.

It is assumed that the slopes within the I levels of factor A are homogeneous. This assumption is tested by using the conventional homogeneity of regression test on the I levels of factor A (not on the IJ cells). If this test reveals heterogeneity of slopes for the different levels of factor A, follow-up homogeneity of regression tests on the pairs of levels of factor A should be run in the case that I is greater than two. Use ANCOVA to test the adjusted effects on factor A for those levels having homogeneous slopes. Use the Johnson–Neyman technique to analyze comparisons involving levels of A with heterogeneous slopes. The same procedure is recommended for factor B. The general rule to keep in mind is that homogeneity of slopes should be present for the levels or cells being compared with ANCOVA. If this is not the case for any comparison of interest, use the Johnson–Neyman technique for those comparisons that involve heterogeneous regression slopes. Two other approaches to the analysis of two-factor designs with heterogeneous slopes have been described by Reeder and Carter (1976) and Searle (1979).

13.7 INTERPRETATION PROBLEMS

The Johnson–Neyman procedure yields results that are generally no more difficult to interpret than are ANCOVA results. There are three issues however, that should be kept in mind when interpreting the outcome of a $J - N$ analysis: (1) the extrapolation problem, (2) the possibility of having heterogeneous slopes but no subjects for which differences are significant, and (3) simultaneous inference problems in the situations where two or more regions are computed. These points are discussed in the remainder of this section.

The Extrapolation Problem

It is important to restrict the generalization of results to the range of values of the covariate(s) included in the sample. Consider the data illustrated in Figure 13.2. The $J - N$ analysis indicates that X scores below 6.04 and above 7.06 are associated with significant differences between the two treatments. Notice, however, that the range of X included in the study is 1 through 11. Suppose we want to know if the effects of the two treatments are different for subjects with covariate scores of 20. It is possible to test this hypothesis by using the Johnson–Neyman procedure, but, strictly speaking, no inferential statements should be made concerning treatment effects at levels of X outside the range of the scores included in the available data. If there are no data on subjects with X scores near 20, we have no justification for making statements concerning treatment effects for subjects with such extreme X scores. It might not be unreasonable to extrapolate results to X scores of 12 or 13, but extrapolation to more extreme scores would be poor practice. If the effects of the treatments for subjects with X scores of 20 is important, the investigator should make sure to include subjects with this score in the experiment. Extrapolation is useful for hypothesis generation, but it can be dangerous if it is not recognized as a form of speculation.

Inconsistencies in Outcomes of Homogeneity of Regression and Johnson–Neyman Tests

A large homogeneity of regression slopes F ratio is usually a sign that the Johnson–Neyman procedure will reveal regions of significance and nonsignificance. It is possible, however, to obtain a significant homogeneity of regression F ratio, but no region of significance *within the range of the sample data*. This will generally occur only when the heterogeneity of slopes is slight and the associated F is just barely significant. The regions of significance may lie outside the range of covariate scores included in the sample. As has already been pointed out, a researcher should be very careful about making statements of significant differences for ranges of the covariate(s) for which there are no data.

Simultaneous Inference

The probability of making a Type I error in a two-group problem is the probability of rejecting H_0 when the true population difference between means *at a specified point on X* (in the one-covariate case) is zero. Hence if several points on X are selected and a statement concerning the population difference at each of these points is made, the probability of making a Type I error is α *for each statement*. Suppose, however, that we want to hold the probability of making a Type I error at α for all statements simultaneously. In other words, if we state that differences are significant for many different values of X, we want α to be the probability that one or more of the whole set of statements is incorrect.

Alternatively, if a 95% confidence interval is computed for *each* of many points on X using the conventional Johnson–Neyman procedure, we can state for each of the many confidence intervals that the probability is .95 that the obtained confidence interval will span the population mean difference. The probability that the whole collection of confidence intervals simultaneously includes all population differences is *not* .95. The exact probability can not be stated in this case. It is possible, however, to employ a procedure slightly different from the conventional Johnson–Neyman technique to obtain simultaneous tests and confidence intervals. The conventional and modified formulas for the two group case are presented as follows.

Conventional Johnson–Neyman Formulas

Individual Significance Tests	Individual Confidence Intervals
$F = (x_s' D D' x_s)/(x_s' V x_s)$ Critical value $= F_{[\alpha, 1, N - J(C+1)]}$	$x_s' D \pm \sqrt{F_{[\alpha, 1, N - J(C+1)]}} \sqrt{x_s' V x_s}$

Modified Formulas

Simultaneous Significance Tests	Simultaneous Confidence Intervals
Significance tests are computed by using conventional $J - N$ formula, but critical value is $(C+1) F_{[\alpha, C+1, N - J(C+1)]}$	$x_s' D \pm \sqrt{(C+1) F_{[\alpha, C+1, N - J(C+1)]}} \sqrt{x_s' V x_s}$

If more than two groups are involved, the Bonferroni F based on

$$df_1 = C + 1$$
$$df_2 = N - J(C + 1)$$
$$C' = \frac{J(J - 1)}{2}$$

is substituted for the conventional F in the preceding formulas. When this situation is encountered, we can be at least $100(1 - \alpha)$ percent confident that all confidence intervals across all $[J(J - 1)]/2$ group comparisons contain the population mean differences.

The modified formulas presented here are essentially the same as those presented by Potthoff (1964). Potthoff has presented another modified formula for simultaneous confidence intervals based on the same rationale as Scheffé's multiple comparison procedure. The procedure described in this section is generally more powerful than either Potthoff's Scheffé-type procedure (unless sample sizes are *very* small) or Erlander and Gustavsson's (1965) procedure. All these procedures have been compared by Cahen and Linn (1971).

In summary, the conventional Johnson–Neyman technique is appropriate if the experimenter wants to maintain the probability of making a Type I error at α for each individual point on X. If, on the other hand, the objective is to make a probability statement about a whole set of differences, the modified procedure can be employed. This will allow the statement that the probability of one or more false rejections of H_0 in the whole collection or set of statements is less than or equal to α.

Consider again the data analyzed in Sections 13.2 and 13.4. The first significance test of Section 13.4 yielded an F_{obt} of 23.71 for the covariate score 8. The critical value in that section was based on the conventional F statistic with 1 and $N - J(C + 1)$ degrees of freedom. It was appropriate to use this critical value because the experimenter wanted to hold the probability of making a Type I error at .05 for this particular comparison. Suppose, however, that the experimenter now wants to make a statement for each value of X in the experiment. Further, he or she wants the probability of making one or more Type I errors in the whole collection of tests to be less than .05. The modified formula for the critical value yields

$$2(3.37) = 6.74$$

Each test is evaluated with this critical value.

13.8 MULTIPLE DEPENDENT VARIABLES

If multiple dependent variables are employed, the experimenter has a choice of error rates to consider. Since this same problem has been discussed in Chapter 11 it is not repeated here in depth.

The experimenter may decide to treat each dependent variable as a separate experiment. In this case the conventional Johnson–Neyman procedure is applied to each dependent variable, and a separate statement (or set of statements if simultaneous tests are run on several levels of X) is made for each dependent variable. On the other hand, if the experimenter decides to treat the whole family of dependent variables as one experiment, he or she may want the probability of making a Type I error to be equal to or less than α for the whole family of dependent variables. The formulas for simultaneous significance tests and confidence intervals are changed by substituting the Bonferroni F critical value based on

$$df_1 = 1$$
$$df_2 = N - J(C + 1)$$
$$C' = p = \text{number of dependent variables}$$

for the conventional F statistic.

Suppose that five dependent variables rather than one had been involved in the two-group one-covariate problem presented in Section 13.2. The conventional $J - N$ analysis is performed on each dependent variable; the critical value (using a specified x_s) for each of the five tests or confidence intervals is

$$F_{B[\alpha, C', 1, N - J(C+1)]} = F_{B(.05, 5, 1, 26)} = 7.72.$$

If we want the probability of Type I error to be less than say .05 for all dependent variables *and* all levels of X simultaneously, the critical value is

$$(C + 1)F_{B[\alpha, C', C+1, N - J(C+1)]}$$

Hence if there are 30 subjects, two groups, one covariate, five dependent variables, and $\alpha = .05$, the critical value for each test is

$$(2)F_{B[.05, 5, 2, 26]} = (2)5.53 = 11.06$$

for each test.

13.9 MULTIPLE GROUPS, MULTIPLE DEPENDENT VARIABLES, ANY NUMBER OF COVARIATES

The formulas described in Section 13.8 are appropriate only for the two-group case. If the experiment involves multiple groups, multiple dependent variables, and any number of covariates and if the experimenter wants to state that the probability of Type I error is less than α for all tests across all groups and all dependent variables for *a specified* x_s, the critical value appropriate for each test is

$$F_{B[\alpha, C', 1, N - J(C+1)]}$$

where $C' = (p)[J(J-1)/2] = $ Number of "comparisons" employed in entering the Bonferroni table and p is the number of dependent variables.
Hence if there are 30 subjects, four groups, one covariate, five dependent variables, and $\alpha = .05$, the critical value is

$$F_{B(.05, 30, 1, 22)} = 12.83.$$

This approach is appropriate unless the experimenter wants the probability of Type I error to be less than .05 simultaneously for statements made for *all levels of X*, as well as all group comparisons, all dependent variables, and any number of covariates. In this case the following critical value should be employed.

$$(C+1)F_{B[\alpha, C', C+1, N - J(C+1)]}$$

where

$$C' = (p)[J(J-1)/2]$$

If we again use $N = 30$, $J = 4$, $C = 1$, $p = 5$, and $\alpha = .05$, the critical value is

$$(2)F_{B(.05, 30, 2, 22)} = (2)8.26 = 16.52.$$

13.10 NONLINEAR JOHNSON–NEYMAN ANALYSIS

The Johnson–Neyman technique, like the analysis of covariance, is based on the assumption that the relationship between the covariate and the dependent variable is linear. It is possible, however, to modify the analysis to deal with the situation in which the XY relationship is nonlinear. This modification of the Johnson–Neyman technique and a computer program for handling the rather complex computations involved have been presented in Wunderlich and Borich (1974) and Borich et al. (1976).

Before resorting to the Wunderlich–Borich approach, the nature of the nonlinearity should be identified. If the nonlinearity is marked but of monotonic form, a transformation of the original data may yield linearity. The transformed data can then be substituted for the original data in a conventional Johnson–Neyman analysis. In the unlikely situation where the nonlinear relationship is not monotonic, the Wunderlich–Borich modifications should be considered.

13.11 CORRELATED SAMPLES

The conventional Johnson–Neyman technique is appropriate for two independent samples. Where one sample is observed under two conditions, or where matched pairs are randomly assigned to two conditions, the independent sample formula is inappropriate. Dayton (1978) has worked out the correlated sample Johnson–Neyman formula.

The three terms A, B, and C that were defined previously for the independent sample case are redefined for the correlated sample case as follows:

$$A = \frac{\left[b_1^{(1)} - b_1^{(2)} \right]^2}{\left\{ \left[\left[\Sigma_{i=1}^{n}(e_{i1} - e_{i2})^2 / n - 2 \right] F_{(\alpha, 1, n-2)} \right] \right\} - \left(1 / \Sigma_{i=1}^{n} x^2 \right)}$$

$$B = \frac{2 \left[b_0^{(1)} - b_0^{(2)} \right] \left[b_1^{(1)} - b_1^{(2)} \right]}{\left\{ \left[\Sigma_{i=1}^{n}(e_{i1} - e_{i2})^2 / n - 2 \right] F_{(\alpha, 1, n-2)} \right\} + 2\bar{X} / \Sigma_{i=1}^{n} x^2}$$

$$C = \frac{\left[b_0^{(1)} - b_0^{(2)} \right]^2}{\left\{ \left[\Sigma_{i=1}^{n}(e_{i1} - e_{i2})^2 / n - 2 \right] F_{(\alpha, 1, n-2)} \right\} - (1/n) - \left[\bar{X}^2 / \Sigma_{i=1}^{n} x^2 \right]}$$

where

$b_0^{(1)}$, $b_0^{(2)}$ = the regression intercepts for groups 1 and 2, respectively

$b_1^{(1)}$, $b_1^{(2)}$ = regression slopes for groups 1 and 2, respectively,

$\qquad e_{i1}$ = prediction error for ith subject under condition 1; that is, $e_{i1} = (Y_{i1} - \hat{Y}_{i1})$, where $\hat{Y}_{i1}$ is value predicted from X_i using regression of Y_1 on X

$\qquad e_{i2}$ = prediction error for ith subject under condition 2; that is, $e_{i2} = (Y_{i2} - \hat{Y}_{i2})$, where $\hat{Y}_{i2}$ is the value predicted from X_i using the regression of Y_2 on X

$F_{(\alpha, 1, n-2)}$ = critical value of F statistic based on 1 and $n-2$ degrees of freedom

n = number of subjects (in one-sample case) or number of pairs of subjects (in matched-pair case),

$\sum_{i=1}^{n} x^2$ = total sum of squares on covariate,

$\overline{X}$ = covariate mean

The values obtained for A, B, and C are entered into the conventional Johnson–Neyman formulas for the region of nonsignificance. The limits of the region of nonsignificance are

$$X_{L1} = \frac{-B - \sqrt{B^2 - AC}}{A}$$

$$X_{L2} = \frac{-B + \sqrt{B^2 - AC}}{A}$$

13.12 COMPUTER PROGRAMS

Computer programs for conventional linear and nonlinear models are available in Borich et al (1976). Rogosa (1977) has developed programs for complex versions of the Johnson–Neyman procedure that have not been described in this chapter. Among them are variants of procedures for one covariate, for quadratic functions, and for two covariates. He also describes a program for dealing with the problem of measurement error. A program for the analysis of the correlated sample case is available from Dayton (1978).

13.13 SUMMARY

The assumption of homogeneity of regression is very important for the correct interpretation of ANCOVA. When this assumption is violated, the Johnson–Neyman procedure is an appropriate alternative analysis. Rather than evaluating the treatment effects at the covariate mean, as is the case with ANCOVA, the Johnson–Neyman technique evaluates the treatment effects as a function of the level of the covariate. Regions of the X dimension are identified that are associated with significant differences between the treatment groups (if such differences exist). Likewise, a region of nonsignificance is identified that is the range of X scores associated with nonsignificant differences. Extensions of the Johnson–Neyman technique are available to include the case of several groups, several covariates, two-factor designs, multiple dependent variables, and correlated samples.

CHAPTER FOURTEEN
Analysis of Nonequivalent Groups: Part I. True-Score ANCOVA

14.1 INTRODUCTION

The issue of measurement error has been discussed in several previous chapters. In this chapter we (1) review the effects of measurement error in X and in Y, (2) discuss the relationship between several research designs and the effects of measurement error and (3) describe the true-score ANCOVA procedure for analyzing certain designs. The major focus here and in Chapter 15 is on the problem of analyzing the nonequivalent-group pretest–posttest design. The large amount of space devoted to this design is not an indication of our preference for this design. Indeed, the reason for the detailed discussion is because the analysis of this design is so messy. Randomized-group designs are, of course, highly preferred, but the real world is such that some version of the nonrandomized design is often the only alternative to no design at all. In some applied areas, such as program evaluation, it appears that this is the most frequently employed design. Even the discussion here and in Chapter 15 only scratches the surface of the issues involved in analyzing this weak design.

14.2 EFFECTS OF MEASUREMENT ERROR IN DEPENDENT VARIABLE AND IN COVARIATE: RANDOMIZED-GROUP CASE

In this section we briefly review the effects of error in the measurement of the dependent variable and the covariate for the case of randomized group experiments. The effects of measurement error in the case of nonequivalent-group pretest–posttest designs are described in Section 14.3.

Measurement error in the dependent variable is *not* a major problem because the slope associated with the regression of Y on X is not affected. Suppose that the regression of Y on X yields a slope of 1.0 as is shown in Figure 14.1a. Notice that the data are perfectly fit by a straight line and that

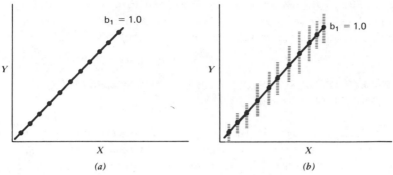

Fig. 14.1 Effect of adding measurement error to dependent variable scores: (*a*) no error in *Y*; (*b*) error in *Y*.

there is no variability around the regression line. Figure 14.1*b* illustrates the nature of the distribution when measurement error is added to the dependent variable scores. The slope is still 1.0 after measurement error has been added, but the regression line no longer is a perfect fit. There is now greater variability on *Y*, and this additional variability is error variability that will contribute to the error mean square (MS). In the case of ANCOVA, where two or more groups are involved, the effect of error in *Y* is an increase in the MS residual within groups, but the pooled slope (b_w) is not affected. This means that the power of the *F* test is decreased and that the width of the confidence intervals is increased. But since the slope is not affected, the adjusted means are not affected and no bias is introduced to the adjusted mean difference.

When measurement error is present in the covariate(s), the effects depend on the experimental design. In the case of randomized-group designs, if the covariate is measured before the treatments are administered (and is thus not affected by them), the power of ANCOVA is reduced relative to what it would be if no error were present, but treatment effects are not biased. Hence the effect of measurement error on either the covariate or the dependent variable is the same (i.e., reduced power) in the case of randomized experiments where the treatment does not affect the covariate. With other designs the effects of measurement error in *X* are likely to be serious. If a randomized design is employed but the covariate is affected by the treatments, biased treatment effects are likely to result. Likewise, if subjects are neither randomly assigned to treatments nor assigned to treatments on the basis of the covariate score (as is done with the biased assignment design), biased treatment effects are likely to result from measurement error in *X*. With the exception of the biased assignment experiment, the experimenter should seriously question the use of ANCOVA if the expected mean difference between covariate means is not zero. The use of true-score ANCOVA should be considered in this case. There are varieties of nonrandomized experiments in which this analysis will yield unbiased treatment effects. Other varieties of nonrandomized experiments are not appropriately analyzed with either conventional ANCOVA or true-score ANCOVA. Distinctions among varieties of this design must be made to select appropriate analyses.

14.3 EFFECT OF MEASUREMENT ERROR IN COVARIATE: NONEQUIVALENT-GROUP CASE

Two situations were described in previous chapters (Sections 6.5 and 7.2) in which measurement error in X often results in bias in the estimate of the adjusted mean difference. The first situation involves a randomized-group design where the covariate is affected by the treatments. The second situation involves the nonequivalent group design. Both designs will generally yield biased treatment effects with ANCOVA if the measurements on the covariate do not contain the information required to appropriately adjust the treatment effects. In the case of the treatment affecting the covariate in randomized designs, the use of the method described later in this chapter—true-score ANCOVA, is generally recommended. In the case of nonequivalent-group pretest–posttest designs, true-score ANCOVA is one of several alternatives that may be appropriate. We attempt to provide an overview of some of the issues that should be considered in analyzing these designs in the remainder of this section.

The question of greatest relevance in attempting to conceptualize the problems in the analysis of nonequivalent-group pretest–posttest designs is, "What would the between-group difference on the posttest be if there were no treatment effect?" Unfortunately, the answer to this question is not always simple, because, to a large extent, it depends on how the subjects are selected into the two (or more) groups. If subjects are selected (or assigned) into groups on the basis of (1) randomization or (2) covariate score, there is little ambiguity about how to answer the question. If the process through which subjects end up in the two groups is *not* known, there is no way that is known to be unbiased to provide an answer to the question. The best that can be done in this case is to conjecture about the nature of the selection process (i.e., postulate some selection models) and then estimate what the posttest difference would be if the guesses were correct.

The reason why it is essential to know the expected between-group difference on the posttest under the condition of no treatment effect is quite straightforward. If we don't know what the difference would be when there is no treatment effect, there is no way to unambiguously interpret the mean difference. The key issue here is finding a way to separate the effects of selection (which are generally associated with nonequivalent groups) from the effects of the treatments. When the expected posttest difference is known, that difference can be compared with the obtained difference. If the obtained difference after applying treatments is much larger or smaller than the difference that is expected when treatments are not applied, a test of the obtained against the expected difference is employed to identify treatment effects. For example, if a 10-point difference is expected between two means based on samples from two different populations but a 25-point difference is obtained in an experiment, 10 of the 25 points are explained by selection, and 15 of the 25 points are described as the difference apparently due to treatments and sampling error. An alternative approach to the problem is to adjust the posttest difference in such a way that the expected difference is set equal to zero and to then directly evaluate any nonzero difference as a reflection of the treatment effects and sampling error. The methods used to arrive at zero differences between the expected adjusted means are the topic of this chapter and Chapter 15.

Varieties of Pretest–Posttest Design

Before moving on to the details of various analysis strategies, we briefly review several experimental designs that have been described previously. Some new varieties of the nonequivalent-group design are then described. A comparison of these designs should clarify the nature of some of the analysis issues that must be considered.

Randomized Groups.   The randomized-group pretest–posttest design involves the random assignment of subjects to treatment conditions and the measurement of all subjects before and after the treatments are administered. Analysis of covariance is the appropriate method of analysis for this design. As can be seen in Table 14.1, the expected mean difference on both the X and the Y variables is zero with this design, thus suggesting that the adjusted population mean difference will appropriately be zero when the null hypothesis is true.

Randomized Groups: Covariate Affected by Treatment.   The second design listed in the table is also a randomized-group design, but the expected adjusted mean difference is not zero because the covariate is affected by the treatment. (Note that this is the only design in Table 14.1 that is not a pretest–posttest design). When this occurs, the expected difference between covariate means is generally not zero, and, unless the slope (β) is zero, the adjusted mean difference is affected by the covariate mean difference. True-score ANCOVA is generally appropriate in this situation. When true-score ANCOVA is appropriately employed with this design, the expected adjusted mean difference is zero.

   This recommendation to employ true-score ANCOVA is based on the assumption that the measurement error in X is independent of the measurement error in Y. This situation must be discriminated from the case in which measurement error in X and Y are correlated. If measurement error is perfectly correlated in X and Y, the appropriate analysis is conventional ANCOVA. Since the degree of correlation between measurement error in X and measurement error in Y is not known, the selection of the most appropriate analysis is difficult. It is reasonable to compute both conventional ANCOVA and true-score ANCOVA with this design. The correct estimate of

Table 14.1 Generally Expected Posttest Difference Using ANCOVA with Six Designs When No Treatment Effects on Y Are Present

Design	Expected Pretest Mean Difference	Expected Posttest Mean Difference	Expected Posttest Adjusted Mean Difference
Randomized groups: covariate unaffected by treatment	0.0	0.0	0.0
Randomized groups: covariate affected by treatment[a]	($\neq$0.0)	0.0	$\neq$0.0
Biased assignment: groups formed on basis of pretest (covariate) scores	$\neq$0.0	$\neq$0.0	0.0
Nonequivalent groups: formed on basis of covariate and additional relevant data	$\neq$0.0	$\neq$0.0	$\neq$0.0
Nonequivalent groups: formed on basis of selection from different populations	$\neq$0.0	$\neq$0.0	$\neq$0.0
Nonequivalent groups: formed on basis of unknown selection factors	$\neq$0.0	$\neq$0.0	$\neq$0.0

[a]This is not a pretest–posttest design.

the adjusted mean difference should fall somewhere between the estimates provided by these analyses.

Biased Assignment.　The biased assignment design involves the assignment of subjects to groups on the basis of the magnitude of the covariate score. For example, all subjects with pretest (covariate) scores above the median may be assigned to one treatment condition, and all subjects with pretest scores below the median may be assigned to another treatment condition. Notice in Table 14.1 that the expected difference between the covariate or pretest means is not zero with this design, but that the expected difference between adjusted means *is* zero. This implies that ANCOVA is the appropriate method to use in analyzing this design.

Nonequivalent Groups: Selection Based on Covariate and Additional Covariate Relevant Information.　The next design can be viewed as a biased assignment design gone wrong. In this situation the subjects are initially assigned to treatments on the basis of their covariate scores; then exceptions are made concerning the criteria for selection. Suppose that the biased assignment approach is involved in the assignment to treatment 1 of all subjects with X

scores below the median and assignment to treatment 2 of all subjects with X scores at or above the median; then a second stage of assignment is initiated in which it is decided that certain additional information will be used to make the assignment decision. This second stage creates problems; notice in Table 14.1 that the expected difference between adjusted means is not zero. The reason why the use of additional information in the assignment process causes problems in attempting to estimate the expected adjusted mean difference is rather subtle and elusive. A hypothetical example might help to clarify the issue.

Let us say that a compensatory reading program is to be evaluated. Subjects are pretested on a short and fairly unreliable reading test before the program begins. The investigator then assigns students to the compensatory reading program if they score below the median on the pretest; those who score at or above the median are assigned to the traditional program. Then, for some reason, it is decided to consider additional information concerning students' reading skills by an examination of (1) teachers' evaluations, (2) parents' evaluations, and (3) students' self-evaluations. These various sources of information are considered in deciding which students are assigned to the compensatory program and which are assigned to the traditional program. There is not, however, a clear-cut rule for combining the information from the various sources. If ANCOVA is applied by using the originally administered pretest as the covariate and another reading test as the dependent variable, this analysis is unlikely to yield an unbiased result. This design is not a biased assignment design because the pretest is not the only basis for assigning subjects to treatments. The analysis problem here is that ANCOVA will adjust the means as they should be adjusted under the condition that the covariate *as measured* is the *only* variable involved in the assignment process. Since subjects were *actually* assigned on the basis of information that was probably more reliable than that provided by the reading pretest alone, the ANCOVA will underadjust the means; the expected mean difference will not be zero when there are no treatment effects. If the de facto assignment is on the basis of true scores, true-score ANCOVA will yield an expected adjusted mean difference of zero.

In general, if the reliability of the measurement used in the assignment of subjects to treatments is higher than the reliability of the measure used as the covariate, ANCOVA will be biased. The problem is illustrated in Figure 14.2, where the two graphs represent hypothetical regression slopes that are associated with the example described previously. The left-hand graph illustrates the biased assignment design in which the covariate is the only information used to assign subjects to treatments. Notice that the adjusted means are equal in this case. The right-hand graph has the same covariate means and the same slopes as does the left-hand one, but the adjusted means are very different. The explanation for the difference in outcome illustrated in the two graphs can be found in the difference in the reliability of the measure used to assign subjects to treatments in the left-hand graph relative to the

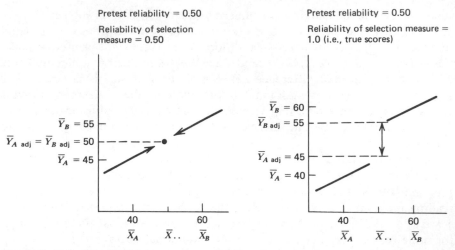

Fig. 14.2 Expected adjusted means when reliability of covariate is same as and different from reliability of selection measure (H_0 true).

reliability of the information used in the right-hand graph. It is necessary to remember how the regression artifact comes about if these graphs are to be understood.

We leave the realistic example illustrated in Figure 14.2 for a few paragraphs to first explain a less realistic but more easily understood example.

Suppose that the reading pretest has zero correlation with the posttest because the reliability of the pretest is zero. If the reliability is zero, all the pretest variability is viewed as random fluctuation. If a subject obtains a pretest score that is very high, for example, there is no reason to expect the posttest score to be very high; or if the pretest score is very low, there is no reason to expect the posttest score to also be very low. In terms of standard scores, the same idea holds. If a subject has a high standard score on an unreliable pretest, there is no reason to expect a high standard score on the posttest. The lack of relationship between standard score deviation on the pretest and the posttest can be seen if we employ the following formula for predicting posttest standard scores from pretest standard scores:

$$\hat{Z}_y = r_{xy}(Z_x)$$

In keeping with our example, $r_{xy} = 0.0$. This means that regardless of the pretest standard score, the predicted posttest standard score is zero. The same principle applies to group means. If the pretest means for the two groups in a biased assignment experiment are, in standard score units, -1.0 and 1.0, the expected posttest mean standard score is zero for each group:

$$r_{xy}(\bar{Z}_x) = \hat{\bar{Z}}_y$$
$$0.0(-1.0) = 0.0$$
$$0.0(1.0) = 0.0$$

In other words, the group means are expected to regress to the mean of the single population from which they were selected. Any procedure used to analyze data from a design such as this one should take into account the regression phenomenon; ANCOVA does this. The conventional adjustment formula for ANCOVA does correctly predict the posttest difference in this situation where $r_{xy} = 0.0$. This is shown in the following formulas for an example where the pretest means are 40 and 60, the grand pretest and posttest means are 50, and $r_{xy} = 0.0$:

$$\overline{Y}_i - b_w\left(\overline{X}_i - \overline{X}..\right) = \overline{Y}_{i\,\text{adj}};$$

hence

$$50 - 0.0(40 - 50) = 50$$
$$50 - 0.0(60 - 50) = 50$$

Notice that the posttest means are the same for both groups even though the pretest scores are 40 and 60; this is because the pretest scores are completely unreliable. Also notice that the adjusted means are equal.

The most important point to grasp in this example is that ANCOVA provides the appropriate expected mean difference between the adjusted means. Hence when the subjects are assigned to treatments on the basis of the observed pretest scores, ANCOVA provides the appropriate analysis regardless of the reliability of the pretest or covariate scores. As long as the information used to assign subjects to groups is the *same* information employed as the covariate, there is no problem of bias in the adjusted treatment mean difference. This is the case illustrated in the left-hand graph in Figure 14.2. If the covariate (pretest) is not the *only* information employed to assign subjects to groups, ANCOVA is likely to provide a biased estimate of treatment effects because, in this case, the covariate does not contain all the information concerning the assignment process. The result of this inconsistency between the covariate and the information actually used to form the groups is illustrated in the right-hand graph in Figure 14.2. Some additional explanation of the effect illustrated may be helpful.

Let us say that the covariate is not correlated with the dependent variable, as in the previous example, but that an additional measure that *is* reliable and *is* correlated with the dependent variable is used to assign subjects to treatments. In this case the posttest means are *not* expected to regress to the same value. That is, the expected mean difference is not zero in this case because the measure used to form the two groups systematically measures real differences in the samples that are also reflected on the posttest. If the unreliable pretest is employed as the covariate, the ANCOVA estimate of the posttest difference will be zero when the expected difference is not zero. Analysis of covariance will predict complete regression of the posttest means because the covariate is completely unreliable. The posttest means will not, however, regress as ANCOVA predicts because the measure used to form the groups is not completely unreliable and is correlated with Y. Hence the expected degree of regression predicted by ANCOVA does not take place, and the adjusted means are not unbiased estimates of the expected posttest means. Under a true null hypothesis, the expected adjusted mean difference is not zero.

This example is somewhat extreme because the hypothetical reliability of the covariate was set at zero. But the same basic problem occurs with more realistic reliability coefficients. A more realistic situation is presented in Figure 14.2, where the reliability of the covariate is 0.50. In the case that the covariate (pretest) is the only variable considered in the assignment of subjects to treatments, the adjusted means are the correct estimate of the posttest mean difference. This is shown in the left-hand graph in Figure 14.2. The right-hand graph illustrates the adjusted mean difference that will result when the measure used to assign subjects to groups does not have the same reliability as the covariate. If the measure actually used in the assignment has perfect reliability (i.e., assignment is on the basis of pretest true scores), the analysis is biased by the adjusted difference illustrated in the right-hand graph. It can be seen, then, that it is critical for the investigator to know whether the covariate is the only variable used to assign subjects to treatments. If it is likely that the assignment of subjects has been made on the basis of additional information that yields a much more reliable measure *of the characteristic purportedly measured by the covariate*, two alternatives to conventional ANCOVA are appropriate.

The preferable alternative is to identify the process through which subjects were actually assigned to groups. In the example of the compensatory reading program, it was pointed out that the different sources of information included in the assignment process were (1) the pretest, (2) teacher's evaluations, (3) parents' evaluations, and (4) students' self-evaluations. These sources were combined in some unknown manner to make assignment decisions. What is required is a procedure to capture the decision-making strategy that was employed. The question is, "How were the four sources of information combined in making the assignment decision?" If the information from the four sources was combined into some "composite score" during the decision-making process, each subject's composite score, if identified, can be used as the covariate in a conventional ANCOVA. If the composite contains all the information used to make the assignment decision, the use of this composite with ANCOVA will yield unbiased treatment effect estimates. Unfortunately, a well-defined method of combining the data from different sources is seldom employed. It is possible in this case to model the decision making strategy.

Suppose that each of the four sources of information described in the preceding paragraph yields a score. The scores may be either continuous or dichotomous. If it is possible to obtain a score for each subject on the pretest and each of the three evaluations (teachers', parents', and self-), the following approach will yield an estimate of the assignment model:

1. Construct a group membership identification variable by assigning a "one" to all subjects in the treatment group and a "zero" to all subjects in the control group.

2. Using a multiple linear regression computer program, regress the zero–one group-membership variable on the variables used in the decision making.
3. Compute the predicted group membership score for each subject from the regression equation, that is,

$$\hat{Y} = b_0 + b_1 X_1 + b_2 X_2 + \cdots + b_m X_m$$

where
   $\hat{Y}$ = predicted group-membership score
   b_0 = estimated slope,
   $b_1 - b_m$ = estimated partial regression coefficients
   $X_1 - X_m$ = scores used in decision-making process
4. Use the predicted group-membership scores as the covariate and posttest scores as the dependent variable in a conventional ANCOVA.

If the predicted group-membership scores are all below one point for one group and above that point for the other group, an approximation of the selection process has been identified and the analysis based on this procedure can be expected to yield more accurate estimates of treatment effects than would be obtained using only the pretest as the covariate. But if the predicted group membership scores for the two groups overlap, the estimated treatment effects are likely to be biased. Overlap is an indication of failure to model the selection process; the process of assigning subjects to groups is not completely explained by a linear combination of the sources of information that are believed to explain the assignments. If the sample sizes are large and the degree of overlap of predicted group-membership scores is low, those subjects whose group membership is not correctly predicted by the regression equation might be deleted from the analysis before ANCOVA is carried out. The results of the ANCOVA based on only those subjects who fit the assignment model suggested by the regression equation do not, of course, generalize to subjects like those deleted from the analysis.

When there are no data available on the various sources of information employed in the assignment process other than the pretest, the approach described in the preceding paragraph cannot be used. In this case it is suggested that two versions of true-score ANCOVA be employed, based on (1) true-scores estimated by using a measure of internal consistency (e.g., coefficient alpha) and (2) true-scores estimated by using the pooled within-group pretest–posttest correlation as the reliability estimate. (Computational details are presented in Section 14.4) Two confidence intervals can then be constructed, with one based on each true-score analysis. It is likely that the population mean difference will be contained between the lower of the two lower limits and the higher of the two upper limits. The validity of the true-score ANCOVA approach is dependent on whether the subjects are assigned to groups on the basis of information that, in effect, is the covariate true-score dimension.

The assignment information may be more reliable than observed covariate scores but less reliable than true covariate scores. In this case neither ANCOVA nor true-score ANCOVA will yield unbiased treatment effects. Suppose that the reliability of the observed scores is 0.50 and the reliability of the variable used in the selection process is 0.80. The correct estimate of treatment effects will fall between the estimates associated with ANCOVA and the estimates associated with true-score ANCOVA—unless there are additional factors that invalidate both analyses.

If the assignment to groups is made on the basis of information (regardless of its reliability) that is *not* related to the dimension measured by the pretest, there is no reason to believe that true-score ANCOVA will provide appropriate estimates of treatment effects. Unfortunately, it is often impossible to know the basis of the assignment or selection decisions. In this case it is reasonable to go ahead and compute true-score ANCOVA as one of several analyses and to then state that the results of this approach are what is expected when the subjects are assigned to treatments on the basis of true scores. These results can then be compared with the results of other analyses that are based on different assumptions concerning the selection or assignment process. Two other varieties of the pretest–posttest design, based on different selection models, are described next.

Nonequivalent Groups: Selected from Different Populations. Suppose that subjects are selected from two populations that are known to differ on the response measure, such as low and high socioeconomic groups. If these subjects are pretested, given different treatments, and then posttested, it can be seen that this is another type of nonequivalent-group pretest–posttest design. Clearly, the process through which the subjects have ended up in the two groups can be considered neither a random process nor a biased assignment process. That is, with this design the groups are formed by selecting subjects from different populations rather than by randomly assigning subjects or assigning subjects to groups on the basis of the pretest scores. This is a design that should never be used if there is a possibility of employing a randomized or biased assignment design instead. The basic problem in analyzing this design—as with other nonequivalent-group designs – is to separate treatment effects from differences that are due to the nonequivalence of the groups. The analysis of this design is described in Chapter 15.

Nonequivalent Groups: Unknown Selection Factors. The last variety of nonequivalent group design that we describe is where the groups have been formed on the basis of unknown selection factors. Unlike the previous design, where the groups were selected from defined populations, this design involves two groups on which pretest and posttest scores are available, but there is no information concerning the process through which subjects have ended up in the two groups. In the complete absence of information concerning the selection process, the best analysis strategy that can be suggested is to apply different analyses that are based on different assumptions concerning the selection model and to then compare the various results. If the different analyses yield essentially the same results, the investigator will have some confidence that the overall conclusion is valid. If the analysis based on one selection model (e.g., selection based on true scores) is not consistent with the analysis based on another selection model (e.g., selection from different populations), the inconsistency should be reported. It should be pointed out that the estimates depend on how the selection process is viewed. There is *no* analysis procedure that will yield unambiguous results with these designs. Even if many different analyses yield the same conclusion, all may be biased in the same direction.

The following analyses are suggested for this unknown selection design if the only available data are pretest and posttest scores:

1. Conventional ANCOVA (pretest as covariate).
2. True-score ANCOVA (using both internal consistency and pooled within-group pretest–posttest correlation to estimate true scores).
3. Standardized change-score analysis (described in Chapter 15) and/or gain-score ANOVA.

ANALYSES BASED ON ADDITIONAL INFORMATION.   Sometimes the "shotgun" approach recommended previously can be improved on by employing information in addition to the pretest and posttest data that may clarify the nature of the selection process. When additional data help to describe the selection process, one or more of the analyses suggested earlier can be eliminated, and hence fewer estimates of treatment effects need be considered. Such information includes (1) pre-pretest scores, (2) background characteristics, and (3) other variables that describe the selection process. Methods of employing each of these types of information are described next.

Suppose that the data available on the two nonequivalent groups include scores that have been collected at two different points in time before treatments are administered (i.e., pretest and pre-pretest scores), as well as posttest scores. Also, suppose that the amount of time between the pre-pretest and the pretest is the same as the amount of time between the pretest and the posttest. We use X_1 to denote the pre-pretest, X_2 to denote the pretest, and Y to denote the posttest. The availability of the pre-pretest scores makes it possible to carry out an approach sometimes called "dry-run" experimentation (Campbell 1974; Director 1974). The idea here is quite straightforward. First, ANCOVA is carried out by using X_1 as the covariate and X_2 as the dependent variable. If this dry-run analysis yields a very small and nonsignificant difference between adjusted means, ANCOVA is considered acceptable for the "real" analysis using X_2 as the covariate and Y as the dependent variable. But if the dry-run analysis yields a significant "treatment" effect, it is concluded that ANCOVA is biased since there has been no treatment administered in this dry-run experiment. If the difference between adjusted means in this first analysis is significant and is not near zero, true-score ANCOVA should be computed next. True-score ANCOVA based on the pooled within-group correlation between X_1 and X_2 and coefficient alpha or some other measure of internal consistency should both be carried out. If either of the true-score ANCOVA approaches yield a nonsignificant result with an adjusted mean difference near zero, it is concluded that an appropriate reliability estimate has been identified. The next step is to carry out another true-score ANCOVA by using estimated X_2 true scores as the covariate (where the true scores are estimated by using the reliability coefficient identified in the previous dry-run analysis) and Y as the dependent variable. The results of this analysis are presumed to yield an unbiased estimate of the treatment effect under the assumption that the behavior of the subjects from X_2 to Y is simulated by the behavior from X_1 to X_2.

If, in the analysis of dry-run data (i.e., X_1 and X_2), neither ANCOVA nor true-score ANCOVA yield adjusted mean differences that are near zero and nonsignificant, gain-score ANOVA and/or standardized change-score analysis (described in Chapter 15) should be computed. These procedures are based on other selection models that may fit the dry-run data. If gain-score ANOVA on $X_2 - X_1$ yields group differences that are effectively zero, this analysis should be used on the $Y - X_2$ gains. If standardized change-score analysis yields group differences that are effectively zero, this analysis should be used. If both gain-score ANOVA and standardized change-score analysis yield essentially no difference, gain-score ANOVA should be used (it is simpler).

In summary, when it is not known how subjects have been selected into groups, the use of dry-run analyses can facilitate identification of the selection model. If subjects have been selected from a single population on the basis of observed X scores, ANCOVA is the appropriate analysis. If subjects have been selected from a single population on the basis of true X scores, true-score ANCOVA is appropriate. If subjects have been selected from two different populations, gain-score ANOVA or standardized change-score analysis may be appropriate. If any of these selection models describe the selection process for the two groups from a nonequivalent design, the estimated effect on X_2 scores should be zero since no treatments have been applied. Any analysis that does not yield a nonsignificant and near-zero difference on X_2 should not be employed in the analysis of Y.

In some nonequivalent-group designs, data available in addition to the pretest and posttest may consist of background information such as race, occupational level, sex, occupation, age, educational level, and income. These variables can be employed in addition to the pretest as multiple covariates in a multiple covariance analysis. The rationale for including both pretest and background variables in the analysis is that both are relevant to the selection process. This general notion can be extended to the cases in which selection is on the basis of either multiple true-score variables or pretest true scores and (fallible) background variables. Analyses for these cases are described in Sections 14.8 and 14.9.

Rather than simply hoping that a collection of variables will remove bias in the comparison of nonequivalent groups, a systematic search and analysis of possible variables can be undertaken. Interview data obtained from, for example, subjects and program directors can provide useful hypotheses concerning the selection factors that are likely to be important in a study. If it is possible to obtain data on these variables after potential selection variables are identified, the following procedure is suggested:

1. Construct a group-membership variable by assigning a "one" to all subjects in the treatment group and a "zero" to all subjects in the control group.
2. Regress the zero–one group membership variable on all the potential selection variables by using multiple regression.
3. Identify the variables in the multiple regression that explain the most variance (e.g., by using the forward addition or stepwise variable selection procedures).
4. Employ the pretest and the variables identified in step 3 as covariates in a multiple ANCOVA.

The use of this procedure will not guarantee an unbiased estimate of the treatment effect since many important selection variables may not be included, but it will generally yield less biased estimates than would be obtained using only the pretest as the covariate. If more than one measurement of each variable in the set is available, multiple covariate true-score ANCOVA should also be computed. If the results of conventional multiple ANCOVA and multiple covariate true-score ANCOVA are not in agreement, it will not be possible to say which of the two results are least biased unless additional information concerning the selection process is known. The only way to know whether true-score estimates or observed scores should be used in the analysis is to know whether the decision concerning a subject entering the treatment or control condition was made on the basis of true or observed scores on the selection variables.

14.4   PORTER'S TRUE-SCORE ANCOVA

An appropriate solution to the problem of how to correct the ANCOVA for unreliability or measurement error was first presented by Lord (1960). Another procedure was suggested a few years later by Porter (1967). Porter's true-score ANCOVA is described here in preference to the Lord procedure because the latter is limited to the two-group case, whereas Porter's can be employed with any number of groups. De Gracie (1968), De Gracie and Fuller (1972), and Stroud (1972) have also proposed multiple group solutions.

    The Porter procedure involves carrying out a standard analysis of covariance by using estimated true covariate scores in place of obtained covariate scores. The true scores are estimated by using

$$\hat{T}_{ij} = \overline{X}_j + r_{xx}\left(X_{ij} - \overline{X}_j\right)$$

where

$\hat{T}$ = estimated true score for ith individual in jth group
X_{ij} = obtained covariate score for ith individual in jth group
$\overline{X}_j$ = covariate mean for subjects in jth group and
r_{xx} = estimated reliability of covariate

Reliability Estimation

There are many complex issues associated with the estimation of reliability for true-score ANCOVA that have not been resolved. Porter (1973), for example, suggests that the parallel forms estimate is the most appropriate. But a strong argument can be made that the pooled within-group correlation between the pretest and the posttest should be used (Campbell and Boruch 1975). Others (e.g., Linn and Werts 1973) have taken the stand that the use of the pooled within-group pretest–posttest correlation is unjustified. Since the basic psychometric issues of reliability estimation are beyond the scope of the present discussion, we refer the reader to the papers mentioned previously and to Cochran (1968) and Nunnally (1978).

Given the complexities associated with appropriately estimating reliability, a simple strategy is to compute more than one true-score ANCOVA with the use of different reliability estimates. First, estimate pretest reliability by using Cronbach's coefficient alpha (or some other measure of internal consistency) to find what is generally considered to be the upper bound of the reliability. [See Nunnally (1970, pp. 550–552) for a full discussion of the computation and interpretation of this estimate.] Then compute the pooled within-group pretest–posttest correlation to estimate the lower bound. Compute two true-score ANCOVA analyses, with the first using coefficient alpha and the second using the pretest–posttest correlation. If the results of both analyses lead to the same conclusions, greater confidence can be placed in the adequacy of the analysis. If the results are not consistent, the inconsistency should be reported to make the problems of the study apparent.

A major advantage of using coefficient alpha and the pretest–posttest correlation as reliability estimates is that only pretest and posttest data are required. Other estimation methods, such as parallel forms and test–retest, require the collection of additional data. If such data are available, it is worthwhile to compute additional reliability estimates from these data. This additional information is especially useful if it appears that the pretest–posttest correlation does not provide an appropriate reliability estimate. It is likely that the pretest–posttest within-group correlation is appropriate if the following assumptions are met:

1. The pretest and the posttest measure the same dimension.
2. The pretest and the posttest have the same reliability.

3. The treatment effects are additive (i.e., are the same for subjects with different covariate scores).

4. The pretest measurement error and the posttest measurement error are uncorrelated.

5. The amount of measurement error in the pretest is not correlated with the true pretest score.

It may seem unrealistic to question the assumption that the pretest and the posttest measure the same dimension since the posttest is simply a re-administration of the same instrument used as the pretest. It is known, however, that the factorial composition of a given instrument is not neces-sarily the same for two administrations; this is especially true if there is a long pretest–posttest lag. The effect of different pretest and posttest factorial composition on the results of the analysis usually is an overadjustment of treatment means.

The problem of different reliabilities for pretest and posttest measurements is also known to be a real concern. Pretest reliability is often different from posttest reliability when young children are observed on measures of aptitude or achievement. If we have parallel measurements for both pretest and posttest periods, it is possible to evaluate whether the reliability is the same during the two points in time. But if we compute the pooled within-group correlation between pretest and posttest measures and use this as the reliabil-ity estimate, we have no way of knowing whether the pretest and posttest reliabilities are the same. The pretest–posttest correlation (i.e., r_{xy_w}) often yields an estimate that falls between the pretest reliability and the posttest reliability if the pretest and posttest reliabilities are not the same. If the pretest reliability is actually much lower than the posttest reliability, the use of r_{xy_w} to estimate true scores will result in an undercorrection when true-score ANCOVA is carried out. On the other hand, if the pretest measurement error is highly correlated with the posttest measurement error (as might happen with very short pretest–posttest lag), true score ANCOVA will overcorrect; conventional ANCOVA is more appropriate in this case. The effect of not meeting the additivity assumption is to lower the correlation (r_{xy_w}), which means that the true-score ANCOVA yields adjusted means that have been overcorrected.

In summary, if only pretest and posttest data are available, it is suggested that both coefficient alpha and the pretest–posttest pooled within-group correlation be employed as reliability estimates in true-score ANCOVA. If additional data are available such as pretest scores on another form of the pretest or a retest, these data should be used to compute additional estimates of reliability. True-score ANCOVA based on several different reliability estimates will provide a range of results that is more useful than a single analysis based on a reliability estimate of unknown adequacy.

The formulas for the parallel form, test–retest, and pretest–posttest esti-mates are as follows.

Parallel-forms Reliability Estimate.

$$r_{x_A x_{B_w}} = \frac{\Sigma x_A x_{B_w}}{\sqrt{(\Sigma x_{A_w}^2)(\Sigma x_{B_w}^2)}}$$

where

$\Sigma x_{A_w}^2$ = pooled within-group sum of squares on form A of pretest
$\Sigma x_{B_w}^2$ = pooled within-group sum of squares on form B of pretest
$\Sigma x_A x_{B_w}$ = pooled within-group sum of deviation cross products for forms A and B

Test–Retest Reliability Estimate.

$$r_{x_1 x_{2_w}} = \frac{\Sigma x_1 x_{2_w}}{\sqrt{(\Sigma x_{1_w}^2)(\Sigma x_{2_w}^2)}}$$

where

$\Sigma x_{1_w}^2$ = pooled within-group sum of squares on first test administration of form A
$\Sigma x_{2_w}^2$ = pooled within-group sum of squares on retest administration of form A (this retest is administered *before* treatments are applied)
$\Sigma x_1 x_{2_w}$ = pooled within-group sum of deviation cross products for test and retest

Pretest–Posttest Pooled Within-group Correlation.

$$r_{xy_w} = \frac{\Sigma xy_w}{\sqrt{(\Sigma x_w^2)(\Sigma y_w^2)}}$$

where

Σx_w^2 = pooled within-group sum of squares on pretest
Σy_w^2 = pooled within-group sum of squares on posttest
Σxy_w = pooled within-group sum of deviation cross products for pretest and posttest

It is important to keep in mind that these estimates are all pooled within groups coefficients. Also note the distinction between the test–retest coefficient and the pretest–posttest coefficient. The test–retest measurements are both pretest measurements, either of which can be employed as the covariate. Both the test and the retest are obtained before the treatments are administered. This differs from the pretest–posttest coefficient, which is, of course, based on measures obtained before and after treatment.

14.5 TRUE-SCORE ANCOVA—COMPUTATIONAL EXAMPLE

Suppose that an experimenter is interested in evaluating the effectiveness of a remedial mathematics program in a high school. Randomized-group and biased assignment pretest–posttest designs are ruled out by the school administration. The experimenter is allowed to administer two forms of a short pretest to all students. A volunteer system is then employed to obtain students for the two groups. Students self-select the conventional (control) or remedial (experimental) treatment on the basis of their evaluations of their own true mathematical ability. Suppose, to keep the computation simple, that five subjects volunteer for each treatment. Treatments are carried out and posttest data are collected. There actually is no population difference in the effects of the two treatments, but the experimenter does not, of course, know this. The data (based on unrealistically small samples) follow.

Raw Data

Group 1—Experimental			Group 2—Control		
Posttest Y	Pretest $X_A{}^a$	Pretest X_B	Posttest Y	Pretest $X_A{}^b$	Pretest X_B
2	3	3	12	16	21
2	6	4	15	17	17
6	4	6	13	21	19
5	7	9	17	21	24
4	9	6	14	25	23
	$^a\bar{X}_{1_A} = 5.8.$			$^b\bar{X}_{2_A} = 20.$	

Reliability Estimation

Sums of Squares and Cross Products on Pretests

Group 1	Group 2
$\Sigma x_A^2 = 22.8$	$\Sigma x_A^2 = 52.0$
$\Sigma x_B^2 = 21.2$	$\Sigma x_B^2 = 32.8$
$\Sigma x_A x_B = 11.6$	$\Sigma x_A x_B = 23.0$

Pooled Within

$$\Sigma x_{A_w}^2 = 74.8$$
$$\Sigma x_{B_w}^2 = 54.0$$
$$\Sigma x_A x_{B_w} = 34.6$$

$$\text{Parallel-forms reliability estimate} = \frac{34.6}{\sqrt{(74.8)(54)}} = 0.54$$

Estimation of True Scores on Pretest (Form A)

Group 1		Group 2	
Subject	$\bar{X}_{1_A} + r_{x_A x_{B_w}}(X_i - \bar{X}_{1_A}) = \hat{T}$	Subject	$\bar{X}_{2_A} + r_{x_A x_{B_w}}(X_i - \bar{X}_{2_A}) = \hat{T}$
1	$5.8 + 0.54(3 - 5.8) = 4.288$	6	$20 + 0.54(16 - 20) = 17.84$
2	$5.8 + 0.54(6 - 5.8) = 5.908$	7	$20 + 0.54(17 - 20) = 18.38$
3	$5.8 + 0.54(4 - 5.8) = 4.828$	8	$20 + 0.54(21 - 20) = 20.54$
4	$5.8 + 0.54(7 - 5.8) = 6.448$	9	$20 + 0.54(21 - 20) = 20.54$
5	$5.8 + 0.54(9 - 5.8) = 7.528$	10	$20 + 0.54(25 - 20) = 22.70$

Data Layout for True-score ANCOVA

Y	D	$\hat{T}$
2	1	4.288
2	1	5.908
6	1	4.828
5	1	6.448
4	1	7.528
12	0	17.84
15	0	18.38
13	0	20.54
17	0	20.54
14	0	22.70

True-score ANCOVA

$$R^2_{yD,\hat{T}} = 0.91169$$
$$R^2_{y\hat{T}} = 0.89486$$
$$F_{obt} = \frac{(0.91169 - 0.89486)/1}{(1 - 0.91169)/7} = 1.33$$

The probability value associated with an F of 1.33 based on 1 and 7 degrees of freedom is .29.

A standard ANCOVA using the obtained X values rather than estimated true-score values as the covariate yields an F of 6.27. The probability value associated with this F is .04. Since the hypothetical population situation is one in which the treatment has no effect, we can see that only true-score ANCOVA results in the correct conclusion. If an α of .05 is employed, the standard ANCOVA results in the incorrect rejection of the null hypothesis concerning adjusted population treatment effects.

14.6 ASSUMPTIONS

Three factors, in addition to the assumptions for the ordinary ANCOVA (excluding random assignment), should be considered in employing true-score ANCOVA: (1) sample size, (2) reliability of the covariate, and (3) correlation of the dependent variable with the covariate. Porter's (1967) Monte Carlo investigation suggests that sample sizes of 20 or more are required for the true-score ANCOVA F statistic to be in close agreement with the theoretical F distribution. This appears to be true as long as the reliability is between 0.5 and 0.9, the correlation between the covariate and the dependent variable is between 0.6 and 0.9, and the number of groups is not over 4. If true-score ANCOVA is employed when $n < 20$, $\rho_{xx} < 0.5$ or > 0.9, or $\rho_{xy} < 0.6$ or > 0.9, the probability of Type I error may differ from the nominal α value by a considerable amount. Other investigations of Porter's procedure by Hughes (1973) and McLean (1975) generally support the use of this method to correct for measurement error in the covariate. Their results are consistent with Porter's.

If the two groups employed in the analysis are quite different from each other, the reliability coefficient (r_{xx}), which is a pooled within-group coefficient, may not be representative of the reliability of either group. In this case separate coefficients can be computed for each group and entered in the conventional formula used to estimate true scores.

An additional point that should be kept in mind is that the true-score ANCOVA procedure is based on the assumption that the errors of measurement in the covariate are distributed randomly and have a mean of zero within each treatment group.

14.7   EMPIRICAL EXAMPLE OF UNKNOWN SELECTION

Wortman et al. (1978) evaluated an education voucher program in which each student received a voucher that was paid directly to the school chosen by the parents. The idea underlying the voucher plan is that competition among schools for students will lead to improved education and schools that are more responsive to the needs of students. One specific purpose of the study was to evaluate the effects of the voucher schools relative to nonvoucher schools on students' test performance on the "cooperative primary reading tests." Test scores were obtained on the same groups of students across 3 years. The first 2 years (first and second grades) preceded implementation of the voucher plan; the first year of the plan was during the third grade. Data from five schools in the voucher plan and three schools not in the voucher plan were analyzed. The voucher schools were broken down into 11 "traditional" academic orientation minischools ($n = 150$) and five "nontraditional" academic orientation minischools ($n = 84$). The three nonvoucher schools ($n = 120$) were all classified as having a traditional academic orientation.

We classify this study as an example of a nonequivalent-group pretest–posttest design in which the selection into groups is based on unknown factors. It is clear with this design that subjects were not randomly assigned to voucher and nonvoucher schools and were not assigned on the basis of pretest scores. Since the selection factors were not known, the authors

Table 14.2   Comparison of Four Analyses of Voucher Study

Analysis	Estimated Mean Difference (Voucher–Nonvoucher)	F	Probability
Dry-run Analysis (Based on Pre-pretest and Pretest Scores)			
Gain-score ANOVA	0.68	< 1.0	—
ANCOVA	0.66	< 1.0	—
True-score ANCOVA (based on a pretest reliability estimate)	0.45	< 1.0	—
True-score ANCOVA (based on pretest–posttest correlation)	-0.33	< 1.0	—
Main Analysis (Based on Pretest and Posttest Scores)			
Gain-score ANOVA	-2.61	8.46	$<.01$
ANCOVA	-1.88	5.38	$<.05$
True-score ANCOVA (based on a pretest reliability estimate)	-2.07	6.52	$<.05$
True-score ANCOVA (based on pretest–posttest correlation)	-2.64	10.44	$<.01$

Source: Wortman et al. (1978).

computed multiple analyses. Comparisons of voucher and nonvoucher groups on third-grade reading performance yielded the results shown in Table 14.2. Notice that the same four analytic procedures were applied to the data from years 1 and 2 (the dry-run experiment) as were applied to years 2 and 3 (the main analysis). The dry run analyses reveal no significant effects, thus suggesting that none of these procedures will yield highly biased results when applied to the posttest data. The main analysis reveals general consistency among the four analytic procedures. The conclusion of all four analyses is that the voucher schools are inferior to the nonvoucher schools. (Additional analyses, not reported here, suggest that the lower overall performance of the voucher group was caused by a few low performing nontraditional subgroups or minischools.)

14.8   TRUE-SCORE ANALYSIS FOR MULTIPLE COVARIATE CASE

If subjects are assigned (or self-selected) to treatment groups on the basis of true-score information on two or more variables, Porter's true-score procedure is not appropriate since it handles the case of only one covariate. Several solutions to the problem of developing a true-score analysis procedure for multiple covariates have been proposed. The solutions can be classified into two general categories.

The first category consists of solutions that are essentially generalizations of the Lord (1960) and Porter (1967) approaches. Pravalpruk and Porter (1974), McGaw and Cummings (1975), and Misselt (1977) have proposed solutions that fall into this category. It appears, on the basis of Monte Carlo results provided by each of these investigators, that Misselt's procedure provides the most satisfactory solution among these three.

The second set of solutions is based on the structural equation model approach developed by Jöreskog (1970a, 1970b, 1973, 1974, 1977). The procedures described by Keesling and Wiley (1977), Magidson (1977), Rindskopf (1978), and Sörbom (1978) fall into this group. Sörbom's approach represents the state of the art in the analysis of the nonequivalent group design. His approach and the associated computer program are not, however, recommended for the casual data analyzer. Those who employ Sörbom's approach should be acquainted with structural equation models described in, for example, Goldberger and Duncan (1973) and Kenny (1979).

A simple approach to the problem of true-score analysis with multiple covariates is presented as follows:

1. Obtain pretest measures on m pretests (or any other variables to be used as covariates).
2. Compute the intercorrelation of all pretests; if all measures are positively correlated, proceed to step 3. If measures are negatively correlated, change the scoring of measures to yield positive intercorrelations among all pretests; proceed to step 3.
3. Transform all pretest measure scores to standard scores.
4. Compute the mean standard score for each subject:

$$\frac{Z_{1i} + Z_{2i} + \cdots + Z_{mi}}{m} = \bar{Z}_{Ai}$$

5. Obtain parallel form or retest measures on the m pretest variables.
6. Transform this second set of pretest scores to standard scores.
7. Compute the mean standard score on the second set of measures for each subject:

$$\frac{Z_{1i} + Z_{2i} + \cdots + Z_{mi}}{m} = \bar{Z}_{Bi}$$

8. Compute the pooled within-group correlation between the two sets of composite scores (i.e., $r_{\bar{Z}_A, \bar{Z}_B}$).
9. Carry out Porter's true-score ANCOVA, using $r_{\bar{Z}_A, \bar{Z}_B}$ as the reliability estimate and either Z_A or Z_B as the covariate.

Studies of the adequacy of this procedure relative to other multiple true-score ANCOVA procedures have not been conducted.

14.9 TRUE-SCORE ANCOVA WITH CONTROL FOR BACKGROUND VARIABLES

It is possible that the best guess of the selection model is one in which selection is based on true pretest scores *and* background variables (e.g., race, income, age). In this case an analysis can be performed that removes variability associated with the background variables from the estimated true scores (Kenny 1979). The following approach is suggested:

1. Regress the pretest on the group membership dummy variable and the background variable(s).
2. Predict X scores from the regression equation of step 1:

$$\hat{X}_{ij} = b_0 + b_1 d + b_2 BG_1 + b_3 BG_2 + \cdots + b_m BG_{m-1}$$

where

$\hat{X}_{ij}$ = predicted pretest score for ith individual in jth group
d = zero–one group membership dummy variable
$BG_1 - BG_{m-1} = m-1$ background variables

3. Estimate the reliability of the difference $(X - \hat{X})$ by using the methods of Section 14.4.
4. Estimate background corrected pretest true scores by using

$$\hat{T}_{.BG_{ij}} = r_{xx}(X_{ij} - \hat{X}_{ij}) + \hat{X}_{ij}$$

where

$\hat{T}_{.BG_{ij}}$ = background corrected true-score estimate for ith subject in jth group
r_{xx} = estimated reliability of difference $(X - \hat{X})$
X_{ij} = pretest score for ith subject in jth group

5. Compute a conventional multiple ANCOVA by using $\hat{T}_{.BG}$ and the background variables as covariates and the posttest as the dependent variable.

Other Sources of Bias in Nonequivalent-group Pretest–Posttest Designs

Measurement error is not the only reason for underadjustment of treatment means in the covariance analysis of these designs. There are often pretest–posttest fluctuations in the relative standing of subjects within groups that are uncorrelated with either the posttest behavior or the selection factors. This type of within-group fluctuation is a reflection of true changes that take place, but the effect of such fluctuation on the slope (b_w) is the same as with measurement error. Consequently, ANCOVA yields adjusted means that are biased. Suppose that the heart rate of two groups of subjects is measured at two points in time. Heart rate can be measured very reliably, but it is unlikely that a perfect correlation would be found between the pretest and posttest measurements, even with zero measurement error. Differences in the true relative standing of subjects within groups can be expected to occur on the posttest relative to the pretest. These differences may be the result of changes in exercise patterns, weight increases or decreases, diet, and so on. But the effect of these changes is to reduce the slope associated with the regression of posttest heart rate on pretest heart rate. Since there is a source of variation in the pretest heart rate that is related to neither the selection of subjects into groups nor the posttest scores, it can be expected to result in an underadjustment of the means (Cronbach et al. 1977).

There are several other problems that should be considered in the interpretation and analysis of nonequivalent-group designs. Adjusted posttest differences may not be zero in the absence of treatment effects if there are environmental changes between pretesting and posttesting that affect one group but not the other, if the groups are not developing at the same rate, or if there are instrumentation problems (e.g., ceiling and floor effects). These issues and several others are described in greater detail in Chapter 15.

14.10 SUMMARY

Measurement error has an effect on all ANCOVA designs but does not lead to bias in treatment estimates in all cases. It can be shown that measurement error in the dependent variable does not, regardless of the experimental design, lead to bias in the estimates of the adjusted means. The major consequence of measurement error in the dependent variable is a reduction in the power of the ANCOVA F test.

Measurement error in the covariate can cause serious problems with certain designs. In the case of completely randomized and biased assignment designs, the effect is simply a reduction in the power of the ANCOVA F test. But with most nonequivalent group designs, treatment effects are biased; this bias is often serious.

True-score ANCOVA provides an estimate of the outcome that ANCOVA would yield if hypothetical true (error-free) scores were employed as the covariate rather than observed (fallible) scores. Such a procedure is appropriate in the analysis of certain experimental designs.

Five varieties of the pretest–posttest design are: (1) randomized group, (2) biased assignment, (3) nonequivalent group where selection is based on the covariate and additional information, (4) nonequivalent group where groups are selected from different populations, and (5) nonequivalent group where selection is based on unknown factors. Varieties 1 and 2 are appropriately analyzed by using ANCOVA. It is suggested that variety 3 be analyzed by using both ANCOVA and true-score ANCOVA. When subjects are selected into groups on the basis of covariate-relevant information that is more reliable than the observed covariate scores, ANCOVA yields biased adjusted means. True-score ANCOVA will also yield biased adjusted means if the reliability of the variable used in the selection process is less than perfect. The range of adjusted mean differences associated with ANCOVA and true-score ANCOVA provides an interval within which the correct (but unknown) adjusted treatment effect is likely to fall. It is sometimes possible to analyze this design with less ambiguity (than is associated with the use of both ANCOVA and true-score ANCOVA) if information on the specific variables used in the selection process is available. It may be possible to identify the selection process and to use information concerning the process as a covariate in a conventional ANCOVA.

The nonequivalent-group design with selection from two different populations should not generally be analyzed with ANCOVA. Standardized change-score analysis or gain-score ANOVA is appropriate for some cases of this design. When nonequivalent groups have been selected by using an unknown selection process, multiple analyses are necessary. If the only available data are pretest and posttest scores, it is suggested that the following analyses be computed: (1) ANCOVA, (2) true-score ANCOVA using both internal consistency and pretest–posttest reliability estimates, and (3) gain-score ANOVA and/or standardized change-score analysis. Since these analyses produce estimates of treatment effects that are based on different selection models, the various estimates may span the results that are associated with the actual but unknown selection process. If the results of all analyses are generally consistent, some confidence can be placed in the conclusions. However, it is always possible that all the analyses are biased in the same direction by additional confounding variables. If the various analyses yield very different results, little can be done to clarify the interpretation unless additional data are available. Three types of data are helpful in narrowing down the number of analyses that should be attended to. First, if pre-pretest data are available in addition to the pretest and posttest data, it may be possible to evaluate the adequacy of the different analyses by running each analysis on the pre-pretest and pretest data. If the pre-pretest is conceptualized as the covariate and the pretest is conceptualized as the dependent variable, it is reasonable to expect an appropriate analysis procedure to yield a treatment-effect estimate that is effectively zero, because no treatment has been applied. Analyses that do not produce treatment-effect estimates of approximately zero in this type of dry-run experiment should not be employed in the main analysis on the dependent variable. Two other types of additional information that can be used to clarify the nature of the selection model are (1) questionnaire or interviewer data from subjects or others involved in the research project concerning reasons for selection and (2) general background data that may be associated with treatment group membership.

CHAPTER FIFTEEN
Analysis of Nonequivalent Groups: Part II. Standardized Change-Score Analysis

15.1 INTRODUCTION

Several varieties of the pretest–posttest design have been described in previous chapters. It has been pointed out that the analysis of covariance is the appropriate method of analysis for this design when the experimenter has control of the assignment of subjects to groups. This occurs when the experimenter either (1) randomly assigns subjects to treatments or (2) assigns subjects to treatments on the basis of the covariate scores. In the latter case the groups are not equivalent on the pretest (covariate) or the posttest, but, since the selection (assignment) of subjects is completely explained by the covariate, the posttest means are appropriately adjusted to remove the bias associated with the nonequivalence of the groups.

Analysis of covariance fails to provide the appropriate adjustment of posttest means when employed with nonequivalent-group designs in which the covariate does not contain all the relevant information used in the selection of subjects into groups. Alternatives to conventional ANCOVA are required in this case. The choice of an alternative analysis is dictated by the nature of the information available concerning the selection process. A summary of the pretest–posttest designs and analyses described previously is contained in Table 15.1.

Several additional varieties of the nonequivalent-group pretest–posttest design are described in this chapter along with some suggested analysis procedures. Before describing these designs and analyses, we must reiterate previous warnings about the use of nonequivalent-group designs. When the experimenter has a choice, these designs should be avoided if there is any way to do so. Randomized designs are not just a slight preference—they are *so much* better. When randomized designs are not possible, there are many

Table 15.1   Pretest–Posttest Designs and Suggested Analysis Procedures

Type of Pretest–Posttest Design	Suggested Analysis Procedure
Randomized group	ANCOVA
Biased assignment	ANCOVA
Comparative–developmental	ANOVA on gains
Nonequivalent group: selection based on pretest and additional covariate–relevant information (approximate true scores)	True-score ANCOVA
Nonequivalent group: selection based on pretest and other fallible selection and background variables	Multiple ANCOVA
Nonequivalent group: selection based on multiple true-score variables	Multiple true-score ANCOVA
Nonequivalent group: selection based on unknown factors	Combination of ANCOVA, true-score ANCOVA, gain-score ANOVA, standardized change-score analysis and, if additional data are available, multiple ANCOVA and multiple true-score ANCOVA and/or dry-run analyses

strong alternatives to the nonequivalent-group design that should be considered. The classic work of Campbell and Stanley (1966) and, more recently, Cook and Campbell (1979) should be consulted for descriptions of many useful alternatives to both randomized- and nonequivalent-group designs. An especially strong useful alternative is some variety of time-series design. Certain forms of this design are very flexible, easily interpreted, and practical to implement. Useful references on this design include Cook and Campbell (1979), Glass et al. (1975), and Hersen and Barlow (1976).

All but one of the versions of the nonequivalent-group design described in the remainder of this chapter and summarized in Table 15.2 involve the selection of subjects from different populations. For example, subjects from male and female populations may be selected. Other examples of this type of selection include sampling subjects from two different census tracts, different organizations, or volunteer and nonvolunteer populations (Kenny 1975). We may be interested in simply observing samples from two (or more) populations at two points in time to compare developmental or maturational rates. Recall that when the groups are not subjected to differential treatments, we call this a *comparative–developmental* design. An example of this type of design was described in Chapter 7, where the investigator had interest in comparing the weight gains for male and female subjects both exposed to the same diet. Different populations were involved, but differential treatments

Table 15.2 Versions of Nonequivalent-group Pretest–Posttest Design Based on Selection from Different Populations or Selection Not at Pretest

Design Characteristics	Suggested Analysis
Selection from different populations, parallel mean growth curves, $\sigma^2_{pre} = \sigma^2_{post}$ or $\sigma^2_{pre} \neq \sigma^2_{post}$	Gain-score ANOVA
Selection from different populations, fan-spread growth model, $\sigma^2_{pre} \neq \sigma^2_{post}$	Standardized change-score analysis
Selection from different populations, fan-spread growth model, observations collected at three equally spaced points in time	ANCOVA using $(X_2 - X_1)$ as covariate and $(Y - X_2)$ as dependent variable
Selection (from a single population) not at point of pretesting	Standardized change-score analysis

were not. If the subjects from the male and female populations had been exposed to different diets (treatments), we would have a nonequivalent treatment-group design based on selection from different populations.

Among the many issues of relevance in the choice of an appropriate analysis of such designs are the selection model and the growth model. The importance of the selection model (a topic thoroughly explored by Kenny 1975) has been mentioned repeatedly in earlier chapters. The growth model (Bryk and Weisberg 1977) has not been mentioned previously. Both the selection model and the growth model are important in selecting an analysis procedure because these models describe the processes underlying the generation of the data. If nothing is known about how the data were generated, it is impossible to specify the appropriate method of analysis.

Knowledge of growth models is not necessarily critical in experiments based on observations made at only one point in time. But with nonequivalent-group pretest–posttest designs, it is very important to consider the nature of change from pretest to posttest periods when no treatments are applied. With some pretest–posttest designs, such as the randomized-group version, knowledge of the nature of the pretest to posttest growth or change is not as important as in the case of nonequivalent groups selected from different populations. This is because the pretest-to-posttest change, whatever its nature, is expected to be the same across groups that are probabilistically equivalent. But when groups from *different* populations are observed at two points in time, the change for one group may be quite different than the change for another group. In other words, if the gain scores for two groups from different populations are compared, we must know the expected gains (under a true null hypothesis) for each group if an evaluation of treatment effects is to be made. Suppose that the natural growth or gain for one group is twice the growth for the other group. If the experimenter does not know that the two groups are gaining at different rates (in the absence of treatments) and administers different treatments, the mean difference in gain scores is not a reflection of the treatment effects. Rather, it is a combined measure of treatment and differential growth effects. The two are confounded. Clearly, the extent to which the mean gain difference is attributable to treatments is dictated by the nature of the growth model.

There are many possible growth models; only two are described in this chapter. The parallel mean growth model and the fan-spread model are considered in the next two sections. Other growth models are described in Bryk (1977) and Bryk and Weisberg (1977).

15.2   ANALYSIS OF DATA THAT CONFORM TO PARALLEL MEAN GROWTH MODEL

An illustration of data that conform to the parallel mean growth model is provided in Figure 15.1a, b. Notice that both groups are gaining at the same rate under the condition that the null hypothesis concerning treatment effects is true. Since the difference between the two groups on the pretest is the same at the posttest, the mean gain scores are equal for the two groups. When the data conform to this model, the appropriate method of analysis is ANOVA on gain scores. If the subjects within the groups are heterogeneous in terms of background variables such as age, sex, and socioeconomic level, the power of the analysis can be improved by using multiple ANCOVA. The analysis is carried out by using gain scores as the dependent variable and scores on background variables as the covariates.

15.3   FAN-SPREAD MODEL

Campbell (1971) has pointed out that ordinary gain-score analysis is inappropriate for many studies in which the pretest population means are not equal. The gain for subjects with high pretest scores is frequently higher than for subjects with low pretest scores. That is, subjects with high pretest scores are often likely to develop or mature at a higher rate than are low pretest subjects. This difference in maturational rate can be expected to occur between groups, thus producing what Campbell calls a *selection by maturation interaction*. The mean gain for groups with different pretest means is not expected to be the same; hence the mean difference between groups is expected to increase as the subjects develop.

Associated with the increase in the separation of means across time is an increase in the variance within groups. The concomitant increase in both mean differences and within-group variances across the developmental period defines Campbell's "fan-spread hypothesis." Kenny (1975) points out that in

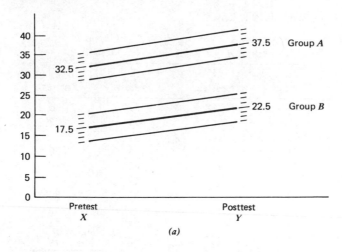

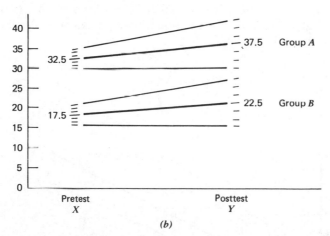

Fig. 15.1   Parallel mean growth (H_0 true), with (a) homogeneous pretest and posttest variance and (b) heterogeneous pretest and posttest variance.

its strictest form, the fan-spread hypothesis stipulates that the mean difference between groups relative to the pooled within-groups standard deviation is constant over time.

Consider the following example. Ten subjects from school A are taught mathematics using method A materials. Ten subjects from school B are taught mathematics by the same teacher using method B materials. A common measure of mathematics achievement is employed as the pretest and posttest. The investigator wants to know whether the effectiveness of the two methods differs. The data follow:

Group A Method A		Group B Method B	
X Pretest	Y Posttest	X Pretest	Y Posttest
79	158	53	106
77	154	51	102
73	146	47	94
71	142	45	90
70	140	44	88
70	140	44	88
69	138	43	86
67	134	41	82
63	126	37	74
61	122	35	70
$\bar{X}_A = 70$	$\bar{Y}_A = 140$	$\bar{X}_B = 44$	$\bar{Y}_B = 88$

These data are illustrated in Figure 15.2. (The values in this example were selected to illustrate the fan-spread notion as clearly as possible; they are not, however, realistic.)

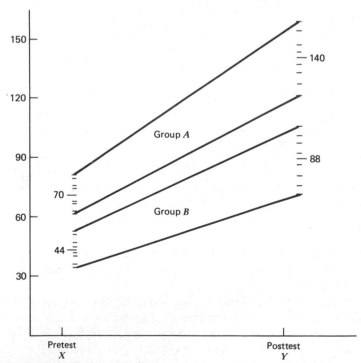

Fig. 15.2 Fan-spread data from pretest–posttest nonequivalent treatment groups selected from different populations (true H_0).

The pretest pooled within-group standard deviation is

$$s_{x_w} = \sqrt{\frac{\Sigma x_A^2 + \Sigma x_B^2}{n_A + n_B - 2}} = \sqrt{\frac{280 + 280}{10 + 10 - 2}} = 5.58$$

and the mean pretest difference is

$$\bar{X}_A - \bar{X}_B = 70 - 44 = 26$$

The posttest pooled within-group standard deviation is

$$s_{y_w} = \sqrt{\frac{\Sigma y_A^2 + \Sigma y_B^2}{n_A + n_B - 2}} = \sqrt{\frac{1120 + 1120}{10 + 10 - 2}} = 11.16$$

and the mean posttest difference is

$$\bar{Y}_A - \bar{Y}_B = 140 - 88 = 52$$

Hence the data agree with the fan-spread hypothesis because

$$\frac{26}{5.58} \sim \frac{52}{11.16}$$

Keep in mind the fact that the increase in both mean difference and variance occurs when there is *no* treatment effect according to the fan-spread hypothesis.

Since there is much confusion concerning an acceptable analysis of this type of very weak design, several clearly inappropriate analyses are shown in Figure 15.3. The first analysis involves the computation of a correlated sample t test on the gain scores for each group. The gains for group A are significant and the gains for group B are significant. This approach is not considered appropriate because it does not attend to the research questions of interest. The t tests tell us that posttest scores are significantly higher than pretest scores for both groups; these tests do not tell us whether methods A and B are differentially effective, nor whether there are, in fact, any treatment effects at all. Treatment and maturation effects are confounded for both t tests.

The second method of analysis involves computation of an independent sample t or ANOVA F on the group A gains versus the group B gains. This approach yields a significant difference in gain scores for the two methods. Hence this analysis, unlike the first method of analysis, does involve a direct comparison of the two groups. Unfortunately, this comparison is not as easily interpreted as it might appear. There is more than one reason for a difference in gains for the two groups; the two most likely explanations are treatment effects and differential growth or maturation effects. This analysis, therefore,

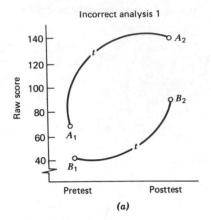

(a)

Raw gain-score correlated sample t test for groups A and B separately:

$$\bar{Y}_A - \bar{X}_A = 140 - 70 = 70 \text{ and}$$
$$t_{obt} = 39.69 \text{ (Critical value} = t_{(.05, 9)} = 2.26).$$
$$\bar{Y}_B - \bar{X}_B = 88 - 44 = 44 \text{ and}$$
$$t_{obt} = 24.95 \text{ (Critical value} = t_{(.05, 9)} = 2.26).$$

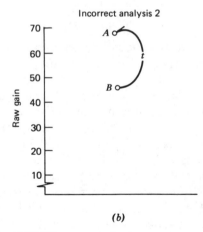

(b)

1. Independent sample t (or ANOVA) on group A gains versus group B gains:

$$A \text{ gains} - B \text{ gains} = 70 - 44 = 26 \quad \text{and}$$
$$t_{obt} = 10.42 \text{ (critical value} = t_{(.05, 18)} = 2.1) \quad \text{and (equivalently)},$$

2. Two-factor repeated-measurement ANOVA F test on trials by treatments interaction $= 108.6$ (critical value $= F_{(.05, 1, 18)} = 4.41$)

Fig. 15.3 Three incorrect analyses of a pretest–posttest nonequivalent treatment-group design based on selection from different populations.

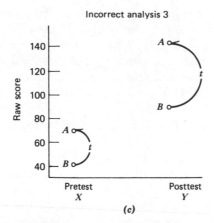

Incorrect analysis 3

Decisions (significant or nonsignificant) based on independent sample t (or ANOVA) on group A versus group B for both pretest and postobservations are compared:

$$\bar{X}_A - \bar{X}_B = 70 - 44 = 26 \qquad \text{and}$$
$$t_{obt} = 10.42 \ (\text{critical value} = t_{.05, 18} = 2.1)$$
$$\bar{Y}_A - \bar{Y}_B = 140 - 88 = 52 \qquad \text{and}$$
$$t_{obt} = 10.42 \ (\text{critical value} = t_{.05, 18} = 2.1)$$

Fig. 15.3 *Continued*

is considered inadequate because a significant test statistic does not necessarily imply that there are differential treatment effects. Maturation or natural growth should be considered along with treatment effects as an explanation of a significant test statistic. This design is sometimes analyzed by using the trials-by-treatments interaction F test from a two-factor repeated measures analysis. Recall from Chapter 7 that this analysis is conceptually and algebraically equivalent to t or F on gains; it is thus inappropriate for the same reasons.

The third incorrect method of analysis involves the computation of an independent sample t or ANOVA F on group A versus group B for the pretest and also for the posttest. Both tests are significant. We may be tempted to conclude that there are no treatment effects because both tests are significant. In this particular example we would be correct, but in general we should not base a decision concerning treatment effects on whether the pre- and posttest statistics yield significant or nonsignificant results. Suppose, for example, that the pretest t is 2.08 and the posttest t is 2.11. A critical value of 2.10 will lead us to conclude that the pretest difference between groups A and B is not significant and the posttest difference is significant (using $\alpha = .05$). If we then follow the inappropriate procedure of comparing the pretest decision (not significant) with the posttest decision (significant), we conclude that significant differences in treatment effects exist where the data do not warrant such a conclusion. The agreement or disagreement of the decisions reached on the basis of tests on A versus B on the pretest and A versus B on the posttest is *not* a test of the significance of the difference in the effects at the pretest and posttest periods. What is required is a test of the difference between the two t values t_{pre} and t_{post}. Standardized change-score analysis provides such a test. The rationale and computational details of this analysis are described in Section 15.4.

Analysis of Fan Spread When Data from Two Pretest Points Are Available

If measures are obtained at three equally spaced points in time and the treatment occurs at the second point in time, the data include: (1) pre-pretest scores (X_1), (2) pretest scores (X_2), and (3) posttest scores (Y). An appropriate way to efficiently employ all these data in one analysis is to (1) use the gain scores $(X_2 - X_1)$ as the covariate and (2) use the gain scores $(Y - X_2)$ as the dependent variable in a conventional analysis of covariance.

15.4 COMPUTATION AND RATIONALE FOR SCSA

The general idea underlying the fan-spread hypothesis is that variability within and between groups increases over time because growth is cumulative. It has been shown that if the posttest variance exceeds the pretest variance —as suggested by the fan-spread hypothesis—the conventional gain-score analysis is inappropriate. That is, significant results will frequently be obtained in the absence of real treatment effects. Since the bias is associated with the difference between the pretest and posttest variances, some method of transforming the data to yield homogeneous variances is in order. One method is to standardize all scores by dividing each (1) pretest score by the pooled within-group pretest standard deviation and (2) posttest score by the pooled within-group posttest standard deviation. This will force pretest scores and posttest scores to have exactly the same variance (i.e., $s_x^2 = s_y^2$). Rather than directly analyzing the pretest–posttest change in the standardized scores, other methods of carrying out standardized change-score analysis (SCSA) due to Kenny (1975) are recommended. It is suggested, however, that as a preliminary step the adequacy of the fan-spread hypothesis be evaluated to decide whether to employ an ordinary gain-score analysis or SCSA.

The adequacy of the fan-spread hypothesis can be evaluated by searching the literature for longitudinal studies that deal with the kinds of populations and variables of interest. Kenny (1975) points out that fan spread is generally found on cognitive tests when groups are formed on the basis of race or social class. A review of the literature in many areas of the behavioral and biological sciences can be expected to reveal the fan-spread pattern. This pattern will never be found, however, when the raw-score pretest and posttest units have been subjected to some type of standardization or ranking transformation.

If previously collected data are not available to confirm the fan-spread pattern, but if the researcher must make a choice between gain-score analysis and SCSA, a test of the pooled within-group variances for pre- and posttests is in order. We suggest the following test:

$$\frac{\left(s_{y_w}^2 - s_{x_w}^2\right)\sqrt{N-4}}{2s_{x_w}s_{y_w}\sqrt{1-r_{xy_w}^2}} = t$$

where

s_{x_w}, s_{y_w} = pooled within-group standard deviations on pretest and posttest measures, respectively (see Section 15.3 for formulas)

r_{xy_w} = pooled within-group correlation between pretest and posttest measures; that is,

$$\frac{\left(\Sigma xy_A + \Sigma xy_B\right)}{\sqrt{\left(\Sigma x_A^2 + \Sigma x_B^2\right)\left(\Sigma y_A^2 + \Sigma y_B^2\right)}}$$

N = total number of subjects in the two groups combined

It is recommended that the obtained t statistic be compared with the critical value of t for $\alpha = .30$ based on $N-4$ degrees of freedom. If the obtained value does not exceed the critical value, the hypothesis of equal pretest and posttest variances is retained and ANOVA or t on gain scores is probably appropriate.

A difference between pretest and posttest variances does not *necessarily* mean that fan spread exists. Notice in Figure 15.1*b* that the pretest and posttest variances are heterogeneous but fan spread is not present because the increase in variance is not accompanied by an increase in the mean difference between groups. Conventional ANOVA on gains is appropriate here. It is also possible to have homogeneous pretest and posttest variances but different growth rates for each group. In this case both ANOVA on gains and SCSA are inappropriate as tests of treatment effects. Clearly, then, the test on the equality of the pretest and posttest variances is not a definitive test of the fan-spread hypothesis. It tests a necessary but not a sufficient fan-spread condition. If there are no longitudinal data on the two groups when the same (or a similar) response measure is used, there is no way to adequately test the fan-spread hypothesis.

If the pretest and posttest variances are homogeneous, both gain-score analysis and SCSA test the same hypothesis, but the former is preferred because (1) the computation is simpler, (2) the results of the t or ANOVA on gains are understood by most researchers, and (3) the descriptive statistics associated with gain-score analysis (e.g., raw gain means) are easily understood by both researchers and nonresearchers.

Computation Procedures

Standardized change score analysis can be performed using either of two different procedures (Kenny 1975, 1976). One procedure involves testing the significance of the difference between two correlation coefficients. The second approach involves an analysis of variance on the difference between the group means on an adjusted gain score. Both procedures are described in this unit. Since investigations of the relative power, robustness, and other characteristics of these procedures have not yet been undertaken, the choice between the two is arbitrary.

SCSA through Correlational Analysis. The correlational approach to SCSA requires the following steps:

1. Construct a design matrix in which the column of pretest scores is denoted X, the column of posttest scores is denoted Y, and the group membership dummy variable is denoted d.
2. Compute the correlation between the pretest scores and the dummy variable (i.e., r_{xd}).
3. Compute the correlation between the posttest scores and the dummy variable (i.e., r_{yd}).
4. Test the significance of the difference between r_{xd} and r_{yd}, using the Hotelling–Williams procedure for testing the difference between two correlated correlation coefficients (i.e., test H_0: $\rho_{xd} = \rho_{yd}$). If the hypothesis is rejected, it is concluded that treatment effects exist.
5. Compute the adjusted mean gain-score difference by using the following formula:

$$\left(\overline{Y}_1 - \overline{Y}_2 \right) - \frac{s_{y_w}}{s_{x_w}} \left(\overline{X}_1 - \overline{X}_2 \right)$$

This provides a descriptive measure of the posttest mean difference due to treatments.

Hotelling–Williams Test. Kenny (1975) points out that the well-known Hotelling test (1940) of the difference between two correlated correlation coefficients is one method that can be employed in SCSA. We recommend that Williams' modification (1959) of the Hotelling test be employed instead because it has been shown that the Type I error rate associated with the Hotelling test is unacceptably high (Neill and Dunn 1975). Neil and Dunn's Monte Carlo work indicates that the Hotelling–Williams test is the most satisfactory among nine tests evaluated. The computational formula for this test is

$$\left(r_{xd} - r_{yd} \right) \sqrt{ \frac{(N-1)(1+r_{xy})}{2 \left[(N-1)/(N-3) \right] |R| + \bar{r}^2 (1 - r_{xy})^3 } } = t$$

where

r_{xd} = correlation between pretest and dummy variable
r_{yd} = correlation between posttest and dummy variable
r_{xy} = correlation between pretest and posttest
$\bar{r} = (r_{xd} + r_{yd})/2$
$|R|$ = determinant of correlation matrix

$$\begin{bmatrix} 1 & r_{xd} & r_{yd} \\ r_{xd} & 1 & r_{xy} \\ r_{yd} & r_{xy} & 1 \end{bmatrix}$$

(using scalar notation, the determinant of this matrix is

$$1 - r_{xy}^2 - r_{xd}^2 - r_{yd}^2 + 2r_{xy}r_{xd}r_{yd})$$

N = total number of subjects in study, and t_{obt} is compared with critical value of t based on $N - 3$ degrees of freedom

RATIONALE FOR EMPLOYING HOTELLING–WILLIAMS TEST.   If we compute F (or t) on the difference between two group means on the pretest and also compute F on the difference between the two group means on the posttest, we need a method for testing the difference between F_{pre} and F_{post} (or t_{pre} and t_{post}). There are no standard tests for this purpose, but it turns out that testing the difference between two correlated correlation coefficients is equivalent to testing the difference between two correlated F values. This is true because the F ratio is related to the correlation coefficient as follows:

$$r = \sqrt{\frac{F}{F + [N - 2]}}$$

Suppose that the difference between two means on the pretest yields an ANOVA F of 10. Further, the ANOVA F on the posttest is 30. If N (the total number of subjects in both groups combined) is 12, we find two correlations:

$$r = \sqrt{\frac{10}{10 + (12 - 2)}} = 0.707$$

$$r = \sqrt{\frac{30}{30 + (12 - 2)}} = 0.866.$$

These correlations are r_{xd} and r_{yd}, the correlations between the pretest and the dummy variable and the posttest and the dummy variable. Recall that the F test on the difference between two means on variable Y is equivalent to the F

test on the significance of a correlation between a dummy variable and Y:

$$\text{ANOVA } F = \frac{r_{yd}^2}{(1 - r_{yd}^2)/N - 2}$$

If it is clear that there is a relationship between r and F, it should seem reasonable to test the difference between r_{xd} and r_{yd} to evaluate whether the pretest standard score difference is equal to the posttest standard score difference.

The difference between the pretest and posttest correlations is of interest because:

1. The squared correlation between the pretest and the dummy variable (r_{xd}^2) provides an estimate of the proportion of the variability on the pretest accounted for by group differences that exist *before* treatments are administered.

2. The squared correlation between the posttest and the dummy variable (r_{yd}^2) provides an estimate of the proportion of the variability on the posttest accounted for by group differences that exist *after* treatments are administered.

3. The difference $|r_{xd}^2 - r_{yd}^2|$ provides an estimate of the proportion of the variability on the posttest that is accounted for by treatment effects independent of pretest group differences. If there is no treatment effect, we expect r_{xd} to equal r_{yd}. (If r_{xd} and r_{yd} have different signs, the rank order of the means on the pretest is reversed on the posttest; $|r_{xd}^2 - r_{yd}^2|$ should not be employed as a descriptive measure in this case).

If the Hotelling–Williams test yields a significant value, it is concluded that treatment effects are present. Notice that it is not necessary for r_{yd}^2 to *exceed* r_{xd}^2 to suggest that there are treatment effects. If r_{xd}^2 exceeds r_{yd}^2, the treatments have decreased the standard score difference between the two groups.

Since the example values mentioned previously are $r_{xd} = 0.707$ and $r_{yd} = 0.866$, we conclude that (1) the difference between the standard score sample means increased after treatments, and (2) the proportion of the variability in the posttest accounted for by treatments independent of pretest group differences is

$$|0.707^2 - 0.866^2| = |0.50 - 0.75| = 0.25$$

The Hotelling–Williams test should be carried out to determine whether the apparent treatment effect is statistically significant.

SCSA through Adjusted Gain-Score ANOVA. The second SCSA analysis approach involves the following steps:

1. Compute adjusted gain scores by using the formula

$$Y - \frac{s_{y_w}}{s_{x_w}}(X)$$

where

  Y = posttest score

  X = pretest score

  s_{y_w} = pooled within-group standard deviation on posttest

  s_{x_w} = pooled within-group standard deviation on pretest

2. Compute an independent sample ANOVA F on the adjusted gain scores, but use degrees of freedom $= 1, N-3$ rather than the conventional $1, N-2$.

Computational Example

Raw Data

	Control Group A		Treatment Group B	
Pretest X	Posttest Y	Pretest X	Posttest Y
2	3	4	10
3	7	5	14
4	7	6	14
5	11	8	20
6	11	10	22
$\bar{X}_A = 4$	$\bar{Y}_A = 7.8$	$\bar{X}_B = 6.6$	$\bar{Y}_B = 16$

HOMOGENEITY OF VARIANCE TEST.

$$\Sigma x_A^2 = 10 \qquad \Sigma x_w^2 = 33.2$$

$$\Sigma y_A^2 = 44.8 \qquad \Sigma y_w^2 = 140.8$$

$$\left. \begin{array}{ll} \Sigma x_B^2 = 23.2 & \Sigma xy_A = 20 \\ \Sigma y_B^2 = 96 & \Sigma xy_B = 46 \end{array} \right\} \quad \Sigma xy_w = 66$$

$$s_{x_w} = \sqrt{\frac{10 + 23.2}{8}} = 2.037$$

$$s_{y_w} = \sqrt{\frac{44.8 + 96}{8}} = 4.195$$

$$r_{xy_w} = \frac{66}{\sqrt{(33.2)(140.8)}} = 0.9653$$

$$t = \frac{(17.6 - 4.15)\sqrt{6}}{2(2.037)(4.195)\sqrt{1 - (0.9653)^2}} = 7.38$$

The critical value of t for $\alpha = .30$ using 6 degrees of freedom is 1.134; since the absolute value of the obtained t exceeds this value, it is concluded that the population pretest and posttest variances differ. Hence SCSA appears to be more appropriate than ordinary gain-score analysis in this situation. The computation of SCSA based on (1) the Hotelling–Williams test and (2) gain-score ANOVA on

$$\left[Y - (s_{y_w}/s_{x_w})(X) \right]$$

is shown in the following paragraphs.

HOTELLING–WILLIAMS TEST

Design Matrix

Pretest X	Posttest Y	Dummy d
2	3	1
3	7	1
4	7	1
5	11	1
6	11	1
4	10	0
5	14	0
6	14	0
8	20	0
10	22	0

The Hotelling–Williams test is as follows:

$$r_{xy} = 0.959$$

$$r_{xd} = -0.5808$$

$$r_{yd} = -0.7377$$

$$|R| = \begin{vmatrix} 1 & -0.5808 & -0.7377 \\ -0.5808 & 1 & 0.959 \\ -0.7377 & 0.959 & 1 \end{vmatrix} = 0.020568$$

$$\bar{r} = \left[(-0.5808) + (-0.7377) \right]/2 = -0.65925$$

$$\bar{r}^2 = 0.43461$$

$$t = (0.1569)\sqrt{\frac{(9)(1.959)}{2\left(\frac{9}{7}\right)(0.020568) + (0.43461)(1 - 0.959)^3}} = 2.86$$

The critical value of t for $\alpha = .05$ is 2.365 for $N - 3 = 7$ degrees of freedom. The proportion of the variability on the posttest that appears to be due to treatment effects is

$$|(-0.5808)^2 - (-0.7377)^2| = 0.21$$

Since the squared pretest correlation is exceeded by the squared posttest correlation, it is concluded that the treatment has increased the standard score difference between posttest means. The adjusted mean difference, in the metric of the posttest, is

$$(7.8 - 16) - 2.059(4 - 6.6) = -2.85.$$

This is the estimated mean difference corrected for both pretest differences between groups and fan spread.

ADJUSTED GAIN-SCORE ANALYSIS ON $Y - (s_{y_w}/s_{x_w})(X)$

Design Matrix

Group	Posttest Y	Adjusted Pretest $\frac{s_{y_w}}{s_{x_w}}(X)$	Adjusted Gain $Y - \frac{s_{y_w}}{s_{x_w}}(X)$	Dummy d
A	3	2.059(2) = 4.118	-1.118	1
	7	" (3) = 6.177	0.823	1
	7	" (4) = 8.236	-1.236	1
	11	" (5) = 10.295	0.705	1
	11	" (6) = 12.354	-1.354	1
B	10	" (4) = 8.236	1.764	0
	14	" (5) = 10.295	3.705	0
	14	" (6) = 12.354	1.646	0
	20	" (8) = 16.472	3.528	0
	22	" (10) = 20.590	1.410	0

Analysis of variance on adjusted gains and the critical value are

$$\text{ANOVA } F = \frac{R^2_{y_{adj}d}}{\left(1 - R^2_{y_{adj}d}\right)/N - 3} = 14.53$$

Critical value $= F_{(.05, 1, 7)} = 5.59$

Adjusted Mean Gains

Group A	Group B	$A - B$ Difference
-0.44	2.41	-2.85

15.5 STANDARDIZED CHANGE-SCORE ANALYSIS WITH CONTROL FOR BACKGROUND VARIABLES

The subjects involved in SCSA studies may be heterogeneous with respect to background variables such as age, income, socioeconomic level, experience, sex, and aptitude. When experimenters design experiments, they often attempt to select subjects who are homogeneous with respect to these characteristics. Applied researchers, however, often find that they must deal with very heterogeneous groups of subjects. If the subjects in the nonequivalent groups are very heterogeneous on these background variables, the researcher may want to reduce this variability to (1) increase the precision of the statistical analysis or (2) reduce possible confounding of the treatment effect with these factors.

Aspects of conventional ANCOVA and SCSA are combined to remove the effects of background characteristics from the analysis. First, the adjusted gain scores are computed using the adjustment formula

$$Y^* = Y - \frac{S_{y|d,BG}}{S_{x|d,BG}} X$$

where

$\quad\quad Y = $ posttest

$S_{y|d,BG} = $ standard error of estimate obtained from regression of Y on dummy variable (d) and background variables (BG)

$S_{x|d,BG} = $ standard error of estimate obtained from regression of X on dummy variable (d) and background variables (BG)

$\quad\quad X = $ pretest

Multiple ANCOVA is then carried out by using the adjusted gain scores as the dependent variable and measures of the background characteristics as the covariates. Degrees of freedom for the two-group case are 1 and $N-3-C$, where C is the number of background variables.

Computational Example

The data employed in the computational example at the end of Section 15.4 are reanalyzed here, using sex as a background variable. As can be seen in the remainder of this section, the design matrix contains a background variable column. Values of one and zero are used to identify male and female subjects, respectively.

$$S_{y|d,BG} = 4.328$$

$$S_{x|d,BG} = 2.172$$

Posttest Y	Adjusted Pretest $\frac{S_{y\|d,BG}}{S_{x\|d,BG}} X$	Adjusted Gain $Y^* = Y - \frac{S_{y\|d,BG}}{S_{x\|d,BG}} X$	Group-membership Dummy d	Background Variable (Sex) BG
3	1.993(2) = 3.986	−0.986	1	0
7	" (3) = 5.979	1.021	1	1
7	" (4) = 7.972	−0.972	1	0
11	" (5) = 9.965	1.035	1	1
11	" (6) = 11.958	−0.958	1	0
10	" (4) = 7.972	2.028	0	1
14	" (5) = 9.965	4.035	0	1
14	" (6) = 11.958	2.042	0	0
20	" (8) = 15.944	4.056	0	1
22	" (10) = 19.930	2.070	0	1

$$F=\frac{\left(R_{y\cdot d,\,BG}^{2}-R_{y\cdot BG}^{2}\right)}{\left(1-R_{y\cdot d,\,BG}^{2}\right)/N-3-C}=\frac{(0.86214-0.49838)}{(0.13786)/6}=\frac{0.36376}{0.02298}=15.83$$

Critical value $=F_{(.05,\,1,\,6)}=5.99$

Adjusted Mean Gains; Sex as Background Variable

Group A	Group B	$A-B$ Difference
0.148	2.53	-2.382

It can be seen that these results using sex as a control variable differ little from those of the previous analysis that did not include this variable.

15.6 NONEQUIVALENT TREATMENT-GROUP DESIGN BASED ON SELECTION AFTER PRETESTING

It has been assumed with all the previously described nonrandomized pre-test–posttest designs that the occasion for selection or assignment to groups is at the pretest. Kenny (1975) describes another selection model in which the occasion for selection occurs some time after the pretesting has taken place. It turns out that ANCOVA is generally inappropriate in this case, but SCSA yields an appropriate solution. A comparison of the expected posttest mean difference under (1) biased assignment, (2) selection from different popula-tions, and (3) selection after pretesting designs may facilitate understanding of why SCSA should be employed when selection takes place after pretesting.

Recall that the basic reason for employing ANCOVA with the biased assignment design is to predict the amount of between-group regression effect that will take place when treatment effects do not exist. That is, through the adjustment procedure associated with ANCOVA, the posttest means are adjusted by the amount that is expected to result from regression. It can be seen in the adjustment formula

$$\overline{Y}_{j\,\text{adj}}=\overline{Y}_{j}-b_{\text{w}}\left(\overline{X}_{j}-\overline{X}..\right)$$

that the predicted amount of between-group regression (i.e., the extent to which posttest means are expected to regress toward each other) is a function of the within-group regression coefficient b_{w}. If the treatment groups are formed on the basis of random assignment or the pretest scores, the coefficient b_{w} correctly predicts the amount by which the posttest means will regress when the ANCOVA assumptions are met. This effect is illustrated in Figure 15.4a. The actual difference between the posttest means is half the difference between the pretest means, and this regression effect is correctly predicted from the adjustment formula because the pooled within-group regression coefficient b_{w} is 0.50. Hence ANCOVA is correct with biased

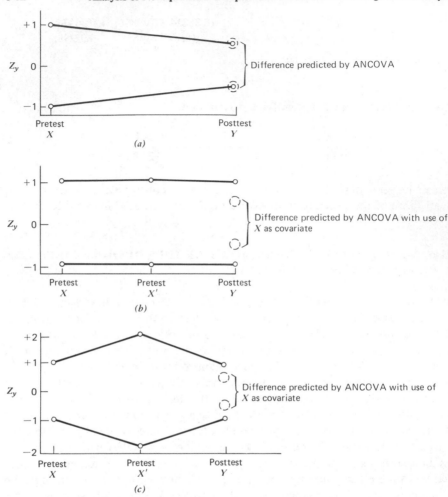

Fig. 15.4 Comparison of actual and predicted posttest differences with biased assignment, selection from different populations, and selection after pretesting designs with use of X as covariate, no treatment effects present: (a) biased assignment design ($b_w = 0.5$); (b) selection from different populations design ($b_w = 0.5$); (c) selection after pretesting design ($b_w = 0.5$).

assignment (as well as randomized-group) designs because the extent to which the posttest group means are expected to regress toward the grand posttest mean is appropriately estimated by the extent the posttest scores within groups regress toward their group means.

Suppose, however, that the design involves the following steps:

1. Pretest X is administered.
2. Pretest X' is administered 6 months later, subjects are assigned to treatment conditions on the basis of X' scores, and treatments are applied.
3. Posttest Y is administered 6 months after treatments begin.

If ANCOVA is carried out by using X (not X') as the covariate and Y as the dependent variable, the means will generally be adjusted inappropriately. This is because b_w is not a correct estimate of the amount of between-group regression to be expected when the pretesting and the selection do not occur at the same time. If we compute the standard-score mean difference between groups on X and on Y, we expect to find that they are equal. But the difference on X' is larger than the differences on X and Y. This is because the difference between standard-score means can be expected to attenuate when measurements are taken at points in time removed from the selection point unless the groups are comprised of subjects from different populations. If the subjects are not from different populations, the standard-score mean difference between groups is likely to be a function of the time separating the selection and the measurement. Hence if the selection takes place midway between the pretest X and the posttest Y (or averages out midway), the standard-score mean differences should be equal. Standardized change-score analysis rather than ANCOVA or true-score ANCOVA should be employed for this design. If the selection of treatments takes place at a point other than midway between the pretest and the posttest, SCSA is likely to be biased. This is not true if the groups are from different populations because between-group standard-score differences are not generally expected to change across time under this model. It should be remembered, however, that a model is just a convenient fiction; real data from different populations may not conform to this model. Also, it is often very difficult to decide whether subjects are from distinctly different but stable populations (where differences between groups should remain relatively constant across time) or from a single population (where differences are expected to attenuate across time).

The expected posttest differences for fixed pretest differences and $b_w = 0.50$ are illustrated in Figure 15.4 for the biased assignment, selection from different populations, and selection after pretest designs. Notice that ANCOVA correctly predicts the posttest difference for the biased assignment experiment only. If ANCOVA is applied in the situations shown in Figure 15.4b or 15.4c, the difference predicted is not the actual difference: ANCOVA is incorrect. The situation depicted in Figure 15.4b is correctly analyzed by using SCSA because the difference between standard-score means is constant when the groups are selected from different populations. It can be seen that ANCOVA incorrectly predicts that the difference between posttest means will be less than the pretest difference. Analysis of covariance also incorrectly predicts the posttest difference in the situation shown in Figure 15.4c. Standardized change-score analysis applied in this situation yields the correct analysis.

The situation shown in Figure 15.4c differs from the biased assignment design in that the variable used as the covariate (i.e., X) is not the same as the variable used to assign subjects to treatments. If X' had been used as the covariate rather than X, the analysis of covariance would have been the correct analysis. But if selection or assignment takes place at X' and scores on X' are not available (or if the selection into treatments at this point is on the basis of unmeasured factors operating between X and Y), SCSA is appropriate. It is not necessary that subjects be assigned to treatments on the basis of some known selection factor. It *is* necessary that the selection into treatments take place midway between the pretest and the posttest unless it is known that the groups are from different populations and that the difference in standard-score means will remain constant from pretest to posttest in the absence of treatment effects. Notice in Figure 15.5*a* that the standard-score difference is constant regardless of the point of selection into treatments because subjects from different populations have been selected; therefore, no between-group regression takes place. Standardized change-score analysis is appropriate with such data regardless of the point of selection.

In Figure 15.5*b* the subjects were assigned to treatments from a single population on the basis of some factor related to the response measure used as pretest and posttest. The selection did not take place midway between X and Y; instead, selection was much closer to the point of pretesting than to the point of posttesting. In this case the difference at X is much larger than at Y even though there are no treatment effects; thus SCSA is biased. The analysis of covariance is also biased in this situation. If a treatment is applied to the group with the lower pretest mean, the SCSA bias is in such a direction as to make it appear that the treatment is effective when it is not; ANCOVA

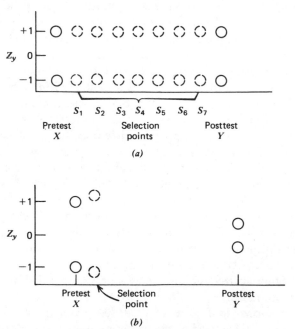

Fig. 15.5 Effect of employing SCSA when selection takes place at various points between pretest and posttest, no treatment effects present: (*a*) subjects from different populations select treatments; (*b*) subjects from same population assigned to treatment on basis of information related to response measure.

is biased in such a direction as to make it appear that the treatment effect is harmful. There does not appear to be a satisfactory analysis in this situation. However, if both SCSA and ANCOVA yield the same decision concerning treatment effects, the interpretation is less ambiguous. Since it is often difficult to decide whether between-group regression should occur it is reasonable to carry out several different analyses routinely with nonequivalent-group designs. If the results do not agree, it may be necessary to collect additional longitudinal data to evaluate the adequacy of the various models.

15.7 INTERPRETATION ISSUES

It has been pointed out that a significant difference between r_{xd} and r_{yd} suggests that treatment effects are present. It should be kept in mind, however, that alternative explanations must always be considered in the evaluation of the outcome of any experiment. Unfortunately, the number of rival explanations associated with nonequivalent-group designs analyzed with the use of SCSA is far greater than the number of rival explanations associated with the randomized-group pretest–posttest design. An excellent description of the problems associated with the latter design can be found in the work of Campbell and Stanley (1966) and Cook and Campbell (1979). Some of the issues relevant to the interpretation of standardized change-score analysis are presented in the remainder of this section.

Ceiling and Floor Effects

Any explanation of a difference between r_{xd} and r_{yd} should be considered in evaluating the outcome of SCSA. The adequacy of the method of measuring the dependent variable should be one of the first considerations. If the pretest measure yields no difference between groups because it is not sensitive enough to measure differences that actually exist, we may have a ceiling or a floor effect. Suppose a measure of mathematics achievement is administered to two groups of third graders who differ in terms of actual mathematics achievement. If the test of achievement is much too difficult for both groups, the pretest means may both be the same (i.e., chance scores) even though the true achievement difference between groups is large. In this case r_{xd} will be zero even though it should be nonzero. If the posttest is administered after the groups have developed to the point that the difference in achievement is reflected in the test scores, the correlation r_{yd} will not be zero. Hence even though no differential treatment effects are present, r_{xd} and r_{yd} will be different as a result of the floor effect.

Ceiling effects result in a similar problem. If the response measure is a psychometric instrument on which almost all subjects obtain the maximum score or a physiological recording instrument that records responses at a rate below the rate of the highest performing subjects, ceiling effects may be present. The problem here, as with floor effects, is that true differences between groups will not show up, even though they are present. If the instrument is adequate at the pretest but has insufficient "ceiling" at the posttest, a difference between r_{xd} and r_{yd} will result independently of any real treatment effects. On the other hand, if the instrumentation has insufficient ceiling during both the pretest and posttest administrations, real treatment effects may be masked. The problems of ceiling and floor effects are not unique to nonequivalent treatment-group analyses, but they are somewhat more likely to occur. This is because nonequivalent groups more often differ to the extent that the range of sensitivity of the instruments is exceeded.

A simple method of identifying ceiling and floor effects is to plot the distributions of the pretest and posttest scores. If skewed distributions are observed, the adequacy of the instrumentation should be questioned. Also, a review of the literature concerning the nature of the populations for which the instrumentation was developed may suggest the range of response levels that is most appropriate.

Measurement Conditions

Since the reliability of a measure affects the size of the correlation between that measure and any other variable, it is important that the conditions affecting reliability be similar for both groups during pretesting and posttesting. For example, if young children are pretested during the first day or two of school, when chaos is likely, the reliability of the measurement will probably be lower than the reliability of a posttest administered during relative calmness. Similarity of environmental conditions does not, however, guarantee that the reliability of measurement will be the same during pretest and posttest administrations.

Subject Characteristics—Reliability Shift

Just as environmental characteristics can have an effect on reliability, the characteristics of the subjects themselves can increase or decrease reliability estimates. Measurement error is often higher with young organisms than with mature organisms, thus suggesting that great care should be taken in interpreting the results of SCSA if the lag between pre- and posttesting of young organisms is substantial.

A method of attempting to eliminate "reliability shift" as an explanation of the difference between r_{xd} and r_{yd} involves estimating the reliability of both the pretest and the posttest. If the estimated reliability (coefficient α or the parallel forms pooled within-groups estimate described in Section 14.4 is appropriate here) is the same for both the pretest and the posttest, the difference between correlations must be due to something other than a reliability shift. If the reliability estimates are different, the correlations should be corrected to eliminate that portion of the difference between them that is accounted for by the reliability shift.

There are several ways of making these corrections. The following approach has been suggested by Kenny (1976):

$$*r_{xd} = r_{xd}\left(\frac{r_{yy_w}}{r_{xx_w}}\right)^{1/4}$$

$$*r_{yd} = r_{yd}\left(\frac{r_{xx_w}}{r_{yy_w}}\right)^{1/4}$$

where

$*r_{xd}$ = estimate of expected correlation between dummy variable and pretest if pretest and posttest had the same reliability

$*r_{yd}$ = estimate of expected correlation between dummy variable and posttest if pretest and posttest had the same reliability

r_{xx_w} = estimated reliability of pretest

r_{yy_w} = estimated reliability of posttest

Suppose that the following correlations were based on two nonequivalent groups of children observed during the first (pretest) and third (posttest) grades. One group was exposed to a conventional reading program and the other group, to an experimental reading program. The difference between the two correlations,

$$r_{xd} = 0.707$$

$$r_{yd} = 0.866$$

is, according to the initial analysis, statistically significant, and the proportion of the variability in the posttest accounted for by the treatment is 0.25. The reliability estimates for the pretest and posttest measures are then considered:

$$r_{xx_w} = 0.60$$

$$r_{yy_w} = 0.90$$

Since these estimates are not the same, it is apparent that the difference $(0.707 - 0.866)$ is at least partly a function of the reliability shift. The corrected correlations are

$$*r_{xd} = 0.707\left(\frac{0.90}{0.60}\right)^{1/4} = 0.78$$

$$*r_{yd} = 0.866\left(\frac{0.60}{0.90}\right)^{1/4} = 0.80$$

Since the corrected correlations are essentially the same, we have no evidence that treatment effects are present. The difference between r_{xd} and r_{yd} is almost completely explained by the reliability shift from 0.60 for the pretest to 0.90 for the posttest. The interpretation of the initial analysis is incorrect. Unless the difference between $*r_{xd}$ and $*r_{yd}$ is obviously not significant (as in the example just described), the Hotelling–Williams test should be applied to these corrected correlations.

An alternative approach to testing for treatment effects when the pretest and posttest reliabilities are different involves a slight modification of the gain-score approach mentioned in the latter part of Section 15.4. Rather than analyzing the adjusted posttest measure

$$ Y - \frac{s_{y_w}}{s_{x_w}}(X) $$

which is appropriate when the pretest and posttest reliabilities are equal, the following measure is analyzed (Kenny 1975):

$$ Y - \frac{s_{y_w}}{s_{x_w}}\left(\frac{r_{yy_w}}{r_{xx_w}}\right)(X) $$

It can be seen that this formula reduces to the equal reliability formula when there is no difference between r_{xx_w} and r_{yy_w}. We suggest that ANOVA on scores based on this transformation be computed by using numerator and denominator degrees of freedom equal to 1 and $N-3$, respectively.

Subject Characteristics—Factorial Composition Shift

Another explanation for a difference between r_{xd} and r_{yd} is a change in the factorial composition of the response variable from the pretest to the posttest period. It has been shown that measures of complex motor responses, for example, measure different dimensions during pretesting or early trials than they do during posttesting or later trials (Fleishman and Hempel 1954 and 1955). Since the same characteristics are not necessarily being measured during pretesting and posttesting, the difference between the pretest and posttest correlations can be explained by the shift in the factorial composition.

A simple descriptive procedure for evaluating whether the factorial composition is the same during pretesting and posttesting involves the following steps:

1. Compute the within-group correlation between the pretest and the posttest.
2. Correct the within-group correlation for attenuation due to unreliability in both the pretest and the posttest by using the following formula:

$$ *r_{xy} = \frac{r_{xy}}{\sqrt{(r_{xx})(r_{yy})}} $$

where

$*r_{xy}$ = disattenuated pretest–posttest correlation
r_{xy} = pretest–posttest correlation
r_{xx} = estimated reliability of pretest
r_{yy} = estimated reliability of posttest

If $*r_{xy}$ is near 1.0, the factorial composition will not have shifted. A problem with this simple procedure is to decide how much discrepancy from 1.0 should be expected with sample data. What is needed here is a test of the hypothesis that the population correlation corrected for attenuation (i.e., disattenuated) equals 1.0. Procedures for this purpose include those developed by Villegas (1964), McNemar (1958), Forsyth and Feldt (1970), Werts et al. (1976), and Jörskog (1978). The Villegas procedure is described in Lord (1973). The Forsyth–Feldt procedure, which is a modification of the McNemar procedure, is described in the text that follows.

McNemar–Forsyth–Feldt Test. The quantity

$$\frac{\dfrac{2 + r_{x_1x_2} + r_{y_1y_2} - r_{x_1y_1} - r_{x_1y_2} - r_{x_2y_1} - r_{x_2y_2}}{2 - r_{x_1y_2} - r_{y_1y_2}}}{1 + \dfrac{2\left[\bar{r}_{rel} - \sqrt{(r_{x_1x_2})(r_{y_1y_2})}\,\right]}{1 - \bar{r}_{rel}}} = F$$

where

x_1, x_2 = parallel forms of pretest
y_1, y_2 = parallel forms of posttest
$\bar{r}_{rel}$ = arithmetic mean of the two reliability estimates, $(r_{x_1x_2} + r_{y_1y_2})/2$
yields an F value that is compared with the critical value of F based on $N - 1$ and $2N - 2$ degrees of freedom. If the obtained F exceeds $F_{(\alpha, N-1, 2N-2)}$ the disattenuated population correlation between the pretest and the posttest will not be 1.0. This means that the shift in factorial composition from the pretest to the posttest should be retained as a possible explanation of the difference between r_{xd} and r_{yd}.

We suggest that this test be computed separately for the experimental and the control groups because the factorial composition is generally more likely to shift for treated subjects than for the untreated control subjects. If either or both tests suggest a shift in factorial composition, the interpretation of the difference between r_{xd} and r_{yd} becomes ambiguous. If the disattenuated pretest–posttest correlation for the control group and the disattenuated pretest–posttest correlation for the experimental group do not differ, a single McNemar–Forsyth–Feldt test applied to the pooled within-group values is appropriate.

An obvious problem with this test, in addition to the computational complexity, is the need for parallel form reliability estimates for both the pretest and the posttest. A simple alternative is to compute coefficient alpha (or another measure of internal consistency such as a corrected split-half estimate) for both the pretest and the posttest and then to apply these estimates in the correction for attenuation formula. If the disattenuated correlation is far from 1.0 (say, below 0.90), the lack of similarity of the pretest and posttest factorial composition should be considered as a viable explanation for the difference between r_{xd} and r_{yd}. A more complete discussion of the general problem of estimating the relationship between the true scores of two measures can be found in Kristof (1973) and Lord and Novick (1968).

Misspecification of Growth Model

Two growth models, parallel mean growth and fan spread, have been described in previous sections. It was pointed out that the appropriate analysis for the parallel mean growth model (i.e., simple gain-score ANOVA) yields biased estimates of treatment effects when the data conform to the fan-spread model. Hence the specification of parallel mean growth when fan spread is the case is an example of misspecification of the growth model that leads to the incorrect analysis. There are many other growth models that may describe the nature of the data and, consequently, many possible types of misspecification. Methods of analyzing data that conform to growth models not described here are being developed (e.g., Bryk and Weisberg 1976).

Misspecification of Selection Model

The major emphasis in this chapter and in Chapter 14 has been on the importance of the selection model. When the analysis procedure is based on a selection model that is not consistent with the nature of the data, the estimates of treatment effects may be biased; the degree of bias can be very high. Since the consequences of misspecifying the selection model are potentially severe, the suggested strategy is to apply analysis procedures based on several different selection models. The number of analyses that should be computed is then inversely related to the amount of information that is available on the selection process. At one end of the selection-information continuum all relevant selection information is known; this is the case when randomized-group and biased assignment designs are employed and ANCOVA is the appropriate analysis. At the other end of the continuum, where there is no information on the selection process, analyses based on all realistic selection models may be relevant. The consequence of not being able to specify the selection model is not being able to state which, if any, of the analyses yield the appropriate solution.

Unknown Events between the Pretest and Posttest

Regardless of the adequacy of the statistical analysis, the estimates of the treatment effects are meaningless if unknown events other than the treatment occur between the pretest and the posttest. This source of invalidity (classified as "history" by Campbell and Stanley 1966) is a problem with all pretest–posttest designs, but when subjects have been selected from two different populations, the chance of different events affecting each is high.

15.8 STATUS OF NONEQUIVALENT-GROUP STUDIES

The current work in the area of nonequivalent treatment-group designs is focused on (1) the conceptual specification of the analysis problems (e.g., Cronbach et al. 1977, Kenny 1975 and Reichardt 1979), (2) structural model solutions (e.g., Jöreskog and Sörbom 1976, 1978; Linn and Werts, 1977; Sörbom 1978), (3) growth models (e.g., Bryk and Weisberg 1974, 1976, Strenio et al. 1977, Weisberg et al. 1979), and (4) attempts to convince researchers that nonequivalent group designs frequently can and should be avoided because, in many cases, stronger research designs can be implemented (e.g., Campbell and Boruch 1975). If work in all these areas is fruitful, there should be fewer nonequivalent-group designs to analyze in the future, but more satisfactory analyses of those that are carried out. A comprehensive discussion of interpretation issues associated with this design can be found in Cook and Campbell (1979).

Recent work on the analysis of observational studies in which there is no formal treatment administration under the control of an experimenter is somewhat scattered (e.g., Cochran 1969, Cochran and Rubin 1973, McKinlay 1975, Meehl 1970, Rindskopf 1978), but it is clear that the trend in this area is toward structural-equation models. The work of Jöreskog (1970a, 1970b, 1973, 1974, 1977, 1978) and Sörbom (1978) is particularly relevant here. They have developed a sophisticated computer program (Jöreskog and Sörbom 1978) that is very useful for a wide variety of problems in the structural-equation area. A structural (or causal) model consists of a set of equations relating effects to causes. If models of the causes, measurement models, and design models are posited, it is possible, if appropriate data are available, to evaluate the adequacy of the explanations (causes). An excellent introduction to structural-equation models has been written by Kenny (1979).

15.9 SUMMARY

Nonequivalent-group designs in which the subjects are (1) selected from different populations or (2) assigned to treatments from a single population midway between pretesting and posttesting can be analyzed by using SCSA.

This analysis is recommended over conventional change-score (gain-score) analysis when the data conform to the fan-spread model. If the natural growth process is different for two groups, a conventional analysis is likely to confound differential growth with treatment effects. Conventional change-score analysis is appropriate when the parallel mean growth model fits the data. It is possible to test the equality of the pretest and posttest variances as a possible indication that fan spread exists. This test does not, however, necessarily identify fan spread.

There are two methods of computing SCSA: (1) testing of the difference between r_{xd} (the correlation between the pretest and the group membership dummy variable) and r_{yd} (the correlation between the posttest and the dummy variable) and (2) computation of ANOVA on gain scores that have been corrected for the difference between the pretest and posttest variances. It is possible to employ background information for control purposes within the SCSA framework.

Interpretation issues that are common to all pretest–posttest designs, but some of which are potentially more troublesome with nonequivalent groups from different populations, include: ceiling and floor effects, differential measurement conditions, reliability shift, factorial composition shift, mis-specification of growth model, misspecification of selection model, and un-known confounding events between pretest and posttest.

Nonequivalent-group designs are very weak, easily misinterpreted, and difficult to analyze. They should be avoided whenever possible. It appears that the future of the analysis of observational studies lies with structural equation methods.

APPENDIX A
Tables

Table A.1. Random numbers

	1 2 3 4 5	6 7 8 9 10	11 12 13 14 15	16 17 18 19 20	21 22 23 24 25
1	10 27 53 96 23	71 50 54 36 23	54 31 04 82 98	04 14 12 15 09	26 78 25 47 47
2	28 41 50 61 88	64 85 27 20 18	83 36 36 05 56	39 71 65 09 62	94 76 62 11 89
3	34 21 42 57 02	59 19 18 97 48	80 30 03 30 98	05 24 67 70 07	84 97 50 87 46
4	61 81 77 23 23	82 82 11 54 08	53 28 70 58 96	44 07 39 55 43	42 34 43 39 28
5	61 15 18 13 54	16 86 20 26 88	90 74 80 55 09	14 53 90 51 17	52 01 63 01 59
6	91 76 21 64 64	44 91 13 32 97	75 31 62 66 54	84 80 32 75 77	56 08 25 70 29
7	00 97 79 08 06	37 30 28 59 85	53 56 68 53 40	01 74 39 59 73	30 19 99 85 48
8	36 46 18 34 94	75 20 80 27 77	78 91 69 16 00	08 43 18 73 68	67 69 61 34 25
9	88 98 99 60 50	65 95 79 42 94	93 62 40 89 96	43 56 47 71 66	46 76 29 67 02
10	04 37 59 87 21	05 02 03 24 17	47 97 81 56 51	92 34 86 01 82	55 51 33 12 91
11	63 62 06 34 41	94 21 78 55 09	72 76 45 16 94	29 95 81 83 83	79 88 01 97 30
12	78 47 23 53 90	34 41 92 45 71	09 23 70 70 07	12 38 92 79 43	14 85 11 47 23
13	87 68 62 15 43	53 14 36 59 25	54 47 33 70 15	59 24 48 40 35	50 03 42 99 36
14	47 60 92 10 77	88 59 53 11 52	66 25 69 07 04	48 68 64 71 06	61 65 70 22 12
15	56 88 87 59 41	65 28 04 67 53	95 79 88 37 31	50 41 06 94 76	81 83 17 16 33
16	02 57 45 86 67	73 43 07 34 48	44 26 87 93 29	77 09 61 67 84	06 69 44 77 75
17	31 54 14 13 17	48 62 11 90 60	68 12 93 64 28	46 24 79 16 76	14 60 25 51 01
18	28 50 16 43 36	28 97 85 58 99	67 22 52 76 23	24 70 36 54 54	59 28 61 71 96
19	63 29 62 66 50	02 63 45 52 38	67 63 47 54 75	83 24 78 43 20	92 63 13 47 48
20	45 65 58 26 51	76 96 59 38 72	86 57 45 71 46	44 67 76 14 55	44 88 01 62 12
21	39 65 36 63 70	77 45 85 50 51	74 13 39 35 22	30 53 36 02 95	49 34 88 73 61
22	73 71 98 16 04	29 18 94 51 23	76 51 94 84 86	79 93 96 38 63	08 58 25 58 94
23	72 20 56 20 11	72 65 71 08 86	79 57 95 13 91	97 48 72 66 48	09 71 17 24 89
24	75 17 26 99 76	89 37 20 70 01	77 31 61 95 46	26 97 05 73 51	53 33 18 72 87
25	37 48 60 82 29	81 30 15 39 14	48 38 75 93 29	06 87 37 78 48	45 56 00 84 47
26	68 08 02 80 72	83 71 46 30 49	89 17 95 88 29	02 39 56 03 46	97 74 06 56 17
27	14 23 98 61 67	70 52 85 01 50	01 84 02 78 43	10 62 98 19 41	18 83 99 47 99
28	49 08 96 21 44	25 27 99 41 28	07 41 08 34 66	19 42 74 39 91	41 96 53 78 72
29	78 37 06 08 43	63 61 62 42 29	39 68 95 10 96	09 24 23 00 62	56 12 80 73 16
30	37 21 34 17 68	68 96 83 23 56	32 84 60 15 31	44 73 67 34 77	91 15 79 74 58
31	14 29 09 34 04	87 83 07 55 07	76 58 30 83 64	87 29 25 58 84	86 50 60 00 25
32	58 43 28 06 36	49 52 83 51 14	47 56 91 29 34	05 87 31 06 95	12 45 57 09 09
33	10 43 67 29 70	80 62 80 03 42	10 80 21 38 84	90 56 35 03 09	43 12 74 49 14
34	44 38 88 39 54	86 97 37 44 22	00 95 01 31 76	17 16 29 56 63	38 78 94 49 81
35	90 69 59 19 51	85 39 52 85 13	07 28 37 07 61	11 16 36 27 03	78 86 72 04 95
36	41 47 10 25 62	97 05 31 03 61	20 26 36 31 62	68 69 86 95 44	84 95 48 46 45
37	91 94 14 63 19	75 89 11 47 11	31 56 34 19 09	79 57 92 36 59	14 93 87 81 40
38	80 06 54 18 66	09 18 94 06 19	98 40 07 17 81	22 45 44 84 11	24 62 20 42 31
39	67 72 77 63 48	84 08 31 55 58	24 33 45 77 58	80 45 67 93 82	75 70 16 08 24
40	59 40 24 13 27	79 26 88 86 30	01 31 60 10 39	53 58 47 70 93	85 81 56 39 38
41	05 90 35 89 95	01 61 16 96 94	50 78 13 69 36	37 68 53 37 31	71 26 35 03 71
42	44 43 80 69 98	46 68 05 14 82	90 78 50 05 62	77 79 13 57 44	59 60 10 39 66
43	61 81 31 96 82	00 57 25 60 59	46 72 60 18 77	55 66 12 62 11	08 99 55 64 57
44	42 88 07 10 05	24 98 65 63 21	47 21 61 88 32	27 80 30 21 60	10 92 35 36 12
45	77 94 30 05 39	28 10 99 00 27	12 73 73 99 12	49 99 57 94 82	96 88 57 17 91
46	78 83 19 76 16	94 11 68 84 26	23 54 20 86 85	23 86 66 99 07	36 37 34 92 09
47	87 76 59 61 81	43 63 64 61 61	65 76 36 95 90	18 48 27 45 68	27 23 65 30 72
48	91 43 05 96 47	55 78 99 95 24	37 55 85 78 78	01 48 41 19 10	35 19 54 07 73
49	84 97 77 72 73	09 62 06 65 72	87 12 49 03 60	41 15 20 76 27	50 47 02 29 16
50	87 41 60 76 83	44 88 96 07 80	83 05 83 38 96	73 70 66 81 90	30 56 10 48 59

Table A.1. *Continued*

	1	2	3	4	5		6	7	8	9	10		11	12	13	14	15		16	17	18	19	20		21	22	23	24	25
1	22	17	68	65	84		68	95	23	92	35		87	02	22	57	51		61	09	43	95	06		58	24	82	03	47
2	19	36	27	59	46		13	79	93	37	55		39	77	32	77	09		85	52	05	30	62		47	83	51	62	74
3	16	77	23	02	77		09	61	87	25	21		28	06	24	25	93		16	71	13	59	78		23	05	47	47	25
4	78	43	76	71	61		20	44	90	32	64		97	67	63	99	61		46	38	03	93	22		69	81	21	99	21
5	03	28	28	26	08		73	37	32	04	05		69	30	16	09	05		88	69	58	28	99		35	07	44	75	47
6	93	22	53	64	39		07	10	63	76	35		87	03	04	79	88		08	13	13	85	51		55	34	57	72	69
7	78	76	58	54	74		92	38	70	96	92		52	06	79	79	45		82	63	18	27	44		69	66	92	19	09
8	23	68	35	26	00		99	53	93	61	28		52	70	05	48	34		56	65	05	61	86		90	92	10	70	80
9	15	39	25	70	99		93	86	52	77	65		15	33	59	05	28		22	87	26	07	47		86	96	98	29	06
10	58	71	96	30	24		18	46	23	34	27		85	13	99	24	44		49	18	09	79	49		74	16	32	23	02
11	57	35	27	33	72		24	53	63	94	09		41	10	76	47	91		44	04	95	49	66		39	60	04	59	81
12	48	50	86	54	48		22	06	34	72	52		82	21	15	65	20		33	29	94	71	11		15	91	29	12	03
13	61	96	48	95	03		07	16	39	33	66		98	56	10	56	79		77	21	30	27	12		90	49	22	23	62
14	36	93	89	41	26		29	70	83	63	51		99	74	20	52	36		87	09	41	15	09		98	60	16	03	03
15	18	87	00	42	31		57	90	12	02	07		23	47	37	17	31		54	08	01	88	63		39	41	88	92	10
16	88	56	53	27	59		33	35	72	67	47		77	34	55	45	70		08	18	27	38	90		16	95	86	70	75
17	09	72	95	84	29		49	41	31	06	70		42	38	06	45	18		64	84	73	31	65		52	53	37	97	15
18	12	96	88	17	31		65	19	69	02	83		60	75	86	90	68		24	64	19	35	51		56	61	87	39	12
19	85	94	57	24	16		92	09	84	38	76		22	00	27	69	85		29	81	94	78	70		21	94	47	90	12
20	38	64	43	59	98		98	77	87	68	07		91	51	67	62	44		40	98	05	93	78		23	32	65	11	18
21	53	44	09	42	72		00	41	86	79	79		68	47	22	00	20		35	55	31	51	51		00	83	63	22	55
22	40	76	66	26	84		57	99	99	90	37		36	63	32	08	58		37	40	13	68	97		87	64	81	07	83
23	02	17	79	18	05		12	59	52	57	02		22	07	90	47	03		28	14	11	30	79		20	69	22	40	98
24	95	17	82	06	53		31	51	10	96	46		92	06	88	07	77		56	11	50	81	69		40	23	72	51	39
25	35	76	22	42	92		96	11	83	44	80		34	68	35	48	77		33	42	40	90	60		73	96	53	97	86
26	26	29	13	56	41		85	47	04	66	08		34	72	57	59	13		82	43	80	46	15		38	26	61	70	04
27	77	80	20	75	82		72	82	32	99	90		63	95	73	76	63		89	73	44	99	05		48	67	26	43	18
28	46	40	66	44	52		91	36	74	43	53		30	82	13	54	00		78	45	63	98	35		55	03	36	67	68
29	37	56	08	18	09		77	53	84	46	47		31	91	18	95	58		24	16	74	11	53		44	10	13	85	57
30	61	65	61	68	66		37	27	47	39	19		84	83	70	07	48		53	21	40	06	71		95	06	79	88	54
31	93	43	69	64	07		34	18	04	52	35		56	27	09	24	86		61	85	53	83	45		19	90	70	99	00
32	21	96	60	12	99		11	20	99	45	18		48	13	93	55	34		18	37	79	49	90		65	97	38	20	46
33	95	20	47	97	97		27	37	83	28	71		00	06	41	41	74		45	89	09	39	84		51	67	11	52	49
34	97	86	21	78	73		10	65	81	92	59		58	76	17	14	97		04	76	62	16	17		17	95	70	45	80
35	69	92	06	34	13		59	71	74	17	32		27	55	10	24	19		23	71	82	13	74		63	52	52	01	41
36	04	31	17	21	56		33	73	99	19	87		26	72	39	27	67		53	77	57	68	93		60	61	97	22	61
37	61	06	98	03	91		87	14	77	43	96		43	00	65	98	50		45	60	33	01	07		98	99	46	50	47
38	85	93	85	86	88		72	87	08	62	40		16	06	10	89	20		23	21	34	74	97		76	38	03	29	63
39	21	74	32	47	45		73	96	07	94	52		09	65	90	77	47		25	76	16	19	33		53	05	70	53	30
40	15	69	53	82	80		79	96	23	53	10		65	39	07	16	29		45	33	02	43	70		02	87	40	41	45
41	02	89	08	04	49		20	21	14	68	86		87	63	93	95	17		11	29	01	95	80		35	14	97	35	33
42	87	18	15	89	79		85	43	01	72	73		08	61	74	51	69		89	74	39	82	15		94	51	33	41	67
43	98	83	71	94	22		59	97	50	99	52		08	52	85	08	40		87	80	61	65	31		91	51	80	32	44
44	10	08	58	21	66		72	68	49	29	31		89	85	84	46	06		59	73	19	85	23		65	09	29	75	63
45	47	90	56	10	08		88	02	84	27	83		42	29	72	23	19		66	56	45	65	79		20	71	53	20	25
46	22	85	61	68	90		49	64	92	85	44		16	40	12	89	88		50	14	49	81	06		01	82	77	45	12
47	67	80	43	79	33		12	83	11	41	16		25	58	19	68	70		77	02	54	00	52		53	43	37	15	26
48	27	62	50	96	72		79	44	61	40	15		14	53	40	65	39		27	31	58	50	28		11	39	03	34	25
49	33	78	80	87	15		38	30	06	38	21		14	47	47	07	26		54	96	87	53	32		40	36	40	96	76
50	13	13	92	66	99		47	24	49	57	74		32	25	43	62	17		10	97	11	69	84		99	63	22	32	98

Source: Taken from Table 3 of Fisher and Yates, *Statistical Tables for Biological, Agricultural and Medical Research*, published by Oliver and Boyd Ltd., Edinburgh, by permission of the authors and publishers.

Table A.2. Critical Values of F for .05 (Roman), .01 (*Italic*), and .001 (Boldface) Levels of Significance

n_2 \ n_1	1	2	3	4	5	6	8	12	24	∞
1	161	200	216	225	230	234	239	244	249	254
	4052	*4999*	*5403*	*5625*	*5724*	*5859*	*5981*	*6106*	*6234*	*6366*
	405284	**500000**	**540379**	**562500**	**576405**	**585937**	**598144**	**610667**	**623497**	**636619**
2	18.51	19.00	19.16	19.25	19.30	19.33	19.37	19.41	19.45	19.50
	98.49	*99.01*	*99.17*	*99.25*	*99.30*	*99.33*	*99.36*	*99.42*	*99.46*	*99.50*
	998.5	**999.0**	**999.2**	**999.2**	**999.3**	**999.3**	**999.4**	**999.4**	**999.5**	**999.5**
3	10.13	9.55	9.28	9.12	9.01	8.94	8.84	8.74	8.64	8.53
	34.12	*30.81*	*29.46*	*28.71*	*28.24*	*27.91*	*27.49*	*27.05*	*26.60*	*26.12*
	167.5	**148.5**	**141.1**	**137.1**	**134.6**	**132.8**	**130.6**	**128.3**	**125.9**	**123.5**
4	7.71	6.94	6.59	6.39	6.26	6.16	6.04	5.91	5.77	5.63
	21.20	*18.00*	*16.69*	*15.98*	*15.52*	*15.21*	*14.80*	*14.37*	*13.93*	*13.46*
	74.14	**61.25**	**56.18**	**53.44**	**51.71**	**50.53**	**49.00**	**47.41**	**45.77**	**44.05**
5	6.61	5.79	5.41	5.19	5.05	4.95	4.82	4.68	4.53	4.36
	16.26	*13.27*	*12.06*	*11.39*	*10.97*	*10.67*	*10.27*	*9.89*	*9.47*	*9.02*
	47.04	**36.61**	**33.20**	**31.09**	**29.75**	**28.84**	**27.64**	**26.42**	**25.14**	**23.78**
6	5.99	5.14	4.76	4.53	4.39	4.28	4.15	4.00	3.84	3.67
	13.74	*10.92*	*9.78*	*9.15*	*8.75*	*8.47*	*8.10*	*7.72*	*7.31*	*6.88*
	35.51	**27.00**	**23.70**	**21.90**	**20.81**	**20.03**	**19.03**	**17.99**	**16.89**	**15.75**
7	5.59	4.74	4.35	4.12	3.97	3.87	3.73	3.57	3.41	3.23
	12.25	*9.55*	*8.45*	*7.85*	*7.46*	*7.19*	*6.84*	*6.47*	*6.07*	*5.65*
	29.22	**21.69**	**18.77**	**17.19**	**16.21**	**15.52**	**14.63**	**13.71**	**12.73**	**11.69**
8	5.32	4.46	4.07	3.84	3.69	3.58	3.44	3.28	3.12	2.93
	11.26	*8.65*	*7.59*	*7.01*	*6.63*	*6.37*	*6.03*	*5.67*	*5.28*	*4.86*
	25.42	**18.49**	**15.83**	**14.39**	**13.49**	**12.86**	**12.04**	**11.19**	**10.30**	**9.34**
9	5.12	4.26	3.86	3.63	3.48	3.37	3.23	3.07	2.90	2.71
	10.56	*8.02*	*6.99*	*6.42*	*6.06*	*5.80*	*5.47*	*5.11*	*4.73*	*4.31*
	22.86	**16.39**	**13.90**	**12.56**	**11.71**	**11.13**	**10.37**	**9.57**	**8.72**	**7.81**
10	4.96	4.10	3.71	3.48	3.33	3.22	3.07	2.91	2.74	2.54
	10.04	*7.56*	*6.55*	*5.99*	*5.64*	*5.39*	*5.06*	*4.71*	*4.33*	*3.91*
	21.04	**14.91**	**12.55**	**11.28**	**10.48**	**9.92**	**9.20**	**8.45**	**7.64**	**6.76**
11	4.84	3.98	3.59	3.36	3.20	3.09	2.95	2.79	2.61	2.40
	9.65	*7.20*	*6.22*	*5.67*	*5.32*	*5.07*	*4.74*	*4.40*	*4.02*	*3.60*
	19.69	**13.81**	**11.56**	**10.35**	**9.58**	**9.05**	**8.35**	**7.63**	**6.85**	**6.00**
12	4.75	3.88	3.49	3.26	3.11	3.00	2.85	2.69	2.50	2.30
	9.33	*6.93*	*5.95*	*5.41*	*5.06*	*4.82*	*4.50*	*4.16*	*3.78*	*3.36*
	18.64	**12.97**	**10.80**	**9.63**	**8.89**	**8.38**	**7.71**	**7.00**	**6.25**	**5.42**

Table A.2. *Continued*

n_2 \ n_1	1	2	3	4	5	6	8	12	24	∞
13	4.67	3.80	3.41	3.18	3.02	2.92	2.77	2.60	2.42	2.21
	9.07	*6.70*	*5.74*	*5.20*	*4.86*	*4.62*	*4.30*	*3.96*	*3.59*	*3.16*
	17.81	**12.31**	**10.21**	**9.07**	**8.35**	**7.86**	**7.21**	**6.52**	**5.78**	**4.97**
14	4.60	3.74	3.34	3.11	2.96	2.85	2.70	2.53	2.35	2.13
	8.86	*6.51*	*5.56*	*5.03*	*4.69*	*4.46*	*4.14*	*3.80*	*3.43*	*3.00*
	17.14	**11.78**	**9.73**	**8.62**	**7.92**	**7.43**	**6.80**	**6.13**	**5.41**	**4.60**
15	4.54	3.68	3.29	3.06	2.90	2.79	2.64	2.48	2.29	2.07
	8.68	*6.36*	*5.42*	*4.89*	*4.56*	*4.32*	*4.00*	*3.67*	*3.29*	*2.87*
	16.59	**11.34**	**9.34**	**8.25**	**7.57**	**7.09**	**6.47**	**5.81**	**5.10**	**4.31**
16	4.49	3.63	3.24	3.01	2.85	2.74	2.59	2.42	2.24	2.01
	8.53	*6.23*	*5.29*	*4.77*	*4.44*	*4.20*	*3.89*	*3.55*	*3.18*	*2.75*
	16.12	**10.97**	**9.00**	**7.94**	**7.27**	**6.81**	**6.19**	**5.55**	**4.85**	**4.06**
17	4.45	3.59	3.20	2.96	2.81	2.70	2.55	2.38	2.19	1.96
	8.40	*6.11*	*5.18*	*4.67*	*4.34*	*4.10*	*3.79*	*3.45*	*3.08*	*2.65*
	15.72	**10.66**	**8.73**	**7.68**	**7.02**	**6.56**	**5.96**	**5.32**	**4.63**	**3.85**
18	4.41	3.55	3.16	2.93	2.77	2.66	2.51	2.34	2.15	1.92
	8.28	*6.01*	*5.09*	*4.58*	*4.25*	*4.01*	*3.71*	*3.37*	*3.00*	*2.57*
	15.38	**10.39**	**8.49**	**7.46**	**6.81**	**6.35**	**5.76**	**5.13**	**4.45**	**3.67**
19	4.38	3.52	3.13	2.90	2.74	2.63	2.48	2.31	2.11	1.88
	8.18	*5.93*	*5.01*	*4.50*	*4.17*	*3.94*	*3.63*	*3.30*	*2.92*	*2.49*
	15.08	**10.16**	**8.28**	**7.26**	**6.61**	**6.18**	**5.59**	**4.97**	**4.29**	**3.52**
20	4.35	3.49	3.10	2.87	2.71	2.60	2.45	2.28	2.08	1.84
	8.10	*5.85*	*4.94*	*4.43*	*4.10*	*3.87*	*3.56*	*3.23*	*2.86*	*2.42*
	14.82	**9.95**	**8.10**	**7.10**	**6.46**	**6.02**	**5.44**	**4.82**	**4.15**	**3.38**
21	4.32	3.47	3.07	2.84	2.68	2.57	2.42	2.25	2.05	1.81
	8.02	*5.78*	*4.87*	*4.37*	*4.04*	*3.81*	*3.51*	*3.17*	*2.80*	*2.36*
	14.59	**9.77**	**7.94**	**6.95**	**6.32**	**5.88**	**5.31**	**4.70**	**4.03**	**3.26**
22	4.30	3.44	3.05	2.82	2.66	2.55	2.40	2.23	2.03	1.78
	7.94	*5.72*	*4.82*	*4.31*	*3.99*	*3.76*	*3.45*	*3.12*	*2.75*	*2.31*
	14.38	**9.61**	**7.80**	**6.81**	**6.19**	**5.76**	**5.19**	**4.58**	**3.92**	**3.15**
23	4.28	3.42	3.03	2.80	2.64	2.53	2.38	2.20	2.00	1.76
	7.88	*5.66*	*4.76*	*4.26*	*3.94*	*3.71*	*3.41*	*3.07*	*2.70*	*2.26*
	14.19	**9.47**	**7.67**	**6.69**	**6.08**	**5.65**	**5.09**	**4.48**	**3.82**	**3.05**
24	4.26	3.40	3.01	2.78	2.62	2.51	2.36	2.18	1.98	1.73
	7.82	*5.61*	*4.72*	*4.22*	*3.90*	*3.67*	*3.36*	*3.03*	*2.66*	*2.21*
	14.03	**9.34**	**7.55**	**6.59**	**5.98**	**5.55**	**4.99**	**4.39**	**3.74**	**2.97**

Table A.2. *Continued*

n_2 \ n_1	1	2	3	4	5	6	8	12	24	∞
25	4.24	3.38	2.99	2.76	2.60	2.49	2.34	2.16	1.96	1.71
	7.77	5.57	4.68	4.18	3.86	3.63	3.32	2.99	2.62	2.17
	13.88	9.22	7.45	6.49	5.88	5.46	4.91	4.31	3.66	2.89
26	4.22	3.37	2.98	2.74	2.59	2.47	2.32	2.15	1.95	1.69
	7.72	5.53	4.64	4.14	3.82	3.59	3.29	2.96	2.58	2.13
	13.74	9.12	7.36	6.41	5.80	5.38	4.83	4.24	3.59	2.82
27	4.21	3.35	2.96	2.73	2.57	2.46	2.30	2.13	1.93	1.67
	7.68	5.49	4.60	4.11	3.78	3.56	3.26	2.93	2.55	2.10
	13.61	9.02	7.27	6.33	5.73	5.31	4.76	4.17	3.52	2.75
28	4.20	3.34	2.95	2.71	2.56	2.44	2.29	2.12	1.91	1.65
	7.64	5.45	4.57	4.07	3.75	3.53	3.23	2.90	2.52	2.06
	13.50	8.93	7.19	6.25	5.66	5.24	4.69	4.11	3.46	2.70
29	4.18	3.33	2.93	2.70	2.54	2.43	2.28	2.10	1.90	1.64
	7.60	5.42	4.54	4.04	3.73	3.50	3.20	2.87	2.49	2.03
	13.39	8.85	7.12	6.19	5.59	5.18	4.64	4.05	3.41	2.64
30	4.17	3.32	2.92	2.69	2.53	2.42	2.27	2.09	1.89	1.62
	7.56	5.39	4.51	4.02	3.70	3.47	3.17	2.84	2.47	2.01
	13.29	8.77	7.05	6.12	5.53	5.12	4.58	4.00	3.36	2.59
40	4.08	3.23	2.84	2.61	2.45	2.34	2.18	2.00	1.79	1.51
	7.31	5.18	4.31	3.83	3.51	3.29	2.99	2.66	2.29	1.80
	12.61	8.25	6.60	5.70	5.13	4.73	4.21	3.64	3.01	2.23
60	4.00	3.15	2.76	2.52	2.37	2.25	2.10	1.92	1.70	1.39
	7.08	4.98	4.13	3.65	3.34	3.12	2.82	2.50	2.12	1.60
	11.97	7.76	6.17	5.31	4.76	4.37	3.87	3.31	2.69	1.90
120	3.92	3.07	2.68	2.45	2.29	2.17	2.02	1.83	1.61	1.25
	6.85	4.79	3.95	3.48	3.17	2.96	2.66	2.34	1.95	1.38
	11.38	7.31	5.79	4.95	4.42	4.04	3.55	3.02	2.40	1.56
∞	3.84	2.99	2.60	2.37	2.21	2.09	1.94	1.75	1.52	1.00
	6.64	4.60	3.78	3.32	3.02	2.80	2.51	2.18	1.79	1.00
	10.83	6.91	5.42	4.62	4.10	3.74	3.27	2.74	2.13	1.00

Source: Taken from Table 3 of Fisher and Yates, *Statistical Tables for Biological, Agricultural and Medical Research*, published by Oliver and Boyd Ltd., Edinburgh, by permission of the authors and publishers.

Table A.3. Critical Values of the t Statistic (α_1 = One Tail; α_2 = Two Tail)

df	α_1 / α_2	.25 / .50	.20 / .40	.15 / .30	.10 / .20	.05 / .10	.025 / .05	.01 / .02	.005 / .01	.0005 / .001
1		1.000	1.376	1.963	3.078	6.314	12.706	31.821	63.657	636.619
2		.816	1.061	1.386	1.886	2.920	4.303	6.965	9.925	31.598
3		.765	.978	1.250	1.638	2.353	3.182	4.541	5.841	12.924
4		.741	.941	1.190	1.533	2.132	2.776	3.747	4.604	8.610
5		.727	.920	1.156	1.476	2.015	2.571	3.365	4.032	6.869
6		.718	.906	1.134	1.440	1.943	2.447	3.143	3.707	5.959
7		.711	.896	1.119	1.415	1.895	2.365	2.998	3.499	5.408
8		.706	.889	1.108	1.397	1.860	2.306	2.896	3.355	5.041
9		.703	.883	1.100	1.383	1.833	2.262	2.821	3.250	4.781
10		.700	.879	1.093	1.372	1.812	2.228	2.764	3.169	4.587
11		.697	.876	1.088	1.363	1.796	2.201	2.718	3.106	4.437
12		.695	.873	1.083	1.356	1.782	2.179	2.681	3.055	4.318
13		.694	.870	1.079	1.350	1.771	2.160	2.650	3.012	4.221
14		.692	.868	1.076	1.345	1.761	2.145	2.624	2.977	4.140
15		.691	.866	1.074	1.341	1.753	2.131	2.602	2.947	4.073
16		.690	.865	1.071	1.337	1.746	2.120	2.583	2.921	4.015
17		.689	.863	1.069	1.333	1.740	2.110	2.567	2.898	3.965
18		.688	.862	1.067	1.330	1.734	2.101	2.552	2.878	3.922
19		.688	.861	1.066	1.328	1.729	2.093	2.539	2.861	3.883
20		.687	.860	1.064	1.325	1.725	2.086	2.528	2.845	3.850
21		.686	.859	1.063	1.323	1.721	2.080	2.518	2.831	3.819
22		.686	.858	1.061	1.321	1.717	2.074	2.508	2.819	3.792
23		.685	.858	1.060	1.319	1.714	2.069	2.500	2.807	3.767
24		.685	.857	1.059	1.318	1.711	2.064	2.492	2.797	3.745
25		.684	.856	1.058	1.316	1.708	2.060	2.485	2.787	3.725
26		.684	.856	1.058	1.315	1.706	2.056	2.479	2.779	3.707
27		.684	.855	1.057	1.314	1.703	2.052	2.473	2.771	3.690
28		.683	.855	1.056	1.313	1.701	2.048	2.467	2.763	3.674
29		.683	.854	1.055	1.311	1.699	2.045	2.462	2.756	3.659
30		.683	.854	1.055	1.310	1.697	2.042	2.457	2.750	3.646
40		.681	.851	1.050	1.303	1.684	2.021	2.423	2.704	3.551
60		.679	.848	1.046	1.296	1.671	2.000	2.390	2.660	3.460
120		.677	.845	1.041	1.289	1.658	1.980	2.358	2.617	3.373
∞		.674	.842	1.036	1.282	1.654	1.960	2.326	2.576	3.291

Source: Taken from Table 3 of Fisher and Yates, *Statistical Tables for Biological, Agricultural and Medical Research*, published by Oliver and Boyd Ltd., Edinburgh, by permission of the authors and publishers.

Table A.4. Critical Values of Bryant–Paulson Generalized Studentized Range Statistic Q_p

					$\alpha = .05$						
					J						
ν	2	3	4	5	6	7	8	10	12	16	20

$C = 1$

ν	2	3	4	5	6	7	8	10	12	16	20
2	7.96	11.00	12.99	14.46	15.61	16.56	17.36	18.65	19.68	21.23	22.40
3	5.42	7.18	8.32	9.17	9.84	10.39	10.86	11.62	12.22	13.14	13.83
4	4.51	5.84	6.69	7.32	7.82	8.23	8.58	9.15	9.61	10.30	10.82
5	4.06	5.17	5.88	6.40	6.82	7.16	7.45	7.93	8.30	8.88	9.32
6	3.79	4.78	5.40	5.86	6.23	6.53	6.78	7.20	7.53	8.04	8.43
7	3.62	4.52	5.09	5.51	5.84	6.11	6.34	6.72	7.03	7.49	7.84
8	3.49	4.34	4.87	5.26	5.57	5.82	6.03	6.39	6.67	7.10	7.43
10	3.32	4.10	4.58	4.93	5.21	5.43	5.63	5.94	6.19	6.58	6.87
12	3.22	3.95	4.40	4.73	4.98	5.19	5.37	5.67	5.90	6.26	6.53
14	3.15	3.85	4.28	4.59	4.83	5.03	5.20	5.48	5.70	6.03	6.29
16	3.10	3.77	4.19	4.49	4.72	4.91	5.07	5.34	5.55	5.87	6.12
18	3.06	3.72	4.12	4.41	4.63	4.82	4.98	5.23	5.44	5.75	5.98
20	3.03	3.67	4.07	4.35	4.57	4.75	4.90	5.15	5.35	5.65	5.88
24	2.98	3.61	3.99	4.26	4.47	4.65	4.79	5.03	5.22	5.51	5.73
30	2.94	3.55	3.91	4.18	4.38	4.54	4.69	4.91	5.09	5.37	5.58
40	2.89	3.49	3.84	4.09	4.29	4.45	4.58	4.80	4.97	5.23	5.43
60	2.85	3.43	3.77	4.01	4.20	4.35	4.48	4.69	4.85	5.10	5.29
120	2.81	3.37	3.70	3.93	4.11	4.26	4.38	4.58	4.73	4.97	5.15

$C = 2$

ν	2	3	4	5	6	7	8	10	12	16	20
2	9.50	13.18	15.59	17.36	18.75	19.89	20.86	22.42	23.66	25.54	26.94
3	6.21	8.27	9.60	10.59	11.37	12.01	12.56	13.44	14.15	15.22	16.02
4	5.04	6.54	7.51	8.23	8.80	9.26	9.66	10.31	10.83	11.61	12.21
5	4.45	5.68	6.48	7.06	7.52	7.90	8.23	8.76	9.18	9.83	10.31
6	4.10	5.18	5.87	6.37	6.77	7.10	7.38	7.84	8.21	8.77	9.20
7	3.87	4.85	5.47	5.92	6.28	6.58	6.83	7.24	7.57	8.08	8.46
8	3.70	4.61	5.19	5.61	5.94	6.21	6.44	6.82	7.12	7.59	7.94
10	3.49	4.31	4.82	5.19	5.49	5.73	5.93	6.27	6.54	6.95	7.26
12	3.35	4.12	4.59	4.93	5.20	5.43	5.62	5.92	6.17	6.55	6.83
14	3.26	3.99	4.44	4.76	5.01	5.22	5.40	5.69	5.92	6.27	6.54
16	3.19	3.90	4.32	4.63	4.88	5.07	5.24	5.52	5.74	6.07	6.33
18	3.14	3.82	4.24	4.54	4.77	4.96	5.13	5.39	5.60	5.92	6.17
20	3.10	3.77	4.17	4.46	4.69	4.88	5.03	5.29	5.49	5.81	6.04
24	3.04	3.69	4.08	4.35	4.57	4.75	4.90	5.14	5.34	5.63	5.86
30	2.99	3.61	3.98	4.25	4.46	4.62	4.77	5.00	5.18	5.46	5.68
40	2.93	3.53	3.89	4.15	4.34	4.50	4.64	4.86	5.04	5.30	5.50
60	2.88	3.46	3.80	4.05	4.24	4.39	4.52	4.73	4.89	5.14	5.33
120	2.82	3.38	3.72	3.95	4.13	4.28	4.40	4.60	4.75	4.99	5.17

						$\alpha = .05$						
						J						
ν	2	3	4	5	6	7	8	10	12	16	20
						$C=3$						
2	10.83	15.06	17.82	19.85	21.45	22.76	23.86	25.66	27.08	29.23	30.83
3	6.92	9.23	10.73	11.84	12.72	13.44	14.06	15.05	15.84	17.05	17.95
4	5.51	7.18	8.25	9.05	9.67	10.19	10.63	11.35	11.92	12.79	13.45
5	4.81	6.16	7.02	7.66	8.17	8.58	8.94	9.52	9.98	10.69	11.22
6	4.38	5.55	6.30	6.84	7.28	7.64	7.94	8.44	8.83	9.44	9.90
7	4.11	5.16	5.82	6.31	6.70	7.01	7.29	7.73	8.08	8.63	9.03
8	3.91	4.88	5.49	5.93	6.29	6.58	6.83	7.23	7.55	8.05	8.42
10	3.65	4.51	5.05	5.44	5.75	6.01	6.22	6.58	6.86	7.29	7.62
12	3.48	4.28	4.78	5.14	5.42	5.65	5.85	6.17	6.43	6.82	7.12
14	3.37	4.13	4.59	4.93	5.19	5.41	5.59	5.89	6.13	6.50	6.78
16	3.29	4.01	4.46	4.78	5.03	5.23	5.41	5.69	5.92	6.27	6.53
18	3.23	3.93	4.35	4.66	4.90	5.10	5.27	5.54	5.76	6.09	6.34
20	3.18	3.86	4.28	4.57	4.81	5.00	5.16	5.42	5.63	5.96	6.20
24	3.11	3.76	4.16	4.44	4.67	4.85	5.00	5.25	5.45	5.75	5.98
30	3.04	3.67	4.05	4.32	4.53	4.70	4.85	5.08	5.27	5.56	5.78
40	2.97	3.57	3.94	4.20	4.40	4.56	4.70	4.92	5.10	5.37	5.57
60	2.90	3.49	3.83	4.08	4.27	4.43	4.56	4.77	4.93	5.19	5.38
120	2.84	3.40	3.73	3.97	4.15	4.30	4.42	4.62	4.77	5.01	5.19

						$\alpha = .01$						
						J						
ν	2	3	4	5	6	7	8	10	12	16	20
						$C=1$						
2	19.09	26.02	30.57	33.93	36.58	38.76	40.60	43.59	45.95	49.55	52.24
3	10.28	13.32	15.32	16.80	17.98	18.95	19.77	21.12	22.19	23.82	25.05
4	7.68	9.64	10.93	11.89	12.65	13.28	13.82	14.70	15.40	16.48	17.29
5	6.49	7.99	8.97	9.70	10.28	10.76	11.17	11.84	12.38	13.20	13.83
6	5.83	7.08	7.88	8.48	8.96	9.36	9.70	10.25	10.70	11.38	11.90
7	5.41	6.50	7.20	7.72	8.14	8.48	8.77	9.26	9.64	10.24	10.69
8	5.12	6.11	6.74	7.20	7.58	7.88	8.15	8.58	8.92	9.46	9.87
10	4.76	5.61	6.15	6.55	6.86	7.13	7.35	7.72	8.01	8.47	8.82
12	4.54	5.31	5.79	6.15	6.43	6.67	6.87	7.20	7.46	7.87	8.18
14	4.39	5.11	5.56	5.89	6.15	6.36	6.55	6.85	7.09	7.47	7.75
16	4.28	4.96	5.39	5.70	5.95	6.15	6.32	6.60	6.83	7.18	7.45
18	4.20	4.86	5.26	5.56	5.79	5.99	6.15	6.42	6.63	6.96	7.22
20	4.14	4.77	5.17	5.45	5.68	5.86	6.02	6.27	6.48	6.80	7.04
24	4.05	4.65	5.02	5.29	5.50	5.68	5.83	6.07	6.26	6.56	6.78
30	3.96	4.54	4.89	5.14	5.34	5.50	5.64	5.87	6.05	6.32	6.53
40	3.88	4.43	4.76	5.00	5.19	5.34	5.47	5.68	5.85	6.10	6.30
60	3.79	4.32	4.64	4.86	5.04	5.18	5.30	5.50	5.65	5.89	6.07
120	3.72	4.22	4.52	4.73	4.89	5.03	5.14	5.32	5.47	5.69	5.85

Table A.4. *Continued*

					$\alpha = .01$						
					J						
v	2	3	4	5	6	7	8	10	12	16	20
					$C=2$						
2	23.11	31.55	37.09	41.19	44.41	47.06	49.31	52.94	55.82	60.20	63.47
3	11.97	15.56	17.91	19.66	21.05	22.19	23.16	24.75	26.01	27.93	29.38
4	8.69	10.95	12.43	13.54	14.41	15.14	15.76	16.77	17.58	18.81	19.74
5	7.20	8.89	9.99	10.81	11.47	12.01	12.47	13.23	13.84	14.77	15.47
6	6.36	7.75	8.64	9.31	9.85	10.29	10.66	11.28	11.77	12.54	13.11
7	5.84	7.03	7.80	8.37	8.83	9.21	9.53	10.06	10.49	11.14	11.64
8	5.48	6.54	7.23	7.74	8.14	8.48	8.76	9.23	9.61	10.19	10.63
10	5.02	5.93	6.51	6.93	7.27	7.55	7.79	8.19	8.50	8.99	9.36
12	4.74	5.56	6.07	6.45	6.75	7.00	7.21	7.56	7.84	8.27	8.60
14	4.56	5.31	5.78	6.13	6.40	6.63	6.82	7.14	7.40	7.79	8.09
16	4.42	5.14	5.58	5.90	6.16	6.37	6.55	6.85	7.08	7.45	7.73
18	4.32	5.00	5.43	5.73	5.98	6.18	6.35	6.63	6.85	7.19	7.46
20	4.25	4.90	5.31	5.60	5.84	6.03	6.19	6.46	6.67	7.00	7.25
24	4.14	4.76	5.14	5.42	5.63	5.81	5.96	6.21	6.41	6.71	6.95
30	4.03	4.62	4.98	5.24	5.44	5.61	5.75	5.98	6.16	6.44	6.66
40	3.93	4.48	4.82	5.07	5.26	5.41	5.54	5.76	5.93	6.19	6.38
60	3.83	4.36	4.68	4.90	5.08	5.22	5.35	5.54	5.70	5.94	6.12
120	3.73	4.24	4.54	4.75	4.91	5.05	5.16	5.35	5.49	5.71	5.88
					$C=3$						
2	26.54	36.26	42.64	47.36	51.07	54.13	56.71	60.90	64.21	69.25	73.01
3	13.45	17.51	20.17	22.15	23.72	25.01	26.11	27.90	29.32	31.50	33.13
4	9.59	12.11	13.77	15.00	15.98	16.79	17.47	18.60	19.50	20.87	21.91
5	7.83	9.70	10.92	11.82	12.54	13.14	13.65	14.48	15.15	16.17	16.95
6	6.85	8.36	9.34	10.07	10.65	11.13	11.54	12.22	12.75	13.59	14.21
7	6.23	7.52	8.36	8.98	9.47	9.88	10.23	10.80	11.26	11.97	12.51
8	5.81	6.95	7.69	8.23	8.67	9.03	9.33	9.84	10.24	10.87	11.34
10	5.27	6.23	6.84	7.30	7.66	7.96	8.21	8.63	8.96	9.48	9.88
12	4.94	5.80	6.34	6.74	7.05	7.31	7.54	7.90	8.20	8.65	9.00
14	4.72	5.51	6.00	6.36	6.65	6.89	7.09	7.42	7.69	8.10	8.41
16	4.56	5.30	5.76	6.10	6.37	6.59	6.77	7.08	7.33	7.71	8.00
18	4.44	5.15	5.59	5.90	6.16	6.36	6.54	6.83	7.06	7.42	7.69
20	4.35	5.03	5.45	5.75	5.99	6.19	6.36	6.63	6.85	7.19	7.45
24	4.22	4.86	5.25	5.54	5.76	5.94	6.10	6.35	6.55	6.87	7.11
30	4.10	4.70	5.06	5.33	5.54	5.71	5.85	6.08	6.27	6.56	6.78
40	3.98	4.54	4.88	5.13	5.32	5.48	5.61	5.83	6.00	6.27	6.47
60	3.86	4.39	4.72	4.95	5.12	5.27	5.39	5.59	5.75	6.00	6.18
120	3.75	4.25	4.55	4.77	4.94	5.07	5.18	5.37	5.51	5.74	5.90

Source: Reprinted from Bryant and Paulson (1976), by permission of the authors and the publishers.

Table A.5. Critical Values of Studentized Range Statistic

Error		$J = number\ of\ means$									
df	α	2	3	4	5	6	7	8	9	10	11
5	.05	3.64	4.60	5.22	5.67	6.03	6.33	6.58	6.80	6.99	7.17
5	.01	5.70	6.98	7.80	8.42	8.91	9.32	9.67	9.97	10.24	10.48
6	.05	3.46	4.34	4.90	5.30	5.63	5.90	6.12	6.32	6.49	6.65
6	.01	5.24	6.33	7.03	7.56	7.97	8.32	8.61	8.87	9.10	9.30
7	.05	3.34	4.16	4.68	5.06	5.36	5.61	5.82	6.00	6.16	6.30
7	.01	4.95	5.92	6.54	7.01	7.37	7.68	7.94	8.17	8.37	8.55
8	.05	3.26	4.04	4.53	4.89	5.17	5.40	5.60	5.77	5.92	6.05
8	.01	4.75	5.64	6.20	6.62	6.96	7.24	7.47	7.68	7.86	8.03
9	.05	3.20	3.95	4.41	4.76	5.02	5.24	5.43	5.59	5.74	5.87
9	.01	4.60	5.43	5.96	6.35	6.66	6.91	7.13	7.33	7.49	7.65
10	.05	3.15	3.88	4.33	4.65	4.91	5.12	5.30	5.46	5.60	5.72
10	.01	4.48	5.27	5.77	6.14	6.43	6.67	6.87	7.05	7.21	7.36
11	.05	3.11	3.82	4.26	4.57	4.82	5.03	5.20	5.35	5.49	5.61
11	.01	4.39	5.15	5.62	5.97	6.25	6.48	6.67	6.84	6.99	7.13
12	.05	3.08	3.77	4.20	4.51	4.75	4.95	5.12	5.27	5.39	5.51
12	.01	4.32	5.05	5.50	5.84	6.10	6.32	6.51	6.67	6.81	6.94
13	.05	3.06	3.73	4.15	4.45	4.69	4.88	5.05	5.19	5.32	5.43
13	.01	4.26	4.96	5.40	5.73	5.98	6.19	6.37	6.53	6.67	6.79
14	.05	3.03	3.70	4.11	4.41	4.64	4.83	4.99	5.13	5.25	5.36
14	.01	4.21	4.89	5.32	5.63	5.88	6.08	6.26	6.41	6.54	6.66
15	.05	3.01	3.67	4.08	4.37	4.59	4.78	4.94	5.08	5.20	5.31
15	.01	4.17	4.84	5.25	5.56	5.80	5.99	6.16	6.31	6.44	6.55
16	.05	3.00	3.65	4.05	4.33	4.56	4.74	4.90	5.03	5.15	5.26
16	.01	4.13	4.79	5.19	5.49	5.72	5.92	6.08	6.22	6.35	6.46
17	.05	2.98	3.63	4.02	4.30	4.52	4.70	4.86	4.99	5.11	5.21
17	.01	4.10	4.74	5.14	5.43	5.66	5.85	6.01	6.15	6.27	6.38
18	.05	2.97	3.61	4.00	4.28	4.49	4.67	4.82	4.96	5.07	5.17
18	.01	4.07	4.70	5.09	5.38	5.60	5.79	5.94	6.08	6.20	6.31
19	.05	2.96	3.59	3.98	4.25	4.47	4.65	4.79	4.92	5.04	5.14
19	.01	4.05	4.67	5.05	5.33	5.55	5.73	5.89	6.02	6.14	6.25
20	.05	2.95	3.58	3.96	4.23	4.45	4.62	4.77	4.90	5.01	5.11
20	.01	4.02	4.64	5.02	5.29	5.51	5.69	5.84	5.97	6.09	6.19
24	.05	2.92	3.53	3.90	4.17	4.37	4.54	4.68	4.81	4.92	5.01
24	.01	3.96	4.55	4.91	5.17	5.37	5.54	5.69	5.81	5.92	6.02
30	.05	2.89	3.49	3.85	4.10	4.30	4.46	4.60	4.72	4.82	4.92
30	.01	3.89	4.45	4.80	5.05	5.24	5.40	5.54	5.65	5.76	5.85
40	.05	2.86	3.44	3.79	4.04	4.23	4.39	4.52	4.63	4.73	4.82
40	.01	3.82	4.37	4.70	4.93	5.11	5.26	5.39	5.50	5.60	5.69
60	.05	2.83	3.40	3.74	3.98	4.16	4.31	4.44	4.55	4.65	4.73
60	.01	3.76	4.28	4.59	4.82	4.99	5.13	5.25	5.36	5.45	5.53
120	.05	2.80	3.36	3.68	3.92	4.10	4.24	4.36	4.47	4.56	4.64
120	.01	3.70	4.20	4.50	4.71	4.87	5.01	5.12	5.21	5.30	5.37
∞	.05	2.77	3.31	3.63	3.86	4.03	4.17	4.29	4.39	4.47	4.55
∞	.01	3.64	4.12	4.40	4.60	4.76	4.88	4.99	5.08	5.16	5.23

Table A.5. *Continued*

<table>
<tr><th colspan="9">J = number of means</th><th colspan="2">Error</th></tr>
<tr><th>12</th><th>13</th><th>14</th><th>15</th><th>16</th><th>17</th><th>18</th><th>19</th><th>20</th><th>α</th><th>df</th></tr>
<tr><td>7.32</td><td>7.47</td><td>7.60</td><td>7.72</td><td>7.83</td><td>7.93</td><td>8.03</td><td>8.12</td><td>8.21</td><td>.05</td><td>5</td></tr>
<tr><td>10.70</td><td>10.89</td><td>11.08</td><td>11.24</td><td>11.40</td><td>11.55</td><td>11.68</td><td>11.81</td><td>11.93</td><td>.01</td><td>5</td></tr>
<tr><td>6.79</td><td>6.92</td><td>7.03</td><td>7.14</td><td>7.24</td><td>7.34</td><td>7.43</td><td>7.51</td><td>7.59</td><td>.05</td><td>6</td></tr>
<tr><td>9.48</td><td>9.65</td><td>9.81</td><td>9.95</td><td>10.08</td><td>10.21</td><td>10.32</td><td>10.43</td><td>10.54</td><td>.01</td><td>6</td></tr>
<tr><td>6.43</td><td>6.55</td><td>6.66</td><td>6.76</td><td>6.85</td><td>6.94</td><td>7.02</td><td>7.10</td><td>7.17</td><td>.05</td><td>7</td></tr>
<tr><td>8.71</td><td>8.86</td><td>9.00</td><td>9.12</td><td>9.24</td><td>9.35</td><td>9.46</td><td>9.55</td><td>9.65</td><td>.01</td><td>7</td></tr>
<tr><td>6.18</td><td>6.29</td><td>6.39</td><td>6.48</td><td>6.57</td><td>6.65</td><td>6.73</td><td>6.80</td><td>6.87</td><td>.05</td><td>8</td></tr>
<tr><td>8.18</td><td>8.31</td><td>8.44</td><td>8.55</td><td>8.66</td><td>8.76</td><td>8.85</td><td>8.94</td><td>9.03</td><td>.01</td><td>8</td></tr>
<tr><td>5.98</td><td>6.09</td><td>6.19</td><td>6.28</td><td>6.36</td><td>6.44</td><td>6.51</td><td>6.58</td><td>6.64</td><td>.05</td><td>9</td></tr>
<tr><td>7.78</td><td>7.91</td><td>8.03</td><td>8.13</td><td>8.23</td><td>8.33</td><td>8.41</td><td>8.49</td><td>8.57</td><td>.01</td><td>9</td></tr>
<tr><td>5.83</td><td>5.93</td><td>6.03</td><td>6.11</td><td>6.19</td><td>6.27</td><td>6.34</td><td>6.40</td><td>6.47</td><td>.05</td><td>10</td></tr>
<tr><td>7.49</td><td>7.60</td><td>7.71</td><td>7.81</td><td>7.91</td><td>7.99</td><td>8.08</td><td>8.15</td><td>8.23</td><td>.01</td><td>10</td></tr>
<tr><td>5.71</td><td>5.81</td><td>5.90</td><td>5.98</td><td>6.06</td><td>6.13</td><td>6.20</td><td>6.27</td><td>6.33</td><td>.05</td><td>11</td></tr>
<tr><td>7.25</td><td>7.36</td><td>7.46</td><td>7.56</td><td>7.65</td><td>7.73</td><td>7.81</td><td>7.88</td><td>7.95</td><td>.01</td><td>11</td></tr>
<tr><td>5.61</td><td>5.71</td><td>5.80</td><td>5.88</td><td>5.95</td><td>6.02</td><td>6.09</td><td>6.15</td><td>6.21</td><td>.05</td><td>12</td></tr>
<tr><td>7.06</td><td>7.17</td><td>7.26</td><td>7.36</td><td>7.44</td><td>7.52</td><td>7.59</td><td>7.66</td><td>7.73</td><td>.01</td><td>12</td></tr>
<tr><td>5.53</td><td>5.63</td><td>5.71</td><td>5.79</td><td>5.86</td><td>5.93</td><td>5.99</td><td>6.05</td><td>6.11</td><td>.05</td><td>13</td></tr>
<tr><td>6.90</td><td>7.01</td><td>7.10</td><td>7.19</td><td>7.27</td><td>7.35</td><td>7.42</td><td>7.48</td><td>7.55</td><td>.01</td><td>13</td></tr>
<tr><td>5.46</td><td>5.55</td><td>5.64</td><td>5.71</td><td>5.79</td><td>5.85</td><td>5.91</td><td>5.97</td><td>6.03</td><td>.05</td><td>14</td></tr>
<tr><td>6.77</td><td>6.87</td><td>6.96</td><td>7.05</td><td>7.13</td><td>7.20</td><td>7.27</td><td>7.33</td><td>7.39</td><td>.01</td><td>14</td></tr>
<tr><td>5.40</td><td>5.49</td><td>5.57</td><td>5.65</td><td>5.72</td><td>5.78</td><td>5.85</td><td>5.90</td><td>5.96</td><td>.05</td><td>15</td></tr>
<tr><td>6.66</td><td>6.76</td><td>6.84</td><td>6.93</td><td>7.00</td><td>7.07</td><td>7.14</td><td>7.20</td><td>7.26</td><td>.01</td><td>15</td></tr>
<tr><td>5.35</td><td>5.44</td><td>5.52</td><td>5.59</td><td>5.66</td><td>5.73</td><td>5.79</td><td>5.84</td><td>5.90</td><td>.05</td><td>16</td></tr>
<tr><td>6.56</td><td>6.66</td><td>6.74</td><td>6.82</td><td>6.90</td><td>6.97</td><td>7.03</td><td>7.09</td><td>7.15</td><td>.01</td><td>16</td></tr>
<tr><td>5.31</td><td>5.39</td><td>5.47</td><td>5.54</td><td>5.61</td><td>5.67</td><td>5.73</td><td>5.79</td><td>5.84</td><td>.05</td><td>17</td></tr>
<tr><td>6.48</td><td>6.57</td><td>6.66</td><td>6.73</td><td>6.81</td><td>6.87</td><td>6.94</td><td>7.00</td><td>7.05</td><td>.01</td><td>17</td></tr>
<tr><td>5.27</td><td>5.35</td><td>5.43</td><td>5.50</td><td>5.57</td><td>5.63</td><td>5.69</td><td>5.74</td><td>5.79</td><td>.05</td><td>18</td></tr>
<tr><td>6.41</td><td>6.50</td><td>6.58</td><td>6.65</td><td>6.73</td><td>6.79</td><td>6.85</td><td>6.91</td><td>6.97</td><td>.01</td><td>18</td></tr>
<tr><td>5.23</td><td>5.31</td><td>5.39</td><td>5.46</td><td>5.53</td><td>5.59</td><td>5.65</td><td>5.70</td><td>5.75</td><td>.05</td><td>19</td></tr>
<tr><td>6.34</td><td>6.43</td><td>6.51</td><td>6.58</td><td>6.65</td><td>6.72</td><td>6.78</td><td>6.84</td><td>6.89</td><td>.01</td><td>19</td></tr>
<tr><td>5.20</td><td>5.28</td><td>5.36</td><td>5.43</td><td>5.49</td><td>5.55</td><td>5.61</td><td>5.66</td><td>5.71</td><td>.05</td><td>20</td></tr>
<tr><td>6.28</td><td>6.37</td><td>6.45</td><td>6.52</td><td>6.59</td><td>6.65</td><td>6.71</td><td>6.77</td><td>6.82</td><td>.01</td><td>20</td></tr>
<tr><td>5.10</td><td>5.18</td><td>5.25</td><td>5.32</td><td>5.38</td><td>5.44</td><td>5.49</td><td>5.55</td><td>5.59</td><td>.05</td><td>24</td></tr>
<tr><td>6.11</td><td>6.19</td><td>6.26</td><td>6.33</td><td>6.39</td><td>6.45</td><td>6.51</td><td>6.56</td><td>6.61</td><td>.01</td><td>24</td></tr>
<tr><td>5.00</td><td>5.08</td><td>5.15</td><td>5.21</td><td>5.27</td><td>5.33</td><td>5.38</td><td>5.43</td><td>5.47</td><td>.05</td><td>30</td></tr>
<tr><td>5.93</td><td>6.01</td><td>6.08</td><td>6.14</td><td>6.20</td><td>6.26</td><td>6.31</td><td>6.36</td><td>6.41</td><td>.01</td><td>30</td></tr>
<tr><td>4.90</td><td>4.98</td><td>5.04</td><td>5.11</td><td>5.16</td><td>5.22</td><td>5.27</td><td>5.31</td><td>5.36</td><td>.05</td><td>40</td></tr>
<tr><td>5.76</td><td>5.83</td><td>5.90</td><td>5.96</td><td>6.02</td><td>6.07</td><td>6.12</td><td>6.16</td><td>6.21</td><td>.01</td><td>40</td></tr>
<tr><td>4.81</td><td>4.88</td><td>4.94</td><td>5.00</td><td>5.06</td><td>5.11</td><td>5.15</td><td>5.20</td><td>5.24</td><td>.05</td><td>60</td></tr>
<tr><td>5.60</td><td>5.67</td><td>5.73</td><td>5.78</td><td>5.84</td><td>5.89</td><td>5.93</td><td>5.97</td><td>6.01</td><td>.01</td><td>60</td></tr>
<tr><td>4.71</td><td>4.78</td><td>4.84</td><td>4.90</td><td>4.95</td><td>5.00</td><td>5.04</td><td>5.09</td><td>5.13</td><td>.05</td><td>120</td></tr>
<tr><td>5.44</td><td>5.50</td><td>5.56</td><td>5.61</td><td>5.66</td><td>5.71</td><td>5.75</td><td>5.79</td><td>5.83</td><td>.01</td><td>120</td></tr>
<tr><td>4.62</td><td>4.68</td><td>4.74</td><td>4.80</td><td>4.85</td><td>4.89</td><td>4.93</td><td>4.97</td><td>5.01</td><td>.05</td><td>∞</td></tr>
<tr><td>5.29</td><td>5.35</td><td>5.40</td><td>5.45</td><td>5.49</td><td>5.54</td><td>5.57</td><td>5.61</td><td>5.65</td><td>.01</td><td>∞</td></tr>
</table>

Source: Abridged from Table 29 in Pearson and Hartley (1958).

364

Table A.6. Critical Values of Dunn–Bonferroni t_{DB} Statistic

BONFERRONI T CRITICAL VALUES FOR FAMILYWISE ALPHA=.05
DEGREES OF FREEDOM

COMPARISONS	3	4	5	6	7	8
2	4.177	3.495	3.163	2.969	2.841	2.752
3	4.857	3.961	3.534	3.287	3.128	3.016
4	5.392	4.315	3.810	3.521	3.335	3.206
5	5.841	4.604	4.032	3.707	3.499	3.355
6	6.232	4.851	4.219	3.863	3.636	3.479
7	6.580	5.068	4.382	3.997	3.753	3.584
8	6.895	5.261	4.526	4.115	3.855	3.677
9	7.185	5.437	4.655	4.221	3.947	3.759
10	7.453	5.598	4.773	4.317	4.029	3.833
11	7.704	5.747	4.882	4.405	4.105	3.900
12	7.940	5.885	4.983	4.486	4.174	3.962
13	8.162	6.015	5.076	4.561	4.239	4.019
14	8.374	6.138	5.164	4.632	4.299	4.072
15	8.575	6.254	5.247	4.698	4.355	4.122
16	8.768	6.364	5.326	4.760	4.408	4.169
17	8.952	6.469	5.400	4.820	4.459	4.214
18	9.129	6.570	5.471	4.876	4.506	4.256
19	9.300	6.666	5.539	4.929	4.551	4.296
20	9.465	6.758	5.604	4.981	4.595	4.334
22	9.778	6.933	5.726	5.077	4.676	4.405
24	10.073	7.096	5.840	5.166	4.750	4.470
26	10.352	7.248	5.945	5.248	4.819	4.531
28	10.617	7.392	6.045	5.326	4.884	4.587
30	10.869	7.529	6.138	5.398	4.945	4.640
35	11.453	7.842	6.352	5.563	5.082	4.759
40	11.984	8.122	6.541	5.709	5.202	4.864
45	12.471	8.376	6.713	5.840	5.310	4.957
50	12.924	8.610	6.869	5.959	5.408	5.041
60	13.745	9.029	7.146	6.169	5.580	5.189
70	14.479	9.398	7.389	6.351	5.728	5.316
80	15.145	9.729	7.604	6.512	5.859	5.428
90	15.757	10.030	7.798	6.657	5.976	5.527
100	16.326	10.306	7.976	6.788	6.082	5.617
150	18.708	11.438	8.692	7.314	6.502	5.973
200	20.605	12.312	9.236	7.708	6.814	6.234
250	22.205	13.034	9.677	8.024	7.063	6.442
300	23.599	13.652	10.054	8.292	7.272	6.616
400	25.984	14.687	10.674	8.730	7.613	6.896
500	27.992	15.543	11.177	9.083	7.885	7.121

Table A.6. *Continued*

```
          BONFERRONI T CRITICAL VALUES FOR FAMILYWISE ALPHA=.05
                           DEGREES OF FREEDOM
COMPARISONS      9         10        11        12        13        14
```

COMPARISONS	9	10	11	12	13	14
2	2.685	2.634	2.593	2.560	2.533	2.510
3	2.933	2.870	2.820	2.779	2.746	2.718
4	3.111	3.038	2.981	2.934	2.896	2.864
5	3.250	3.169	3.106	3.055	3.012	2.977
6	3.364	3.277	3.208	3.153	3.107	3.069
7	3.462	3.368	3.295	3.236	3.187	3.146
8	3.547	3.448	3.370	3.308	3.256	3.214
9	3.622	3.518	3.437	3.371	3.318	3.273
10	3.690	3.581	3.497	3.428	3.372	3.326
11	3.751	3.639	3.551	3.480	3.422	3.374
12	3.808	3.691	3.600	3.527	3.467	3.417
13	3.860	3.740	3.646	3.571	3.509	3.458
14	3.909	3.785	3.689	3.611	3.548	3.495
15	3.954	3.827	3.728	3.649	3.584	3.530
16	3.997	3.867	3.765	3.684	3.618	3.562
17	4.037	3.904	3.800	3.717	3.649	3.593
18	4.075	3.939	3.833	3.749	3.679	3.621
19	4.111	3.973	3.865	3.778	3.708	3.649
20	4.146	4.005	3.895	3.807	3.735	3.675
22	4.210	4.064	3.950	3.859	3.785	3.723
24	4.269	4.118	4.001	3.907	3.830	3.767
26	4.324	4.168	4.048	3.951	3.873	3.807
28	4.374	4.215	4.091	3.992	3.912	3.845
30	4.422	4.259	4.132	4.031	3.948	3.880
35	4.529	4.357	4.223	4.117	4.030	3.958
40	4.622	4.442	4.303	4.192	4.101	4.026
45	4.706	4.518	4.373	4.258	4.164	4.086
50	4.781	4.587	4.437	4.318	4.221	4.140
60	4.912	4.706	4.548	4.422	4.319	4.234
70	5.025	4.809	4.642	4.510	4.403	4.314
80	5.124	4.898	4.724	4.587	4.475	4.383
90	5.211	4.977	4.798	4.655	4.540	4.444
100	5.291	5.049	4.863	4.716	4.597	4.499
150	5.602	5.329	5.120	4.955	4.822	4.712
200	5.830	5.533	5.306	5.128	4.984	4.865
250	6.010	5.694	5.453	5.263	5.111	4.985
300	6.160	5.827	5.574	5.375	5.215	5.084
400	6.401	6.042	5.768	5.554	5.382	5.240
500	6.594	6.211	5.921	5.694	5.513	5.363

Table A.6. *Continued*

```
        BONFERRONI T CRITICAL VALUES FOR FAMILYWISE ALPHA=.05
                          DEGREES OF FREEDOM
COMPARISONS      15        16        17        18        19        20
```

COMPARISONS	15	16	17	18	19	20
2	2.490	2.473	2.458	2.445	2.433	2.423
3	2.694	2.673	2.655	2.639	2.625	2.613
4	2.837	2.813	2.793	2.775	2.759	2.744
5	2.947	2.921	2.898	2.878	2.861	2.845
6	3.036	3.008	2.984	2.963	2.944	2.927
7	3.112	3.082	3.056	3.034	3.014	2.996
8	3.177	3.146	3.119	3.095	3.074	3.055
9	3.235	3.202	3.173	3.149	3.127	3.107
10	3.286	3.252	3.222	3.197	3.174	3.153
11	3.333	3.297	3.267	3.240	3.216	3.195
12	3.375	3.339	3.307	3.279	3.255	3.233
13	3.414	3.377	3.344	3.316	3.291	3.268
14	3.450	3.412	3.378	3.349	3.323	3.301
15	3.484	3.444	3.410	3.380	3.354	3.331
16	3.515	3.475	3.440	3.409	3.383	3.359
17	3.545	3.504	3.468	3.437	3.409	3.385
18	3.573	3.531	3.494	3.463	3.435	3.410
19	3.599	3.556	3.519	3.487	3.459	3.433
20	3.624	3.581	3.543	3.510	3.481	3.455
22	3.670	3.626	3.587	3.553	3.523	3.497
24	3.713	3.667	3.627	3.592	3.561	3.534
26	3.752	3.705	3.664	3.628	3.597	3.569
28	3.788	3.740	3.698	3.661	3.629	3.601
30	3.822	3.772	3.730	3.692	3.659	3.630
35	3.897	3.846	3.801	3.762	3.727	3.697
40	3.963	3.909	3.862	3.822	3.786	3.754
45	4.021	3.965	3.917	3.874	3.837	3.804
50	4.073	4.015	3.965	3.922	3.883	3.850
60	4.163	4.102	4.049	4.003	3.963	3.928
70	4.239	4.175	4.121	4.073	4.031	3.993
80	4.305	4.239	4.182	4.133	4.089	4.051
90	4.364	4.296	4.237	4.186	4.141	4.101
100	4.417	4.346	4.286	4.233	4.187	4.146
150	4.620	4.542	4.475	4.416	4.365	4.319
200	4.766	4.682	4.609	4.546	4.491	4.443
250	4.880	4.791	4.714	4.648	4.590	4.538
300	4.974	4.880	4.800	4.731	4.671	4.617
400	5.122	5.023	4.937	4.863	4.798	4.741
500	5.239	5.134	5.044	4.965	4.898	4.837

Table A.6. *Continued*

```
        BONFERRONI T CRITICAL VALUES FOR FAMILYWISE ALPHA=.05
                              DEGREES OF FREEDOM
COMPARISONS      21        22        23        24        25        26
```

COMPARISONS	21	22	23	24	25	26
2	2.414	2.405	2.398	2.391	2.385	2.379
3	2.601	2.591	2.582	2.574	2.566	2.559
4	2.732	2.720	2.710	2.700	2.692	2.684
5	2.831	2.819	2.807	2.797	2.787	2.779
6	2.912	2.899	2.886	2.875	2.865	2.856
7	2.980	2.965	2.952	2.941	2.930	2.920
8	3.038	3.023	3.009	2.997	2.986	2.975
9	3.090	3.074	3.059	3.046	3.035	3.024
10	3.135	3.119	3.104	3.091	3.078	3.067
11	3.176	3.159	3.144	3.130	3.117	3.106
12	3.214	3.196	3.181	3.166	3.153	3.141
13	3.248	3.230	3.214	3.199	3.186	3.174
14	3.280	3.262	3.245	3.230	3.216	3.204
15	3.310	3.291	3.274	3.258	3.244	3.231
16	3.337	3.318	3.301	3.285	3.270	3.257
17	3.363	3.344	3.326	3.310	3.295	3.282
18	3.388	3.368	3.350	3.333	3.318	3.304
19	3.411	3.390	3.372	3.355	3.340	3.326
20	3.432	3.412	3.393	3.376	3.361	3.346
22	3.473	3.452	3.432	3.415	3.399	3.384
24	3.510	3.488	3.468	3.450	3.434	3.419
26	3.544	3.521	3.501	3.483	3.466	3.451
28	3.575	3.552	3.531	3.513	3.495	3.480
30	3.604	3.581	3.560	3.540	3.523	3.507
35	3.669	3.645	3.623	3.603	3.584	3.567
40	3.726	3.700	3.677	3.656	3.637	3.620
45	3.775	3.749	3.725	3.703	3.684	3.666
50	3.819	3.792	3.768	3.745	3.725	3.707
60	3.896	3.867	3.841	3.818	3.797	3.777
70	3.960	3.931	3.904	3.879	3.857	3.837
80	4.016	3.985	3.958	3.932	3.909	3.888
90	4.066	4.034	4.005	3.979	3.955	3.934
100	4.110	4.077	4.048	4.021	3.996	3.974
150	4.279	4.243	4.210	4.181	4.154	4.129
200	4.399	4.361	4.326	4.294	4.265	4.239
250	4.493	4.452	4.415	4.382	4.352	4.324
300	4.569	4.527	4.488	4.454	4.422	4.393
400	4.690	4.645	4.604	4.567	4.533	4.503
500	4.784	4.736	4.693	4.654	4.619	4.587

Table A.6. *Continued*

```
          BONFERRONI T CRITICAL VALUES FOR FAMILYWISE ALPHA=.05
                          DEGREES OF FREEDOM
COMPARISONS       27        28        29        30        32        34
```

COMPARISONS	27	28	29	30	32	34
2	2.373	2.368	2.364	2.360	2.352	2.345
3	2.552	2.546	2.541	2.536	2.526	2.518
4	2.676	2.669	2.663	2.657	2.647	2.638
5	2.771	2.763	2.756	2.750	2.738	2.728
6	2.847	2.839	2.832	2.825	2.812	2.802
7	2.911	2.902	2.894	2.887	2.874	2.863
8	2.966	2.957	2.949	2.941	2.927	2.915
9	3.014	3.004	2.996	2.988	2.974	2.961
10	3.057	3.047	3.038	3.030	3.015	3.002
11	3.095	3.085	3.076	3.067	3.052	3.039
12	3.130	3.120	3.110	3.102	3.086	3.072
13	3.162	3.152	3.142	3.133	3.117	3.103
14	3.192	3.181	3.171	3.162	3.145	3.131
15	3.219	3.208	3.198	3.189	3.172	3.157
16	3.245	3.234	3.223	3.214	3.197	3.181
17	3.269	3.258	3.247	3.237	3.220	3.204
18	3.292	3.280	3.269	3.259	3.241	3.226
19	3.313	3.301	3.290	3.280	3.262	3.246
20	3.333	3.321	3.310	3.300	3.281	3.265
22	3.371	3.359	3.347	3.336	3.317	3.301
24	3.405	3.392	3.381	3.370	3.350	3.333
26	3.436	3.423	3.411	3.400	3.380	3.362
28	3.465	3.452	3.440	3.428	3.408	3.390
30	3.492	3.479	3.466	3.454	3.433	3.415
35	3.552	3.538	3.525	3.513	3.491	3.471
40	3.604	3.589	3.575	3.563	3.540	3.520
45	3.649	3.634	3.620	3.607	3.583	3.563
50	3.690	3.674	3.659	3.646	3.622	3.601
60	3.759	3.743	3.728	3.714	3.688	3.666
70	3.818	3.801	3.785	3.771	3.744	3.721
80	3.869	3.851	3.835	3.820	3.792	3.769
90	3.914	3.896	3.879	3.863	3.835	3.810
100	3.954	3.935	3.918	3.902	3.873	3.848
150	4.107	4.086	4.067	4.049	4.018	3.990
200	4.215	4.193	4.172	4.154	4.120	4.090
250	4.299	4.275	4.254	4.234	4.198	4.167
300	4.367	4.343	4.320	4.300	4.262	4.230
400	4.474	4.449	4.425	4.403	4.363	4.329
500	4.558	4.530	4.505	4.482	4.441	4.405

Table A.6. *Continued*

```
            BONFERRONI T CRITICAL VALUES FOR FAMILYWISE ALPHA=.05
                            DEGREES OF FREEDOM
COMPARISONS      36          38         40         42         44         46
```

COMPARISONS	36	38	40	42	44	46
2	2.339	2.334	2.329	2.325	2.321	2.317
3	2.511	2.505	2.499	2.494	2.489	2.485
4	2.629	2.622	2.616	2.610	2.605	2.600
5	2.719	2.712	2.704	2.698	2.692	2.687
6	2.792	2.783	2.776	2.769	2.763	2.757
7	2.853	2.844	2.836	2.828	2.822	2.816
8	2.905	2.895	2.887	2.879	2.872	2.866
9	2.950	2.940	2.931	2.924	2.916	2.910
10	2.990	2.980	2.971	2.963	2.956	2.949
11	3.027	3.016	3.007	2.998	2.991	2.984
12	3.060	3.049	3.039	3.031	3.023	3.016
13	3.090	3.079	3.069	3.060	3.052	3.045
14	3.118	3.107	3.096	3.087	3.079	3.071
15	3.144	3.132	3.122	3.112	3.104	3.096
16	3.168	3.156	3.145	3.136	3.127	3.119
17	3.191	3.178	3.168	3.158	3.149	3.141
18	3.212	3.199	3.188	3.179	3.170	3.161
19	3.232	3.219	3.208	3.198	3.189	3.181
20	3.251	3.238	3.227	3.216	3.207	3.199
22	3.286	3.273	3.261	3.250	3.241	3.232
24	3.318	3.304	3.292	3.281	3.272	3.263
26	3.347	3.333	3.321	3.310	3.300	3.291
28	3.374	3.360	3.347	3.336	3.326	3.317
30	3.399	3.385	3.372	3.360	3.350	3.340
35	3.455	3.440	3.426	3.414	3.403	3.394
40	3.503	3.487	3.473	3.461	3.449	3.439
45	3.545	3.529	3.514	3.501	3.490	3.479
50	3.582	3.566	3.551	3.538	3.526	3.515
60	3.647	3.629	3.614	3.600	3.588	3.576
70	3.701	3.683	3.667	3.653	3.640	3.628
80	3.748	3.729	3.713	3.698	3.684	3.672
90	3.789	3.770	3.753	3.737	3.724	3.711
100	3.826	3.806	3.788	3.773	3.758	3.746
150	3.966	3.944	3.925	3.907	3.892	3.878
200	4.064	4.041	4.021	4.002	3.985	3.970
250	4.140	4.116	4.094	4.075	4.057	4.042
300	4.201	4.177	4.154	4.134	4.116	4.099
400	4.299	4.272	4.248	4.227	4.208	4.190
500	4.374	4.346	4.321	4.298	4.278	4.260

Table A.6. *Continued*

COMPARISONS	48	50	DEGREES OF FREEDOM 55	60	65	70
2	2.314	2.311	2.304	2.299	2.295	2.291
3	2.481	2.477	2.469	2.463	2.458	2.453
4	2.595	2.591	2.583	2.575	2.569	2.564
5	2.682	2.678	2.668	2.660	2.654	2.648
6	2.752	2.747	2.737	2.729	2.721	2.715
7	2.810	2.805	2.794	2.785	2.778	2.771
8	2.860	2.855	2.844	2.834	2.826	2.820
9	2.904	2.898	2.887	2.877	2.869	2.862
10	2.943	2.937	2.925	2.915	2.906	2.899
11	2.977	2.972	2.959	2.948	2.940	2.932
12	3.009	3.003	2.990	2.979	2.970	2.962
13	3.038	3.032	3.018	3.007	2.998	2.990
14	3.065	3.058	3.045	3.033	3.024	3.016
15	3.089	3.083	3.069	3.057	3.048	3.039
16	3.112	3.106	3.091	3.080	3.070	3.061
17	3.134	3.127	3.113	3.101	3.090	3.082
18	3.154	3.147	3.132	3.120	3.110	3.101
19	3.173	3.166	3.151	3.139	3.128	3.119
20	3.191	3.184	3.169	3.156	3.146	3.137
22	3.224	3.217	3.202	3.189	3.178	3.168
24	3.255	3.247	3.231	3.218	3.207	3.197
26	3.282	3.275	3.258	3.245	3.233	3.224
28	3.308	3.300	3.284	3.270	3.258	3.248
30	3.332	3.324	3.307	3.293	3.281	3.271
35	3.385	3.376	3.358	3.344	3.331	3.321
40	3.430	3.421	3.403	3.388	3.375	3.364
45	3.470	3.461	3.442	3.426	3.413	3.402
50	3.505	3.496	3.476	3.460	3.447	3.435
60	3.566	3.556	3.536	3.519	3.505	3.493
70	3.617	3.607	3.586	3.568	3.553	3.541
80	3.661	3.651	3.629	3.610	3.595	3.582
90	3.699	3.689	3.666	3.648	3.632	3.619
100	3.734	3.723	3.700	3.681	3.665	3.651
150	3.865	3.853	3.828	3.807	3.789	3.774
200	3.957	3.944	3.917	3.895	3.876	3.860
250	4.027	4.014	3.986	3.962	3.942	3.926
300	4.085	4.071	4.041	4.017	3.997	3.979
400	4.174	4.160	4.129	4.102	4.081	4.063
500	4.244	4.229	4.196	4.169	4.146	4.127

Table A.6. *Continued*

BONFERRONI T CRITICAL VALUES FOR FAMILYWISE ALPHA=.05
DEGREES OF FREEDOM

COMPARISONS	75	80	85	90	95	100
2	2.287	2.284	2.282	2.280	2.277	2.276
3	2.449	2.445	2.442	2.440	2.437	2.435
4	2.559	2.555	2.552	2.549	2.546	2.544
5	2.643	2.639	2.635	2.632	2.629	2.626
6	2.710	2.705	2.701	2.698	2.695	2.692
7	2.766	2.761	2.757	2.753	2.750	2.747
8	2.814	2.809	2.804	2.800	2.797	2.793
9	2.855	2.850	2.845	2.841	2.838	2.834
10	2.892	2.887	2.882	2.878	2.874	2.871
11	2.926	2.920	2.915	2.911	2.907	2.903
12	2.956	2.950	2.945	2.940	2.936	2.933
13	2.983	2.977	2.972	2.967	2.963	2.960
14	3.009	3.003	2.997	2.992	2.988	2.984
15	3.032	3.026	3.020	3.016	3.011	3.007
16	3.054	3.048	3.042	3.037	3.033	3.029
17	3.074	3.068	3.062	3.057	3.053	3.049
18	3.094	3.087	3.081	3.076	3.072	3.067
19	3.112	3.105	3.099	3.094	3.089	3.085
20	3.129	3.122	3.116	3.111	3.106	3.102
22	3.160	3.153	3.147	3.142	3.137	3.133
24	3.189	3.182	3.176	3.170	3.165	3.161
26	3.215	3.208	3.202	3.196	3.191	3.186
28	3.240	3.232	3.226	3.220	3.214	3.210
30	3.262	3.254	3.248	3.242	3.236	3.232
35	3.312	3.304	3.297	3.291	3.285	3.280
40	3.354	3.346	3.339	3.333	3.327	3.322
45	3.392	3.383	3.376	3.369	3.363	3.358
50	3.425	3.416	3.409	3.402	3.396	3.390
60	3.482	3.473	3.465	3.458	3.452	3.446
70	3.530	3.520	3.512	3.505	3.498	3.492
80	3.571	3.561	3.553	3.545	3.538	3.532
90	3.607	3.597	3.588	3.580	3.574	3.567
100	3.639	3.629	3.620	3.612	3.605	3.598
150	3.761	3.750	3.740	3.731	3.723	3.716
200	3.846	3.834	3.824	3.814	3.806	3.799
250	3.911	3.899	3.888	3.878	3.869	3.862
300	3.964	3.951	3.940	3.930	3.921	3.913
400	4.047	4.033	4.021	4.010	4.001	3.992
500	4.110	4.096	4.084	4.072	4.062	4.053

Table A.6. *Continued*

```
          BONFERRONI T CRITICAL VALUES FOR FAMILYWISE ALPHA=.05
                            DEGREES OF FREEDOM
COMPARISONS      150       200       250       300       500      1000
```

COMPARISONS	150	200	250	300	500	1000
2	2.264	2.258	2.255	2.253	2.248	2.245
3	2.421	2.414	2.410	2.407	2.402	2.398
4	2.528	2.521	2.516	2.513	2.507	2.502
5	2.609	2.601	2.596	2.592	2.586	2.581
6	2.674	2.665	2.659	2.656	2.649	2.644
7	2.728	2.718	2.712	2.709	2.701	2.696
8	2.774	2.764	2.758	2.754	2.746	2.740
9	2.814	2.803	2.797	2.793	2.785	2.779
10	2.849	2.839	2.832	2.828	2.820	2.813
11	2.881	2.870	2.864	2.859	2.850	2.844
12	2.910	2.899	2.892	2.887	2.879	2.872
13	2.936	2.925	2.918	2.913	2.904	2.897
14	2.960	2.949	2.942	2.937	2.928	2.921
15	2.983	2.971	2.964	2.959	2.949	2.942
16	3.004	2.992	2.984	2.979	2.970	2.962
17	3.023	3.011	3.003	2.998	2.989	2.981
18	3.042	3.029	3.021	3.016	3.006	2.999
19	3.059	3.046	3.038	3.033	3.023	3.015
20	3.075	3.062	3.054	3.049	3.039	3.031
22	3.105	3.092	3.084	3.079	3.068	3.060
24	3.133	3.119	3.111	3.105	3.094	3.086
26	3.158	3.144	3.135	3.130	3.118	3.110
28	3.181	3.166	3.158	3.152	3.141	3.132
30	3.202	3.187	3.179	3.173	3.161	3.153
35	3.249	3.234	3.225	3.219	3.207	3.198
40	3.290	3.274	3.264	3.258	3.246	3.236
45	3.325	3.309	3.299	3.293	3.280	3.270
50	3.357	3.340	3.330	3.323	3.310	3.300
60	3.411	3.393	3.383	3.376	3.362	3.352
70	3.456	3.437	3.427	3.419	3.405	3.395
80	3.494	3.476	3.464	3.457	3.442	3.432
90	3.528	3.509	3.498	3.490	3.475	3.464
100	3.558	3.539	3.527	3.519	3.504	3.492
150	3.673	3.651	3.638	3.630	3.613	3.601
200	3.752	3.729	3.716	3.707	3.689	3.676
250	3.813	3.789	3.775	3.765	3.747	3.733
300	3.862	3.838	3.823	3.813	3.794	3.779
400	3.939	3.913	3.897	3.887	3.867	3.851
500	3.998	3.971	3.955	3.944	3.922	3.907

Table A.6. *Continued*

COMPARISONS	3	4	DEGREES OF FREEDOM 5	6	7	8
2	7,453	5,598	4,773	4,317	4,029	3,833
3	8,575	6,254	5,247	4,698	4,355	4,122
4	9,465	6,758	5,604	4,981	4,595	4,334
5	10,215	7,173	5,893	5,208	4,785	4,501
6	10,869	7,529	6,138	5,398	4,945	4,640
7	11,453	7,842	6,352	5,563	5,082	4,759
8	11,984	8,122	6,541	5,709	5,202	4,864
9	12,471	8,376	6,713	5,840	5,310	4,957
10	12,924	8,610	6,869	5,959	5,408	5,041
11	13,347	8,827	7,013	6,068	5,497	5,118
12	13,745	9,029	7,146	6,169	5,580	5,189
13	14,122	9,219	7,271	6,263	5,656	5,255
14	14,479	9,398	7,389	6,351	5,728	5,316
15	14,819	9,568	7,499	6,434	5,795	5,374
16	15,145	9,729	7,604	6,512	5,859	5,428
17	15,458	9,883	7,703	6,586	5,919	5,479
18	15,757	10,030	7,798	6,657	5,976	5,527
19	16,047	10,171	7,889	6,724	6,030	5,573
20	16,326	10,306	7,976	6,788	6,082	5,617
22	16,858	10,563	8,139	6,909	6,179	5,700
24	17,358	10,802	8,292	7,021	6,268	5,775
26	17,831	11,026	8,433	7,125	6,351	5,846
28	18,282	11,237	8,567	7,223	6,429	5,911
30	18,708	11,438	8,692	7,314	6,502	5,973
35	19,704	11,899	8,980	7,523	6,668	6,112
40	20,605	12,312	9,236	7,708	6,814	6,234
45	21,433	12,688	9,466	7,874	6,945	6,343
50	22,205	13,034	9,677	8,024	7,063	6,442
60	23,599	13,652	10,054	8,292	7,272	6,616
70	24,847	14,200	10,380	8,523	7,452	6,765
80	25,984	14,687	10,674	8,730	7,613	6,896
90	27,028	15,134	10,936	8,914	7,755	7,013
100	27,992	15,543	11,177	9,083	7,885	7,121
150	32,048	17,219	12,156	9,757	8,398	7,539
200	35,282	18,517	12,892	10,263	8,784	7,849
250	38,015	19,587	13,497	10,667	9,089	8,097
300	40,383	20,508	14,004	11,012	9,346	8,303
400	44,448	22,047	14,852	11,570	9,763	8,644
500	47,885	23,312	15,541	12,030	10,105	8,908

Table A.6. *Continued*

```
          BONFERRONI T CRITICAL VALUES FOR FAMILYWISE ALPHA=.01
                           DEGREES OF FREEDOM
COMPARISONS        9         10         11         12         13         14
```

COMPARISONS	9	10	11	12	13	14
2	3.690	3.581	3.497	3.428	3.372	3.326
3	3.954	3.827	3.728	3.649	3.584	3.530
4	4.146	4.005	3.895	3.807	3.735	3.675
5	4.297	4.144	4.025	3.930	3.852	3.787
6	4.422	4.259	4.132	4.031	3.948	3.880
7	4.529	4.357	4.223	4.117	4.030	3.958
8	4.622	4.442	4.303	4.192	4.101	4.026
9	4.706	4.518	4.373	4.258	4.164	4.086
10	4.781	4.587	4.437	4.318	4.221	4.140
11	4.849	4.649	4.495	4.372	4.272	4.189
12	4.912	4.706	4.548	4.422	4.319	4.234
13	4.971	4.759	4.597	4.467	4.362	4.275
14	5.025	4.809	4.642	4.510	4.403	4.314
15	5.076	4.855	4.685	4.550	4.440	4.350
16	5.124	4.898	4.724	4.587	4.475	4.383
17	5.169	4.939	4.762	4.622	4.508	4.414
18	5.211	4.977	4.798	4.655	4.540	4.444
19	5.252	5.014	4.831	4.687	4.569	4.472
20	5.291	5.049	4.863	4.716	4.597	4.499
22	5.363	5.114	4.923	4.772	4.650	4.549
24	5.429	5.174	4.978	4.823	4.698	4.595
26	5.491	5.229	5.029	4.871	4.743	4.637
28	5.548	5.281	5.076	4.914	4.784	4.676
30	5.602	5.329	5.120	4.955	4.822	4.712
35	5.723	5.438	5.219	5.048	4.909	4.794
40	5.830	5.533	5.306	5.128	4.984	4.865
45	5.924	5.618	5.383	5.199	5.050	4.929
50	6.010	5.694	5.453	5.263	5.111	4.985
60	6.160	5.827	5.574	5.375	5.215	5.084
70	6.289	5.941	5.678	5.470	5.304	5.167
80	6.401	6.042	5.768	5.554	5.382	5.240
90	6.502	6.130	5.848	5.628	5.451	5.305
100	6.594	6.211	5.921	5.694	5.513	5.363
150	6.952	6.527	6.204	5.955	5.755	5.590
200	7.216	6.757	6.410	6.143	5.928	5.752
250	7.422	6.938	6.575	6.292	6.064	5.880
300	7.598	7.090	6.710	6.415	6.179	5.987
400	7.875	7.334	6.925	6.608	6.358	6.153
500	8.100	7.528	7.100	6.763	6.500	6.288

Table A.6. *Continued*

	DEGREES OF FREEDOM					
COMPARISONS	15	16	17	18	19	20
2	3.286	3.252	3.222	3.197	3.174	3.153
3	3.484	3.444	3.410	3.380	3.354	3.331
4	3.624	3.581	3.543	3.510	3.481	3.455
5	3.733	3.686	3.646	3.611	3.579	3.552
6	3.822	3.772	3.730	3.692	3.659	3.630
7	3.897	3.846	3.801	3.762	3.727	3.697
8	3.963	3.909	3.862	3.822	3.786	3.754
9	4.021	3.965	3.917	3.874	3.837	3.804
10	4.073	4.015	3.965	3.922	3.883	3.850
11	4.120	4.060	4.009	3.964	3.925	3.890
12	4.163	4.102	4.049	4.003	3.963	3.928
13	4.202	4.140	4.086	4.039	3.998	3.962
14	4.239	4.175	4.121	4.073	4.031	3.993
15	4.273	4.208	4.152	4.104	4.061	4.023
16	4.305	4.239	4.182	4.133	4.089	4.051
17	4.335	4.268	4.210	4.160	4.116	4.077
18	4.364	4.296	4.237	4.186	4.141	4.101
19	4.391	4.322	4.262	4.210	4.164	4.124
20	4.417	4.346	4.286	4.233	4.187	4.146
22	4.464	4.392	4.330	4.276	4.229	4.187
24	4.508	4.434	4.371	4.315	4.267	4.224
26	4.548	4.473	4.408	4.352	4.302	4.258
28	4.585	4.509	4.443	4.385	4.335	4.290
30	4.620	4.542	4.475	4.416	4.365	4.319
35	4.698	4.617	4.547	4.486	4.433	4.386
40	4.766	4.682	4.609	4.546	4.491	4.443
45	4.826	4.739	4.665	4.600	4.543	4.493
50	4.880	4.791	4.714	4.648	4.590	4.538
60	4.974	4.880	4.800	4.731	4.671	4.617
70	5.054	4.957	4.874	4.802	4.739	4.683
80	5.122	5.023	4.937	4.863	4.798	4.741
90	5.184	5.081	4.993	4.917	4.851	4.792
100	5.239	5.134	5.044	4.965	4.898	4.837
150	5.453	5.338	5.239	5.154	5.079	5.013
200	5.608	5.483	5.379	5.288	5.208	5.139
250	5.727	5.598	5.487	5.394	5.310	5.236
300	5.827	5.691	5.578	5.478	5.393	5.317
400	5.983	5.843	5.720	5.616	5.525	5.443
500	6.106	5.959	5.834	5.722	5.627	5.541

Table A.6. *Continued*

```
            BONFERRONI T CRITICAL VALUES FCR FAMILYWISE ALPHA=.01
                              DEGREES OF FREEDOM
COMPARISONS      21        22        23        24        25        26
```

COMPARISONS	21	22	23	24	25	26
2	3.135	3.119	3.104	3.091	3.078	3.067
3	3.310	3.291	3.274	3.258	3.244	3.231
4	3.432	3.412	3.393	3.376	3.361	3.346
5	3.527	3.505	3.485	3.467	3.450	3.435
6	3.604	3.581	3.560	3.540	3.523	3.507
7	3.669	3.645	3.623	3.603	3.584	3.567
8	3.726	3.700	3.677	3.656	3.637	3.620
9	3.775	3.749	3.725	3.703	3.684	3.666
10	3.819	3.792	3.768	3.745	3.725	3.707
11	3.859	3.831	3.806	3.783	3.763	3.744
12	3.896	3.867	3.841	3.818	3.797	3.777
13	3.929	3.900	3.874	3.850	3.828	3.808
14	3.960	3.931	3.904	3.879	3.857	3.837
15	3.989	3.959	3.932	3.907	3.884	3.864
16	4.016	3.985	3.958	3.932	3.909	3.888
17	4.042	4.010	3.982	3.956	3.933	3.912
18	4.066	4.034	4.005	3.979	3.955	3.934
19	4.088	4.056	4.027	4.000	3.976	3.954
20	4.110	4.077	4.048	4.021	3.996	3.974
22	4.149	4.116	4.086	4.058	4.034	4.011
24	4.186	4.152	4.121	4.093	4.067	4.044
26	4.219	4.184	4.153	4.124	4.098	4.075
28	4.250	4.215	4.183	4.154	4.127	4.103
30	4.279	4.243	4.210	4.181	4.154	4.129
35	4.344	4.306	4.272	4.242	4.214	4.188
40	4.399	4.361	4.326	4.294	4.265	4.239
45	4.449	4.409	4.373	4.340	4.311	4.284
50	4.493	4.452	4.415	4.382	4.352	4.324
60	4.569	4.527	4.488	4.454	4.422	4.393
70	4.634	4.590	4.550	4.514	4.482	4.452
80	4.690	4.645	4.604	4.567	4.533	4.503
90	4.739	4.693	4.651	4.613	4.579	4.547
100	4.784	4.736	4.693	4.654	4.619	4.587
150	4.955	4.902	4.856	4.813	4.775	4.740
200	5.077	5.022	4.973	4.928	4.887	4.850
250	5.171	5.114	5.062	5.015	4.973	4.933
300	5.249	5.190	5.137	5.088	5.044	5.002
400	5.373	5.308	5.251	5.201	5.154	5.113
500	5.467	5.403	5.343	5.289	5.239	5.195

Table A.6. *Continued*

```
             BONFERRONI T CRITICAL VALUES FOR FAMILYWISE ALPHA=.01
                                 DEGREES OF FREEDOM
 COMPARISONS        27        28        29        30        32        34
```

COMPARISONS	27	28	29	30	32	34
2	3.057	3.047	3.038	3.030	3.015	3.002
3	3.219	3.208	3.198	3.189	3.172	3.157
4	3.333	3.321	3.310	3.300	3.281	3.265
5	3.421	3.408	3.396	3.385	3.365	3.348
6	3.492	3.479	3.466	3.454	3.433	3.415
7	3.552	3.538	3.525	3.513	3.491	3.471
8	3.604	3.589	3.575	3.563	3.540	3.520
9	3.649	3.634	3.620	3.607	3.583	3.563
10	3.690	3.674	3.659	3.646	3.622	3.601
11	3.726	3.710	3.695	3.681	3.657	3.635
12	3.759	3.743	3.728	3.714	3.688	3.666
13	3.790	3.773	3.758	3.743	3.717	3.695
14	3.818	3.801	3.785	3.771	3.744	3.721
15	3.845	3.827	3.811	3.796	3.769	3.746
16	3.869	3.851	3.835	3.820	3.792	3.769
17	3.892	3.874	3.858	3.842	3.814	3.790
18	3.914	3.896	3.879	3.863	3.835	3.810
19	3.934	3.916	3.899	3.883	3.854	3.830
20	3.954	3.935	3.918	3.902	3.873	3.848
22	3.990	3.971	3.953	3.936	3.907	3.881
24	4.023	4.003	3.985	3.968	3.938	3.912
26	4.053	4.033	4.015	3.997	3.967	3.940
28	4.081	4.061	4.042	4.024	3.993	3.966
30	4.107	4.086	4.067	4.049	4.018	3.990
35	4.165	4.143	4.124	4.105	4.072	4.044
40	4.215	4.193	4.172	4.154	4.120	4.090
45	4.259	4.237	4.216	4.196	4.161	4.131
50	4.299	4.275	4.254	4.234	4.198	4.167
60	4.367	4.343	4.320	4.300	4.262	4.230
70	4.425	4.399	4.376	4.355	4.317	4.283
80	4.474	4.449	4.425	4.403	4.363	4.329
90	4.518	4.492	4.467	4.445	4.404	4.369
100	4.558	4.530	4.505	4.482	4.441	4.405
150	4.709	4.679	4.652	4.627	4.582	4.543
200	4.816	4.784	4.756	4.730	4.682	4.641
250	4.899	4.866	4.836	4.808	4.759	4.716
300	4.968	4.932	4.901	4.872	4.821	4.777
400	5.075	5.038	5.006	4.974	4.922	4.874
500	5.156	5.118	5.085	5.055	4.998	4.948

Table A.6. *Continued*

COMPARISONS	36	38	40	42	44	46
2	2.990	2.980	2.971	2.963	2.956	2.949
3	3.144	3.132	3.122	3.112	3.104	3.096
4	3.251	3.238	3.227	3.216	3.207	3.199
5	3.333	3.319	3.307	3.296	3.286	3.277
6	3.399	3.385	3.372	3.360	3.350	3.340
7	3.455	3.440	3.426	3.414	3.403	3.394
8	3.503	3.487	3.473	3.461	3.449	3.439
9	3.545	3.529	3.514	3.501	3.490	3.479
10	3.582	3.566	3.551	3.538	3.526	3.515
11	3.616	3.599	3.584	3.570	3.558	3.547
12	3.647	3.629	3.614	3.600	3.588	3.576
13	3.675	3.657	3.642	3.627	3.615	3.603
14	3.701	3.683	3.667	3.653	3.640	3.628
15	3.725	3.707	3.691	3.676	3.663	3.651
16	3.748	3.729	3.713	3.698	3.684	3.672
17	3.769	3.750	3.733	3.718	3.704	3.692
18	3.789	3.770	3.753	3.737	3.724	3.711
19	3.808	3.788	3.771	3.755	3.741	3.729
20	3.826	3.806	3.788	3.773	3.758	3.746
22	3.859	3.838	3.821	3.805	3.790	3.777
24	3.889	3.868	3.850	3.833	3.819	3.805
26	3.916	3.895	3.877	3.860	3.845	3.831
28	3.942	3.921	3.902	3.885	3.869	3.855
30	3.966	3.944	3.925	3.907	3.892	3.878
35	4.018	3.996	3.976	3.958	3.942	3.927
40	4.064	4.041	4.021	4.002	3.985	3.970
45	4.104	4.081	4.060	4.041	4.024	4.008
50	4.140	4.116	4.094	4.075	4.057	4.042
60	4.201	4.177	4.154	4.134	4.116	4.099
70	4.254	4.228	4.205	4.184	4.165	4.148
80	4.299	4.272	4.248	4.227	4.208	4.190
90	4.338	4.311	4.286	4.264	4.245	4.227
100	4.374	4.346	4.321	4.298	4.278	4.260
150	4.509	4.478	4.451	4.427	4.405	4.386
200	4.605	4.572	4.544	4.519	4.495	4.474
250	4.678	4.645	4.615	4.588	4.565	4.543
300	4.739	4.704	4.673	4.647	4.621	4.598
400	4.832	4.796	4.764	4.735	4.710	4.687
500	4.907	4.868	4.834	4.805	4.778	4.754

Table A.6. *Continued*

```
          BONFERRONI T CRITICAL VALUES FOR FAMILYWISE ALPHA=.01
                            DEGREES OF FREEDOM
COMPARISONS     48        50        55        60        65        70
```

COMPARISONS	48	50	55	60	65	70
2	2.943	2.937	2.925	2.915	2.906	2.899
3	3.089	3.083	3.069	3.057	3.048	3.039
4	3.191	3.184	3.169	3.156	3.146	3.137
5	3.269	3.261	3.245	3.232	3.220	3.211
6	3.332	3.324	3.307	3.293	3.281	3.271
7	3.385	3.376	3.358	3.344	3.331	3.321
8	3.430	3.421	3.403	3.388	3.375	3.364
9	3.470	3.461	3.442	3.426	3.413	3.402
10	3.505	3.496	3.476	3.460	3.447	3.435
11	3.537	3.528	3.508	3.491	3.477	3.465
12	3.566	3.556	3.536	3.519	3.505	3.493
13	3.592	3.583	3.562	3.545	3.530	3.518
14	3.617	3.607	3.586	3.568	3.553	3.541
15	3.640	3.630	3.608	3.590	3.575	3.562
16	3.661	3.651	3.629	3.610	3.595	3.582
17	3.681	3.670	3.648	3.630	3.614	3.601
18	3.699	3.689	3.666	3.648	3.632	3.619
19	3.717	3.707	3.684	3.665	3.649	3.635
20	3.734	3.723	3.700	3.681	3.665	3.651
22	3.765	3.754	3.730	3.710	3.694	3.680
24	3.793	3.782	3.758	3.738	3.721	3.707
26	3.819	3.807	3.783	3.762	3.745	3.731
28	3.843	3.831	3.806	3.785	3.768	3.753
30	3.865	3.853	3.828	3.807	3.789	3.774
35	3.914	3.902	3.876	3.854	3.836	3.820
40	3.957	3.944	3.917	3.895	3.876	3.860
45	3.994	3.981	3.953	3.930	3.911	3.895
50	4.027	4.014	3.986	3.962	3.942	3.926
60	4.085	4.071	4.041	4.017	3.997	3.979
70	4.133	4.119	4.088	4.063	4.042	4.024
80	4.174	4.160	4.129	4.102	4.081	4.063
90	4.211	4.196	4.164	4.138	4.116	4.096
100	4.244	4.229	4.196	4.169	4.146	4.127
150	4.368	4.352	4.316	4.287	4.263	4.242
200	4.456	4.438	4.401	4.371	4.345	4.323
250	4.523	4.505	4.467	4.434	4.408	4.385
300	4.578	4.560	4.519	4.487	4.459	4.435
400	4.664	4.646	4.604	4.568	4.539	4.516
500	4.730	4.712	4.668	4.631	4.601	4.574

Table A.6. *Continued*

BONFERRONI T CRITICAL VALUES FOR FAMILYWISE ALPHA=.01

DEGREES OF FREEDOM

COMPARISONS	75	80	85	90	95	100
2	2.892	2.887	2.882	2.878	2.874	2.871
3	3.032	3.026	3.020	3.016	3.011	3.007
4	3.129	3.122	3.116	3.111	3.106	3.102
5	3.202	3.195	3.189	3.183	3.178	3.174
6	3.262	3.254	3.248	3.242	3.236	3.232
7	3.312	3.304	3.297	3.291	3.285	3.280
8	3.354	3.346	3.339	3.333	3.327	3.322
9	3.392	3.383	3.376	3.369	3.363	3.358
10	3.425	3.416	3.409	3.402	3.396	3.390
11	3.455	3.446	3.438	3.431	3.425	3.420
12	3.482	3.473	3.465	3.458	3.452	3.446
13	3.507	3.498	3.490	3.482	3.476	3.470
14	3.530	3.520	3.512	3.505	3.498	3.492
15	3.551	3.542	3.533	3.526	3.519	3.513
16	3.571	3.561	3.553	3.545	3.538	3.532
17	3.590	3.580	3.571	3.563	3.556	3.550
18	3.607	3.597	3.588	3.580	3.574	3.567
19	3.624	3.613	3.604	3.597	3.590	3.583
20	3.639	3.629	3.620	3.612	3.605	3.598
22	3.668	3.657	3.648	3.640	3.633	3.626
24	3.694	3.684	3.674	3.666	3.658	3.652
26	3.718	3.707	3.698	3.689	3.682	3.675
28	3.740	3.729	3.720	3.711	3.703	3.696
30	3.761	3.750	3.740	3.731	3.723	3.716
35	3.807	3.795	3.785	3.776	3.768	3.761
40	3.846	3.834	3.824	3.814	3.806	3.799
45	3.881	3.868	3.858	3.848	3.840	3.832
50	3.911	3.899	3.888	3.878	3.869	3.862
60	3.964	3.951	3.940	3.930	3.921	3.913
70	4.009	3.995	3.983	3.973	3.964	3.955
80	4.047	4.033	4.021	4.010	4.001	3.992
90	4.080	4.066	4.054	4.043	4.033	4.025
100	4.110	4.096	4.084	4.072	4.062	4.053
150	4.224	4.209	4.195	4.183	4.173	4.163
200	4.304	4.288	4.274	4.261	4.249	4.240
250	4.365	4.349	4.334	4.320	4.309	4.299
300	4.416	4.398	4.383	4.369	4.357	4.347
400	4.493	4.476	4.460	4.445	4.432	4.421
500	4.554	4.534	4.517	4.504	4.491	4.478

Table A.6. *Continued*

```
          BONFERRONI T CRITICAL VALUES FOR FAMILYWISE ALPHA=.01
                            DEGREES OF FREEDOM
```

COMPARISONS	150	200	250	300	500	1000
2	2,849	2,839	2,832	2,828	2,820	2,813
3	2,983	2,971	2,964	2,959	2,949	2,942
4	3,075	3,062	3,054	3,049	3,039	3,031
5	3,145	3,131	3,123	3,118	3,107	3,098
6	3,202	3,187	3,179	3,173	3,161	3,153
7	3,249	3,234	3,225	3,219	3,207	3,198
8	3,290	3,274	3,264	3,258	3,246	3,236
9	3,325	3,309	3,299	3,293	3,280	3,270
10	3,357	3,340	3,330	3,323	3,310	3,300
11	3,385	3,368	3,358	3,351	3,337	3,327
12	3,411	3,393	3,383	3,376	3,362	3,352
13	3,434	3,416	3,406	3,398	3,384	3,374
14	3,456	3,437	3,427	3,419	3,405	3,395
15	3,476	3,457	3,446	3,439	3,424	3,414
16	3,494	3,476	3,464	3,457	3,442	3,432
17	3,512	3,493	3,482	3,474	3,459	3,448
18	3,528	3,509	3,498	3,490	3,475	3,464
19	3,544	3,524	3,513	3,505	3,490	3,478
20	3,558	3,539	3,527	3,519	3,504	3,492
22	3,585	3,565	3,553	3,545	3,530	3,518
24	3,610	3,590	3,577	3,569	3,553	3,541
26	3,633	3,612	3,599	3,591	3,575	3,562
28	3,653	3,632	3,620	3,611	3,595	3,582
30	3,673	3,651	3,638	3,630	3,613	3,601
35	3,715	3,693	3,680	3,671	3,654	3,641
40	3,752	3,729	3,716	3,707	3,689	3,676
45	3,784	3,761	3,747	3,738	3,719	3,706
50	3,813	3,789	3,775	3,765	3,747	3,733
60	3,862	3,838	3,823	3,813	3,794	3,779
70	3,904	3,878	3,863	3,853	3,833	3,818
80	3,939	3,913	3,897	3,887	3,867	3,851
90	3,970	3,943	3,927	3,917	3,896	3,881
100	3,998	3,971	3,955	3,944	3,922	3,907
150	4,103	4,074	4,056	4,045	4,022	4,006
200	4,176	4,146	4,128	4,116	4,091	4,073
250	4,233	4,201	4,182	4,169	4,144	4,125
300	4,279	4,246	4,227	4,214	4,187	4,168
400	4,351	4,315	4,295	4,282	4,256	4,235
500	4,406	4,369	4,347	4,334	4,308	4,285

Source: Taken from Huitema, *Bonferroni Statistics*: *Quick and Dirty Methods of Simultaneous Statistical Inference*. (In preparation.)

Table A.7. Upper Percentage Points of the χ^2 Distribution

df	0.99	0.98	0.95	0.90	0.80	0.70	0.50	0.30	0.20	0.10	0.05	0.02	0.01	0.001
1	0.0^3157	0.0^3628	0.00393	0.0158	0.0642	0.148	0.455	1.074	1.642	2.706	3.841	5.412	6.635	10.827
2	0.0201	0.0404	0.103	0.211	0.446	0.713	1.386	2.408	3.219	4.605	5.991	7.824	9.210	13.815
3	0.115	0.185	0.352	0.584	1.005	1.424	2.366	3.665	4.642	6.251	7.815	9.837	11.345	16.266
4	0.297	0.429	0.711	1.064	1.649	2.195	3.357	4.878	5.989	7.779	9.488	11.668	13.277	18.467
5	0.554	0.752	1.145	1.610	2.343	3.000	4.351	6.064	7.289	9.236	11.070	13.388	15.086	20.515
6	0.872	1.134	1.635	2.204	3.070	3.828	5.348	7.231	8.558	10.645	12.592	15.033	16.812	22.457
7	1.239	1.564	2.167	2.833	3.822	4.671	6.346	8.383	9.803	12.017	14.067	16.622	18.475	24.322
8	1.646	2.032	2.733	3.490	4.594	5.527	7.344	9.524	11.030	13.362	15.507	18.168	20.090	26.125
9	2.088	2.532	3.325	4.168	5.380	6.393	8.343	10.656	12.242	14.684	16.919	19.679	21.666	27.877
10	2.558	3.059	3.940	4.865	6.179	7.267	9.342	11.781	13.442	15.987	18.307	21.161	23.209	29.588
11	3.053	3.609	4.575	5.578	6.989	8.148	10.341	12.899	14.631	17.275	19.675	22.618	24.725	31.264
12	3.571	4.178	5.226	6.304	7.807	9.034	11.340	14.011	15.812	18.549	21.026	24.054	26.217	32.909
13	4.107	4.765	5.892	7.042	8.634	9.926	12.340	15.119	16.985	19.812	22.362	25.472	27.688	34.528
14	4.660	5.368	6.571	7.790	9.467	10.821	13.339	16.222	18.151	21.064	23.685	26.873	29.141	36.123
15	5.229	5.985	7.261	8.547	10.307	11.721	14.339	17.322	19.311	22.307	24.996	28.259	30.578	37.697

Table A.7. *Continued*

df	0.99	0.98	0.95	0.90	0.80	0.70	0.50	0.30	0.20	0.10	0.05	0.02	0.01	0.001
16	5.812	6.614	7.962	9.312	11.152	12.624	15.338	18.418	20.465	23.542	26.296	29.633	32.000	39.252
17	6.408	7.255	8.672	10.085	12.002	13.531	16.338	19.511	21.615	24.769	27.587	30.995	33.409	40.790
18	7.015	7.906	9.390	10.865	12.857	14.440	17.338	20.601	22.760	25.989	28.869	32.346	34.805	42.312
19	7.633	8.567	10.117	11.651	13.716	15.352	18.338	21.689	23.900	27.204	30.144	33.687	36.191	43.820
20	8.260	9.237	10.851	12.443	14.578	16.266	19.337	22.775	25.038	28.412	31.410	35.020	37.566	45.315
21	8.897	9.915	11.591	13.240	15.445	17.182	20.337	23.858	26.171	29.615	32.671	36.343	38.932	46.797
22	9.542	10.600	12.338	14.041	16.314	18.101	21.337	24.939	27.301	30.813	33.924	37.659	40.289	48.268
23	10.196	11.293	13.091	14.848	17.187	19.021	22.337	26.018	28.429	32.007	35.172	38.968	41.638	49.728
24	10.856	11.992	13.848	15.659	18.062	19.943	23.337	27.096	29.553	33.196	36.415	40.270	42.980	51.179
25	11.524	12.697	14.611	16.473	18.940	20.867	24.337	28.172	30.675	34.382	37.652	41.566	44.314	52.620
26	12.198	13.409	15.379	17.292	19.820	21.792	25.336	29.246	31.795	35.563	38.885	42.856	45.642	54.052
27	12.879	14.125	16.151	18.114	20.703	22.719	26.336	30.319	32.912	36.741	40.113	44.140	46.963	55.476
28	13.565	14.847	16.928	18.939	21.588	23.647	27.336	31.391	34.027	37.916	41.337	45.419	48.278	56.893
29	14.256	15.574	17.708	19.768	22.475	24.577	28.336	32.461	35.139	39.087	42.557	46.693	49.588	58.302
30	14.953	16.306	18.493	20.599	23.364	25.508	29.336	33.530	36.250	40.256	43.773	47.962	50.892	59.703

For $\nu > 30$, the expression $\sqrt{2\chi^2} - \sqrt{2\nu - 1}$ may be used as a normal deviate with unit variance.

Source: Taken from Table 4 of Fisher and Yates by permission of the authors and the publishers.

Table A.8. Critical Values of Bonferroni F_B Statistic

```
BONFERRONI F CRITICAL VALUES FOR FAMILYWISE ALPHA= .050
```

COMPARISONS	DF1 = 1 DF2 = 3	1 4	1 5	1 6	1 7	1 8	1 9
2	17.44	12.21	10.01	8.81	8.07	7.57	7.21
3	23.60	15.69	12.49	10.81	9.78	9.10	8.60
4	29.08	18.62	14.52	12.40	11.12	10.28	9.68
5	34.13	21.19	16.26	13.74	12.25	11.26	10.56
6	38.82	23.53	17.79	14.92	13.22	12.10	11.31
7	43.27	25.68	19.21	15.98	14.09	12.85	11.98
8	47.55	27.69	20.48	16.93	14.86	13.52	12.58
9	51.61	29.56	21.68	17.82	15.58	14.12	13.11
10	55.58	31.32	22.77	18.63	16.24	14.68	13.62
11	59.33	33.01	23.83	19.39	16.84	15.21	14.07
12	63.03	34.63	24.83	20.13	17.42	15.69	14.50
13	66.63	36.17	25.78	20.79	17.97	16.16	14.91
14	70.07	37.70	26.66	21.46	18.48	16.59	15.28
15	73.58	39.13	27.54	22.07	18.97	16.98	15.64
16	76.86	40.48	28.38	22.68	19.44	17.38	15.97
17	80.12	41.84	29.18	23.23	19.87	17.75	16.29
18	83.38	43.14	29.94	23.76	20.30	18.11	16.60
19	86.50	44.42	30.69	24.30	20.71	18.46	16.90
20	89.60	45.67	31.41	24.82	21.11	18.78	17.20
25	104.28	51.47	34.73	27.11	22.91	20.26	18.45
30	118.22	56.64	37.69	29.13	24.47	21.53	19.55
40	143.54	65.97	42.82	32.59	27.08	23.65	21.38
50	167.06	74.15	47.16	35.52	29.26	25.42	22.86
100	266.48	106.20	63.63	46.11	36.99	31.58	27.97
250	492.84	169.80	93.69	64.39	49.90	41.49	36.13
500	783.54	241.46	124.92	82.50	62.17	50.71	43.49

COMPARISONS	DF1 = 1 DF2 = 10	1 11	1 12	1 13	1 14	1 15	1 16
2	6.94	6.72	6.55	6.41	6.30	6.20	6.12
3	8.24	7.95	7.72	7.54	7.39	7.26	7.15
4	9.23	8.88	8.61	8.39	8.20	8.05	7.91
5	10.04	9.65	9.33	9.07	8.86	8.68	8.53
6	10.74	10.29	9.94	9.65	9.42	9.22	9.05
7	11.35	10.86	10.47	10.16	9.90	9.68	9.50
8	11.89	11.36	10.95	10.60	10.33	10.09	9.90
9	12.37	11.81	11.37	11.01	10.71	10.46	10.25
10	12.82	12.23	11.75	11.37	11.06	10.80	10.57
11	13.24	12.60	12.11	11.71	11.38	11.10	10.87
12	13.62	12.96	12.44	12.02	11.67	11.39	11.14
13	13.98	13.30	12.75	12.31	11.95	11.66	11.40
14	14.32	13.61	13.04	12.59	12.21	11.91	11.64
15	14.64	13.90	13.32	12.84	12.46	12.14	11.87
16	14.96	14.18	13.58	13.08	12.69	12.36	12.07
17	15.24	14.45	13.82	13.32	12.90	12.56	12.27
18	15.52	14.69	14.06	13.53	13.11	12.76	12.46
19	15.79	14.94	14.27	13.75	13.30	12.95	12.64
20	16.03	15.17	14.50	13.94	13.51	13.13	12.83
25	17.17	16.20	15.43	14.83	14.34	13.93	13.59
30	18.13	17.08	16.24	15.59	15.05	14.61	14.23
40	19.72	18.50	17.58	16.83	16.20	15.70	15.27
50	21.04	19.69	18.66	17.80	17.15	16.59	16.12
100	25.49	23.65	22.23	21.14	20.25	19.49	18.88
250	32.42	29.75	27.71	26.12	24.83	23.81	22.95
500	38.58	35.06	32.44	30.39	28.77	27.45	26.35

Table A.8. *Continued*

```
            BONFERRONI F CRITICAL VALUES FOR FAMILYWISE ALPHA= .050
```

COMPARISONS	DF1 = DF2 =	1 17	1 18	1 19	1 20	1 22	1 24	1 26
2		6.04	5.98	5.92	5.87	5.79	5.72	5.66
3		7.05	6.97	6.89	6.83	6.71	6.62	6.55
4		7.80	7.70	7.61	7.53	7.40	7.29	7.20
5		8.40	8.29	8.18	8.10	7.95	7.82	7.72
6		8.91	8.78	8.67	8.57	8.40	8.27	8.15
7		9.34	9.20	9.09	8.98	8.79	8.65	8.52
8		9.73	9.58	9.45	9.34	9.14	8.98	8.85
9		10.07	9.91	9.78	9.65	9.45	9.28	9.15
10		10.38	10.22	10.07	9.95	9.73	9.55	9.41
11		10.67	10.50	10.34	10.21	9.98	9.79	9.65
12		10.93	10.76	10.59	10.46	10.21	10.02	9.87
13		11.18	10.99	10.83	10.69	10.43	10.23	10.07
14		11.41	11.21	11.05	10.89	10.64	10.43	10.26
15		11.63	11.42	11.24	11.09	10.83	10.61	10.44
16		11.84	11.63	11.44	11.28	11.01	10.78	10.61
17		12.03	11.81	11.62	11.46	11.18	10.95	10.77
18		12.21	11.99	11.79	11.63	11.34	11.11	10.91
19		12.39	12.16	11.96	11.79	11.50	11.26	11.07
20		12.54	12.32	12.11	11.95	11.64	11.40	11.20
25		13.29	13.04	12.82	12.62	12.29	12.01	11.80
30		13.91	13.62	13.40	13.18	12.83	12.53	12.29
40		14.91	14.60	14.33	14.09	13.69	13.36	13.10
50		15.71	15.38	15.08	14.83	14.37	14.02	13.74
100		18.37	17.93	17.53	17.20	16.63	16.16	15.80
250		22.22	21.60	21.07	20.62	19.82	19.22	18.69
500		25.44	24.64	24.00	23.42	22.43	21.65	21.03

COMPARISONS	DF1 = DF2 =	1 28	1 30	1 40	1 60	1 120	1 500	1 1000
2		5.61	5.57	5.43	5.28	5.15	5.05	5.04
3		6.48	6.43	6.25	6.07	5.90	5.77	5.75
4		7.13	7.06	6.84	6.63	6.43	6.28	6.26
5		7.63	7.56	7.31	7.08	6.85	6.69	6.66
6		8.06	7.98	7.71	7.44	7.20	7.02	6.99
7		8.43	8.33	8.04	7.76	7.49	7.29	7.26
8		8.74	8.65	8.33	8.03	7.74	7.54	7.51
9		9.02	8.93	8.59	8.28	7.98	7.76	7.72
10		9.29	9.18	8.83	8.50	8.18	7.95	7.91
11		9.52	9.41	9.04	8.69	8.36	8.13	8.09
12		9.73	9.62	9.24	8.87	8.54	8.28	8.24
13		9.94	9.82	9.42	9.05	8.69	8.43	8.39
14		10.12	10.00	9.59	9.20	8.83	8.57	8.53
15		10.29	10.16	9.75	9.35	8.97	8.70	8.66
16		10.46	10.33	9.90	9.48	9.10	8.82	8.77
17		10.61	10.48	10.04	9.61	9.22	8.93	8.89
18		10.76	10.62	10.17	9.74	9.33	9.04	9.00
19		10.90	10.76	10.29	9.85	9.44	9.14	9.10
20		11.04	10.89	10.41	9.96	9.54	9.23	9.19
25		11.62	11.46	10.94	10.44	9.99	9.65	9.60
30		12.11	11.93	11.37	10.84	10.35	9.99	9.94
40		12.88	12.69	12.06	11.47	10.93	10.54	10.47
50		13.50	13.28	12.62	11.96	11.38	10.96	10.89
100		15.48	15.21	14.36	13.55	12.81	12.28	12.19
250		18.27	17.91	16.76	15.71	14.73	14.04	13.94
500		20.53	20.08	18.66	17.39	16.22	15.39	15.25

386

Table A.8. *Continued*

BONFERRONI F CRITICAL VALUES FOR FAMILYWISE ALPHA= .050

COMPARISONS	DF1 = DF2 =	2 3	2 4	2 5	2 6	2 7	2 8	2 9
2		16.04	10.65	8.43	7.26	6.54	6.06	5.71
3		21.49	13.49	10.36	8.74	7.78	7.13	6.68
4		26.35	15.89	11.93	9.93	8.74	7.96	7.42
5		30.82	18.00	13.27	10.92	9.55	8.65	8.02
6		34.99	19.91	14.47	11.80	10.24	9.24	8.54
7		38.95	21.66	15.55	12.58	10.86	9.76	8.99
8		42.71	23.30	16.54	13.29	11.42	10.23	9.40
9		46.32	24.83	17.45	13.94	11.93	10.65	9.77
10		49.80	26.28	18.31	14.54	12.41	11.04	10.11
11		53.16	27.66	19.12	15.11	12.84	11.41	10.42
12		56.43	28.99	19.89	15.65	13.25	11.74	10.71
13		59.61	30.25	20.62	16.15	13.64	12.06	10.98
14		62.70	31.47	21.31	16.63	14.01	12.36	11.24
15		65.72	32.64	21.98	17.08	14.36	12.65	11.48
16		68.68	33.78	22.62	17.52	14.69	12.92	11.71
17		71.57	34.87	23.24	17.94	15.01	13.18	11.93
18		74.41	35.95	23.83	18.34	15.31	13.42	12.14
19		77.19	36.99	24.41	18.73	15.60	13.66	12.35
20		79.94	38.00	24.96	19.10	15.89	13.89	12.54
25		93.00	42.73	27.53	20.81	17.16	14.92	13.41
30		105.20	46.99	29.80	22.30	18.27	15.80	14.14
40		127.76	54.57	33.74	24.85	20.13	17.27	15.38
50		148.51	61.25	37.12	27.00	21.69	18.50	16.39
100		236.58	87.45	49.77	34.79	27.21	22.75	19.87
250		437.12	139.44	72.92	48.29	36.39	29.63	25.37
500		694.45	197.99	97.02	61.64	45.14	35.99	30.34

COMPARISONS	DF1 = DF2 =	2 10	2 11	2 12	2 13	2 14	2 15	2 16
2		5.46	5.26	5.10	4.97	4.86	4.77	4.69
3		6.34	6.08	5.87	5.70	5.56	5.45	5.35
4		7.01	6.70	6.45	6.26	6.09	5.95	5.83
5		7.56	7.21	6.93	6.70	6.51	6.36	6.23
6		8.03	7.63	7.33	7.08	6.87	6.70	6.55
7		8.43	8.01	7.67	7.40	7.18	6.99	6.84
8		8.80	8.34	7.98	7.69	7.45	7.26	7.09
9		9.13	8.64	8.26	7.95	7.70	7.49	7.31
10		9.43	8.91	8.51	8.19	7.92	7.70	7.51
11		9.70	9.16	8.74	8.40	8.13	7.90	7.70
12		9.96	9.40	8.96	8.60	8.32	8.07	7.87
13		10.20	9.62	9.16	8.79	8.49	8.24	8.03
14		10.43	9.82	9.35	8.97	8.66	8.40	8.18
15		10.65	10.02	9.52	9.13	8.81	8.55	8.32
16		10.85	10.20	9.69	9.29	8.96	8.68	8.45
17		11.04	10.37	9.85	9.44	9.10	8.82	8.58
18		11.23	10.54	10.00	9.58	9.23	8.94	8.70
19		11.40	10.70	10.15	9.71	9.35	9.06	8.81
20		11.57	10.85	10.29	9.84	9.48	9.17	8.92
25		12.33	11.53	10.90	10.41	10.01	9.68	9.40
30		12.97	12.10	11.43	10.89	10.46	10.10	9.80
40		14.04	13.04	12.28	11.68	11.19	10.79	10.45
50		14.91	13.81	12.98	12.31	11.78	11.34	10.97
100		17.86	16.40	15.30	14.43	13.73	13.16	12.69
250		22.47	20.38	18.81	17.60	16.63	15.85	15.20
500		26.54	23.85	21.85	20.31	19.09	18.11	17.30

Table A.8. *Continued*

BONFERRONI F CRITICAL VALUES FOR FAMILYWISE ALPHA= .050

COMPARISONS	DF1 = DF2 =	2 17	2 18	2 19	2 20	2 22	2 24	2 26
2		4.62	4.56	4.51	4.46	4.38	4.32	4.27
3		5.26	5.18	5.12	5.06	4.96	4.88	4.81
4		5.73	5.65	5.57	5.50	5.38	5.29	5.21
5		6.11	6.01	5.93	5.85	5.72	5.61	5.53
6		6.43	6.32	6.22	6.14	6.00	5.88	5.79
7		6.70	6.58	6.48	6.39	6.24	6.11	6.01
8		6.94	6.82	6.71	6.61	6.45	6.32	6.21
9		7.16	7.03	6.91	6.81	6.64	6.50	6.38
10		7.35	7.21	7.09	6.99	6.81	6.66	6.54
11		7.53	7.39	7.26	7.15	6.96	6.81	6.68
12		7.70	7.55	7.41	7.30	7.10	6.95	6.82
13		7.85	7.69	7.56	7.44	7.24	7.07	6.94
14		7.99	7.83	7.69	7.57	7.36	7.19	7.05
15		8.13	7.96	7.82	7.69	7.48	7.30	7.16
16		8.26	8.08	7.94	7.80	7.58	7.41	7.26
·17		8.37	8.20	8.05	7.91	7.69	7.50	7.35
18		8.49	8.31	8.15	8.01	7.78	7.60	7.44
19		8.60	8.41	8.25	8.11	7.88	7.69	7.53
20		8.70	8.51	8.35	8.21	7.97	7.77	7.61
25		9.16	8.95	8.77	8.62	8.35	8.14	7.97
30		9.54	9.32	9.13	8.96	8.68	8.45	8.26
40		10.16	9.92	9.70	9.51	9.20	8.95	8.74
50		10.66	10.39	10.16	9.95	9.61	9.34	9.12
100		12.29	11.94	11.65	11.38	10.95	10.61	10.33
250		14.65	14.19	13.79	13.44	12.86	12.40	12.03
500		16.62	16.05	15.55	15.12	14.41	13.85	13.40

COMPARISONS	DF1 = DF2 =	2 28	2 30	2 40	2 60	2 120	2 500	2 1000
2		4.22	4.18	4.05	3.92	3.80	3.72	3.70
3		4.76	4.71	4.54	4.39	4.24	4.13	4.11
4		5.15	5.09	4.90	4.72	4.55	4.42	4.40
5		5.45	5.39	5.18	4.98	4.79	4.65	4.63
6		5.71	5.64	5.41	5.19	4.98	4.83	4.81
7		5.93	5.85	5.61	5.37	5.15	4.99	4.97
8		6.12	6.04	5.78	5.53	5.30	5.13	5.10
9		6.29	6.21	5.93	5.67	5.42	5.25	5.22
10		6.44	6.35	6.07	5.80	5.54	5.36	5.33
11		6.58	6.49	6.19	5.91	5.64	5.45	5.42
12		6.71	6.62	6.30	6.01	5.74	5.54	5.51
13		6.83	6.73	6.41	6.11	5.83	5.62	5.59
14		6.94	6.84	6.51	6.20	5.91	5.70	5.67
15		7.04	6.94	6.60	6.28	5.98	5.77	5.74
16		7.14	7.03	6.69	6.36	6.05	5.84	5.80
17		7.23	7.12	6.77	6.43	6.12	5.90	5.86
18		7.32	7.21	6.84	6.50	6.18	5.96	5.92
19		7.40	7.29	6.92	6.57	6.24	6.01	5.98
20		7.48	7.36	6.99	6.63	6.30	6.06	6.03
25		7.82	7.70	7.29	6.90	6.55	6.29	6.25
30		8.11	7.98	7.54	7.13	6.75	6.48	6.44
40		8.57	8.42	7.94	7.49	7.07	6.77	6.73
50		8.93	8.77	8.25	7.77	7.32	7.00	6.96
100		10.09	9.90	9.25	8.65	8.10	7.72	7.66
250		11.73	11.47	10.62	9.85	9.15	8.66	8.59
500		13.03	12.72	11.70	10.78	9.96	9.38	9.29

Table A.8. *Continued*

COMPARISONS	DF1 = 3 DF2 = 3	3 4	3 5	3 6	3 7	3 8	3 9
2	15.44	9.98	7.76	6.60	5.89	5.42	5.08
3	20.61	12.58	9.47	7.89	6.94	6.32	5.88
4	25.22	14.77	10.87	8.91	7.77	7.02	6.49
5	29.46	16.69	12.06	9.78	8.45	7.59	6.99
6	33.42	18.44	13.12	10.54	9.05	8.09	7.42
7	37.17	20.04	14.08	11.21	9.57	8.52	7.80
8	40.74	21.53	14.95	11.83	10.05	8.91	8.13
9	44.17	22.93	15.77	12.39	10.48	9.27	8.44
10	47.47	24.26	16.53	12.92	10.88	9.60	8.72
11	50.66	25.52	17.25	13.41	11.26	9.90	8.98
12	53.76	26.72	17.93	13.87	11.61	10.18	9.22
13	56.77	27.88	18.57	14.31	11.93	10.45	9.44
14	59.71	28.99	19.19	14.72	12.25	10.70	9.65
15	62.57	30.06	19.78	15.12	12.54	10.94	9.86
16	65.38	31.10	20.35	15.49	12.83	11.17	10.05
17	68.12	32.11	20.89	15.86	13.10	11.38	10.23
18	70.81	33.08	21.42	16.21	13.35	11.59	10.40
19	73.46	34.03	21.93	16.54	13.60	11.79	10.57
20	76.06	34.96	22.43	16.87	13.84	11.98	10.73
25	88.45	39.27	24.70	18.35	14.93	12.84	11.44
30	100.04	43.16	26.71	19.64	15.87	13.57	12.05
40	121.44	50.08	30.20	21.84	17.45	14.81	13.07
50	141.11	56.18	33.20	23.70	18.77	15.83	13.90
100	224.72	80.10	44.42	30.45	23.46	19.39	16.77
250	415.01	127.55	64.94	42.14	31.26	25.14	21.30
500	659.58	181.01	86.28	53.69	38.68	30.46	25.40

COMPARISONS	DF1 = 3 DF2 = 10	3 11	3 12	3 13	3 14	3 15	3 16
2	4.83	4.63	4.47	4.35	4.24	4.15	4.08
3	5.55	5.30	5.10	4.94	4.81	4.69	4.60
4	6.10	5.81	5.57	5.38	5.23	5.10	4.98
5	6.55	6.22	5.95	5.74	5.56	5.42	5.29
6	6.93	6.56	6.27	6.04	5.85	5.69	5.55
7	7.27	6.87	6.55	6.30	6.09	5.92	5.77
8	7.57	7.14	6.80	6.53	6.31	6.13	5.97
9	7.84	7.38	7.02	6.74	6.50	6.31	6.14
10	8.08	7.60	7.23	6.93	6.68	6.48	6.30
11	8.31	7.80	7.41	7.10	6.84	6.63	6.45
12	8.52	7.99	7.59	7.26	6.99	6.77	6.58
13	8.72	8.17	7.75	7.41	7.13	6.90	6.71
14	8.90	8.34	7.90	7.55	7.26	7.03	6.83
15	9.08	8.49	8.04	7.68	7.39	7.14	6.94
16	9.24	8.64	8.18	7.80	7.50	7.25	7.04
17	9.40	8.78	8.30	7.92	7.61	7.35	7.14
18	9.55	8.92	8.42	8.03	7.71	7.45	7.23
19	9.70	9.04	8.54	8.14	7.81	7.54	7.32
20	9.83	9.17	8.65	8.24	7.91	7.63	7.40
25	10.45	9.71	9.15	8.70	8.33	8.03	7.78
30	10.98	10.18	9.56	9.08	8.69	8.36	8.09
40	11.84	10.94	10.25	9.70	9.26	8.90	8.60
50	12.55	11.56	10.80	10.21	9.73	9.34	9.01
100	14.97	13.65	12.66	11.89	11.27	10.76	10.34
250	18.71	16.86	15.47	14.40	13.55	12.87	12.30
500	22.04	19.66	17.90	16.55	15.49	14.64	13.93

Table A.8. *Continued*

BONFERRONI F CRITICAL VALUES FOR FAMILYWISE ALPHA= .050

COMPARISONS	DF1 = DF2 =	3 17	3 18	3 19	3 20	3 22	3 24	3 26
2		4,01	3,95	3,90	3,86	3,78	3,72	3,67
3		4,52	4,44	4,38	4,32	4,23	4,15	4,09
4		4,89	4,80	4,73	4,67	4,56	4,47	4,40
5		5,19	5,09	5,01	4,94	4,82	4,72	4,64
6		5,43	5,33	5,24	5,16	5,03	4,93	4,84
7		5,65	5,54	5,44	5,36	5,22	5,10	5,01
8		5,84	5,72	5,62	5,53	5,38	5,26	5,16
9		6,00	5,88	5,77	5,68	5,52	5,39	5,29
10		6,16	6,03	5,92	5,82	5,65	5,52	5,41
11		6,30	6,16	6,05	5,94	5,77	5,63	5,52
12		6,42	6,29	6,17	6,06	5,88	5,74	5,62
13		6,54	6,40	6,28	6,17	5,98	5,83	5,71
14		6,65	6,51	6,38	6,27	6,08	5,92	5,80
15		6,76	6,61	6,48	6,36	6,16	6,01	5,88
16		6,86	6,70	6,57	6,45	6,25	6,09	5,95
17		6,95	6,79	6,65	6,53	6,33	6,16	6,03
18		7,04	6,88	6,73	6,61	6,40	6,23	6,09
19		7,12	6,96	6,81	6,68	6,47	6,30	6,16
20		7,21	7,04	6,89	6,76	6,54	6,36	6,22
25		7,56	7,38	7,21	7,07	6,84	6,65	6,49
30		7,86	7,66	7,49	7,34	7,08	6,88	6,71
40		8,34	8,12	7,93	7,76	7,48	7,26	7,07
50		8,73	8,49	8,28	8,10	7,80	7,55	7,36
100		9,99	9,69	9,42	9,20	8,82	8,51	8,27
250		11,82	11,42	11,07	10,77	10,27	9,87	9,55
500		13,35	12,85	12,42	12,05	11,44	10,96	10,58

COMPARISONS	DF1 = DF2 =	3 28	3 30	3 40	3 60	3 120	3 500	3 1000
2		3,63	3,59	3,46	3,34	3,23	3,14	3,13
3		4,04	3,99	3,84	3,69	3,55	3,44	3,43
4		4,33	4,28	4,10	3,93	3,77	3,66	3,64
5		4,57	4,51	4,31	4,13	3,95	3,82	3,80
6		4,76	4,70	4,49	4,28	4,09	3,96	3,93
7		4,93	4,86	4,63	4,42	4,22	4,07	4,05
8		5,07	5,00	4,76	4,53	4,32	4,17	4,14
9		5,20	5,13	4,87	4,64	4,41	4,25	4,23
10		5,32	5,24	4,98	4,73	4,50	4,33	4,31
11		5,42	5,34	5,07	4,81	4,57	4,40	4,37
12		5,52	5,44	5,15	4,89	4,64	4,46	4,44
13		5,61	5,52	5,23	4,96	4,71	4,52	4,49
14		5,69	5,60	5,30	5,03	4,76	4,58	4,55
15		5,77	5,68	5,37	5,09	4,82	4,63	4,60
16		5,84	5,75	5,44	5,14	4,87	4,67	4,64
17		5,91	5,82	5,50	5,20	4,92	4,72	4,69
18		5,98	5,88	5,55	5,25	4,96	4,76	4,73
19		6,04	5,94	5,61	5,30	5,01	4,80	4,77
20		6,10	6,00	5,66	5,34	5,05	4,84	4,81
25		6,36	6,25	5,88	5,54	5,23	5,00	4,97
30		6,58	6,46	6,07	5,71	5,37	5,13	5,10
40		6,92	6,79	6,36	5,97	5,60	5,34	5,30
50		7,19	7,05	6,59	6,17	5,78	5,51	5,46
100		8,07	7,89	7,33	6,81	6,34	6,01	5,96
250		9,29	9,06	8,34	7,68	7,09	6,68	6,61
500		10,26	9,99	9,13	8,35	7,66	7,18	7,11

Table A.8. *Continued*

```
              BONFERRONI F CRITICAL VALUES FOR FAMILYWISE ALPHA= .050
```

COMPARISONS	DF1 = 4 DF2 = 3	4 4	4 5	4 6	4 7	4 8	4 9
2	15.10	9.60	7.39	6.23	5.52	5.05	4.72
3	20.12	12.07	8.98	7.41	6.48	5.86	5.43
4	24.59	14.15	10.28	8.35	7.22	6.49	5.98
5	28.71	15.98	11.39	9.15	7.85	7.01	6.42
6	32.56	17.63	12.38	9.84	8.39	7.45	6.80
7	36.20	19.15	13.27	10.46	8.86	7.84	7.14
8	39.66	20.57	14.09	11.03	9.29	8.19	7.44
9	42.99	21.90	14.85	11.55	9.69	8.51	7.71
10	46.19	23.15	15.56	12.03	10.05	8.81	7.96
11	49.29	24.35	16.22	12.48	10.39	9.08	8.19
12	52.30	25.49	16.86	12.90	10.71	9.33	8.40
13	55.22	26.59	17.46	13.30	11.00	9.57	8.60
14	58.07	27.64	18.03	13.68	11.29	9.80	8.79
15	60.86	28.66	18.58	14.05	11.56	10.01	8.97
16	63.58	29.64	19.11	14.39	11.81	10.21	9.14
17	66.25	30.60	19.62	14.73	12.06	10.41	9.30
18	68.86	31.53	20.11	15.05	12.29	10.59	9.45
19	71.43	32.43	20.59	15.35	12.52	10.77	9.60
20	73.95	33.30	21.05	15.65	12.73	10.94	9.74
25	85.98	37.39	23.16	17.01	13.72	11.71	10.38
30	97.22	41.09	25.04	18.19	14.57	12.37	10.92
40	118.00	47.65	28.29	20.22	16.00	13.48	11.82
50	137.10	53.43	31.09	21.92	17.20	14.39	12.56
100	218.27	76.13	41.54	28.11	21.44	17.58	15.11
250	402.94	121.15	60.63	38.83	28.50	22.73	19.13
500	640.45	171.93	80.53	49.42	35.22	27.49	22.77

COMPARISONS	DF1 = 4 DF2 = 10	4 11	4 12	4 13	4 14	4 15	4 16
2	4.47	4.27	4.12	4.00	3.89	3.80	3.73
3	5.11	4.86	4.67	4.51	4.38	4.27	4.18
4	5.60	5.31	5.08	4.90	4.74	4.62	4.51
5	5.99	5.67	5.41	5.20	5.04	4.89	4.77
6	6.33	5.97	5.69	5.47	5.28	5.13	4.99
7	6.63	6.24	5.94	5.69	5.49	5.33	5.18
8	6.89	6.47	6.15	5.89	5.68	5.50	5.35
9	7.13	6.69	6.34	6.07	5.85	5.66	5.50
10	7.34	6.88	6.52	6.23	6.00	5.80	5.64
11	7.54	7.06	6.68	6.38	6.14	5.93	5.76
12	7.73	7.23	6.83	6.52	6.27	6.06	5.88
13	7.90	7.38	6.97	6.65	6.39	6.17	5.98
14	8.07	7.53	7.11	6.77	6.50	6.27	6.08
15	8.22	7.66	7.23	6.89	6.61	6.37	6.18
16	8.37	7.79	7.35	6.99	6.71	6.47	6.27
17	8.51	7.91	7.46	7.10	6.80	6.56	6.35
18	8.64	8.03	7.56	7.19	6.89	6.64	6.43
19	8.77	8.14	7.67	7.28	6.97	6.72	6.50
20	8.89	8.25	7.76	7.37	7.06	6.80	6.58
25	9.43	8.73	8.19	7.77	7.42	7.13	6.90
30	9.89	9.14	8.56	8.10	7.73	7.42	7.16
40	10.66	9.80	9.15	8.64	8.22	7.88	7.60
50	11.28	10.35	9.63	9.07	8.62	8.25	7.94
100	13.41	12.18	11.25	10.52	9.95	9.48	9.08
250	16.70	14.98	13.69	12.69	11.91	11.27	10.75
500	19.63	17.42	15.79	14.55	13.57	12.78	12.14

Table A.8. *Continued*

BONFERRONI F CRITICAL VALUES FOR FAMILYWISE ALPHA= .050

COMPARISONS	DF1 = DF2 =	4 17	4 18	4 19	4 20	4 22	4 24	4 26
2		3.67	3.61	3.56	3.51	3.44	3.38	3.33
3		4.10	4.03	3.97	3.91	3.82	3.74	3.68
4		4.42	4.33	4.26	4.20	4.10	4.01	3.94
5		4.67	4.58	4.50	4.43	4.31	4.22	4.14
6		4.88	4.78	4.70	4.62	4.49	4.39	4.31
7		5.06	4.96	4.87	4.79	4.65	4.54	4.45
8		5.22	5.11	5.02	4.93	4.79	4.67	4.57
9		5.37	5.25	5.15	5.06	4.91	4.79	4.69
10		5.50	5.37	5.27	5.17	5.02	4.89	4.79
11		5.62	5.49	5.38	5.28	5.12	4.98	4.88
12		5.73	5.59	5.48	5.38	5.21	5.07	4.96
13		5.83	5.69	5.57	5.47	5.29	5.15	5.04
14		5.92	5.78	5.66	5.55	5.37	5.23	5.11
15		6.01	5.87	5.74	5.63	5.45	5.30	5.18
16		6.10	5.95	5.82	5.71	5.52	5.36	5.24
17		6.17	6.02	5.89	5.78	5.58	5.43	5.30
18		6.25	6.09	5.96	5.84	5.65	5.49	5.36
19		6.32	6.16	6.03	5.91	5.70	5.54	5.41
20		6.39	6.23	6.09	5.97	5.76	5.60	5.46
25		6.69	6.52	6.37	6.23	6.01	5.83	5.69
30		6.95	6.76	6.60	6.45	6.22	6.03	5.87
40		7.36	7.15	6.97	6.81	6.55	6.34	6.17
50		7.68	7.46	7.27	7.10	6.81	6.59	6.41
100		8.75	8.47	8.23	8.02	7.67	7.39	7.16
250		10.31	9.94	9.62	9.34	8.88	8.52	8.22
500		11.60	11.14	10.75	10.41	9.86	9.42	9.07

COMPARISONS	DF1 = DF2 =	4 28	4 30	4 40	4 60	4 120	4 500	4 1000
2		3.29	3.25	3.13	3.01	2.89	2.81	2.80
3		3.63	3.59	3.43	3.29	3.15	3.05	3.04
4		3.88	3.83	3.66	3.49	3.34	3.22	3.21
5		4.07	4.02	3.83	3.65	3.48	3.36	3.34
6		4.24	4.17	3.97	3.78	3.60	3.46	3.44
7		4.37	4.31	4.09	3.89	3.69	3.55	3.53
8		4.49	4.43	4.20	3.98	3.78	3.63	3.61
9		4.60	4.53	4.29	4.07	3.85	3.70	3.68
10		4.70	4.62	4.37	4.14	3.92	3.76	3.74
11		4.79	4.71	4.45	4.21	3.98	3.82	3.79
12		4.87	4.79	4.52	4.27	4.04	3.87	3.84
13		4.94	4.86	4.58	4.33	4.09	3.92	3.89
14		5.01	4.93	4.64	4.38	4.13	3.96	3.93
15		5.08	4.99	4.70	4.43	4.18	4.00	3.97
16		5.14	5.05	4.75	4.48	4.22	4.04	4.01
17		5.19	5.10	4.80	4.52	4.26	4.07	4.04
18		5.25	5.16	4.85	4.56	4.30	4.10	4.08
19		5.30	5.21	4.89	4.60	4.33	4.14	4.11
20		5.35	5.25	4.93	4.64	4.36	4.17	4.14
25		5.56	5.46	5.12	4.80	4.50	4.30	4.26
30		5.74	5.63	5.27	4.93	4.62	4.40	4.37
40		6.03	5.91	5.51	5.14	4.80	4.57	4.53
50		6.25	6.12	5.70	5.31	4.95	4.69	4.66
100		6.97	6.82	6.30	5.82	5.39	5.09	5.04
250		7.98	7.78	7.12	6.52	5.98	5.61	5.56
500		8.78	8.54	7.76	7.06	6.44	6.01	5.94

Table A.8. *Continued*

BONFERRONI F CRITICAL VALUES FOR FAMILYWISE ALPHA= .050

COMPARISONS	DF1 = DF2 =	5 3	5 4	5 5	5 6	5 7	5 8	5 9
2		14.89	9.36	7.15	5.99	5.29	4.82	4.48
3		19.81	11.75	8.67	7.11	6.18	5.57	5.14
4		24.20	13.75	9.90	8.00	6.88	6.15	5.65
5		28.24	15.52	10.97	8.75	7.46	6.63	6.06
6		32.01	17.12	11.91	9.40	7.96	7.04	6.41
7		35.58	18.59	12.76	9.99	8.41	7.41	6.72
8		38.98	19.96	13.54	10.52	8.81	7.73	6.99
9		42.25	21.24	14.26	11.01	9.18	8.03	7.24
10		45.39	22.46	14.94	11.46	9.52	8.30	7.47
11		48.43	23.61	15.58	11.89	9.84	8.55	7.68
12		51.38	24.72	16.18	12.29	10.13	8.79	7.88
13		54.25	25.77	16.75	12.67	10.41	9.01	8.06
14		57.05	26.79	17.30	13.03	10.68	9.22	8.24
15		59.78	27.78	17.83	13.37	10.93	9.42	8.40
16		62.45	28.73	18.33	13.70	11.17	9.61	8.56
17		65.07	29.65	18.82	14.01	11.40	9.79	8.71
18		67.63	30.54	19.29	14.31	11.62	9.96	8.85
19		70.15	31.41	19.74	14.60	11.83	10.13	8.99
20		72.62	32.26	20.18	14.88	12.03	10.28	9.12
25		84.42	36.21	22.20	16.16	12.95	11.00	9.70
30		95.46	39.78	23.99	17.28	13.74	11.61	10.20
40		115.85	46.12	27.09	19.19	15.09	12.64	11.03
50		134.59	51.71	29.75	20.80	16.21	13.48	11.71
100		214.20	73.63	39.72	26.65	20.17	16.44	14.06
250		395.49	117.10	57.95	36.76	26.77	21.22	17.77
500		628.44	166.13	76.89	46.74	33.06	25.64	21.11

COMPARISONS	DF1 = DF2 =	5 10	5 11	5 12	5 13	5 14	5 15	5 16
2		4.24	4.04	3.89	3.77	3.66	3.58	3.50
3		4.83	4.58	4.39	4.23	4.10	4.00	3.90
4		5.27	4.99	4.76	4.58	4.43	4.31	4.20
5		5.64	5.32	5.06	4.86	4.69	4.56	4.44
6		5.95	5.59	5.32	5.10	4.92	4.76	4.64
7		6.22	5.84	5.54	5.30	5.11	4.94	4.81
8		6.46	6.05	5.74	5.48	5.27	5.10	4.96
9		6.67	6.24	5.91	5.64	5.42	5.24	5.09
10		6.87	6.42	6.07	5.79	5.56	5.37	5.21
11		7.06	6.59	6.22	5.93	5.69	5.49	5.32
12		7.23	6.74	6.36	6.05	5.80	5.60	5.43
13		7.39	6.88	6.48	6.17	5.91	5.70	5.52
14		7.54	7.01	6.60	6.28	6.01	5.79	5.61
15		7.68	7.13	6.71	6.38	6.11	5.88	5.69
16		7.81	7.25	6.82	6.48	6.20	5.97	5.77
17		7.94	7.36	6.92	6.57	6.28	6.05	5.85
18		8.06	7.47	7.02	6.66	6.36	6.12	5.92
19		8.18	7.57	7.11	6.74	6.44	6.19	5.99
20		8.29	7.67	7.20	6.82	6.51	6.26	6.05
25		8.79	8.11	7.59	7.17	6.84	6.57	6.34
30		9.21	8.48	7.92	7.47	7.12	6.82	6.57
40		9.91	9.08	8.46	7.96	7.56	7.24	6.96
50		10.48	9.58	8.89	8.35	7.92	7.57	7.27
100		12.42	11.24	10.36	9.66	9.11	8.66	8.29
250		15.44	13.79	12.56	11.62	10.87	10.27	9.77
500		18.12	16.02	14.47	13.29	12.36	11.62	11.01

Table A.8. *Continued*

```
              BONFERRONI F CRITICAL VALUES FOR FAMILYWISE ALPHA= .050
```

COMPARISONS	DF1 = 5 DF2 = 17	5 18	5 19	5 20	5 22	5 24	5 26
2	3.44	3.38	3.33	3.29	3.21	3.15	3.10
3	3.83	3.76	3.70	3.64	3.55	3.48	3.42
4	4.11	4.03	3.96	3.90	3.79	3.71	3.64
5	4.34	4.25	4.17	4.10	3.99	3.89	3.82
6	4.52	4.43	4.35	4.27	4.15	4.05	3.96
7	4.69	4.59	4.50	4.42	4.29	4.18	4.09
8	4.83	4.72	4.63	4.55	4.41	4.29	4.20
9	4.96	4.85	4.75	4.66	4.51	4.39	4.30
10	5.07	4.96	4.85	4.76	4.61	4.49	4.38
11	5.18	5.06	4.95	4.86	4.70	4.57	4.46
12	5.28	5.15	5.04	4.94	4.78	4.65	4.54
13	5.37	5.24	5.12	5.02	4.85	4.72	4.61
14	5.45	5.32	5.20	5.10	4.92	4.78	4.67
15	5.53	5.39	5.27	5.17	4.99	4.84	4.73
16	5.61	5.47	5.34	5.23	5.05	4.90	4.78
17	5.68	5.53	5.41	5.29	5.11	4.96	4.83
18	5.75	5.60	5.47	5.35	5.16	5.01	4.88
19	5.81	5.66	5.53	5.41	5.21	5.06	4.93
20	5.87	5.72	5.58	5.46	5.26	5.11	4.98
25	6.14	5.97	5.83	5.70	5.48	5.31	5.17
30	6.37	6.19	6.03	5.89	5.67	5.48	5.34
40	6.73	6.53	6.36	6.21	5.96	5.76	5.60
50	7.02	6.81	6.62	6.46	6.19	5.98	5.80
100	7.98	7.71	7.48	7.27	6.94	6.68	6.46
250	9.36	9.00	8.70	8.44	8.00	7.66	7.39
500	10.50	10.07	9.70	9.39	8.87	8.46	8.13

COMPARISONS	DF1 = 5 DF2 = 28	5 30	5 40	5 60	5 120	5 500	5 1000
2	3.06	3.03	2.90	2.79	2.67	2.59	2.58
3	3.36	3.32	3.17	3.03	2.90	2.80	2.78
4	3.58	3.53	3.36	3.20	3.05	2.94	2.93
5	3.75	3.70	3.51	3.34	3.17	3.05	3.04
6	3.90	3.84	3.64	3.45	3.27	3.14	3.12
7	4.02	3.95	3.74	3.54	3.36	3.22	3.20
8	4.12	4.06	3.83	3.62	3.43	3.29	3.26
9	4.22	4.15	3.91	3.70	3.49	3.34	3.32
10	4.30	4.23	3.99	3.76	3.55	3.40	3.37
11	4.38	4.30	4.05	3.82	3.60	3.44	3.42
12	4.45	4.37	4.11	3.87	3.65	3.49	3.46
13	4.51	4.43	4.17	3.92	3.69	3.53	3.50
14	4.57	4.49	4.22	3.97	3.73	3.56	3.54
15	4.63	4.55	4.27	4.01	3.77	3.59	3.57
16	4.68	4.60	4.31	4.05	3.80	3.63	3.60
17	4.73	4.65	4.35	4.08	3.83	3.66	3.63
18	4.78	4.69	4.40	4.12	3.87	3.68	3.66
19	4.82	4.73	4.43	4.15	3.89	3.71	3.68
20	4.87	4.78	4.47	4.19	3.92	3.73	3.71
25	5.06	4.96	4.63	4.32	4.04	3.84	3.81
30	5.21	5.11	4.76	4.44	4.14	3.93	3.90
40	5.46	5.35	4.97	4.62	4.30	4.07	4.03
50	5.66	5.53	5.13	4.76	4.42	4.18	4.14
100	6.28	6.14	5.64	5.20	4.79	4.51	4.46
250	7.16	6.97	6.35	5.79	5.29	4.94	4.89
500	7.86	7.63	6.90	6.25	5.67	5.27	5.21

Table A.8. *Continued*

BONFERRONI F CRITICAL VALUES FOR FAMILYWISE ALPHA= .050

COMPARISONS	DF1 = 6 DF2 = 3	6 4	6 5	6 6	6 7	6 8	6 9
2	14.73	9.20	6.98	5.82	5.12	4.65	4.32
3	19.59	11.52	8.45	6.89	5.97	5.37	4.94
4	23.93	13.48	9.65	7.75	6.64	5.92	5.41
5	27.91	15.21	10.67	8.47	7.19	6.37	5.80
6	31.63	16.77	11.58	9.10	7.67	6.76	6.13
7	35.16	18.20	12.40	9.66	8.10	7.10	6.42
8	38.52	19.54	13.16	10.17	8.48	7.41	6.68
9	41.74	20.79	13.86	10.64	8.83	7.69	6.92
10	44.84	21.97	14.51	11.07	9.16	7.95	7.13
11	47.84	23.10	15.13	11.48	9.46	8.19	7.33
12	50.75	24.18	15.71	11.86	9.74	8.41	7.52
13	53.58	25.21	16.27	12.23	10.00	8.62	7.69
14	56.34	26.21	16.80	12.57	10.26	8.82	7.86
15	59.03	27.17	17.31	12.90	10.50	9.01	8.01
16	61.67	28.09	17.79	13.21	10.72	9.19	8.16
17	64.25	28.99	18.26	13.52	10.94	9.36	8.30
18	66.78	29.87	18.72	13.81	11.15	9.52	8.43
19	69.27	30.72	19.15	14.08	11.35	9.68	8.56
20	71.71	31.54	19.58	14.35	11.55	9.83	8.68
25	83.36	35.40	21.53	15.58	12.42	10.50	9.23
30	94.25	38.89	23.26	16.65	13.18	11.08	9.70
40	114.36	45.07	26.26	18.48	14.46	12.06	10.49
50	132.84	50.52	28.83	20.03	15.52	12.86	11.13
100	211.43	71.91	38.47	25.63	19.30	15.66	13.34
250	390.09	114.37	56.09	35.33	25.58	20.18	16.82
500	620.25	162.25	74.43	44.91	31.56	24.36	19.98

COMPARISONS	DF1 = 6 DF2 = 10	6 11	6 12	6 13	6 14	6 15	6 16
2	4.07	3.88	3.73	3.60	3.50	3.41	3.34
3	4.63	4.38	4.19	4.04	3.91	3.80	3.71
4	5.04	4.76	4.54	4.36	4.21	4.09	3.98
5	5.39	5.07	4.82	4.62	4.46	4.32	4.20
6	5.68	5.33	5.06	4.84	4.66	4.51	4.38
7	5.93	5.55	5.26	5.03	4.84	4.68	4.54
8	6.15	5.76	5.44	5.20	4.99	4.82	4.68
9	6.36	5.94	5.61	5.35	5.13	4.95	4.80
10	6.54	6.10	5.76	5.48	5.26	5.07	4.91
11	6.72	6.25	5.89	5.61	5.37	5.18	5.02
12	6.88	6.39	6.02	5.72	5.48	5.28	5.11
13	7.03	6.53	6.14	5.83	5.58	5.37	5.20
14	7.17	6.65	6.25	5.93	5.67	5.46	5.28
15	7.30	6.77	6.35	6.03	5.76	5.54	5.36
16	7.42	6.88	6.45	6.12	5.84	5.62	5.43
17	7.54	6.98	6.55	6.20	5.92	5.69	5.50
18	7.66	7.08	6.64	6.28	6.00	5.76	5.56
19	7.77	7.18	6.72	6.36	6.07	5.83	5.62
20	7.87	7.27	6.80	6.44	6.14	5.89	5.68
25	8.34	7.67	7.17	6.76	6.44	6.17	5.95
30	8.73	8.02	7.47	7.04	6.69	6.41	6.17
40	9.39	8.59	7.97	7.49	7.10	6.79	6.52
50	9.93	9.05	8.38	7.86	7.44	7.09	6.80
100	11.75	10.60	9.74	9.07	8.53	8.10	7.74
250	14.57	12.98	11.79	10.88	10.16	9.58	9.10
500	17.08	15.05	13.56	12.42	11.54	10.82	10.24

Table A.8. *Continued*

```
          BONFERRONI F CRITICAL VALUES FOR FAMILYWISE ALPHA= .050
```

COMPARISONS	DF1 = DF2 =	6 17	6 18	6 19	6 20	6 22	6 24	6 26
2		3.28	3.22	3.17	3.13	3.05	2.99	2.94
3		3.63	3.56	3.50	3.45	3.36	3.29	3.23
4		3.89	3.82	3.75	3.69	3.58	3.50	3.43
5		4.10	4.01	3.94	3.87	3.76	3.67	3.59
6		4.28	4.18	4.10	4.03	3.90	3.81	3.72
7		4.42	4.32	4.24	4.16	4.03	3.92	3.84
8		4.56	4.45	4.36	4.27	4.14	4.03	3.94
9		4.67	4.56	4.46	4.38	4.24	4.12	4.02
10		4.78	4.66	4.56	4.47	4.32	4.20	4.10
11		4.88	4.76	4.65	4.56	4.40	4.28	4.17
12		4.97	4.84	4.73	4.64	4.48	4.35	4.24
13		5.05	4.92	4.81	4.71	4.54	4.41	4.30
14		5.13	4.99	4.88	4.78	4.61	4.47	4.36
15		5.20	5.06	4.94	4.84	4.67	4.53	4.41
16		5.27	5.13	5.01	4.90	4.72	4.58	4.46
17		5.33	5.19	5.07	4.96	4.78	4.63	4.51
18		5.39	5.25	5.12	5.01	4.82	4.68	4.55
19		5.45	5.30	5.17	5.06	4.87	4.72	4.60
20		5.51	5.36	5.23	5.11	4.92	4.76	4.64
25		5.76	5.59	5.45	5.33	5.12	4.95	4.81
30		5.96	5.79	5.64	5.50	5.28	5.10	4.96
40		6.30	6.10	5.94	5.79	5.55	5.35	5.20
50		6.56	6.36	6.18	6.02	5.76	5.55	5.38
100		7.43	7.18	6.95	6.76	6.44	6.18	5.98
250		8.70	8.36	8.07	7.82	7.40	7.07	6.81
500		9.75	9.33	8.98	8.68	8.18	7.79	7.47

COMPARISONS	DF1 = DF2 =	6 28	6 30	6 40	6 60	6 120	6 500	6 1000
2		2.90	2.87	2.74	2.63	2.52	2.43	2.42
3		3.18	3.13	2.98	2.84	2.71	2.61	2.60
4		3.37	3.32	3.16	3.00	2.85	2.74	2.72
5		3.53	3.47	3.29	3.12	2.96	2.84	2.82
6		3.65	3.60	3.40	3.22	3.04	2.92	2.90
7		3.76	3.70	3.49	3.30	3.12	2.98	2.96
8		3.86	3.79	3.58	3.37	3.18	3.04	3.02
9		3.94	3.88	3.65	3.43	3.24	3.09	3.07
10		4.02	3.95	3.71	3.49	3.28	3.14	3.11
11		4.09	4.02	3.77	3.54	3.33	3.18	3.15
12		4.15	4.08	3.83	3.59	3.37	3.21	3.19
13		4.21	4.13	3.87	3.63	3.41	3.25	3.22
14		4.27	4.19	3.92	3.67	3.44	3.28	3.25
15		4.32	4.24	3.96	3.71	3.48	3.31	3.28
16		4.36	4.28	4.00	3.75	3.51	3.34	3.31
17		4.41	4.32	4.04	3.78	3.54	3.36	3.34
18		4.45	4.37	4.08	3.81	3.56	3.39	3.36
19		4.49	4.41	4.11	3.84	3.59	3.41	3.38
20		4.53	4.44	4.14	3.87	3.61	3.43	3.40
25		4.70	4.60	4.29	3.99	3.72	3.52	3.49
30		4.84	4.74	4.40	4.09	3.80	3.60	3.57
40		5.06	4.95	4.59	4.25	3.94	3.72	3.69
50		5.24	5.12	4.73	4.37	4.04	3.81	3.78
100		5.81	5.66	5.19	4.76	4.37	4.10	4.06
250		6.59	6.41	5.81	5.28	4.80	4.47	4.42
500		7.22	7.00	6.30	5.68	5.13	4.76	4.70

Table A.8. *Continued*

BONFERRONI F CRITICAL VALUES FOR FAMILYWISE ALPHA= .050

COMPARISONS	DF1 = 7 DF2 = 3	7 4	7 5	7 6	7 7	7 8	7 9
2	14.62	9.07	6.85	5.70	4.99	4.53	4.20
3	19.43	11.36	8.29	6.74	5.82	5.21	4.79
4	23.73	13.28	9.45	7.56	6.46	5.74	5.24
5	27.67	14.98	10.46	8.26	6.99	6.18	5.61
6	31.36	16.51	11.34	8.87	7.46	6.55	5.93
7	34.85	17.92	12.14	9.41	7.87	6.88	6.21
8	38.17	19.23	12.88	9.91	8.24	7.18	6.45
9	41.36	20.46	13.56	10.36	8.57	7.45	6.68
10	44.43	21.62	14.20	10.79	8.89	7.69	6.89
11	47.41	22.73	14.80	11.18	9.18	7.92	7.08
12	50.29	23.79	15.37	11.55	9.45	8.14	7.25
13	53.09	24.80	15.91	11.90	9.70	8.34	7.42
14	55.82	25.78	16.43	12.24	9.95	8.53	7.57
15	58.49	26.72	16.92	12.56	10.18	8.71	7.72
16	61.10	27.63	17.40	12.86	10.40	8.88	7.86
17	63.65	28.51	17.86	13.15	10.61	9.04	8.00
18	66.16	29.37	18.30	13.43	10.81	9.20	8.12
19	68.63	30.21	18.72	13.70	11.00	9.35	8.25
20	71.04	31.02	19.14	13.96	11.19	9.49	8.36
25	82.57	34.80	21.04	15.15	12.03	10.14	8.89
30	93.36	38.23	22.73	16.19	12.76	10.70	9.34
40	113.28	44.30	25.65	17.97	13.99	11.63	10.09
50	131.59	49.66	28.16	19.46	15.02	12.40	10.70
100	209.37	70.66	37.55	24.89	18.66	15.08	12.80
250	386.34	112.35	54.73	34.29	24.72	19.42	16.13
500	614.21	159.38	72.61	43.56	30.48	23.42	19.14

COMPARISONS	DF1 = 7 DF2 = 10	7 11	7 12	7 13	7 14	7 15	7 16
2	3.95	3.76	3.61	3.48	3.38	3.29	3.22
3	4.48	4.24	4.05	3.89	3.76	3.66	3.57
4	4.87	4.60	4.37	4.20	4.05	3.93	3.82
5	5.20	4.89	4.64	4.44	4.28	4.14	4.03
6	5.48	5.13	4.86	4.65	4.47	4.32	4.20
7	5.72	5.35	5.06	4.82	4.63	4.48	4.34
8	5.93	5.54	5.23	4.98	4.78	4.61	4.47
9	6.13	5.71	5.38	5.12	4.91	4.74	4.59
10	6.30	5.86	5.52	5.25	5.03	4.85	4.69
11	6.47	6.01	5.65	5.37	5.14	4.95	4.79
12	6.62	6.14	5.77	5.48	5.24	5.04	4.88
13	6.76	6.27	5.89	5.58	5.33	5.13	4.96
14	6.89	6.38	5.99	5.68	5.42	5.21	5.03
15	7.02	6.49	6.09	5.77	5.51	5.29	5.11
16	7.14	6.60	6.18	5.85	5.58	5.36	5.17
17	7.25	6.70	6.27	5.93	5.66	5.43	5.24
18	7.36	6.79	6.36	6.01	5.73	5.49	5.30
19	7.46	6.88	6.44	6.08	5.79	5.56	5.36
20	7.56	6.97	6.51	6.15	5.86	5.62	5.41
25	8.01	7.36	6.85	6.46	6.14	5.88	5.66
30	8.38	7.68	7.14	6.72	6.38	6.10	5.86
40	9.01	8.22	7.62	7.15	6.77	6.45	6.19
50	9.52	8.66	8.00	7.49	7.08	6.74	6.46
100	11.25	10.13	9.28	8.63	8.11	7.68	7.33
250	13.94	12.38	11.22	10.33	9.63	9.07	8.60
500	16.32	14.34	12.89	11.79	10.93	10.23	9.66

Table A.8. *Continued*

BONFERRONI F CRITICAL VALUES FOR FAMILYWISE ALPHA= .050

COMPARISONS	DF1 = DF2 =	7 17	7 18	7 19	7 20	7 22	7 24	7 26
2		3.16	3.10	3.05	3.01	2.93	2.87	2.82
3		3.49	3.42	3.36	3.31	3.22	3.15	3.08
4		3.73	3.65	3.59	3.53	3.42	3.34	3.27
5		3.93	3.84	3.77	3.70	3.59	3.50	3.42
6		4.09	4.00	3.91	3.84	3.72	3.62	3.54
7		4.23	4.13	4.04	3.97	3.84	3.73	3.65
8		4.35	4.25	4.15	4.07	3.94	3.83	3.74
9		4.46	4.35	4.25	4.17	4.03	3.91	3.82
10		4.56	4.44	4.34	4.26	4.11	3.99	3.89
11		4.65	4.53	4.43	4.34	4.18	4.06	3.96
12		4.73	4.61	4.50	4.41	4.25	4.12	4.02
13		4.81	4.68	4.57	4.48	4.31	4.18	4.08
14		4.88	4.75	4.64	4.54	4.37	4.24	4.13
15		4.95	4.82	4.70	4.60	4.43	4.29	4.18
16		5.02	4.88	4.76	4.66	4.48	4.34	4.22
17		5.08	4.94	4.81	4.71	4.53	4.38	4.27
18		5.13	4.99	4.87	4.76	4.57	4.43	4.31
19		5.19	5.04	4.92	4.80	4.62	4.47	4.35
20		5.24	5.09	4.96	4.85	4.66	4.51	4.39
25		5.47	5.31	5.17	5.05	4.84	4.68	4.55
30		5.66	5.49	5.34	5.21	5.00	4.82	4.68
40		5.97	5.79	5.62	5.48	5.24	5.05	4.90
50		6.22	6.02	5.85	5.69	5.44	5.24	5.07
100		7.04	6.78	6.57	6.38	6.07	5.82	5.62
250		8.21	7.89	7.60	7.36	6.95	6.64	6.38
500		9.19	8.79	8.45	8.16	7.68	7.30	6.99

COMPARISONS	DF1 = DF2 =	7 28	7 30	7 40	7 60	7 120	7 500	7 1000
2		2.78	2.75	2.62	2.51	2.39	2.31	2.30
3		3.03	2.99	2.84	2.70	2.57	2.48	2.46
4		3.22	3.17	3.00	2.84	2.70	2.59	2.57
5		3.36	3.30	3.12	2.95	2.79	2.68	2.66
6		3.48	3.42	3.22	3.04	2.87	2.75	2.73
7		3.58	3.51	3.31	3.12	2.94	2.81	2.79
8		3.66	3.60	3.38	3.18	2.99	2.86	2.84
9		3.74	3.67	3.45	3.24	3.04	2.90	2.88
10		3.81	3.74	3.51	3.29	3.09	2.94	2.92
11		3.87	3.80	3.56	3.34	3.13	2.98	2.95
12		3.93	3.86	3.61	3.38	3.16	3.01	2.99
13		3.99	3.91	3.66	3.42	3.20	3.04	3.02
14		4.04	3.96	3.70	3.46	3.23	3.07	3.04
15		4.08	4.00	3.74	3.49	3.26	3.09	3.07
16		4.13	4.05	3.77	3.52	3.29	3.12	3.09
17		4.17	4.09	3.81	3.55	3.31	3.14	3.12
18		4.21	4.12	3.84	3.58	3.34	3.16	3.14
19		4.25	4.16	3.87	3.61	3.36	3.18	3.16
20		4.28	4.19	3.90	3.63	3.38	3.20	3.18
25		4.44	4.34	4.03	3.74	3.47	3.29	3.26
30		4.56	4.47	4.14	3.83	3.55	3.35	3.32
40		4.77	4.66	4.30	3.98	3.67	3.46	3.43
50		4.93	4.82	4.44	4.09	3.77	3.54	3.51
100		5.45	5.31	4.85	4.44	4.06	3.80	3.76
250		6.17	5.99	5.42	4.91	4.45	4.13	4.08
500		6.75	6.54	5.86	5.27	4.74	4.38	4.32

Table A.8. *Continued*

COMPARISONS	DF1 = 8 DF2 = 3	8 4	8 5	8 6	8 7	8 8	8 9
2	14.54	8.98	6.76	5.60	4.90	4.43	4.10
3	19.31	11.23	8.16	6.61	5.70	5.10	4.67
4	23.57	13.13	9.31	7.42	6.32	5.61	5.11
5	27.49	14.80	10.29	8.10	6.84	6.03	5.47
6	31.15	16.31	11.16	8.70	7.29	6.39	5.77
7	34.61	17.70	11.95	9.23	7.69	6.71	6.04
8	37.91	18.99	12.67	9.71	8.05	7.00	6.28
9	41.08	20.20	13.34	10.15	8.38	7.26	6.50
10	44.13	21.35	13.96	10.57	8.68	7.50	6.69
11	47.07	22.44	14.55	10.95	8.96	7.72	6.88
12	49.94	23.49	15.11	11.31	9.22	7.93	7.05
13	52.72	24.49	15.64	11.66	9.47	8.12	7.21
14	55.43	25.45	16.14	11.98	9.71	8.30	7.36
15	58.08	26.38	16.63	12.29	9.93	8.48	7.50
16	60.67	27.28	17.10	12.59	10.15	8.64	7.64
17	63.20	28.15	17.54	12.87	10.35	8.80	7.76
18	65.69	28.99	17.98	13.15	10.55	8.95	7.89
19	68.13	29.82	18.40	13.41	10.73	9.10	8.01
20	70.53	30.62	18.80	13.67	10.91	9.24	8.12
25	81.98	34.35	20.67	14.83	11.73	9.86	8.63
30	92.68	37.72	22.32	15.84	12.44	10.40	9.06
40	112.45	43.72	25.19	17.57	13.64	11.30	9.78
50	130.61	48.99	27.65	19.03	14.63	12.05	10.37
100	207.81	69.71	36.86	24.33	18.17	14.64	12.40
250	383.61	110.82	53.69	33.49	24.05	18.83	15.61
500	609.46	157.06	71.22	42.53	29.65	22.71	18.50

COMPARISONS	DF1 = 8 DF2 = 10	8 11	8 12	8 13	8 14	8 15	8 16
2	3.86	3.66	3.51	3.39	3.29	3.20	3.12
3	4.36	4.12	3.93	3.78	3.65	3.55	3.45
4	4.74	4.47	4.25	4.07	3.92	3.80	3.70
5	5.06	4.74	4.50	4.30	4.14	4.00	3.89
6	5.32	4.98	4.71	4.50	4.32	4.17	4.05
7	5.55	5.19	4.90	4.67	4.48	4.32	4.19
8	5.76	5.37	5.06	4.82	4.62	4.45	4.31
9	5.95	5.53	5.21	4.95	4.74	4.57	4.42
10	6.12	5.68	5.34	5.08	4.86	4.67	4.52
11	6.27	5.82	5.47	5.19	4.96	4.77	4.61
12	6.42	5.95	5.58	5.29	5.06	4.86	4.69
13	6.56	6.07	5.69	5.39	5.15	4.94	4.77
14	6.68	6.18	5.79	5.48	5.23	5.02	4.85
15	6.80	6.29	5.88	5.57	5.31	5.09	4.91
16	6.92	6.39	5.97	5.65	5.38	5.16	4.98
17	7.03	6.48	6.06	5.72	5.45	5.23	5.04
18	7.13	6.57	6.14	5.80	5.52	5.29	5.10
19	7.23	6.66	6.22	5.87	5.58	5.35	5.15
20	7.33	6.74	6.29	5.93	5.64	5.40	5.20
25	7.75	7.11	6.62	6.23	5.91	5.65	5.44
30	8.12	7.42	6.89	6.47	6.14	5.86	5.63
40	8.72	7.94	7.34	6.88	6.51	6.20	5.94
50	9.20	8.35	7.71	7.21	6.80	6.47	6.19
100	10.87	9.76	8.94	8.29	7.78	7.37	7.02
250	13.45	11.92	10.78	9.92	9.23	8.68	8.22
500	15.73	13.80	12.38	11.30	10.46	9.78	9.22

Table A.8. *Continued*

```
                 BONFERRONI F CRITICAL VALUES FOR FAMILYWISE ALPHA= .050
```

COMPARISONS	DF1 = 8 DF2 = 17	8 18	8 19	8 20	8 22	8 24	8 26
2	3.06	3.00	2.96	2.91	2.84	2.78	2.73
3	3.38	3.31	3.25	3.20	3.11	3.03	2.97
4	3.61	3.53	3.46	3.40	3.30	3.22	3.15
5	3.79	3.71	3.63	3.56	3.45	3.36	3.29
6	3.94	3.85	3.77	3.70	3.58	3.48	3.40
7	4.08	3.98	3.89	3.82	3.69	3.58	3.50
8	4.19	4.09	4.00	3.92	3.78	3.68	3.59
9	4.30	4.19	4.09	4.01	3.87	3.75	3.66
10	4.39	4.28	4.18	4.09	3.94	3.83	3.73
11	4.48	4.36	4.26	4.16	4.01	3.89	3.79
12	4.55	4.43	4.33	4.23	4.08	3.95	3.85
13	4.63	4.50	4.39	4.30	4.14	4.01	3.90
14	4.70	4.57	4.45	4.36	4.19	4.06	3.95
15	4.76	4.63	4.51	4.41	4.24	4.11	4.00
16	4.82	4.69	4.57	4.46	4.29	4.15	4.04
17	4.88	4.74	4.62	4.51	4.34	4.20	4.08
18	4.93	4.79	4.67	4.56	4.38	4.24	4.12
19	4.98	4.84	4.71	4.61	4.42	4.28	4.15
20	5.03	4.89	4.76	4.65	4.46	4.31	4.19
25	5.25	5.09	4.95	4.83	4.63	4.47	4.34
30	5.43	5.26	5.12	4.99	4.78	4.61	4.47
40	5.73	5.54	5.38	5.24	5.01	4.82	4.67
50	5.96	5.76	5.59	5.44	5.19	4.99	4.83
100	6.73	6.48	6.27	6.09	5.78	5.54	5.34
250	7.84	7.52	7.24	7.01	6.61	6.30	6.05
500	8.76	8.38	8.04	7.76	7.29	6.92	6.63

COMPARISONS	DF1 = 8 DF2 = 28	8 30	8 40	8 60	8 120	8 500	8 1000
2	2.69	2.65	2.53	2.41	2.30	2.22	2.20
3	2.92	2.88	2.73	2.59	2.46	2.36	2.35
4	3.09	3.04	2.88	2.72	2.57	2.47	2.45
5	3.23	3.17	2.99	2.82	2.66	2.55	2.53
6	3.34	3.28	3.09	2.91	2.73	2.61	2.59
7	3.43	3.37	3.17	2.97	2.79	2.67	2.65
8	3.51	3.45	3.23	3.03	2.85	2.71	2.69
9	3.58	3.52	3.29	3.09	2.89	2.75	2.73
10	3.65	3.58	3.35	3.13	2.93	2.79	2.77
11	3.71	3.64	3.40	3.18	2.97	2.82	2.80
12	3.76	3.69	3.45	3.22	3.00	2.85	2.83
13	3.81	3.74	3.49	3.25	3.03	2.88	2.85
14	3.86	3.78	3.53	3.29	3.06	2.90	2.88
15	3.90	3.82	3.56	3.32	3.09	2.93	2.90
16	3.94	3.86	3.60	3.35	3.11	2.95	2.92
17	3.98	3.90	3.63	3.37	3.14	2.97	2.94
18	4.02	3.94	3.66	3.40	3.16	2.99	2.96
19	4.05	3.97	3.69	3.42	3.18	3.01	2.98
20	4.09	4.00	3.71	3.45	3.20	3.02	3.00
25	4.23	4.14	3.83	3.55	3.29	3.10	3.07
30	4.35	4.25	3.93	3.63	3.36	3.16	3.13
40	4.54	4.44	4.09	3.76	3.47	3.26	3.23
50	4.69	4.58	4.21	3.86	3.55	3.33	3.30
100	5.18	5.04	4.59	4.18	3.82	3.56	3.52
250	5.85	5.68	5.12	4.62	4.17	3.86	3.81
500	6.38	6.18	5.53	4.95	4.43	4.08	4.03

Table A.8. *Continued*

```
                BONFERRONI F CRITICAL VALUES FOR FAMILYWISE ALPHA= .050
```

COMPARISONS	DF1 = 12 / DF2 = 3	12 / 4	12 / 5	12 / 6	12 / 7	12 / 8	12 / 9
2	14.34	8.75	6.52	5.37	4.67	4.20	3.87
3	19.02	10.93	7.86	6.32	5.41	4.81	4.39
4	23.21	12.76	8.95	7.08	5.99	5.28	4.79
5	27.05	14.37	9.89	7.72	6.47	5.67	5.11
6	30.65	15.83	10.72	8.28	6.89	6.00	5.39
7	34.05	17.17	11.47	8.78	7.26	6.29	5.63
8	37.29	18.42	12.15	9.23	7.59	6.56	5.85
9	40.39	19.60	12.79	9.65	7.90	6.80	6.05
10	43.39	20.70	13.38	10.03	8.18	7.01	6.23
11	46.28	21.76	13.94	10.40	8.44	7.22	6.39
12	49.09	22.77	14.48	10.74	8.68	7.41	6.55
13	51.83	23.73	14.98	11.06	8.92	7.59	6.70
14	54.49	24.66	15.46	11.36	9.13	7.76	6.83
15	57.09	25.56	15.93	11.66	9.34	7.92	6.96
16	59.63	26.43	16.37	11.94	9.54	8.07	7.08
17	62.12	27.27	16.80	12.20	9.73	8.21	7.20
18	64.57	28.09	17.21	12.46	9.91	8.35	7.31
19	66.96	28.88	17.61	12.71	10.08	8.49	7.42
20	69.32	29.66	17.99	12.95	10.25	8.61	7.52
25	80.56	33.26	19.77	14.04	11.21	9.19	7.98
30	91.07	36.52	21.35	14.99	11.67	9.68	8.38
40	110.47	42.31	24.07	16.62	12.78	10.51	9.03
50	128.32	47.41	26.42	17.99	13.71	11.19	9.57
100	204.15	67.42	35.19	22.96	16.99	13.58	11.42
250	376.73	107.10	51.22	31.58	22.45	17.43	14.33
500	598.43	151.89	67.90	40.09	27.64	20.98	16.96

COMPARISONS	DF1 = 12 / DF2 = 10	12 / 11	12 / 12	12 / 13	12 / 14	12 / 15	12 / 16
2	3.62	3.43	3.28	3.15	3.05	2.96	2.89
3	4.08	3.84	3.65	3.50	3.37	3.27	3.18
4	4.42	4.15	3.93	3.75	3.61	3.49	3.39
5	4.71	4.40	4.16	3.96	3.80	3.67	3.55
6	4.95	4.61	4.34	4.13	3.96	3.82	3.69
7	5.15	4.79	4.51	4.28	4.10	3.94	3.81
8	5.34	4.95	4.66	4.42	4.22	4.06	3.92
9	5.51	5.10	4.79	4.53	4.33	4.16	4.01
10	5.66	5.24	4.91	4.64	4.43	4.25	4.10
11	5.80	5.36	5.02	4.74	4.52	4.33	4.18
12	5.93	5.47	5.12	4.83	4.60	4.41	4.25
13	6.06	5.58	5.21	4.92	4.68	4.48	4.32
14	6.17	5.68	5.30	5.00	4.75	4.55	4.38
15	6.28	5.78	5.39	5.07	4.82	4.61	4.44
16	6.39	5.87	5.46	5.15	4.89	4.67	4.49
17	6.48	5.95	5.54	5.21	4.95	4.73	4.55
18	6.58	6.03	5.61	5.28	5.01	4.78	4.60
19	6.67	6.11	5.68	5.34	5.06	4.84	4.64
20	6.75	6.18	5.74	5.40	5.12	4.88	4.69
25	7.14	6.51	6.03	5.66	5.35	5.10	4.89
30	7.47	6.79	6.28	5.87	5.55	5.28	5.06
40	8.00	7.25	6.68	6.23	5.87	5.58	5.33
50	8.45	7.63	7.00	6.52	6.13	5.81	5.55
100	9.94	8.88	8.09	7.48	6.99	6.59	6.26
250	12.27	10.81	9.73	8.90	8.25	7.73	7.30
500	14.33	12.49	11.14	10.12	9.32	8.69	8.16

Table A.8. *Continued*

```
            BONFERRONI F CRITICAL VALUES FOR FAMILYWISE ALPHA= .050

              DF1 =   12      12      12      12      12      12      12
COMPARISONS   DF2 =   17      18      19      20      22      24      26

    2                2.82    2.77    2.72    2.68    2.60    2.54    2.49
    3                3.10    3.03    2.97    2.92    2.83    2.76    2.70
    4                3.30    3.22    3.15    3.09    2.99    2.91    2.84
    5                3.46    3.37    3.30    3.23    3.12    3.03    2.96
    6                3.59    3.50    3.42    3.35    3.23    3.13    3.05
    7                3.70    3.60    3.52    3.44    3.32    3.22    3.13
    8                3.80    3.70    3.61    3.53    3.40    3.29    3.21
    9                3.89    3.78    3.69    3.61    3.47    3.36    3.27
   10                3.97    3.86    3.76    3.68    3.53    3.42    3.32
   11                4.04    3.93    3.83    3.74    3.59    3.47    3.38
   12                4.11    3.99    3.89    3.80    3.65    3.52    3.42
   13                4.18    4.05    3.95    3.85    3.70    3.57    3.47
   14                4.23    4.11    4.00    3.90    3.74    3.62    3.51
   15                4.29    4.16    4.05    3.95    3.79    3.66    3.55
   16                4.34    4.21    4.10    4.00    3.83    3.69    3.58
   17                4.39    4.26    4.14    4.04    3.87    3.73    3.62
   18                4.44    4.30    4.18    4.08    3.90    3.76    3.65
   19                4.48    4.34    4.22    4.12    3.94    3.80    3.68
   20                4.53    4.38    4.26    4.15    3.97    3.83    3.71
   25                4.71    4.56    4.43    4.31    4.12    3.96    3.83
   30                4.87    4.71    4.56    4.44    4.24    4.07    3.94
   40                5.12    4.94    4.79    4.65    4.43    4.25    4.11
   50                5.32    5.13    4.97    4.82    4.58    4.39    4.24
  100                5.98    5.75    5.55    5.37    5.08    4.85    4.66
  250                6.94    6.63    6.37    6.15    5.78    5.49    5.25
  500                7.73    7.37    7.05    6.78    6.34    6.00    5.72

              DF1 =   12      12      12      12      12      12      12
COMPARISONS   DF2 =   28      30      40      60     120     500    1000

    2                2.45    2.41    2.29    2.17    2.05    1.97    1.96
    3                2.64    2.60    2.45    2.31    2.18    2.08    2.07
    4                2.79    2.74    2.57    2.42    2.27    2.16    2.14
    5                2.90    2.84    2.66    2.50    2.34    2.22    2.20
    6                2.99    2.93    2.74    2.56    2.39    2.27    2.25
    7                3.06    3.00    2.80    2.62    2.44    2.31    2.29
    8                3.13    3.07    2.86    2.66    2.48    2.34    2.32
    9                3.19    3.13    2.91    2.70    2.51    2.37    2.35
   10                3.25    3.18    2.95    2.74    2.54    2.40    2.38
   11                3.29    3.23    2.99    2.78    2.57    2.43    2.40
   12                3.34    3.27    3.03    2.81    2.60    2.45    2.43
   13                3.38    3.31    3.06    2.83    2.62    2.47    2.45
   14                3.42    3.35    3.09    2.86    2.64    2.49    2.46
   15                3.46    3.38    3.12    2.89    2.66    2.51    2.48
   16                3.49    3.41    3.15    2.91    2.68    2.52    2.50
   17                3.52    3.44    3.18    2.93    2.70    2.54    2.51
   18                3.55    3.47    3.20    2.95    2.72    2.55    2.53
   19                3.58    3.50    3.22    2.97    2.73    2.56    2.54
   20                3.61    3.53    3.25    2.99    2.75    2.58    2.55
   25                3.73    3.64    3.34    3.07    2.81    2.63    2.61
   30                3.83    3.73    3.42    3.13    2.87    2.68    2.65
   40                3.98    3.88    3.54    3.24    2.95    2.75    2.72
   50                4.11    4.00    3.64    3.32    3.02    2.81    2.77
  100                4.51    4.38    3.95    3.57    3.22    2.97    2.94
  250                5.06    4.90    4.37    3.90    3.48    3.19    3.15
  500                5.50    5.31    4.70    4.16    3.68    3.36    3.31
```

Table A.8. *Continued*

COMPARISONS	DF1 = 25 DF2 = 3	25 4	25 5	25 6	25 7	25 8	25 9
2	14.12	8.50	6.27	5.11	4.40	3.94	3.60
3	18.71	10.59	7.53	6.00	5.08	4.49	4.07
4	22.81	12.36	8.56	6.70	5.62	4.91	4.42
5	26.58	13.91	9.45	7.30	6.06	5.26	4.71
6	30.10	15.31	10.23	7.82	6.44	5.56	4.96
7	33.43	16.61	10.94	8.28	6.78	5.83	5.18
8	36.61	17.81	11.59	8.70	7.08	6.07	5.37
9	39.66	18.94	12.19	9.09	7.36	6.28	5.55
10	42.59	20.00	12.76	9.45	7.62	6.48	5.71
11	45.43	21.02	13.29	9.79	7.86	6.67	5.86
12	48.19	21.99	13.79	10.10	8.09	6.84	6.00
13	50.86	22.92	14.27	10.40	8.30	7.00	6.13
14	53.47	23.81	14.72	10.69	8.50	7.15	6.25
15	56.02	24.68	15.16	10.96	8.69	7.30	6.36
16	58.51	25.51	15.58	11.22	8.87	7.43	6.47
17	60.95	26.32	15.98	11.47	9.05	7.57	6.58
18	63.35	27.11	16.37	11.71	9.21	7.69	6.68
19	65.70	27.87	16.75	11.94	9.37	7.81	6.77
20	68.01	28.61	17.11	12.16	9.53	7.93	6.86
25	79.02	32.09	18.79	13.18	10.22	8.45	7.28
30	89.33	35.22	20.28	14.06	10.83	8.89	7.63
40	108.35	40.79	22.86	15.58	11.84	9.64	8.21
50	125.84	45.70	25.08	16.85	12.69	10.26	8.69
100	200.13	64.95	33.37	21.48	15.70	12.41	10.33
250	369.34	103.16	48.53	29.49	20.71	15.89	12.93
500	586.78	146.23	64.31	37.41	25.46	19.10	15.28

COMPARISONS	DF1 = 25 DF2 = 10	25 11	25 12	25 13	25 14	25 15	25 16
2	3.35	3.16	3.01	2.88	2.78	2.69	2.61
3	3.76	3.52	3.33	3.18	3.05	2.94	2.85
4	4.06	3.79	3.57	3.40	3.25	3.13	3.03
5	4.31	4.00	3.76	3.57	3.41	3.28	3.17
6	4.52	4.19	3.93	3.72	3.55	3.40	3.28
7	4.71	4.35	4.07	3.85	3.66	3.51	3.38
8	4.87	4.49	4.20	3.96	3.77	3.60	3.47
9	5.02	4.62	4.31	4.06	3.86	3.69	3.55
10	5.15	4.74	4.41	4.15	3.94	3.77	3.62
11	5.28	4.84	4.51	4.24	4.02	3.84	3.68
12	5.39	4.94	4.59	4.32	4.09	3.90	3.74
13	5.50	5.04	4.68	4.39	4.16	3.96	3.80
14	5.60	5.12	4.75	4.46	4.22	4.02	3.85
15	5.70	5.21	4.82	4.52	4.28	4.07	3.90
16	5.79	5.28	4.89	4.58	4.33	4.12	3.95
17	5.88	5.36	4.96	4.64	4.38	4.17	3.99
18	5.96	5.43	5.02	4.69	4.43	4.21	4.03
19	6.04	5.49	5.08	4.75	4.48	4.26	4.07
20	6.12	5.56	5.13	4.80	4.52	4.30	4.11
25	6.45	5.85	5.38	5.02	4.72	4.48	4.27
30	6.74	6.09	5.59	5.20	4.89	4.63	4.41
40	7.22	6.49	5.94	5.51	5.16	4.87	4.64
50	7.60	6.81	6.22	5.75	5.38	5.07	4.82
100	8.92	7.91	7.15	6.56	6.10	5.72	5.41
250	10.97	9.58	8.56	7.78	7.16	6.67	6.26
500	12.78	11.05	9.78	8.82	8.07	7.47	6.98

Table A.8. *Continued*

```
            BONFERRONI F CRITICAL VALUES FOR FAMILYWISE ALPHA= .050
```

COMPARISONS	DF1 = 25 DF2 = 17	25 18	25 19	25 20	25 22	25 24	25 26
2	2.55	2.49	2.44	2.40	2.32	2.26	2.21
3	2.77	2.70	2.64	2.59	2.50	2.43	2.36
4	2.94	2.86	2.79	2.73	2.63	2.55	2.48
5	3.07	2.98	2.91	2.84	2.73	2.64	2.57
6	3.18	3.09	3.01	2.94	2.82	2.72	2.64
7	3.27	3.17	3.09	3.02	2.89	2.79	2.71
8	3.35	3.25	3.16	3.08	2.95	2.85	2.76
9	3.43	3.32	3.23	3.15	3.01	2.90	2.81
10	3.49	3.38	3.29	3.20	3.06	2.95	2.85
11	3.55	3.44	3.34	3.25	3.11	2.99	2.89
12	3.61	3.49	3.39	3.30	3.15	3.03	2.93
13	3.66	3.54	3.44	3.34	3.19	3.07	2.96
14	3.71	3.59	3.48	3.38	3.23	3.10	3.00
15	3.75	3.63	3.52	3.42	3.26	3.13	3.02
16	3.80	3.67	3.56	3.46	3.29	3.16	3.05
17	3.84	3.71	3.59	3.49	3.32	3.19	3.08
18	3.87	3.74	3.63	3.52	3.35	3.22	3.10
19	3.91	3.78	3.66	3.55	3.38	3.24	3.13
20	3.95	3.81	3.69	3.58	3.41	3.26	3.15
25	4.10	3.95	3.82	3.71	3.52	3.37	3.25
30	4.23	4.07	3.93	3.82	3.62	3.46	3.33
40	4.44	4.26	4.12	3.99	3.77	3.60	3.46
50	4.60	4.42	4.26	4.12	3.89	3.71	3.56
100	5.14	4.92	4.73	4.56	4.28	4.06	3.88
250	5.93	5.64	5.39	5.18	4.83	4.56	4.34
500	6.57	6.23	5.94	5.69	5.28	4.96	4.70

COMPARISONS	DF1 = 25 DF2 = 28	25 30	25 40	25 60	25 120	25 500	25 1000
2	2.16	2.12	1.99	1.87	1.75	1.65	1.64
3	2.31	2.27	2.12	1.97	1.83	1.72	1.71
4	2.42	2.37	2.20	2.04	1.89	1.77	1.76
5	2.51	2.45	2.27	2.10	1.93	1.81	1.79
6	2.58	2.52	2.33	2.14	1.97	1.84	1.82
7	2.64	2.58	2.37	2.18	2.00	1.86	1.84
8	2.69	2.62	2.41	2.21	2.02	1.89	1.87
9	2.73	2.67	2.45	2.24	2.05	1.91	1.88
10	2.77	2.71	2.48	2.27	2.07	1.92	1.90
11	2.81	2.74	2.51	2.29	2.09	1.94	1.91
12	2.85	2.78	2.54	2.31	2.10	1.95	1.93
13	2.88	2.81	2.56	2.33	2.12	1.96	1.94
14	2.91	2.83	2.59	2.35	2.13	1.97	1.95
15	2.94	2.86	2.61	2.37	2.15	1.99	1.96
16	2.96	2.88	2.63	2.39	2.16	2.00	1.97
17	2.99	2.91	2.65	2.40	2.17	2.00	1.98
18	3.01	2.93	2.66	2.41	2.18	2.01	1.99
19	3.03	2.95	2.68	2.43	2.19	2.02	2.00
20	3.05	2.97	2.70	2.44	2.20	2.03	2.00
25	3.14	3.06	2.77	2.50	2.24	2.06	2.04
30	3.22	3.13	2.82	2.54	2.28	2.09	2.06
40	3.34	3.24	2.91	2.61	2.33	2.13	2.10
50	3.43	3.33	2.98	2.67	2.37	2.17	2.14
100	3.74	3.61	3.21	2.84	2.50	2.27	2.23
250	4.16	4.00	3.51	3.07	2.67	2.40	2.36
500	4.49	4.31	3.75	3.24	2.80	2.50	2.45

Table A.8. *Continued*

BONFERRONI F CRITICAL VALUES FOR FAMILYWISE ALPHA= .050

COMPARISONS	DF1 = 50 DF2 = 3	50 4	50 5	50 6	50 7	50 8	50 9
2	14.01	8.38	6.14	4.98	4.28	3.81	3.47
3	18.56	10.43	7.38	5.84	4.93	4.33	3.91
4	22.62	12.17	8.38	6.52	5.43	4.73	4.24
5	26.35	13.69	9.24	7.09	5.86	5.07	4.52
6	29.84	15.07	10.00	7.59	6.22	5.35	4.75
7	33.14	16.33	10.69	8.04	6.55	5.60	4.95
8	36.29	17.51	11.32	8.45	6.84	5.83	5.14
9	39.31	18.62	11.91	8.82	7.11	6.03	5.30
10	42.21	19.67	12.45	9.17	7.35	6.22	5.45
11	45.02	20.66	12.97	9.49	7.58	6.40	5.59
12	47.75	21.61	13.46	9.80	7.80	6.56	5.72
13	50.40	22.53	13.92	10.09	8.00	6.71	5.85
14	52.99	23.41	14.37	10.36	8.19	6.86	5.96
15	55.52	24.25	14.79	10.63	8.38	6.99	6.07
16	57.98	25.07	15.20	10.88	8.55	7.13	6.17
17	60.40	25.87	15.59	11.12	8.72	7.25	6.27
18	62.77	26.64	15.97	11.35	8.87	7.37	6.37
19	65.10	27.39	16.34	11.57	9.03	7.48	6.46
20	67.38	28.12	16.69	11.78	9.17	7.59	6.54
25	78.30	31.52	18.33	12.76	9.84	8.09	6.93
30	88.50	34.60	19.78	13.62	10.42	8.51	7.26
40	107.34	40.07	22.29	15.08	11.39	9.22	7.81
50	124.67	44.88	24.44	16.31	12.20	9.80	8.26
100	198.26	63.77	32.51	20.77	15.08	11.85	9.81
250	365.74	101.25	47.25	28.50	19.87	15.15	12.26
500	581.36	143.48	62.61	36.12	24.42	18.20	14.46

COMPARISONS	DF1 = 50 DF2 = 10	50 11	50 12	50 13	50 14	50 15	50 16
2	3.22	3.03	2.87	2.74	2.64	2.55	2.47
3	3.60	3.36	3.17	3.01	2.89	2.78	2.69
4	3.88	3.61	3.39	3.21	3.07	2.95	2.84
5	4.12	3.81	3.57	3.38	3.22	3.08	2.97
6	4.31	3.98	3.72	3.51	3.34	3.19	3.07
7	4.48	4.13	3.85	3.63	3.44	3.29	3.16
8	4.64	4.26	3.97	3.73	3.54	3.38	3.24
9	4.78	4.38	4.07	3.82	3.62	3.45	3.31
10	4.90	4.49	4.17	3.91	3.70	3.52	3.37
11	5.02	4.59	4.25	3.98	3.77	3.59	3.43
12	5.13	4.68	4.33	4.06	3.83	3.64	3.49
13	5.23	4.77	4.41	4.12	3.89	3.70	3.54
14	5.32	4.85	4.48	4.19	3.95	3.75	3.58
15	5.41	4.92	4.54	4.24	4.00	3.80	3.63
16	5.50	5.00	4.61	4.30	4.05	3.84	3.67
17	5.58	5.06	4.67	4.35	4.10	3.89	3.71
18	5.66	5.13	4.72	4.40	4.14	3.93	3.75
19	5.73	5.19	4.78	4.45	4.18	3.97	3.78
20	5.80	5.25	4.83	4.50	4.22	4.00	3.81
25	6.12	5.52	5.06	4.70	4.41	4.17	3.96
30	6.39	5.74	5.25	4.87	4.56	4.30	4.09
40	6.83	6.12	5.57	5.15	4.80	4.52	4.29
50	7.19	6.42	5.83	5.37	5.00	4.70	4.45
100	8.42	7.43	6.69	6.11	5.66	5.29	4.98
250	10.34	8.98	7.99	7.23	6.63	6.15	5.75
500	12.03	10.34	9.11	8.17	7.45	6.87	6.39

Table A.8. *Continued*

BONFERRONI F CRITICAL VALUES FOR FAMILYWISE ALPHA= .050

COMPARISONS	DF1 = 50 DF2 = 17	50 18	50 19	50 20	50 22	50 24	50 26
2	2.41	2.35	2.30	2.25	2.17	2.11	2.05
3	2.61	2.54	2.48	2.42	2.33	2.25	2.19
4	2.75	2.67	2.61	2.54	2.44	2.36	2.29
5	2.87	2.78	2.71	2.64	2.53	2.44	2.36
6	2.97	2.88	2.80	2.72	2.60	2.51	2.43
7	3.05	2.95	2.87	2.79	2.67	2.57	2.48
8	3.12	3.02	2.93	2.86	2.72	2.62	2.53
9	3.19	3.08	2.99	2.91	2.77	2.66	2.57
10	3.25	3.14	3.04	2.96	2.82	2.70	2.61
11	3.30	3.19	3.09	3.00	2.86	2.74	2.64
12	3.35	3.24	3.13	3.04	2.89	2.77	2.67
13	3.40	3.28	3.17	3.08	2.93	2.80	2.70
14	3.44	3.32	3.21	3.12	2.96	2.83	2.73
15	3.48	3.36	3.25	3.15	2.99	2.86	2.75
16	3.52	3.39	3.28	3.18	3.02	2.89	2.78
17	3.56	3.43	3.31	3.21	3.05	2.91	2.80
18	3.59	3.46	3.34	3.24	3.07	2.93	2.82
19	3.62	3.49	3.37	3.27	3.09	2.95	2.84
20	3.65	3.52	3.40	3.29	3.12	2.98	2.86
25	3.79	3.64	3.52	3.40	3.22	3.07	2.94
30	3.91	3.75	3.61	3.50	3.30	3.14	3.01
40	4.09	3.92	3.77	3.65	3.43	3.26	3.12
50	4.24	4.06	3.90	3.77	3.54	3.36	3.21
100	4.72	4.50	4.31	4.15	3.88	3.66	3.49
250	5.42	5.14	4.90	4.70	4.36	4.09	3.87
500	6.00	5.67	5.38	5.14	4.74	4.43	4.18

COMPARISONS	DF1 = 50 DF2 = 28	50 30	50 40	50 60	50 120	50 500	50 1000
2	2.01	1.97	1.83	1.70	1.56	1.46	1.44
3	2.14	2.09	1.93	1.78	1.63	1.51	1.49
4	2.23	2.18	2.00	1.83	1.67	1.54	1.52
5	2.30	2.24	2.06	1.88	1.70	1.57	1.54
6	2.36	2.30	2.10	1.91	1.73	1.59	1.56
7	2.41	2.35	2.14	1.94	1.75	1.60	1.58
8	2.45	2.39	2.17	1.97	1.77	1.62	1.59
9	2.49	2.43	2.20	1.99	1.78	1.63	1.60
10	2.53	2.46	2.23	2.01	1.80	1.64	1.61
11	2.56	2.49	2.25	2.03	1.81	1.65	1.62
12	2.59	2.52	2.27	2.04	1.82	1.66	1.63
13	2.62	2.54	2.29	2.06	1.83	1.67	1.64
14	2.64	2.57	2.31	2.07	1.84	1.67	1.65
15	2.66	2.59	2.33	2.09	1.85	1.68	1.65
16	2.69	2.61	2.35	2.10	1.86	1.69	1.66
17	2.71	2.63	2.36	2.11	1.87	1.69	1.67
18	2.73	2.65	2.38	2.12	1.88	1.70	1.67
19	2.74	2.66	2.39	2.13	1.89	1.71	1.68
20	2.76	2.68	2.40	2.14	1.89	1.71	1.68
25	2.84	2.75	2.46	2.18	1.92	1.73	1.70
30	2.90	2.81	2.51	2.22	1.95	1.75	1.72
40	3.00	2.91	2.58	2.27	1.99	1.78	1.75
50	3.09	2.98	2.64	2.32	2.02	1.80	1.77
100	3.34	3.22	2.82	2.45	2.11	1.87	1.83
250	3.70	3.55	3.06	2.62	2.23	1.95	1.91
500	3.98	3.81	3.25	2.76	2.32	2.01	1.97

Table A.8. *Continued*

```
              BONFERRONI F CRITICAL VALUES FOR FAMILYWISE ALPHA= .010

              DF1 =    1        1        1        1        1        1        1
COMPARISONS|  DF2 =    3        4        5        6        7        8        9

      2              55.58    31.32    22.77    18.63    16.24    14.68    13.62
      3              73.58    39.13    27.54    22.07    18.97    16.98    15.64
      4              89.60    45.67    31.41    24.82    21.11    18.78    17.20
      5             104.28    51.47    34.73    27.11    22.91    20.26    18.45
      6             118.22    56.64    37.69    29.13    24.47    21.53    19.55
      7             131.08    61.50    40.31    30.96    25.84    22.66    20.52
      8             143.54    65.97    42.82    32.59    27.08    23.65    21.38
      9             155.60    70.14    45.04    34.08    28.18    24.55    22.15
     10             167.06    74.15    47.16    35.52    29.26    25.42    22.86
     11             178.14    77.85    49.14    36.80    30.24    26.18    23.52
     12             188.78    81.46    51.11    38.05    31.11    26.92    24.12
     13             199.52    84.95    52.84    39.25    32.02    27.63    24.69
     14             209.46    88.29    54.59    40.33    32.84    28.26    25.26
     15             219.45    91.50    56.26    41.42    33.61    28.90    25.74
     16             229.48    94.65    57.77    42.41    34.32    29.47    26.25
     17             239.06    97.63    59.37    43.34    35.01    30.01    26.73
     18             248.17   100.59    60.81    44.29    35.68    30.56    27.17
     19             257.31   103.39    62.27    45.24    36.33    31.06    27.57
     20             266.48   106.20    63.63    46.11    36.99    31.58    27.97
     25             309.66   119.17    69.94    50.00    39.81    33.75    29.82
     30             349.73   130.90    75.53    53.47    42.27    35.71    31.40
     40             424.55   151.51    85.34    59.42    46.45    38.86    33.99
     50             492.84   169.80    93.69    64.39    49.90    41.49    36.13
    100             783.54   241.46   124.92    82.50    62.17    50.71    43.49
    250            1445.11   383.64   182.16   113.79    82.61    65.57    55.09
    500            2293.01   543.44   241.53   144.72   102.10    79.35    65.62

              DF1 =    1        1        1        1        1        1        1
COMPARISONS|  DF2 =   10       11       12       13       14       15       16

      2              12.82    12.23    11.75    11.37    11.06    10.80    10.57
      3              14.64    13.90    13.32    12.84    12.46    12.14    11.87
      4              16.03    15.17    14.50    13.94    13.51    13.13    12.83
      5              17.17    16.20    15.43    14.83    14.34    13.93    13.59
      6              18.13    17.08    16.24    15.59    15.05    14.61    14.23
      7              18.99    17.83    16.96    16.23    15.68    15.18    14.79
      8              19.72    18.50    17.58    16.83    16.20    15.70    15.27
      9              20.43    19.12    18.13    17.34    16.69    16.18    15.72
     10              21.04    19.69    18.66    17.80    17.15    16.59    16.12
     11              21.63    20.21    19.10    18.24    17.55    16.97    16.49
     12              22.14    20.67    19.56    18.66    17.93    17.33    16.82
     13              22.67    21.12    19.97    19.04    18.29    17.65    17.13
     14              23.14    21.55    20.33    19.37    18.62    17.96    17.44
     15              23.55    21.96    20.70    19.71    18.93    18.25    17.70
     16              23.97    22.32    21.04    20.02    19.22    18.54    17.96
     17              24.41    22.69    21.35    20.34    19.48    18.80    18.23
     18              24.76    23.01    21.68    20.62    19.74    19.05    18.45
     19              25.12    23.33    21.98    20.88    20.02    19.29    18.68
     20              25.49    23.65    22.23    21.14    20.25    19.49    18.88
     25              27.08    25.03    23.50    22.30    21.31    20.51    19.85
     30              28.39    26.23    24.54    23.23    22.19    21.36    20.62
     40              30.61    28.17    26.31    24.82    23.65    22.71    21.92
     50              32.42    29.75    27.71    26.12    24.83    23.81    22.95
    100              38.58    35.06    32.44    30.39    28.77    27.45    26.35
    250              48.14    43.22    39.59    36.77    34.58    32.80    31.34
    500              56.67    50.40    45.73    42.25    39.53    37.28    35.51
```

Table A.8. *Continued*

```
          BONFERRONI F CRITICAL VALUES FOR FAMILYWISE ALPHA= .010
```

COMPARISONS	DF1 = DF2 =	1 17	1 18	1 19	1 20	1 22	1 24	1 26
2		10.38	10.22	10.07	9.95	9.73	9.55	9.41
3		11.63	11.42	11.24	11.09	10.83	10.61	10.44
4		12.54	12.32	12.11	11.95	11.64	11.40	11.20
5		13.29	13.04	12.82	12.62	12.29	12.01	11.80
6		13.91	13.62	13.40	13.18	12.83	12.53	12.29
7		14.45	14.14	13.90	13.66	13.28	12.98	12.73
8		14.91	14.60	14.33	14.09	13.69	13.36	13.10
9		15.34	15.00	14.73	14.47	14.05	13.71	13.44
10		15.71	15.38	15.08	14.83	14.37	14.02	13.74
11		16.06	15.71	15.40	15.13	14.69	14.32	14.01
12		16.40	16.03	15.71	15.43	14.96	14.57	14.26
13		16.70	16.31	15.99	15.70	15.22	14.82	14.51
14		16.98	16.59	16.26	15.94	15.46	15.05	14.71
15		17.25	16.83	16.49	16.19	15.68	15.27	14.92
16		17.49	17.08	16.72	16.41	15.88	15.46	15.12
17		17.73	17.31	16.94	16.62	16.07	15.66	15.31
18		17.96	17.52	17.13	16.82	16.28	15.84	15.47
19		18.18	17.71	17.34	17.02	16.44	15.99	15.64
20		18.37	17.93	17.53	17.20	16.63	16.16	15.80
25		19.28	18.79	18.36	18.00	17.36	16.90	16.49
30		20.01	19.52	19.06	18.66	18.01	17.48	17.06
40		21.24	20.67	20.18	19.73	19.00	18.43	17.97
50		22.22	21.60	21.07	20.62	19.82	19.22	18.69
100		25.44	24.64	24.00	23.42	22.43	21.65	21.03
250		30.11	29.09	28.19	27.42	26.14	25.15	24.34
500		34.03	32.75	31.66	30.70	29.19	27.98	26.98

COMPARISONS	DF1 = DF2 =	1 28	1 30	1 40	1 60	1 120	1 500	1 1000
2		9.29	9.18	8.83	8.50	8.18	7.95	7.91
3		10.29	10.16	9.75	9.35	8.97	8.70	8.66
4		11.04	10.89	10.41	9.96	9.54	9.23	9.19
5		11.62	11.46	10.94	10.44	9.99	9.65	9.60
6		12.11	11.93	11.37	10.84	10.35	9.99	9.94
7		12.51	12.33	11.74	11.18	10.66	10.28	10.22
8		12.88	12.69	12.06	11.47	10.93	10.54	10.47
9		13.21	13.02	12.35	11.73	11.16	10.76	10.70
10		13.50	13.28	12.62	11.96	11.38	10.96	10.89
11		13.76	13.56	12.85	12.18	11.58	11.15	11.07
12		14.00	13.80	13.07	12.39	11.76	11.30	11.23
13		14.25	14.02	13.27	12.57	11.91	11.45	11.39
14		14.45	14.22	13.44	12.72	12.08	11.59	11.53
15		14.65	14.41	13.62	12.88	12.22	11.72	11.65
16		14.84	14.60	13.79	13.04	12.35	11.84	11.78
17		15.02	14.75	13.95	13.17	12.46	11.97	11.89
18		15.17	14.94	14.09	13.31	12.59	12.08	11.99
19		15.34	15.09	14.22	13.43	12.69	12.19	12.10
20		15.48	15.21	14.36	13.55	12.81	12.28	12.19
25		16.15	15.87	14.93	14.06	13.26	12.70	12.63
30		16.69	16.41	15.40	14.49	13.64	13.05	12.96
40		17.58	17.25	16.17	15.18	14.26	13.60	13.52
50		18.27	17.91	16.76	15.71	14.73	14.04	13.94
100		20.53	20.08	18.66	17.39	16.22	15.39	15.25
250		23.68	23.12	21.30	19.66	18.19	17.18	17.02
500		26.19	25.55	23.37	21.44	19.73	18.56	18.36

Table A.8. *Continued*

```
            BONFERRONI F CRITICAL VALUES FOR FAMILYWISE ALPHA= .010
```

	DF1 =	2	2	2	2	2	2	2
COMPARISONS	DF2 =	3	4	5	6	7	8	9
2		49.80	26.28	18.31	14.54	12.41	11.04	10.11
3		65.72	32.64	21.98	17.08	14.36	12.65	11.48
4		79.94	38.00	24.96	19.10	15.89	13.89	12.54
5		93.00	42.73	27.53	20.81	17.16	14.92	13.41
6		105.20	46.99	29.80	22.30	18.27	15.80	14.14
7		116.74	50.92	31.86	23.63	19.25	16.57	14.80
8		127.78	54.57	33.74	24.85	20.13	17.27	15.38
9		138.31	58.00	35.49	25.97	20.94	17.91	15.90
10		148.51	61.25	37.12	27.00	21.69	18.50	16.39
11		158.33	64.34	38.66	27.97	22.38	19.04	16.84
12		167.89	67.27	40.12	28.88	23.04	19.54	17.25
13		177.19	70.12	41.51	29.74	23.65	20.02	17.64
14		186.19	72.84	42.83	30.56	24.23	20.47	18.01
15		195.05	75.46	44.10	31.34	24.78	20.89	18.35
16		203.71	77.99	45.32	32.09	25.31	21.30	18.69
17		212.19	80.47	46.49	32.80	25.82	21.69	19.00
18		220.44	82.86	47.63	33.49	26.29	22.06	19.30
19		228.58	85.19	48.73	34.15	26.76	22.41	19.59
20		236.58	87.45	49.77	34.79	27.21	22.75	19.87
25		274.80	98.02	54.66	37.71	29.22	24.28	21.10
30		310.53	107.54	58.99	40.27	30.98	25.60	22.16
40		376.51	124.51	66.49	44.63	33.93	27.82	23.92
50		437.12	139.44	72.92	48.29	36.39	29.63	25.37
100		694.45	197.99	97.02	61.64	45.14	35.99	30.34
250		1279.74	314.08	140.97	84.68	59.69	46.29	38.23
500		2028.89	444.36	186.76	107.41	73.49	55.78	45.29

	DF1 =	2	2	2	2	2	2	2
COMPARISONS	DF2 =	10	11	12	13	14	15	16
2		9.43	8.91	8.51	8.19	7.92	7.70	7.51
3		10.65	10.02	9.52	9.13	8.81	8.55	8.32
4		11.57	10.85	10.29	9.84	9.48	9.17	8.92
5		12.33	11.53	10.90	10.41	10.01	9.68	9.40
6		12.97	12.10	11.43	10.89	10.46	10.10	9.80
7		13.54	12.60	11.88	11.31	10.85	10.46	10.14
8		14.04	13.04	12.28	11.68	11.19	10.79	10.45
9		14.49	13.45	12.64	12.01	11.50	11.08	10.72
10		14.91	13.81	12.98	12.31	11.78	11.34	10.97
11		15.29	14.15	13.28	12.59	12.04	11.58	11.20
12		15.64	14.46	13.56	12.85	12.28	11.80	11.41
13		15.98	14.76	13.82	13.09	12.49	12.01	11.61
14		16.29	15.03	14.07	13.31	12.70	12.21	11.79
15		16.59	15.29	14.30	13.52	12.90	12.38	11.96
16		16.87	15.53	14.52	13.72	13.09	12.56	12.12
17		17.13	15.77	14.73	13.91	13.26	12.72	12.27
18		17.39	15.99	14.93	14.09	13.43	12.88	12.42
19		17.63	16.20	15.12	14.27	13.58	13.02	12.56
20		17.86	16.40	15.30	14.43	13.73	13.16	12.69
25		18.91	17.31	16.10	15.16	14.40	13.79	13.27
30		19.80	18.08	16.79	15.78	14.97	14.31	13.76
40		21.26	19.35	17.90	16.78	15.89	15.17	14.56
50		22.47	20.38	18.81	17.60	16.63	15.85	15.20
100		26.54	23.85	21.85	20.31	19.09	18.11	17.30
250		32.89	29.16	26.45	24.36	22.75	21.44	20.36
500		38.53	33.80	30.42	27.84	25.83	24.22	22.93

Table A.8. *Continued*

BONFERRONI F CRITICAL VALUES FOR FAMILYWISE ALPHA= .010

COMPARISONS	DF1 = DF2 =	2 17	2 18	2 19	2 20	2 22	2 24	2 26
2		7.35	7.21	7.09	6.99	6.81	6.66	6.54
3		8.13	7.96	7.82	7.69	7.48	7.30	7.16
4		8.70	8.51	8.35	8.21	7.97	7.77	7.61
5		9.16	8.95	8.77	8.62	8.35	8.14	7.97
6		9.54	9.32	9.13	8.96	8.68	8.45	8.26
7		9.87	9.64	9.43	9.25	8.95	8.71	8.52
8		10.16	9.92	9.70	9.51	9.20	8.95	8.74
9		10.42	10.16	9.94	9.74	9.42	9.15	8.94
10		10.66	10.39	10.16	9.95	9.61	9.34	9.12
11		10.87	10.60	10.36	10.14	9.79	9.51	9.28
12		11.07	10.79	10.54	10.32	9.96	9.67	9.43
13		11.26	10.96	10.71	10.48	10.11	9.81	9.57
14		11.43	11.13	10.87	10.64	10.25	9.95	9.69
15		11.59	11.28	11.01	10.78	10.39	10.07	9.82
16		11.75	11.43	11.16	10.91	10.51	10.19	9.93
17		11.89	11.57	11.29	11.04	10.63	10.30	10.04
18		12.03	11.70	11.41	11.16	10.74	10.41	10.14
19		12.16	11.82	11.53	11.27	10.85	10.51	10.24
20		12.29	11.94	11.65	11.38	10.95	10.61	10.33
25		12.84	12.47	12.15	11.87	11.40	11.03	10.73
30		13.30	12.91	12.57	12.27	11.78	11.38	11.07
40		14.05	13.62	13.24	12.92	12.38	11.95	11.61
50		14.65	14.19	13.79	13.44	12.86	12.40	12.03
100		16.62	16.05	15.55	15.12	14.41	13.85	13.40
250		19.48	18.73	18.08	17.54	16.62	15.90	15.34
500		21.84	20.94	20.17	19.51	18.40	17.55	16.87

COMPARISONS	DF1 = DF2 =	2 28	2 30	2 40	2 60	2 120	2 500	2 1000
2		6.44	6.35	6.07	5.80	5.54	5.36	5.33
3		7.04	6.94	6.60	6.28	5.98	5.77	5.74
4		7.48	7.36	6.99	6.63	6.30	6.06	6.03
5		7.82	7.70	7.29	6.90	6.55	6.29	6.25
6		8.11	7.98	7.54	7.13	6.75	6.48	6.44
7		8.35	8.21	7.75	7.32	6.92	6.64	6.59
8		8.57	8.42	7.94	7.49	7.07	6.77	6.73
9		8.76	8.61	8.10	7.64	7.20	6.90	6.85
10		8.93	8.77	8.25	7.77	7.32	7.00	6.96
11		9.09	8.93	8.39	7.89	7.43	7.10	7.05
12		9.23	9.06	8.51	8.00	7.53	7.19	7.14
13		9.36	9.19	8.62	8.10	7.62	7.27	7.22
14		9.49	9.31	8.73	8.19	7.70	7.35	7.30
15		9.60	9.42	8.83	8.28	7.78	7.42	7.37
16		9.71	9.53	8.92	8.36	7.85	7.49	7.43
17		9.82	9.63	9.01	8.44	7.92	7.55	7.50
18		9.91	9.72	9.09	8.51	7.98	7.61	7.55
19		10.01	9.81	9.17	8.58	8.05	7.66	7.61
20		10.09	9.90	9.25	8.65	8.10	7.72	7.66
25		10.48	10.27	9.57	8.94	8.36	7.95	7.88
30		10.80	10.58	9.84	9.18	8.57	8.14	8.07
40		11.32	11.07	10.28	9.55	8.90	8.43	8.36
50		11.73	11.47	10.62	9.85	9.15	8.66	8.59
100		13.03	12.72	11.70	10.78	9.96	9.38	9.29
250		14.86	14.46	13.18	12.05	11.03	10.33	10.23
500		16.32	15.87	14.36	13.03	11.86	11.07	10.92

Table A.8. *Continued*

COMPARISONS	DF1 = DF2 =	3 3	3 4	3 5	3 6	3 7	3 8	3 9
2		47,47	24,26	16,53	12,92	10,88	9,60	8,72
3		62,57	30,06	19,78	15,12	12,54	10,94	9,86
4		76,06	34,96	22,43	16,87	13,84	11,98	10,73
5		88,45	39,27	24,70	18,35	14,93	12,84	11,44
6		100,04	43,16	26,71	19,64	15,87	13,57	12,05
7		111,00	46,75	28,53	20,79	16,70	14,23	12,59
8		121,44	50,08	30,20	21,84	17,45	14,81	13,07
9		131,45	53,21	31,75	22,81	18,14	15,34	13,50
10		141,11	56,18	33,20	23,70	18,77	15,83	13,90
11		150,44	59,00	34,57	24,54	19,36	16,28	14,27
12		159,50	61,69	35,86	25,33	19,92	16,71	14,61
13		168,32	64,27	37,09	26,08	20,44	17,10	14,94
14		176,90	66,76	38,26	26,79	20,93	17,48	15,24
15		185,29	69,16	39,39	27,46	21,40	17,83	15,53
16		193,50	71,47	40,47	28,11	21,85	18,17	15,80
17		201,49	73,72	41,50	28,73	22,27	18,50	16,06
18		209,39	75,90	42,51	29,32	22,68	18,81	16,31
19		217,12	78,03	43,48	29,90	23,08	19,10	16,54
20		224,72	80,10	44,42	30,45	23,46	19,39	16,77
25		260,93	89,72	48,75	32,98	25,18	20,67	17,79
30		294,86	98,44	52,58	35,19	26,66	21,77	18,66
40		357,37	113,90	59,23	38,96	29,17	23,61	20,11
50		415,01	127,55	64,94	42,14	31,26	25,14	21,30
100		659,58	181,01	86,28	53,69	38,68	30,46	25,40
250		1217,21	286,85	125,38	73,66	51,04	39,06	31,89
500		1934,12	406,03	165,70	93,41	62,82	46,97	37,75

COMPARISONS	DF1 = DF2 =	3 10	3 11	3 12	3 13	3 14	3 15	3 16
2		8,08	7,60	7,23	6,93	6,68	6,48	6,30
3		9,08	8,49	8,04	7,68	7,39	7,14	6,94
4		9,83	9,17	8,65	8,24	7,91	7,63	7,40
5		10,45	9,71	9,15	8,70	8,33	8,03	7,78
6		10,98	10,18	9,56	9,08	8,69	8,36	8,09
7		11,44	10,58	9,93	9,41	8,99	8,65	8,36
8		11,84	10,94	10,25	9,70	9,26	8,90	8,60
9		12,21	11,26	10,54	9,97	9,51	9,13	8,81
10		12,55	11,56	10,80	10,21	9,73	9,34	9,01
11		12,86	11,83	11,05	10,43	9,93	9,52	9,18
12		13,16	12,09	11,27	10,63	10,12	9,70	9,35
13		13,43	12,32	11,48	10,82	10,29	9,86	9,50
14		13,68	12,54	11,68	11,00	10,46	10,01	9,64
15		13,92	12,75	11,87	11,17	10,61	10,16	9,77
16		14,15	12,95	12,04	11,33	10,76	10,29	9,90
17		14,37	13,14	12,21	11,48	10,90	10,42	10,02
18		14,58	13,32	12,37	11,62	11,03	10,54	10,13
19		14,78	13,49	12,52	11,76	11,15	10,65	10,24
20		14,97	13,65	12,66	11,89	11,27	10,76	10,34
25		15,81	14,39	13,31	12,47	11,80	11,25	10,80
30		16,54	15,01	13,85	12,96	12,24	11,66	11,18
40		17,73	16,03	14,75	13,76	12,97	12,33	11,80
50		18,71	16,86	15,47	14,40	13,55	12,87	12,30
100		22,04	19,66	17,90	16,55	15,49	14,64	13,93
250		27,20	23,95	21,56	19,78	18,35	17,24	16,31
500		31,81	27,68	24,74	22,53	20,80	19,43	18,30

Table A.8. *Continued*

```
            BONFERRONI F CRITICAL VALUES FOR FAMILYWISE ALPHA= .010
```

COMPARISONS	DF1 = 3 DF2 = 17	3 18	3 19	3 20	3 22	3 24	3 26
2	6.16	6.03	5.92	5.82	5.65	5.52	5.41
3	6.76	6.61	6.48	6.36	6.16	6.01	5.88
4	7.21	7.04	6.89	6.76	6.54	6.36	6.22
5	7.56	7.38	7.21	7.07	6.84	6.65	6.49
6	7.86	7.66	7.49	7.34	7.08	6.88	6.71
7	8.12	7.90	7.72	7.56	7.29	7.08	6.91
8	8.34	8.12	7.93	7.76	7.48	7.26	7.07
9	8.54	8.31	8.11	7.94	7.65	7.41	7.22
10	8.73	8.49	8.28	8.10	7.80	7.55	7.36
11	8.89	8.65	8.43	8.24	7.93	7.68	7.48
12	9.05	8.79	8.57	8.38	8.06	7.80	7.59
13	9.19	8.93	8.70	8.51	8.18	7.91	7.70
14	9.33	9.06	8.82	8.62	8.28	8.01	7.79
15	9.45	9.18	8.94	8.73	8.39	8.11	7.89
16	9.57	9.29	9.05	8.83	8.48	8.20	7.97
17	9.68	9.40	9.15	8.93	8.57	8.28	8.05
18	9.79	9.50	9.24	9.02	8.66	8.37	8.13
19	9.89	9.59	9.34	9.11	8.74	8.44	8.20
20	9.99	9.69	9.42	9.20	8.82	8.51	8.27
25	10.42	10.09	9.81	9.56	9.16	8.84	8.57
30	10.78	10.43	10.13	9.87	9.44	9.10	8.83
40	11.36	10.98	10.65	10.37	9.90	9.53	9.23
50	11.82	11.42	11.07	10.77	10.27	9.87	9.55
100	13.35	12.85	12.42	12.05	11.44	10.96	10.58
250	15.55	14.91	14.36	13.89	13.12	12.51	12.02
500	17.40	16.61	15.97	15.39	14.49	13.76	13.18

COMPARISONS	DF1 = 3 DF2 = 28	3 30	3 40	3 60	3 120	3 500	3 1000
2	5.32	5.24	4.98	4.73	4.50	4.33	4.31
3	5.77	5.68	5.37	5.09	4.82	4.63	4.60
4	6.10	6.00	5.66	5.34	5.05	4.84	4.81
5	6.36	6.25	5.88	5.54	5.23	5.00	4.97
6	6.58	6.46	6.07	5.71	5.37	5.13	5.10
7	6.76	6.64	6.23	5.85	5.49	5.25	5.21
8	6.92	6.79	6.36	5.97	5.60	5.34	5.30
9	7.06	6.93	6.48	6.07	5.70	5.43	5.39
10	7.19	7.05	6.59	6.17	5.78	5.51	5.46
11	7.31	7.17	6.69	6.26	5.86	5.58	5.53
12	7.42	7.27	6.79	6.34	5.93	5.64	5.59
13	7.52	7.37	6.87	6.41	5.99	5.70	5.65
14	7.61	7.46	6.95	6.48	6.05	5.75	5.70
15	7.70	7.54	7.02	6.54	6.11	5.80	5.75
16	7.78	7.62	7.09	6.60	6.16	5.85	5.80
17	7.86	7.69	7.15	6.66	6.21	5.89	5.84
18	7.93	7.76	7.22	6.71	6.26	5.93	5.88
19	8.00	7.83	7.27	6.76	6.30	5.97	5.92
20	8.07	7.89	7.33	6.81	6.34	6.01	5.96
25	8.36	8.17	7.57	7.02	6.52	6.17	6.12
30	8.60	8.40	7.77	7.19	6.67	6.30	6.25
40	8.98	8.77	8.09	7.47	6.90	6.51	6.45
50	9.29	9.06	8.34	7.68	7.09	6.68	6.61
100	10.26	9.99	9.13	8.35	7.66	7.18	7.11
250	11.63	11.29	10.21	9.26	8.42	7.84	7.76
500	12.71	12.32	11.07	9.97	9.01	8.35	8.25

Table A.8. *Continued*

```
          BONFERRONI F CRITICAL VALUES FOR FAMILYWISE ALPHA= .010
```

COMPARISONS	DF1 = 4 DF2 = 3	4 4	4 5	4 6	4 7	4 8	4 9
2	46.19	23.15	15.56	12.03	10.05	8.81	7.96
3	60.86	28.66	18.58	14.05	11.56	10.01	8.97
4	73.95	33.30	21.05	15.65	12.73	10.94	9.74
5	85.98	37.39	23.16	17.01	13.72	11.71	10.38
6	97.22	41.09	25.04	18.19	14.57	12.37	10.92
7	107.86	44.49	26.74	19.25	15.32	12.95	11.39
8	118.00	47.65	28.29	20.22	16.00	13.48	11.82
9	127.74	50.63	29.73	21.10	16.62	13.95	12.21
10	137.10	53.43	31.09	21.92	17.20	14.39	12.56
11	146.16	56.11	32.36	22.69	17.73	14.80	12.89
12	154.95	58.66	33.56	23.42	18.23	15.18	13.19
13	163.51	61.11	34.70	24.10	18.71	15.53	13.48
14	171.84	63.47	35.80	24.75	19.15	15.87	13.75
15	179.98	65.75	36.85	25.37	19.58	16.19	14.00
16	187.94	67.95	37.85	25.96	19.98	16.49	14.24
17	195.75	70.08	38.82	26.53	20.37	16.78	14.47
18	203.36	72.15	39.75	27.08	20.74	17.06	14.69
19	210.88	74.16	40.66	27.61	21.10	17.32	14.90
20	218.27	76.13	41.54	28.11	21.44	17.58	15.11
25	253.39	85.28	45.57	30.43	23.00	18.73	16.01
30	286.32	93.53	49.13	32.46	24.34	19.71	16.79
40	347.07	108.20	55.31	35.92	26.61	21.37	18.07
50	402.94	121.15	60.63	38.83	28.50	22.73	19.13
100	640.45	171.93	80.53	49.42	35.22	27.49	22.77
250	1181.57	272.71	116.84	67.75	46.43	35.20	28.51
500	1879.47	386.49	154.59	85.70	56.98	42.28	33.72

COMPARISONS	DF1 = 4 DF2 = 10	4 11	4 12	4 13	4 14	4 15	4 16
2	7.34	6.88	6.52	6.23	6.00	5.80	5.64
3	8.22	7.66	7.23	6.89	6.61	6.37	6.18
4	8.89	8.25	7.76	7.37	7.06	6.80	6.58
5	9.43	8.73	8.19	7.77	7.42	7.13	6.90
6	9.89	9.14	8.56	8.10	7.73	7.42	7.16
7	10.30	9.49	8.87	8.38	7.99	7.67	7.39
8	10.66	9.80	9.15	8.64	8.22	7.88	7.60
9	10.98	10.09	9.40	8.86	8.43	8.08	7.78
10	11.28	10.35	9.63	9.07	8.62	8.25	7.94
11	11.56	10.58	9.84	9.26	8.80	8.41	8.10
12	11.81	10.81	10.04	9.44	8.96	8.56	8.24
13	12.05	11.01	10.22	9.60	9.11	8.70	8.36
14	12.28	11.21	10.39	9.76	9.25	8.83	8.49
15	12.49	11.39	10.55	9.90	9.38	8.95	8.60
16	12.69	11.56	10.71	10.04	9.51	9.07	8.71
17	12.88	11.73	10.85	10.17	9.62	9.18	8.81
18	13.06	11.80	10.99	10.29	9.74	9.28	8.90
19	13.24	12.03	11.12	10.41	9.84	9.38	9.00
20	13.41	12.18	11.25	10.52	9.95	9.48	9.08
25	14.16	12.82	11.81	11.02	10.40	9.89	9.47
30	14.79	13.36	12.28	11.45	10.78	10.24	9.80
40	15.85	14.25	13.06	12.14	11.41	10.82	10.33
50	16.70	14.98	13.69	12.69	11.91	11.27	10.75
100	19.63	17.42	15.79	14.55	13.57	12.78	12.14
250	24.18	21.17	18.98	17.32	16.04	15.01	14.17
500	28.22	24.44	21.71	19.70	18.12	16.87	15.86

413

Table A.8. *Continued*

```
        BONFERRONI F CRITICAL VALUES FOR FAMILYWISE ALPHA= .010

              DF1 =    4      4      4      4      4      4      4
COMPARISONS   DF2 =   17     18     19     20     22     24     26

     2               5.50   5.37   5.27   5.17   5.02   4.89   4.79
     3               6.01   5.87   5.74   5.63   5.45   5.30   5.18
     4               6.39   6.23   6.09   5.97   5.76   5.60   5.46
     5               6.69   6.52   6.37   6.23   6.01   5.83   5.69
     6               6.95   6.76   6.60   6.45   6.22   6.03   5.87
     7               7.16   6.97   6.79   6.64   6.39   6.19   6.03
     8               7.36   7.15   6.97   6.81   6.55   6.34   6.17
     9               7.53   7.31   7.12   6.96   6.69   6.47   6.29
    10               7.68   7.46   7.27   7.10   6.81   6.59   6.41
    11               7.83   7.59   7.39   7.22   6.93   6.70   6.51
    12               7.96   7.72   7.51   7.33   7.03   6.80   6.60
    13               8.08   7.83   7.62   7.44   7.13   6.89   6.69
    14               8.19   7.94   7.73   7.54   7.22   6.97   6.77
    15               8.30   8.04   7.82   7.63   7.31   7.05   6.84
    16               8.40   8.14   7.91   7.71   7.39   7.13   6.91
    17               8.50   8.23   8.00   7.80   7.46   7.20   6.98
    18               8.59   8.31   8.08   7.87   7.53   7.26   7.04
    19               8.67   8.40   8.16   7.95   7.60   7.33   7.11
    20               8.75   8.47   8.23   8.02   7.67   7.39   7.16
    25               9.12   8.82   8.56   8.33   7.95   7.66   7.41
    30               9.42   9.10   8.83   8.59   8.19   7.88   7.62
    40               9.92   9.57   9.27   9.01   8.57   8.23   7.96
    50              10.31   9.94   9.62   9.34   8.88   8.52   8.22
   100              11.60  11.14  10.75  10.41   9.86   9.42   9.07
   250              13.47  12.89  12.39  11.96  11.26  10.71  10.27
   500              15.02  14.33  13.74  13.22  12.39  11.75  11.23

              DF1 =    4      4      4      4      4      4      4
COMPARISONS   DF2 =   28     30     40     60    120    500   1000

     2               4.70   4.62   4.37   4.14   3.92   3.76   3.74
     3               5.08   4.99   4.70   4.43   4.18   4.00   3.97
     4               5.35   5.25   4.93   4.64   4.36   4.17   4.14
     5               5.56   5.46   5.12   4.80   4.50   4.30   4.26
     6               5.74   5.63   5.27   4.93   4.62   4.40   4.37
     7               5.89   5.78   5.40   5.04   4.72   4.49   4.45
     8               6.03   5.91   5.51   5.14   4.80   4.57   4.53
     9               6.15   6.02   5.61   5.23   4.88   4.63   4.60
    10               6.25   6.12   5.70   5.31   4.95   4.69   4.66
    11               6.35   6.22   5.78   5.38   5.01   4.75   4.71
    12               6.44   6.30   5.85   5.44   5.06   4.80   4.76
    13               6.52   6.38   5.92   5.50   5.12   4.84   4.80
    14               6.60   6.46   5.99   5.56   5.16   4.89   4.84
    15               6.67   6.53   6.05   5.61   5.21   4.93   4.88
    16               6.74   6.59   6.10   5.66   5.25   4.96   4.92
    17               6.80   6.65   6.15   5.70   5.29   5.00   4.95
    18               6.86   6.71   6.20   5.74   5.32   5.03   4.98
    19               6.92   6.76   6.25   5.78   5.36   5.06   5.02
    20               6.97   6.82   6.30   5.82   5.39   5.09   5.04
    25               7.21   7.05   6.49   5.99   5.54   5.22   5.17
    30               7.41   7.24   6.65   6.13   5.65   5.32   5.27
    40               7.73   7.54   6.91   6.35   5.84   5.49   5.43
    50               7.98   7.78   7.12   6.52   5.98   5.61   5.56
   100               8.78   8.54   7.76   7.06   6.44   6.01   5.94
   250               9.91   9.61   8.64   7.79   7.04   6.52   6.45
   500              10.81  10.46   9.33   8.36   7.50   6.92   6.84
```

Table A.8. *Continued*

BONFERRONI F CRITICAL VALUES FOR FAMILYWISE ALPHA= .010

COMPARISONS	DF1 = 5 DF2 = 3	5 4	5 5	5 6	5 7	5 8	5 9
2	45.39	22.46	14.94	11.46	9.52	8.30	7.47
3	59.78	27.78	17.83	13.37	10.93	9.42	8.40
4	72.62	32.26	20.18	14.88	12.03	10.28	9.12
5	84.42	36.21	22.20	16.16	12.95	11.00	9.70
6	95.46	39.78	23.99	17.28	13.74	11.61	10.20
7	105.89	43.07	25.60	18.28	14.45	12.15	10.64
8	115.85	46.12	27.09	19.19	15.09	12.64	11.03
9	125.38	49.00	28.46	20.03	15.67	13.08	11.39
10	134.59	51.71	29.75	20.80	16.21	13.48	11.71
11	143.47	54.29	30.96	21.53	16.71	13.86	12.02
12	152.11	56.77	32.11	22.21	17.17	14.21	12.30
13	160.49	59.13	33.20	22.86	17.62	14.54	12.56
14	168.66	61.41	34.25	23.47	18.03	14.86	12.81
15	176.65	63.60	35.25	24.06	18.43	15.15	13.04
16	184.45	65.73	36.20	24.62	18.81	15.43	13.26
17	192.11	67.79	37.13	25.15	19.17	15.70	13.48
18	199.61	69.79	38.02	25.67	19.52	15.96	13.68
19	206.97	71.74	38.89	26.16	19.85	16.20	13.87
20	214.20	73.63	39.72	26.65	20.17	16.44	14.06
25	248.75	82.47	43.56	28.83	21.63	17.51	14.89
30	281.01	90.45	46.97	30.74	22.89	18.42	15.61
40	340.65	104.64	52.87	34.01	25.01	19.95	16.79
50	395.49	117.10	57.95	36.76	26.77	21.22	17.77
100	628.44	166.13	76.89	46.74	33.06	25.64	21.11
250	1159.64	263.35	111.62	64.06	43.50	32.78	26.41
500	1842.58	373.92	147.86	81.02	53.41	39.37	31.20

COMPARISONS	DF1 = 5 DF2 = 10	5 11	5 12	5 13	5 14	5 15	5 16
2	6.87	6.42	6.07	5.79	5.56	5.37	5.21
3	7.68	7.13	6.71	6.38	6.11	5.88	5.69
4	8.29	7.67	7.20	6.82	6.51	6.26	6.05
5	8.79	8.11	7.59	7.17	6.84	6.57	6.34
6	9.21	8.48	7.92	7.47	7.12	6.82	6.57
7	9.58	8.80	8.20	7.73	7.35	7.04	6.78
8	9.91	9.08	8.46	7.96	7.56	7.24	6.96
9	10.21	9.34	8.68	8.17	7.75	7.41	7.12
10	10.48	9.58	8.89	8.35	7.92	7.57	7.27
11	10.73	9.80	9.08	8.53	8.08	7.71	7.41
12	10.97	10.00	9.26	8.69	8.22	7.85	7.53
13	11.19	10.18	9.43	8.83	8.36	7.97	7.65
14	11.39	10.36	9.58	8.97	8.49	8.09	7.75
15	11.59	10.53	9.73	9.10	8.60	8.20	7.86
16	11.77	10.69	9.87	9.23	8.72	8.30	7.95
17	11.94	10.83	10.00	9.34	8.82	8.40	8.04
18	12.11	10.98	10.12	9.46	8.92	8.49	8.13
19	12.27	11.11	10.24	9.56	9.02	8.58	8.21
20	12.42	11.24	10.36	9.66	9.11	8.66	8.29
25	13.11	11.83	10.86	10.12	9.52	9.04	8.64
30	13.69	12.32	11.29	10.50	9.86	9.35	8.92
40	14.66	13.13	12.00	11.12	10.42	9.86	9.40
50	15.44	13.79	12.56	11.62	10.87	10.27	9.77
100	18.12	16.02	14.47	13.29	12.36	11.62	11.01
250	22.28	19.43	17.35	15.79	14.58	13.61	12.82
500	25.98	22.41	19.84	17.93	16.44	15.28	14.32

BONFERRONI F CRITICAL VALUES FOR FAMILYWISE ALPHA= .010

COMPARISONS	DF1 = DF2 =	5 17	5 18	5 19	5 20	5 22	5 24	5 26
2		5.07	4.96	4.85	4.76	4.61	4.49	4.38
3		5.53	5.39	5.27	5.17	4.99	4.84	4.73
4		5.87	5.72	5.58	5.46	5.26	5.11	4.98
5		6.14	5.97	5.83	5.70	5.48	5.31	5.17
6		6.37	6.19	6.03	5.89	5.67	5.48	5.34
7		6.56	6.37	6.21	6.06	5.82	5.63	5.47
8		6.73	6.53	6.36	6.21	5.96	5.76	5.60
9		6.88	6.68	6.50	6.34	6.08	5.87	5.70
10		7.02	6.81	6.62	6.46	6.19	5.98	5.80
11		7.15	6.93	6.74	6.57	6.29	6.07	5.89
12		7.27	7.04	6.84	6.67	6.38	6.16	5.97
13		7.37	7.14	6.94	6.76	6.47	6.24	6.05
14		7.48	7.24	7.03	6.85	6.55	6.31	6.12
15		7.57	7.33	7.11	6.93	6.62	6.38	6.18
16		7.66	7.41	7.19	7.01	6.70	6.45	6.25
17		7.74	7.49	7.27	7.08	6.76	6.51	6.30
18		7.83	7.57	7.34	7.15	6.82	6.57	6.36
19		7.90	7.64	7.41	7.21	6.88	6.62	6.41
20		7.98	7.71	7.48	7.27	6.94	6.68	6.46
25		8.30	8.01	7.76	7.55	7.19	6.91	6.68
30		8.57	8.27	8.00	7.78	7.40	7.10	6.86
40		9.01	8.68	8.39	8.14	7.74	7.42	7.16
50		9.36	9.00	8.70	8.44	8.00	7.66	7.39
100		10.50	10.07	9.70	9.39	8.87	8.46	8.13
250		12.17	11.61	11.15	10.75	10.09	9.58	9.17
500		13.55	12.89	12.33	11.85	11.09	10.48	10.01

COMPARISONS	DF1 = DF2 =	5 28	5 30	5 40	5 60	5 120	5 500	5 1000
2		4.30	4.23	3.99	3.76	3.55	3.40	3.37
3		4.63	4.55	4.27	4.01	3.77	3.59	3.57
4		4.87	4.78	4.47	4.19	3.92	3.73	3.71
5		5.06	4.96	4.63	4.32	4.04	3.84	3.81
6		5.21	5.11	4.76	4.44	4.14	3.93	3.90
7		5.34	5.23	4.87	4.53	4.22	4.00	3.97
8		5.46	5.35	4.97	4.62	4.30	4.07	4.03
9		5.56	5.44	5.05	4.69	4.36	4.13	4.09
10		5.66	5.53	5.13	4.76	4.42	4.18	4.14
11		5.74	5.62	5.20	4.82	4.47	4.22	4.18
12		5.82	5.69	5.26	4.87	4.51	4.26	4.22
13		5.89	5.76	5.32	4.92	4.56	4.30	4.26
14		5.96	5.82	5.38	4.97	4.60	4.34	4.30
15		6.02	5.88	5.43	5.01	4.63	4.37	4.33
16		6.08	5.94	5.48	5.05	4.67	4.40	4.36
17		6.13	5.99	5.52	5.09	4.70	4.43	4.39
18		6.19	6.04	5.56	5.13	4.73	4.46	4.41
19		6.24	6.09	5.60	5.16	4.76	4.48	4.44
20		6.28	6.14	5.64	5.20	4.79	4.51	4.46
25		6.49	6.33	5.81	5.34	4.91	4.61	4.57
30		6.67	6.50	5.95	5.46	5.01	4.70	4.65
40		6.94	6.76	6.17	5.64	5.17	4.84	4.79
50		7.16	6.97	6.35	5.79	5.29	4.94	4.89
100		7.86	7.63	6.90	6.25	5.67	5.27	5.21
250		8.83	8.55	7.66	6.86	6.17	5.70	5.63
500		9.61	9.28	8.25	7.34	6.55	6.02	5.94

Table A.8. *Continued*

```
            BONFERRONI F CRITICAL VALUES FOR FAMILYWISE ALPHA= .010

              DF1 =    6       6       6       6       6       6       6
COMPARISONS   DF2 =    3       4       5       6       7       8       9

     2               44.84   21.97   14.51   11.07    9.16    7.95    7.13
     3               59.03   27.17   17.31   12.90   10.50    9.01    8.01
     4               71.71   31.54   19.58   14.35   11.55    9.83    8.68
     5               83.36   35.40   21.53   15.58   12.42   10.50    9.23
     6               94.25   38.89   23.26   16.65   13.18   11.08    9.70
     7              104.55   42.09   24.82   17.61   13.85   11.60   10.12
     8              114.36   45.07   26.26   18.48   14.46   12.06   10.49
     9              123.78   47.88   27.59   19.29   15.01   12.47   10.82
    10              132.84   50.52   28.83   20.03   15.52   12.86   11.13
    11              141.62   53.05   30.01   20.73   16.00   13.21   11.41
    12              150.13   55.45   31.12   21.38   16.44   13.55   11.68
    13              158.40   57.76   32.17   22.00   16.86   13.86   11.92
    14              166.47   59.99   33.18   22.59   17.26   14.16   12.16
    15              174.36   62.13   34.14   23.15   17.64   14.44   12.38
    16              182.07   64.20   35.07   23.69   18.00   14.70   12.59
    17              189.63   66.22   35.96   24.20   18.34   14.96   12.79
    18              197.02   68.17   36.83   24.70   18.67   15.20   12.98
    19              204.25   70.07   37.66   25.17   18.99   15.43   13.16
    20              211.43   71.91   38.47   25.63   19.30   15.66   13.34
    25              245.49   80.54   42.19   27.73   20.68   16.66   14.12
    30              277.36   88.34   45.47   29.56   21.88   17.53   14.79
    40              336.12  102.15   51.19   32.69   23.90   18.98   15.91
    50              390.09  114.37   56.09   35.33   25.58   20.18   16.82
   100              620.25  162.25   74.43   44.91   31.56   24.36   19.98
   250             1143.49  257.30  108.02   61.51   41.50   31.10   24.96
   500             1819.05  364.92  143.10   77.95   50.95   37.35   29.49

              DF1 =    6       6       6       6       6       6       6
COMPARISONS   DF2 =   10      11      12      13      14      15      16

     2                6.54    6.10    5.76    5.48    5.26    5.07    4.91
     3                7.30    6.77    6.35    6.03    5.76    5.54    5.36
     4                7.87    7.27    6.80    6.44    6.14    5.89    5.68
     5                8.34    7.67    7.17    6.76    6.44    6.17    5.95
     6                8.73    8.02    7.47    7.04    6.69    6.41    6.17
     7                9.08    8.32    7.74    7.28    6.91    6.61    6.35
     8                9.39    8.59    7.97    7.49    7.10    6.79    6.52
     9                9.67    8.83    8.19    7.68    7.28    6.95    6.67
    10                9.93    9.05    8.38    7.86    7.44    7.09    6.80
    11               10.16    9.25    8.56    8.01    7.58    7.23    6.93
    12               10.38    9.44    8.72    8.16    7.71    7.35    7.04
    13               10.59    9.61    8.88    8.30    7.84    7.46    7.15
    14               10.78    9.78    9.02    8.43    7.96    7.57    7.25
    15               10.96    9.93    9.15    8.55    8.07    7.67    7.34
    16               11.13   10.08    9.28    8.66    8.17    7.76    7.43
    17               11.30   10.22    9.41    8.77    8.27    7.85    7.51
    18               11.45   10.35    9.52    8.88    8.36    7.94    7.59
    19               11.60   10.48    9.63    8.97    8.45    8.02    7.67
    20               11.75   10.60    9.74    9.07    8.53    8.10    7.74
    25               12.39   11.14   10.21    9.49    8.91    8.44    8.06
    30               12.93   11.60   10.61    9.84    9.23    8.73    8.32
    40               13.84   12.36   11.26   10.41    9.74    9.20    8.75
    50               14.57   12.98   11.79   10.88   10.16    9.58    9.10
   100               17.08   15.05   13.56   12.42   11.54   10.82   10.24
   250               20.97   18.22   16.24   14.74   13.57   12.64   11.89
   500               24.41   20.99   18.54   16.72   15.30   14.18   13.27
```

Table A.8. *Continued*

```
              BONFERRONI F CRITICAL VALUES FOR FAMILYWISE ALPHA= .010
```

COMPARISONS	DF1 = DF2 =	6 17	6 18	6 19	6 20	6 22	6 24	6 26
2		4.78	4.66	4.56	4.47	4.32	4.20	4.10
3		5.20	5.06	4.94	4.84	4.67	4.53	4.41
4		5.51	5.36	5.23	5.11	4.92	4.76	4.64
5		5.76	5.59	5.45	5.33	5.12	4.95	4.81
6		5.96	5.79	5.64	5.50	5.28	5.10	4.96
7		6.14	5.96	5.80	5.66	5.42	5.24	5.09
8		6.30	6.10	5.94	5.79	5.55	5.35	5.20
9		6.44	6.24	6.06	5.91	5.66	5.46	5.29
10		6.56	6.36	6.18	6.02	5.76	5.55	5.38
11		6.68	6.46	6.28	6.12	5.85	5.64	5.46
12		6.79	6.57	6.37	6.21	5.93	5.71	5.53
13		6.88	6.66	6.46	6.29	6.01	5.79	5.60
14		6.98	6.75	6.55	6.37	6.08	5.85	5.67
15		7.06	6.83	6.62	6.45	6.15	5.92	5.73
16		7.15	6.91	6.70	6.52	6.21	5.98	5.78
17		7.22	6.98	6.77	6.58	6.27	6.03	5.83
18		7.30	7.05	6.83	6.64	6.33	6.08	5.88
19		7.37	7.11	6.89	6.70	6.39	6.14	5.93
20		7.43	7.18	6.95	6.76	6.44	6.18	5.98
25		7.73	7.45	7.22	7.01	6.67	6.39	6.17
30		7.98	7.69	7.43	7.22	6.85	6.57	6.34
40		8.38	8.06	7.79	7.55	7.16	6.85	6.60
50		8.70	8.36	8.07	7.82	7.40	7.07	6.81
100		9.75	9.33	8.98	8.68	8.18	7.79	7.47
250		11.27	10.74	10.30	9.91	9.29	8.80	8.41
500		12.54	11.90	11.37	10.92	10.19	9.61	9.17

COMPARISONS	DF1 = DF2 =	6 28	6 30	6 40	6 60	6 120	6 500	6 1000
2		4.02	3.95	3.71	3.49	3.28	3.14	3.11
3		4.32	4.24	3.96	3.71	3.48	3.31	3.28
4		4.53	4.44	4.14	3.87	3.61	3.43	3.40
5		4.70	4.60	4.29	3.99	3.72	3.52	3.49
6		4.84	4.74	4.40	4.09	3.80	3.60	3.57
7		4.96	4.85	4.50	4.17	3.88	3.66	3.63
8		5.06	4.95	4.59	4.25	3.94	3.72	3.69
9		5.16	5.04	4.66	4.31	3.99	3.77	3.73
10		5.24	5.12	4.73	4.37	4.04	3.81	3.78
11		5.32	5.20	4.79	4.43	4.09	3.85	3.82
12		5.39	5.26	4.85	4.47	4.13	3.89	3.85
13		5.45	5.32	4.90	4.52	4.17	3.92	3.88
14		5.51	5.38	4.95	4.56	4.20	3.95	3.91
15		5.57	5.44	5.00	4.60	4.23	3.98	3.94
16		5.62	5.49	5.04	4.63	4.27	4.01	3.97
17		5.67	5.53	5.08	4.67	4.29	4.03	3.99
18		5.72	5.58	5.12	4.70	4.32	4.06	4.02
19		5.76	5.62	5.15	4.73	4.35	4.08	4.04
20		5.81	5.66	5.19	4.76	4.37	4.10	4.06
25		5.99	5.84	5.34	4.89	4.48	4.19	4.15
30		6.15	5.99	5.46	4.99	4.56	4.27	4.22
40		6.39	6.22	5.66	5.15	4.70	4.38	4.34
50		6.59	6.41	5.81	5.28	4.80	4.47	4.42
100		7.22	7.00	6.30	5.68	5.13	4.76	4.70
250		8.09	7.83	6.97	6.23	5.57	5.12	5.06
500		8.79	8.49	7.50	6.65	5.91	5.40	5.33

Table A.8. *Continued*

COMPARISONS	DF1 =	7	7	7	7	7	7	7
	DF2 =	3	4	5	6	7	8	9
2		44.43	21.62	14.20	10.79	8.89	7.69	6.89
3		58.49	26.72	16.92	12.56	10.18	8.71	7.72
4		71.04	31.02	19.14	13.96	11.19	9.49	8.36
5		82.57	34.80	21.04	15.15	12.03	10.14	8.89
6		93.36	38.23	22.73	16.19	12.76	10.70	9.34
7		103.55	41.37	24.25	17.12	13.41	11.19	9.73
8		113.28	44.30	25.65	17.97	13.99	11.63	10.09
9		122.59	47.06	26.95	18.74	14.53	12.03	10.40
10		131.59	49.66	28.16	19.46	15.02	12.40	10.70
11		140.28	52.13	29.30	20.14	15.48	12.74	10.97
12		148.70	54.50	30.39	20.77	15.91	13.06	11.22
13		156.89	56.77	31.42	21.37	16.31	13.36	11.46
14		164.89	58.95	32.40	21.94	16.70	13.64	11.68
15		172.70	61.06	33.34	22.49	17.06	13.91	11.89
16		180.33	63.09	34.24	23.01	17.41	14.17	12.09
17		187.79	65.06	35.11	23.50	17.74	14.41	12.28
18		195.11	66.98	35.96	23.98	18.06	14.64	12.46
19		202.31	68.85	36.77	24.45	18.36	14.87	12.64
20		209.37	70.66	37.55	24.89	18.66	15.08	12.80
25		243.09	79.14	41.18	26.92	19.99	16.05	13.55
30		274.66	86.78	44.38	28.70	21.15	16.88	14.20
40		332.90	100.36	49.94	31.73	23.09	18.27	15.26
50		386.34	112.35	54.73	34.29	24.72	19.42	16.13
100		614.21	159.38	72.61	43.56	30.48	23.42	19.14
250		1132.31	252.81	105.40	59.67	40.09	29.91	23.90
500		1801.35	358.02	139.45	75.57	49.15	35.89	28.16

COMPARISONS	DF1 =	7	7	7	7	7	7	7
	DF2 =	10	11	12	13	14	15	16
2		6.30	5.86	5.52	5.25	5.03	4.85	4.69
3		7.02	6.49	6.09	5.77	5.51	5.29	5.11
4		7.56	6.97	6.51	6.15	5.86	5.62	5.41
5		8.01	7.36	6.85	6.46	6.14	5.88	5.66
6		8.38	7.68	7.14	6.72	6.38	6.10	5.86
7		8.71	7.97	7.39	6.95	6.58	6.29	6.04
8		9.01	8.22	7.62	7.15	6.77	6.45	6.19
9		9.27	8.45	7.82	7.33	6.93	6.60	6.33
10		9.52	8.66	8.00	7.49	7.08	6.74	6.46
11		9.74	8.85	8.17	7.64	7.21	6.87	6.58
12		9.95	9.02	8.32	7.78	7.34	6.98	6.68
13		10.15	9.19	8.47	7.91	7.46	7.09	6.78
14		10.33	9.35	8.61	8.03	7.57	7.19	6.87
15		10.50	9.49	8.73	8.14	7.67	7.28	6.96
16		10.67	9.63	8.85	8.25	7.77	7.37	7.04
17		10.82	9.77	8.97	8.35	7.86	7.46	7.12
18		10.97	9.89	9.08	8.45	7.95	7.53	7.20
19		11.11	10.01	9.18	8.54	8.03	7.61	7.27
20		11.25	10.13	9.28	8.63	8.11	7.68	7.33
25		11.86	10.64	9.73	9.02	8.46	8.01	7.63
30		12.38	11.08	10.10	9.35	8.76	8.28	7.88
40		13.24	11.79	10.72	9.90	9.24	8.72	8.28
50		13.94	12.38	11.22	10.33	9.63	9.07	8.60
100		16.32	14.34	12.89	11.79	10.93	10.23	9.66
250		20.01	17.36	15.42	13.97	12.84	11.94	11.21
500		23.30	19.96	17.61	15.82	14.46	13.38	12.50

Table A.8. *Continued*

BONFERRONI F CRITICAL VALUES FOR FAMILYWISE ALPHA= .010

COMPARISONS	DF1 = DF2 =	7 17	7 18	7 19	7 20	7 22	7 24	7 26
2		4.56	4.44	4.34	4.26	4.11	3.99	3.89
3		4.95	4.82	4.70	4.60	4.43	4.29	4.18
4		5.24	5.09	4.96	4.85	4.66	4.51	4.39
5		5.47	5.31	5.17	5.05	4.84	4.68	4.55
6		5.66	5.49	5.34	5.21	5.00	4.82	4.68
7		5.83	5.65	5.49	5.36	5.13	4.95	4.80
8		5.97	5.79	5.62	5.48	5.24	5.05	4.90
9		6.11	5.91	5.74	5.59	5.35	5.15	4.99
10		6.22	6.02	5.85	5.69	5.44	5.24	5.07
11		6.33	6.12	5.94	5.78	5.52	5.31	5.14
12		6.43	6.22	6.03	5.87	5.60	5.39	5.21
13		6.52	6.30	6.11	5.95	5.67	5.45	5.27
14		6.61	6.38	6.19	6.02	5.74	5.51	5.33
15		6.69	6.46	6.26	6.09	5.80	5.57	5.39
16		6.77	6.53	6.33	6.15	5.86	5.63	5.44
17		6.84	6.60	6.39	6.21	5.92	5.68	5.49
18		6.91	6.66	6.45	6.27	5.97	5.73	5.53
19		6.97	6.73	6.51	6.33	6.02	5.77	5.58
20		7.04	6.78	6.57	6.38	6.07	5.82	5.62
25		7.31	7.04	6.81	6.61	6.28	6.01	5.80
30		7.54	7.26	7.01	6.80	6.45	6.17	5.95
40		7.92	7.61	7.34	7.11	6.73	6.43	6.19
50		8.21	7.89	7.60	7.36	6.95	6.64	6.38
100		9.19	8.79	8.45	8.16	7.68	7.30	6.99
250		10.61	10.10	9.67	9.30	8.70	8.23	7.86
500		11.78	11.18	10.68	10.23	9.52	8.98	8.54

COMPARISONS	DF1 = DF2 =	7 28	7 30	7 40	7 60	7 120	7 500	7 1000
2		3.81	3.74	3.51	3.29	3.09	2.94	2.92
3		4.08	4.00	3.74	3.49	3.26	3.09	3.07
4		4.28	4.19	3.90	3.63	3.38	3.20	3.18
5		4.44	4.34	4.03	3.74	3.47	3.29	3.26
6		4.56	4.47	4.14	3.83	3.55	3.35	3.32
7		4.67	4.57	4.23	3.91	3.62	3.41	3.38
8		4.77	4.66	4.30	3.98	3.67	3.46	3.43
9		4.86	4.74	4.37	4.03	3.72	3.50	3.47
10		4.93	4.82	4.44	4.09	3.77	3.54	3.51
11		5.00	4.88	4.49	4.13	3.81	3.58	3.54
12		5.07	4.95	4.54	4.18	3.84	3.61	3.57
13		5.13	5.00	4.59	4.22	3.88	3.64	3.60
14		5.18	5.05	4.64	4.26	3.91	3.67	3.63
15		5.23	5.10	4.68	4.29	3.94	3.69	3.65
16		5.28	5.15	4.72	4.32	3.96	3.71	3.68
17		5.33	5.19	4.75	4.35	3.99	3.74	3.70
18		5.37	5.23	4.79	4.38	4.01	3.76	3.72
19		5.41	5.27	4.82	4.41	4.04	3.78	3.74
20		5.45	5.31	4.85	4.44	4.06	3.80	3.76
25		5.62	5.47	4.99	4.55	4.15	3.88	3.84
30		5.76	5.61	5.10	4.64	4.23	3.94	3.90
40		5.99	5.82	5.28	4.79	4.35	4.05	4.00
50		6.17	5.99	5.42	4.91	4.45	4.13	4.08
100		6.75	6.54	5.86	5.27	4.74	4.38	4.32
250		7.55	7.29	6.47	5.76	5.13	4.70	4.64
500		8.18	7.89	6.95	6.13	5.42	4.95	4.88

Table A.8. *Continued*

```
            BONFERRONI F CRITICAL VALUES FOR FAMILYWISE ALPHA= .010
```

COMPARISONS	DF1 = 8 DF2 = 17	8 18	8 19	8 20	8 22	8 24	8 26
2	4.39	4.28	4.18	4.09	3.94	3.83	3.73
3	4.76	4.63	4.51	4.41	4.24	4.11	4.00
4	5.03	4.89	4.76	4.65	4.46	4.31	4.19
5	5.25	5.09	4.95	4.83	4.63	4.47	4.34
6	5.43	5.26	5.12	4.99	4.78	4.61	4.47
7	5.59	5.41	5.26	5.12	4.90	4.72	4.58
8	5.73	5.54	5.38	5.24	5.01	4.82	4.67
9	5.85	5.66	5.49	5.35	5.10	4.91	4.75
10	5.96	5.76	5.59	5.44	5.19	4.99	4.83
11	6.06	5.86	5.68	5.53	5.27	5.06	4.90
12	6.16	5.95	5.76	5.61	5.34	5.13	4.96
13	6.25	6.03	5.84	5.68	5.41	5.19	5.02
14	6.33	6.11	5.91	5.75	5.47	5.25	5.07
15	6.40	6.18	5.98	5.81	5.53	5.31	5.12
16	6.48	6.25	6.05	5.87	5.59	5.36	5.17
17	6.54	6.31	6.11	5.93	5.64	5.41	5.22
18	6.61	6.37	6.16	5.98	5.69	5.45	5.26
19	6.67	6.43	6.22	6.04	5.73	5.49	5.30
20	6.73	6.48	6.27	6.09	5.78	5.54	5.34
25	6.99	6.73	6.50	6.30	5.98	5.72	5.51
30	7.21	6.93	6.69	6.48	6.14	5.87	5.65
40	7.56	7.26	7.00	6.77	6.40	6.11	5.88
50	7.84	7.52	7.24	7.01	6.61	6.30	6.05
100	8.76	8.38	8.04	7.76	7.29	6.92	6.63
250	10.10	9.61	9.19	8.83	8.25	7.79	7.43
500	11.21	10.63	10.13	9.71	9.02	8.49	8.07

COMPARISONS	DF1 = 8 DF2 = 28	8 30	8 40	8 60	8 120	8 500	8 1000
2	3.65	3.58	3.35	3.13	2.93	2.79	2.77
3	3.90	3.82	3.56	3.32	3.09	2.93	2.90
4	4.09	4.00	3.71	3.45	3.20	3.02	3.00
5	4.23	4.14	3.83	3.55	3.29	3.10	3.07
6	4.35	4.25	3.93	3.63	3.36	3.16	3.13
7	4.45	4.35	4.01	3.70	3.41	3.21	3.18
8	4.54	4.44	4.09	3.76	3.47	3.26	3.23
9	4.62	4.51	4.15	3.82	3.51	3.30	3.26
10	4.69	4.58	4.21	3.86	3.55	3.33	3.30
11	4.76	4.64	4.26	3.91	3.59	3.36	3.33
12	4.82	4.70	4.31	3.95	3.62	3.39	3.36
13	4.88	4.75	4.35	3.99	3.65	3.42	3.38
14	4.93	4.80	4.39	4.02	3.68	3.44	3.41
15	4.97	4.85	4.43	4.05	3.71	3.47	3.43
16	5.02	4.89	4.47	4.08	3.73	3.49	3.45
17	5.06	4.93	4.50	4.11	3.75	3.51	3.47
18	5.10	4.97	4.53	4.14	3.78	3.53	3.49
19	5.14	5.01	4.56	4.16	3.80	3.54	3.51
20	5.18	5.04	4.59	4.18	3.82	3.56	3.52
25	5.34	5.19	4.72	4.29	3.90	3.63	3.59
30	5.47	5.32	4.82	4.37	3.97	3.69	3.65
40	5.68	5.52	4.99	4.51	4.08	3.79	3.74
50	5.85	5.68	5.12	4.62	4.17	3.86	3.81
100	6.38	6.18	5.53	4.95	4.43	4.08	4.03
250	7.13	6.88	6.09	5.39	4.79	4.38	4.31
500	7.72	7.43	6.53	5.74	5.05	4.60	4.53

Table A.8. *Continued*

```
               BONFERRONI F CRITICAL VALUES FOR FAMILYWISE ALPHA= .010

               DF1 =     8        8        8        8        8        8        8
COMPARISONS    DF2 =     3        4        5        6        7        8        9

      2               44.13    21.35    13.96    10.57     8.68     7.50     6.69
      3               58.08    26.38    16.63    12.29     9.93     8.48     7.50
      4               70.53    30.62    18.80    13.67    10.91     9.24     8.12
      5               81.98    34.35    20.67    14.83    11.73     9.86     8.63
      6               92.68    37.72    22.32    15.84    12.44    10.40     9.06
      7              102.80    40.83    23.82    16.75    13.07    10.88     9.44
      8              112.45    43.72    25.19    17.57    13.64    11.30     9.78
      9              121.70    46.43    26.46    18.33    14.16    11.69    10.09
     10              130.61    48.99    27.65    19.03    14.63    12.05    10.37
     11              139.25    51.44    28.77    19.69    15.08    12.38    10.63
     12              147.61    53.77    29.83    20.31    15.50    12.69    10.87
     13              155.75    56.00    30.84    20.89    15.89    12.97    11.10
     14              163.66    58.16    31.80    21.45    16.26    13.25    11.31
     15              171.41    60.23    32.72    21.98    16.62    13.51    11.52
     16              178.99    62.24    33.61    22.49    16.95    13.76    11.71
     17              186.41    64.19    34.47    22.97    17.28    13.99    11.89
     18              193.70    66.08    35.29    23.44    17.58    14.22    12.07
     19              200.81    67.92    36.09    23.89    17.88    14.43    12.24
     20              207.81    69.71    36.86    24.33    18.17    14.64    12.40
     25              241.32    78.06    40.41    26.31    19.46    15.57    13.12
     30              272.54    85.61    43.56    28.04    20.58    16.38    13.74
     40              330.38    99.00    49.02    31.00    22.48    17.72    14.76
     50              383.61   110.82    53.69    33.49    24.05    18.83    15.61
    100              609.46   157.06    71.22    42.53    29.65    22.71    18.50
    250             1123.86   249.02   103.18    58.22    38.98    28.97    23.08
    500             1787.99   353.41   136.63    73.76    47.80    34.72    27.21

               DF1 =     8        8        8        8        8        8        8
COMPARISONS    DF2 =    10       11       12       13       14       15       16

      2                6.12     5.68     5.34     5.08     4.86     4.67     4.52
      3                6.80     6.29     5.88     5.57     5.31     5.09     4.91
      4                7.33     6.74     6.29     5.93     5.64     5.40     5.20
      5                7.75     7.11     6.62     6.23     5.91     5.65     5.44
      6                8.12     7.42     6.89     6.47     6.14     5.86     5.63
      7                8.43     7.69     7.13     6.69     6.33     6.04     5.80
      8                8.72     7.94     7.34     6.88     6.51     6.20     5.94
      9                8.97     8.16     7.54     7.05     6.66     6.34     6.08
     10                9.20     8.35     7.71     7.21     6.80     6.47     6.19
     11                9.42     8.54     7.87     7.35     6.93     6.59     6.30
     12                9.62     8.71     8.02     7.48     7.05     6.70     6.41
     13                9.81     8.87     8.16     7.60     7.16     6.80     6.50
     14                9.98     9.02     8.29     7.72     7.27     6.90     6.59
     15               10.15     9.16     8.41     7.83     7.36     6.98     6.67
     16               10.31     9.29     8.53     7.93     7.46     7.07     6.75
     17               10.46     9.42     8.64     8.03     7.54     7.15     6.82
     18               10.60     9.54     8.74     8.12     7.63     7.22     6.89
     19               10.74     9.65     8.84     8.21     7.71     7.30     6.96
     20               10.87     9.76     8.94     8.29     7.78     7.37     7.02
     25               11.45    10.26     9.36     8.67     8.12     7.67     7.30
     30               11.95    10.67     9.72     8.98     8.40     7.93     7.54
     40               12.77    11.36    10.31     9.50     8.86     8.34     7.92
     50               13.45    11.92    10.78     9.92     9.23     8.68     8.22
    100               15.73    13.80    12.38    11.30    10.46     9.78     9.22
    250               19.30    16.68    14.79    13.38    12.28    11.40    10.69
    500               22.44    19.18    16.86    15.14    13.82    12.76    11.90
```

Table A.8. *Continued*

```
          BONFERRONI F CRITICAL VALUES FOR FAMILYWISE ALPHA= .010
```

COMPARISONS	DF1 = 12 DF2 = 3	12 4	12 5	12 6	12 7	12 8	12 9
2	43.39	20.70	13.38	10.03	8.18	7.01	6.23
3	57.09	25.56	15.93	11.66	9.34	7.92	6.96
4	69.32	29.66	17.99	12.95	10.25	8.61	7.52
5	80.56	33.26	19.77	14.04	11.01	9.19	7.98
6	91.07	36.52	21.35	14.99	11.67	9.68	8.38
7	101.00	39.52	22.77	15.84	12.26	10.12	8.72
8	110.47	42.31	24.07	16.62	12.78	10.51	9.03
9	119.56	44.93	25.29	17.33	13.26	10.87	9.31
10	128.32	47.41	26.42	17.99	13.71	11.19	9.57
11	136.78	49.77	27.49	18.61	14.12	11.50	9.81
12	144.99	52.02	28.49	19.19	14.51	11.78	10.03
13	152.99	54.18	29.45	19.74	14.87	12.05	10.24
14	160.77	56.26	30.37	20.26	15.22	12.30	10.43
15	168.38	58.27	31.25	20.76	15.55	12.54	10.62
16	175.82	60.20	32.09	21.24	15.86	12.76	10.79
17	183.09	62.09	32.91	21.69	16.16	12.98	10.96
18	190.22	63.92	33.69	22.13	16.45	13.19	11.12
19	197.24	65.69	34.45	22.56	16.72	13.39	11.27
20	204.15	67.42	35.19	22.96	16.99	13.58	11.42
25	236.96	75.49	38.57	24.82	18.19	14.44	12.07
30	267.72	82.78	41.56	26.45	19.23	15.17	12.63
40	324.54	95.72	46.75	29.23	20.99	16.41	13.57
50	376.73	107.10	51.22	31.58	22.45	17.43	14.33
100	598.43	151.89	67.90	40.09	27.64	20.98	16.96
250	1103.42	240.87	98.47	54.85	36.27	26.72	21.14
500	1755.27	341.22	130.19	69.37	44.52	32.07	24.90

COMPARISONS	DF1 = 12 DF2 = 10	12 11	12 12	12 13	12 14	12 15	12 16
2	5.66	5.24	4.91	4.64	4.43	4.25	4.10
3	6.28	5.78	5.39	5.07	4.82	4.61	4.44
4	6.75	6.18	5.74	5.40	5.12	4.88	4.69
5	7.14	6.51	6.03	5.66	5.35	5.10	4.89
6	7.47	6.79	6.28	5.87	5.55	5.28	5.06
7	7.75	7.04	6.49	6.06	5.72	5.44	5.20
8	8.00	7.25	6.68	6.23	5.87	5.58	5.33
9	8.23	7.45	6.85	6.38	6.01	5.70	5.44
10	8.45	7.63	7.00	6.52	6.13	5.81	5.55
11	8.64	7.79	7.15	6.65	6.24	5.91	5.64
12	8.82	7.94	7.28	6.76	6.35	6.01	5.73
13	8.99	8.08	7.40	6.87	6.44	6.10	5.81
14	9.15	8.22	7.52	6.97	6.54	6.18	5.89
15	9.30	8.34	7.62	7.07	6.62	6.26	5.96
16	9.44	8.46	7.73	7.16	6.70	6.33	6.02
17	9.57	8.58	7.82	7.24	6.78	6.40	6.09
18	9.70	8.68	7.92	7.32	6.85	6.47	6.15
19	9.83	8.79	8.01	7.40	6.92	6.53	6.21
20	9.94	8.88	8.09	7.48	6.99	6.59	6.26
25	10.47	9.32	8.47	7.81	7.28	6.85	6.50
30	10.92	9.70	8.79	8.08	7.53	7.08	6.71
40	11.66	10.31	9.31	8.54	7.93	7.44	7.04
50	12.27	10.81	9.73	8.90	8.25	7.73	7.30
100	14.33	12.49	11.14	10.12	9.32	8.69	8.16
250	17.53	15.06	13.28	11.95	10.91	10.09	9.43
500	20.36	17.31	15.12	13.51	12.26	11.27	10.48

Table A.8. *Continued*

```
            BONFERRONI F CRITICAL VALUES FOR FAMILYWISE ALPHA= .010
```

COMPARISONS	DF1 = DF2 =	12 17	12 18	12 19	12 20	12 22	12 24	12 26
2		3.97	3.86	3.76	3.68	3.53	3.42	3.32
3		4.29	4.16	4.05	3.95	3.79	3.66	3.55
4		4.53	4.38	4.26	4.15	3.97	3.83	3.71
5		4.71	4.56	4.43	4.31	4.12	3.96	3.83
6		4.87	4.71	4.56	4.44	4.24	4.07	3.94
7		5.00	4.83	4.68	4.55	4.34	4.17	4.03
8		5.12	4.94	4.79	4.65	4.43	4.25	4.11
9		5.23	5.04	4.88	4.74	4.51	4.33	4.18
10		5.32	5.13	4.97	4.82	4.58	4.39	4.24
11		5.41	5.21	5.04	4.90	4.65	4.45	4.30
12		5.49	5.29	5.12	4.96	4.71	4.51	4.35
13		5.57	5.36	5.18	5.03	4.77	4.56	4.40
14		5.64	5.43	5.24	5.08	4.82	4.61	4.44
15		5.70	5.49	5.30	5.14	4.87	4.66	4.48
16		5.77	5.55	5.36	5.19	4.92	4.70	4.52
17		5.82	5.60	5.41	5.24	4.96	4.74	4.56
18		5.88	5.65	5.46	5.28	5.00	4.78	4.60
19		5.93	5.70	5.50	5.33	5.04	4.81	4.63
20		5.98	5.75	5.55	5.37	5.08	4.85	4.66
25		6.21	5.96	5.74	5.55	5.24	5.00	4.80
30		6.39	6.13	5.90	5.71	5.38	5.12	4.92
40		6.70	6.41	6.16	5.95	5.60	5.33	5.10
50		6.94	6.63	6.37	6.15	5.78	5.49	5.25
100		7.73	7.37	7.05	6.78	6.34	6.00	5.72
250		8.88	8.42	8.03	7.69	7.15	6.72	6.38
500		9.83	9.28	8.82	8.43	7.80	7.31	6.92

COMPARISONS	DF1 = DF2 =	12 28	12 30	12 40	12 60	12 120	12 500	12 1000
2		3.25	3.18	2.95	2.74	2.54	2.40	2.38
3		3.46	3.38	3.12	2.89	2.66	2.51	2.48
4		3.61	3.53	3.25	2.99	2.75	2.58	2.55
5		3.73	3.64	3.34	3.07	2.81	2.63	2.61
6		3.83	3.73	3.42	3.13	2.87	2.68	2.65
7		3.91	3.81	3.49	3.19	2.91	2.72	2.69
8		3.98	3.88	3.54	3.24	2.95	2.75	2.72
9		4.05	3.94	3.60	3.28	2.99	2.78	2.75
10		4.11	4.00	3.64	3.32	3.02	2.81	2.77
11		4.16	4.05	3.68	3.35	3.04	2.83	2.80
12		4.21	4.10	3.72	3.38	3.07	2.85	2.82
13		4.26	4.14	3.76	3.41	3.09	2.87	2.84
14		4.30	4.18	3.79	3.44	3.11	2.89	2.85
15		4.34	4.22	3.82	3.46	3.13	2.91	2.87
16		4.38	4.25	3.85	3.48	3.15	2.92	2.89
17		4.41	4.29	3.88	3.51	3.17	2.94	2.90
18		4.44	4.32	3.90	3.53	3.19	2.95	2.91
19		4.48	4.35	3.93	3.55	3.20	2.96	2.93
20		4.51	4.38	3.95	3.57	3.22	2.97	2.94
25		4.64	4.50	4.05	3.65	3.28	3.03	2.99
30		4.75	4.60	4.13	3.71	3.33	3.07	3.03
40		4.92	4.77	4.27	3.82	3.42	3.14	3.10
50		5.06	4.90	4.37	3.90	3.48	3.19	3.15
100		5.50	5.31	4.70	4.16	3.68	3.36	3.31
250		6.11	5.88	5.15	4.50	3.95	3.57	3.52
500		6.60	6.33	5.49	4.77	4.15	3.73	3.67

Table A.8. *Continued*

```
                 BONFERRONI F CRITICAL VALUES FOR FAMILYWISE ALPHA= .010

                  DF1 =    25       25       25       25       25       25       25
   COMPARISONS    DF2 =     3        4        5        6        7        8        9

        2               42,59    20,00    12,76     9,45     7,62     6,48     5,71
        3               56,02    24,68    15,16    10,96     8,69     7,30     6,36
        4               68,01    28,61    17,11    12,16     9,53     7,93     6,86
        5               79,02    32,09    18,79    13,18    10,22     8,45     7,28
        6               89,33    35,22    20,28    14,06    10,83     8,89     7,63
        7               99,06    38,11    21,63    14,86    11,36     9,29     7,93
        8              108,35    40,79    22,86    15,58    11,84     9,64     8,21
        9              117,26    43,31    24,01    16,24    12,28     9,96     8,46
       10              125,84    45,70    25,08    16,85    12,69    10,26     8,69
       11              134,14    47,97    26,09    17,43    13,07    10,53     8,90
       12              142,18    50,13    27,04    17,97    13,43    10,79     9,10
       13              150,02    52,21    27,95    18,48    13,76    11,03     9,28
       14              157,65    54,21    28,82    18,97    14,08    11,26     9,46
       15              165,11    56,15    29,65    19,43    14,38    11,47     9,62
       16              172,38    58,01    30,45    19,87    14,67    11,68     9,78
       17              179,56    59,82    31,22    20,30    14,94    11,87     9,93
       18              186,53    61,58    31,96    20,71    15,20    12,06    10,07
       19              193,41    63,29    32,68    21,10    15,46    12,24    10,20
       20              200,13    64,95    33,37    21,48    15,70    12,41    10,33
       25              232,34    72,70    36,57    23,21    16,80    13,19    10,92
       30              262,49    79,71    39,40    24,73    17,75    13,85    11,42
       40              318,19    92,19    44,32    27,31    19,36    14,97    12,25
       50              369,34   103,16    48,53    29,49    20,71    15,89    12,93
      100              586,78   146,23    64,31    37,41    25,46    19,10    15,28
      250             1081,95   231,89    93,15    51,09    33,39    24,31    19,00
      500             1721,26   328,19   123,38    64,69    40,91    29,13    22,36

                  DF1 =    25       25       25       25       25       25       25
   COMPARISONS    DF2 =    10       11       12       13       14       15       16

        2                5,15     4,74     4,41     4,15     3,94     3,77     3,62
        3                5,70     5,21     4,82     4,52     4,28     4,07     3,90
        4                6,12     5,56     5,13     4,80     4,52     4,30     4,11
        5                6,45     5,85     5,38     5,02     4,72     4,48     4,27
        6                6,74     6,09     5,59     5,20     4,89     4,63     4,41
        7                6,99     6,30     5,78     5,36     5,03     4,76     4,53
        8                7,22     6,49     5,94     5,51     5,16     4,87     4,64
        9                7,42     6,66     6,08     5,63     5,27     4,98     4,73
       10                7,60     6,81     6,22     5,75     5,38     5,07     4,82
       11                7,78     6,96     6,34     5,86     5,47     5,16     4,89
       12                7,93     7,09     6,45     5,96     5,56     5,24     4,97
       13                8,08     7,21     6,56     6,05     5,64     5,31     5,03
       14                8,22     7,33     6,66     6,14     5,72     5,38     5,10
       15                8,35     7,44     6,75     6,22     5,79     5,44     5,16
       16                8,48     7,54     6,84     6,29     5,86     5,50     5,21
       17                8,60     7,64     6,92     6,37     5,92     5,56     5,26
       18                8,71     7,73     7,00     6,44     5,98     5,62     5,31
       19                8,82     7,82     7,08     6,50     6,04     5,67     5,36
       20                8,92     7,91     7,15     6,56     6,10     5,72     5,41
       25                9,39     8,29     7,48     6,85     6,35     5,94     5,61
       30                9,78     8,62     7,75     7,08     6,55     6,13     5,77
       40               10,44     9,15     8,20     7,47     6,89     6,43     6,05
       50               10,97     9,58     8,56     7,78     7,16     6,67     6,26
      100               12,78    11,05     9,78     8,82     8,07     7,47     6,98
      250               15,60    13,28    11,62    10,37     9,40     8,64     8,02
      500               18,12    15,23    13,19    11,69    10,54     9,62     8,89
```

Table A.8. *Continued*

BONFERRONI F CRITICAL VALUES FOR FAMILYWISE ALPHA= .010

COMPARISONS	DF1 = DF2 =	25 17	25 18	25 19	25 20	25 22	25 24	25 26
2		3.49	3.38	3.29	3.20	3.06	2.95	2.85
3		3.75	3.63	3.52	3.42	3.26	3.13	3.02
4		3.95	3.81	3.69	3.58	3.41	3.26	3.15
5		4.10	3.95	3.82	3.71	3.52	3.37	3.25
6		4.23	4.07	3.93	3.82	3.62	3.46	3.33
7		4.34	4.17	4.03	3.91	3.70	3.53	3.40
8		4.44	4.26	4.12	3.99	3.77	3.60	3.46
9		4.52	4.34	4.19	4.06	3.83	3.66	3.51
10		4.60	4.42	4.26	4.12	3.89	3.71	3.56
11		4.67	4.48	4.32	4.18	3.94	3.76	3.60
12		4.74	4.55	4.38	4.23	3.99	3.80	3.64
13		4.80	4.60	4.43	4.28	4.04	3.84	3.68
14		4.86	4.66	4.48	4.33	4.08	3.88	3.71
15		4.91	4.71	4.53	4.37	4.12	3.91	3.75
16		4.96	4.75	4.57	4.41	4.15	3.95	3.78
17		5.01	4.80	4.61	4.45	4.19	3.98	3.81
18		5.06	4.84	4.65	4.49	4.22	4.01	3.83
19		5.10	4.88	4.69	4.52	4.25	4.03	3.86
20		5.14	4.92	4.73	4.56	4.28	4.06	3.88
25		5.33	5.09	4.88	4.71	4.41	4.18	3.99
30		5.48	5.23	5.01	4.83	4.52	4.28	4.08
40		5.73	5.46	5.23	5.02	4.69	4.43	4.22
50		5.93	5.64	5.39	5.18	4.83	4.56	4.34
100		6.57	6.23	5.94	5.69	5.28	4.96	4.70
250		7.51	7.09	6.72	6.41	5.91	5.52	5.21
500		8.29	7.79	7.36	7.01	6.42	5.97	5.62

COMPARISONS	DF1 = DF2 =	25 28	25 30	25 40	25 60	25 120	25 500	25 1000
2		2.77	2.71	2.48	2.27	2.07	1.92	1.90
3		2.94	2.86	2.61	2.37	2.15	1.99	1.96
4		3.05	2.97	2.70	2.44	2.20	2.03	2.00
5		3.14	3.06	2.77	2.50	2.24	2.06	2.04
6		3.22	3.13	2.82	2.54	2.28	2.09	2.06
7		3.28	3.19	2.87	2.58	2.31	2.11	2.08
8		3.34	3.24	2.91	2.61	2.33	2.13	2.10
9		3.39	3.29	2.95	2.64	2.35	2.15	2.12
10		3.43	3.33	2.98	2.67	2.37	2.17	2.14
11		3.47	3.37	3.01	2.69	2.39	2.18	2.15
12		3.51	3.40	3.04	2.71	2.41	2.19	2.16
13		3.55	3.44	3.07	2.73	2.42	2.21	2.17
14		3.58	3.47	3.09	2.75	2.44	2.22	2.18
15		3.61	3.49	3.11	2.77	2.45	2.23	2.19
16		3.64	3.52	3.13	2.78	2.46	2.24	2.20
17		3.66	3.55	3.15	2.80	2.47	2.24	2.21
18		3.69	3.57	3.17	2.81	2.48	2.25	2.22
19		3.71	3.59	3.19	2.83	2.49	2.26	2.23
20		3.74	3.61	3.21	2.84	2.50	2.27	2.23
25		3.84	3.71	3.28	2.89	2.54	2.30	2.26
30		3.92	3.78	3.34	2.94	2.58	2.33	2.29
40		4.05	3.91	3.43	3.01	2.63	2.37	2.33
50		4.16	4.00	3.51	3.07	2.67	2.40	2.36
100		4.49	4.31	3.75	3.24	2.80	2.50	2.45
250		4.95	4.74	4.07	3.48	2.97	2.62	2.57
500		5.33	5.09	4.32	3.66	3.10	2.72	2.66

Table A.8. *Continued*

```
              BONFERRONI F CRITICAL VALUES FOR FAMILYWISE ALPHA= .010

              DF1 =    50       50       50       50       50       50       50
COMPARISONS   DF2 =     3        4        5        6        7        8        9

    2               42.21    19.67    12.45     9.17     7.35     6.22     5.45
    3               55.52    24.25    14.79    10.63     8.38     6.99     6.07
    4               67.38    28.12    16.69    11.78     9.17     7.59     6.54
    5               78.30    31.52    18.33    12.76     9.84     8.09     6.93
    6               88.50    34.60    19.78    13.62    10.42     8.51     7.26
    7               98.15    37.43    21.08    14.38    10.93     8.88     7.55
    8              107.34    40.07    22.29    15.08    11.39     9.22     7.81
    9              116.17    42.54    23.40    15.71    11.81     9.52     8.04
   10              124.67    44.88    24.44    16.31    12.20     9.80     8.26
   11              132.89    47.11    25.42    16.86    12.56    10.06     8.46
   12              140.87    49.24    26.35    17.38    12.90    10.31     8.64
   13              148.60    51.28    27.23    17.88    13.22    10.54     8.82
   14              156.16    53.24    28.08    18.35    13.53    10.75     8.98
   15              163.55    55.13    28.89    18.79    13.82    10.96     9.14
   16              170.79    56.97    29.66    19.22    14.09    11.15     9.28
   17              177.83    58.75    30.41    19.63    14.35    11.34     9.42
   18              184.77    60.46    31.13    20.02    14.60    11.51     9.56
   19              191.59    62.15    31.83    20.40    14.84    11.68     9.69
   20              198.26    63.77    32.51    20.77    15.08    11.85     9.81
   25              230.18    71.39    35.61    22.44    16.13    12.58    10.36
   30              259.96    78.28    38.37    23.90    17.04    13.21    10.83
   40              318.12    90.51    43.15    26.39    18.58    14.27    11.61
   50              365.74   101.25    47.25    28.50    19.87    15.15    12.26
  100              581.36   143.48    62.61    36.12    24.42    18.20    14.46
  250             1071.95   227.53    90.65    49.36    31.99    23.12    17.96
  500             1703.92   322.47   119.94    62.37    39.22    27.68    21.12

              DF1 =    50       50       50       50       50       50       50
COMPARISONS   DF2 =    10       11       12       13       14       15       16

    2                4.90     4.49     4.17     3.91     3.70     3.52     3.37
    3                5.41     4.92     4.54     4.24     4.00     3.80     3.63
    4                5.80     5.25     4.83     4.50     4.22     4.00     3.81
    5                6.12     5.52     5.06     4.70     4.41     4.17     3.96
    6                6.39     5.74     5.25     4.87     4.56     4.30     4.09
    7                6.62     5.94     5.42     5.02     4.69     4.42     4.20
    8                6.83     6.12     5.57     5.15     4.80     4.52     4.29
    9                7.02     6.27     5.71     5.26     4.91     4.62     4.37
   10                7.19     6.42     5.83     5.37     5.00     4.70     4.45
   11                7.35     6.55     5.94     5.47     5.09     4.78     4.52
   12                7.50     6.67     6.05     5.56     5.17     4.85     4.59
   13                7.64     6.79     6.14     5.64     5.24     4.92     4.65
   14                7.77     6.89     6.23     5.72     5.31     4.98     4.70
   15                7.89     7.00     6.32     5.80     5.38     5.04     4.76
   16                8.01     7.09     6.40     5.87     5.44     5.09     4.81
   17                8.12     7.18     6.48     5.93     5.50     5.15     4.85
   18                8.23     7.27     6.55     6.00     5.55     5.19     4.90
   19                8.33     7.35     6.62     6.06     5.61     5.24     4.94
   20                8.42     7.43     6.69     6.11     5.66     5.29     4.98
   25                8.86     7.79     6.99     6.37     5.88     5.49     5.16
   30                9.23     8.09     7.24     6.59     6.07     5.65     5.31
   40                9.84     8.58     7.65     6.94     6.38     5.93     5.55
   50               10.34     8.98     7.99     7.23     6.63     6.15     5.75
  100               12.03    10.34     9.11     8.17     7.45     6.87     6.39
  250               14.67    12.41    10.80     9.59     8.66     7.92     7.33
  500               17.01    14.22    12.25    10.81     9.69     8.81     8.11
```

427

Table A.8. *Continued*

BONFERRONI F CRITICAL VALUES FOR FAMILYWISE ALPHA= .010

COMPARISONS	DF1 = 50 DF2 = 17	50 18	50 19	50 20	50 22	50 24	50 26
2	3.25	3.14	3.04	2.96	2.82	2.70	2.61
3	3.48	3.36	3.25	3.15	2.99	2.86	2.75
4	3.65	3.52	3.40	3.29	3.12	2.98	2.86
5	3.79	3.64	3.52	3.40	3.22	3.07	2.94
6	3.91	3.75	3.61	3.50	3.30	3.14	3.01
7	4.00	3.84	3.70	3.58	3.37	3.21	3.07
8	4.09	3.92	3.77	3.65	3.43	3.26	3.12
9	4.17	3.99	3.84	3.71	3.49	3.31	3.17
10	4.24	4.06	3.90	3.77	3.54	3.36	3.21
11	4.30	4.12	3.96	3.82	3.58	3.40	3.25
12	4.36	4.17	4.01	3.86	3.63	3.44	3.28
13	4.42	4.22	4.05	3.91	3.66	3.47	3.31
14	4.47	4.27	4.10	3.95	3.70	3.50	3.34
15	4.52	4.31	4.14	3.99	3.73	3.53	3.37
16	4.56	4.36	4.18	4.02	3.77	3.56	3.40
17	4.61	4.40	4.21	4.06	3.80	3.59	3.42
18	4.65	4.43	4.25	4.09	3.82	3.61	3.44
19	4.69	4.47	4.28	4.12	3.85	3.64	3.46
20	4.72	4.50	4.31	4.15	3.88	3.66	3.49
25	4.89	4.65	4.45	4.28	3.99	3.76	3.58
30	5.02	4.78	4.57	4.39	4.08	3.85	3.65
40	5.24	4.98	4.75	4.56	4.24	3.98	3.78
50	5.42	5.14	4.90	4.70	4.36	4.09	3.87
100	6.00	5.67	5.38	5.14	4.74	4.43	4.18
250	6.84	6.43	6.08	5.78	5.29	4.91	4.61
500	7.53	7.05	6.64	6.29	5.73	5.30	4.96

COMPARISONS	DF1 = 50 DF2 = 28	50 30	50 40	50 60	50 120	50 500	50 1000
2	2.53	2.46	2.23	2.01	1.80	1.64	1.61
3	2.66	2.59	2.33	2.09	1.85	1.68	1.65
4	2.76	2.68	2.40	2.14	1.89	1.71	1.68
5	2.84	2.75	2.46	2.18	1.92	1.73	1.70
6	2.90	2.81	2.51	2.22	1.95	1.75	1.72
7	2.96	2.86	2.54	2.25	1.97	1.77	1.73
8	3.00	2.91	2.58	2.27	1.99	1.78	1.75
9	3.05	2.95	2.61	2.30	2.00	1.79	1.76
10	3.09	2.98	2.64	2.32	2.02	1.80	1.77
11	3.12	3.01	2.66	2.33	2.03	1.81	1.77
12	3.15	3.04	2.68	2.35	2.04	1.82	1.78
13	3.18	3.07	2.70	2.37	2.05	1.83	1.79
14	3.21	3.10	2.72	2.38	2.06	1.83	1.80
15	3.23	3.12	2.74	2.39	2.07	1.84	1.80
16	3.26	3.14	2.76	2.41	2.08	1.84	1.81
17	3.28	3.16	2.77	2.42	2.09	1.85	1.81
18	3.30	3.18	2.79	2.43	2.09	1.86	1.82
19	3.32	3.20	2.80	2.44	2.10	1.86	1.82
20	3.34	3.22	2.82	2.45	2.11	1.87	1.83
25	3.43	3.30	2.88	2.49	2.14	1.89	1.85
30	3.50	3.36	2.92	2.53	2.16	1.90	1.86
40	3.61	3.47	3.00	2.58	2.20	1.93	1.89
50	3.70	3.55	3.06	2.62	2.23	1.95	1.91
100	3.98	3.81	3.25	2.76	2.32	2.01	1.97
250	4.37	4.16	3.51	2.94	2.44	2.09	2.04
500	4.68	4.45	3.71	3.08	2.53	2.15	2.10

Source: Taken from Huitema, *Bonferroni Statistics*: *Quick and Dirty Methods of Simultaneous Statistical Inference*. (In preparation.)

APPENDIX B

Design Matrices for Single-classification ANCOVA and Homogeneity of Regression Analysis

Two Groups

	D	X	DX	Y
Group 1	1			
Group 2	0		0	

Three Groups

	D_1	D_2	X	D_1X	D_2X	Y
Group 1	1	0			0	
Group 2	0	1		0		
Group 3	0	0		0	0	

Four Groups

	D_1	D_2	D_3	X	D_1X	D_2X	D_3X	Y
Group 1	1	0	0			0	0	
Group 2	0	1	0		0		0	
Group 3	0	0	1		0	0		
Group 4	0	0	0		0	0	0	

Five Groups

	D_1	D_2	D_3	D_4	X	D_1X	D_2X	D_3X	D_4X	Y
Group 1	1	0	0	0			0	0	0	
Group 2	0	1	0	0		0		0	0	
Group 3	0	0	1	0		0	0		0	
Group 4	0	0	0	1		0	0	0		
Group 5	0	0	0	0		0	0	0	0	

Six Groups

	D_1	D_2	D_3	D_4	D_5	X	D_1X	D_2X	D_3X	D_4X	D_5X	Y
Group 1	1	0	0	0	0			0	0	0	0	
Group 2	0	1	0	0	0		0		0	0	0	
Group 3	0	0	1	0	0		0	0		0	0	
Group 4	0	0	0	1	0		0	0	0		0	
Group 5	0	0	0	0	1		0	0	0	0		
Group 6	0	0	0	0	0		0	0	0	0	0	

Seven Groups

	D_1	D_2	D_3	D_4	D_5	D_6	X	D_1X	D_2X	D_3X	D_4X	D_5X	D_6X	Y
Group 1	1	0	0	0	0	0			0	0	0	0	0	
Group 2	0	1	0	0	0	0		0		0	0	0	0	
Group 3	0	0	1	0	0	0		0	0		0	0	0	
Group 4	0	0	0	1	0	0		0	0	0		0	0	
Group 5	0	0	0	0	1	0		0	0	0	0		0	
Group 6	0	0	0	0	0	1		0	0	0	0	0		
Group 7	0	0	0	0	0	0		0	0	0	0	0	0	

References

Atiqullah, M., "The Robustness of the Covariance Analysis of a One-way Classification," *Biometrika*, **51**, 365–373 (1964).

Bernhardson, C. S., "Type I Error Rates When Multiple Comparison Procedures Follow a Significant *F* Test of ANOVA," *Biometrics*, **31**, 229–232 (1975).

Bock, R. D., " Contributions of Multivariate Statistical Methods to Educational Research," in R. B. Cattell (Ed.), *Handbook of Multivariate Experimental Psychology*, Rand McNally, Chicago, 1966.

Bock, R. D., "Conditional and Unconditional Inference in the Analysis of Repeated Measurements," paper presented at Symposium on the Application of Statistical Techniques to Psychological Research, Canadian Psychological Association, York University, 1969.

Bock, R. D., *Multivariate Statistical Methods in Behavioral Research*, McGraw-Hill, New York, 1975.

Bock, R. D., and Haggard, E. A., "The Use of Multivariate Analysis of Variance in Behavioral Research," in D. K. Whitla (Ed.), *Handbook of Measurement and Assessment in Behavioral Sciences*, Addison-Wesley, Reading, Mass., 1968.

Bohrnstedt, G. W., "Observations on the Measurement of Change," in E. F. Borgatta and G. W. Bohrnstedt (Eds.), *Sociological Methodology—1969*, Jossey-Bass, San Francisco, 1969, pp. 113–133.

Borich, G. D., Godbout, R. D., and Wunderlich, K. W., *The Analysis of Aptitude–Treatment Interactions: Computer Programs and Calculations*, Oasis Press, Austin, Tex., 1976.

Borich, G. D., and Wunderlich, K. W., "A Note on Some Statistical Considerations for Using Johnson–Neyman Regions of Significance," paper presented at Annual Meeting of the American Psychological Association, Montreal, Quebec, 1973.

Box, G. E. P., "Problems in the Analysis of Growth and Wear Curves," *Biometrics*, **6**, 362–389 (1950).

Box, G. E. P., "Some Theories on Quadratic Forms Applied in the Study of Analysis of Variance Problems: II. Effects of Inequality of Variance and Covariance Between Errors in the Two-Way Classification," *Annals of Mathematical Statistics*, **35**, 484–498 (1954).

Box, G. E. P., and Jenkins, G. M., *Time Series Analysis: Forecasting and Control*, Holden, San Francisco, 1970.

Box, G. E. P., and Tiao, G. C., "A Change in Level of a Nonstationary Time Series," *Biometrika*, **52**, 181–192 (1965).

Box, G. E. P., and Tiao, G. C., "Intervention Analysis with Applications to Economic and Environmental Problems," *Journal of the American Statistical Association*, **70**, 70–79 (1975).

Bradley, R. A., and Srivastava, S. S., "Correlation in Polynomial Regression," *American Statistician*, **33**, 11–14 (1979).

Bryant, J. L., and Paulson, A. S., "An Extension of Tukey's Method of Multiple Comparisons to Experimental Designs with Random Concomitant Variables," *Biometrika*, **63**, 631–638 (1976).

Bryk, A. S., and Weisberg, H. I., "A New Approach to Analyzing Quasi-experimental Data: Value-added Analysis," paper presented at Annual Meeting of the American Statistical Association, St. Louis, 1974.

Bryk, A. S. and Weisberg, H. I., "Value-added Analysis: A Dynamic Approach to the Estimation of Treatment Effects," *Journal of Educational Statistics*, **1**, 127–155 (1976).

Bryk, A. D., and Weisberg, H. I., "Use of the Nonequivalent Control Group Design When Subjects are Growing," *Psychological Bulletin*, **85**, 950–962 (1977).

Cahen, L. S., and Linn, R. L., "Regions of Significant Criterion Differences in Aptitude–Treatment-interaction Research," *American Educational Research Journal*, **8**, 521–530 (1971).

Calkins, D. S., "An Empirical Investigation of Some Effects of the Violation of the Assumption that the Covariate in Analysis of Covariance is Measured without Error," paper presented at Annual Meeting of the American Educational Research Association, Chicago, 1974.

Calkins, D. S., and Jennings, E., "An Empirical Investigation of Some Effects of The Violation of the Assumption that the Covariable in Analysis of Covariance is a Mathematical Variable," paper presented at Annual Meeting of the American Educational Research Association, Chicago, 1972.

Campbell, D. T., "Temporal Changes in Treatment–Effect Correlations: A Quasi-experimental Model for Institutional Records and Longitudinal Studies," in *Proceedings of the 1970 Invitational Conference on Testing Problems*, Educational Testing Service, Princeton, N.J., 1971, pp. 93–110.

Campbell, D. T., "Measurement and Experimentation in Social Settings," Psychology Department Technical Report 627b, Northwestern University, Evanston, Ill., 1974.

Campbell, D. T., and Boruch, R. F., "Making the Case for Randomized Assignment to Treatments by Considering the Alternatives: Six Ways in Which Quasi-experimental Evaluations Tend to Underestimate Effects," in C. A. Bennet and A. A. Lumsdaine (Eds.), *Evaluation and Experience: Some Critical Issues in Assessing Social Programs*, Academic, New York, 1975.

Campbell, D. T., and Erlebacher, A. E., "How Regression Artifacts in Quasi-experimental Evaluations can Mistakenly Make Compensatory Education Look Harmful," in J. Hellmuth (Ed.), *Compensatory Education: A National Debate*, (Vol. 3), *Disadvantaged Child*, Brunner/Mazel, New York, 1970.

Campbell, D. T., and Stanley, J. C., *Experimental and Quasi-experimental Designs for Research*, Rand McNally, Chicago, 1966.

Carlson, J. E., and Timm, N. H., "Analysis of Nonorthogonal Fixed-effects Designs," *Psychological Bulletin*, **81**, 563–570 (1974).

Carmer, S. G., and Swanson, M. R., "An Evaluation of Ten Pairwise Multiple Comparison Procedures by Monte Carlo Methods," *Journal of the American Statistical Association*, **68**, 66–74 (1973).

Ceurvorst, R. W., and Stock, W. A., "Comments on the Analysis of Covariance with Repeated Measures Designs," *Multivariate Behavioral Research*, **13**, 509–513 (1978).

Cochran, W. G., "Analysis of Covariance: Its Nature and Uses," *Biometrics*, **13**, 261–281 (1957).

Cochran, W. G., "Errors of Measurement in Statistics," *Technometrics*, **10**, 637–666 (1968).

Cochran, W. G., "The Use of Covariance in Observational Studies," *The Journal of the Royal Statistical Society*, Series C, **18**, 270–275 (1969).

Cochran, W. G., and Rubin, D. B., "Controllable Bias in Observational Studies: A Review," *Sankhya*, Series A, **35**, 417–446 (1973).

Cohen, J., *Statistical Power Analysis for the Behavioral Sciences*, 2nd ed., Academic, New York, 1977.

Cook, T. D., and Campbell, D. T., *Quasi-Experimentation*, Rand McNally, Chicago, 1979.

Cooley, W. W., and Lohnes, P. R., *Multivariate Data Analysis*, Wiley, New York, 1971.

Cramer, E. M., and Appelbaum, M. I., "The Validity of Polynomial Regression in the Random Regression Model," *Review of Educational Research*, **48**, 511–515 (1978).

Cronbach, L. J., and Furby, L., "How Should We Measure 'Change'—or Should We?" *Psychological Bulletin*, **74**, 68–80 (1970).

Cronbach, L. J., Rogosa, D. R., Floden, R. E., and Price, G. G., "Analysis of Covariance in Nonrandomized Experiments: Parameters Affecting Bias," occasional paper, Stanford University, Stanford Evaluation Consortium, Berkeley, Calif., 1977.

Cronbach, L. J., and Snow, R. E., *Aptitudes and Instructional Methods*, Irvington, New York, 1976.

D'Agostino, R. B., "An Omnibus Test of Normality for Moderate and Large Size Samples," *Biometrika*, **58**, 341–348 (1971).

Dayton, C. M., "Johnson–Neyman Regions for Dependent Samples," paper presented at Annual Meeting of the American Statistical Association, San Diego, 1978.

De Gracie, J. S., "Analysis of Covariance When the Concomitant Variable is Measured with Error," doctoral dissertation, University of Iowa, Iowa City, 1968.

De Gracie, J. S., and Fuller, W. A., "Estimation of the Slope and Analysis of Covariance When the Concomitant Variable is Measured with Error," *Journal of the American Statistical Association*, **67**, 930–937 (1972).

Director, S. M., "Underadjustment Bias in the Quasi-experimental Evaluation of Manpower Training," doctoral dissertation, Northwestern University, Evanston, Ill., 1974.

Draper, N. R., and Smith, H., *Applied Regression Analysis*, Wiley, New York, 1966.

Duncan, D. B., "Multiple Range and Multiple F Tests," *Biometrics*, **11**, 1–42 (1955).

Dunn, O. J., "Multiple Comparisons among Means," *Journal of the American Statistical Association*, **56**, 52–64 (1961).

Dunnett, C. W., "A Multiple Comparison Procedure for Comparing Several Treatments with a Control," *Journal of the American Statistical Association*, **50**, 1096–1121 (1955).

Dunnett, C. W., "Pairwise Multiple Comparisons in the Homogeneous Variance, Unequal Sample Size Case," paper presented at Meeting of the Eastern North American Region of the Institute of Mathematical Statistics and the American Statistical Association, New Orleans, Louisiana, 1979.

Dunnett, C. W., and Sabel, M. "A bivariate generalization of Students' t-distribution with tables for certain special cases." *Biometrika*, 1954, **41**, 153–169.

Dyer, H. S., Linn, R. L., and Patton, M. J., "A Comparison of Four Methods of Obtaining Discrepancy Measures Based on Observed and Predicted School System Means on Achievement Tests," *American Educational Research Journal*, **6**, 591–605 (1969).

Edwards, A. L., *Experimental Design in Psychological Research*, Holt, Rinehart and Winston, New York, 1968.

Elashoff, J. D., "Analysis of Covariance: A Delicate Instrument", *American Educational Research Journal*, **6**, 383–401 (1969).

Erlander, S., and Gustavsson, J., "Simultaneous Confidence Regions in Normal Regression Analysis with an Application to Road Accidents," *Review of the International Statistical Institute*, **33**, 364–377 (1965).

Evans, S. H., and Anastasio, E. H., "Misuse of Analysis of Covariance When Treatment Effect and Covariate are Confounded," *Psychological Bulletin*, **69**, 225–234 (1968).

Ezekiel, M., and Fox, K. A., *Methods of Correlation and Regression Analysis*, Wiley, New York, 1959.

Federer, W. T., "Variance and Covariance Analysis for Unbalanced Classifications," *Biometrics*, **13**, 333–361 (1957).

Federer, W. T., and Henderson, H. V., "Covariance Analyses of Designed Experiments Using Statistical Packages," Biometrics Unit paper BU-652-M, Cornell University, Ithaca, N.Y., July 1978a.

Federer, W. T., and Henderson, H. V., "Covariance Analyses of Designed Experiments Using Statistical Packages," in *Proceedings, Statistical Computing Section, American Statistical Association*, San Diego, 1978b, pp. 332–337.

Feldt, L. S., "A Comparison of the Precision of Three Experimental Designs Employing a Concomitant Variable," *Psychometrika*, **23**, 335–354 (1958).

Ferguson, G. A., *Statistical Analysis in Psychology and Education*, 4th ed. McGraw-Hill, New York, 1976.

Fisher, R. A., *Statistical Methods for Research Workers*, 4th ed., Oliver and Boyd, Edinburgh, 1932.

Fisher, R. A., and Yates, F., *Statistical Tables for Biological, Agricultural and Medical Research*, Oliver and Boyd, Edinburgh, 1963.

Fleischman, E. A., and Hempel, W. E., Jr., "Changes in Factor Structure of a Complex Psychomotor Test as a Function of Practice," *Psychometrika*, **19**, 239–252 (1954).

Fleishman, E. A., and Hempel, W. E., Jr., "The Relation Between Abilities and Improvement with Practice in a Visual Discrimination Task," *Journal of Experimental Psychology*, **49**, 301–312 (1955).

Forsyth, R. A., and Feldt, L. S., "Some Theoretical and Empirical Results Related to McNemar's Test that the Population Correlation Coefficient Corrected for Attenuation Equals 1.0," *American Educational Research Journal*, **7**, 197–207 (1970).

Fredericks, H. D., "A Comparison of the Doman-Dalacato Method and a Behavior Modification Method upon the Coordination of Mongoloids" doctoral dissertation, University of Oregon, Eugene, 1969.

Fuller, W. A., and Hidiroglou, M. A., "Regression Estimation after Correction for Attenuation," *Journal of the American Statistical Association*, **73**, 99–104 (1978).

Gaito, J, "The Bolles–Messick Coefficient of Utility," *Psychological Reports*, **4**, 595–598 (1958).

Games, P. A., Winkler, H. B., and Probert, D. A., "Robust Tests for Homogeneity of Variance," *Educational and Psychological Measurement*, **32**, 887–909 (1972).

Gartside, P. S., "A Study of Methods for Comparing Several Variances," *Journal of the American Statistical Association*, **67**, 342–346 (1972).

Geisser, S., and Greenhouse, F. W., "An Extension of Box's Results on the Use of F-Distribution in Multivariate Analysis," *Annals of Mathematical Statistics*, **29**, 885–891 (1958).

Glass, G. V., Peckham, P. D., and Sanders, J. R., "Consequences of Failure to Meet Assumptions Underlying the Fixed Effects Analysis of Variance and Covariance," *Review of Educational Research*, **42**, 237–288 (1972).

Glass, G. V., Willson, V. L., and Gottman, J. M., *Design and Analysis of Time-Series Experiments*, Colorado Associated University Press, Boulder, 1975.

Goldberger, A. S., and Duncan, O. D., *Structural Equation Models in the Social Sciences*, Seminar Press, New York, 1973.

Greenhouse, S. W., and Geisser, S., "On Methods in the Analysis of Profile Data," *Psychometrika*, **24**, 95–112 (1959).

Greenwald, A. G., "Within-Subjects Designs: To Use or Not to Use?" *Psychological Bulletin*, **83**, 314–320 (1976).

Halperin, M., and Greenhouse, S. W., "Note on Multiple Comparisons for Adjusted Means in the Analysis of Covariance," *Biometrika*, **45**, 256–259 (1958).

Hamilton, B. L., "An Empirical Investigation of the Effects of Heterogeneous Regression Slopes in Analysis of Covariance," paper presented at Annual Meeting of the American Educational Research Association, Chicago, 1974.

Hamilton, B. L., "A Monte Carlo Comparison of the Power of Parametric ANCOVA with the Power of Three Nonparametric Techniques," paper presented at Annual Meeting of the American Educational Research Association, Washington, D.C., 1975.

Hamilton, B. L., "A Monte Carlo Test of the Robustness of Parametric and Nonparametric Analysis of Covariance against Unequal Regression Slopes," *Journal of the American Statistical Association*, **71**, 864–869 (1976).

Harris, R. J., *A Primer of Multivariate Statistics*, Academic, New York, 1975.

Hays, W. L., *Statistics for the Social Sciences*, 2nd ed. Holt, Rinehart and Winston, New York, 1973.

Henderson, C. R., Jr., and Henderson, C. R., "Analysis of Covariance in Mixed Models with Unequal Subclass Numbers," *Communications in Statistics*, **A8**, 751–787 (1979).

Hersen, M., and Barlow, D. H., *Single Case Experimental Designs: Strategies for Studying Behavior Change*, Pergamon Press, New York, 1976.

Hollingsworth, H. H., "An Analytical Investigation of the Robustness and Power of ANCOVA with the Presence of Heterogeneous Regression Slopes," paper presented at Annual Meeting of the American Educational Research Association, Washington, D.C., 1976.

Hopkins, K. D., and Glass, G. V., *Basic Statistics for the Behavioral Sciences*, Prentice-Hall, Englewood Cliffs, N.J., 1978.

Hotelling, H., "The Selection of Variates for Use in Prediction with Some Comments on the General Problem of Nuisance Parameters," *Annals of Mathematical Statistics*, **11**, 271–283 (1940).

Hughes, P. F., "The Robustness of the Analysis of Covariance to Departures from the Assumptions of Homogeneous Regression Parameters and Fixed and Reliable Covariate Measures," doctoral dissertation, State University of New York at Albany, 1973.

Huitema, B. E., "The Need for Multivariate Analysis in Experimental Psychology," paper presented at Michigan Academy of Science, Arts and Letters, East Lansing, 1974.

Huitema, B. E., *Bonferroni Statistics: Quick and Dirty Methods of Simultaneous Statistical Inference*, manuscript in preparation.

Huynk, H., and Feldt, L. S., "Conditions under Which Mean Square Ratios in Repeated Measures Designs Have Exact F-Distributions," *Journal of the American Statistical Association*, **65**, 1582–1589 (1979).

Johnson, P. O., and Neyman, J., "Tests of Certain Linear Hypotheses and Their Application to Some Educational Problems," *Statistical Research Memoirs*, **1**, 57–93 (1936).

Jones, L. V., "Analysis of Variance in Its Multivariate Developments," in R. B. Cattell (Ed.), *Handbook of Multivariate Experimental Psychology*. Rand McNally, Chicago, 1966, pp. 244–266.

Jöreskog, K. G., "Estimation and Testing of Simplex Models," *The British Journal of Mathematical and Statistical Psychology*, **23**, 121–145 (1970a).

Jöreskog, K. G., "A General Method for Analysis of Covariance Structures," *Biometrika*, **57**, 239–251 (1970b).

Jöreskog, K. G., "A General Method for Estimating a Linear Structural Equation System," in A. S. Goldberger and O. D. Duncan (Eds.), *Structural Equation Models in the Social Sciences*, McGraw-Hill, New York, 1973.

Jöreskog, K. G., "Analyzing Psychological Data by Structural Analysis of Covariance Matrices," in R. C. Atkinson, D. H. Krantz and P. D. Suppes (Eds.), *Contemporary Developments in Mathematical Psychology*, Vol. 2, Freeman, San Francisco, 1974.

Jöreskog, K. G., "Structural Equation Models in the Social Sciences: Specification, Estimation and Testing," in P. R. Krishnaiah (Ed.), *Application of Statistics*, North Holland, Amsterdam, 1977.

Jöreskog, K. G., "Structural Analysis of Covariance and Correlation Matrices," *Psychometrika*, **43**, 443–477 (1978).

Jöreskog, K. G., and Sörbom, D., "Some Models and Estimation Methods for Analysis of Longitudinal Data," in D. J. Aigner and A. S. Goldberger (Eds.), *Latent Variables in Socioeconomic Models*, North Holland, Amsterdam, 1976.

Jöreskog, K. G., and Sörbom, D., *LISREL IV: Analysis of Linear Structural Relationships by the Method of Maximum Likelihood: User's Guide*, International Educational Services, Chicago, 1978.

Keesling, J. W., and Wiley, D. E., "Measurement Error and the Analysis of Quasi-experimental Data. Mehr Licht!" Studies of Educative Processes, Technical Report, CEMREL, Chicago, July 1977.

Kenny, D. A., "A Quasi-experimental Approach to Assessing Treatment Effects in the Nonequivalent Control Group Design," *Psychological Bulletin*, **82**, 345–362 (1975).

Kenny, D. A., Private communication, August 12, 1976.

Kenny, D. A., *Correlation and Causality*, Wiley-Interscience, New York, 1979.

Keuls, M., "The Use of Studentized Range in Connection with an Analysis of Variance," *Euphytica*, **1**, 112–122 (1952).

Kirk, R. E., *Experimental Design: Procedures for the Behavioral Sciences*, Brooks/Cole, Belmont, Calif., 1968.

Kocher, A. T., "An Investigation of the Effects of Non-homogeneous within-group Regression Coefficients upon the F Test of Analysis of Covariance," paper presented at Annual Meeting of the American Educational Research Association, Chicago, 1974.

Kramer, C. Y., "Extensions of Multiple Range Test to Group Means with Unequal Numbers of Replications," *Biometrics*, **12**, 307–310 (1956).

Kristof, W., "Testing a Linear Relation between True Scores of Two Measures," *Psychometrika*, **38**, 101–111 (1973).

Layard, M. W. J., "Robust Large-sample Tests for Homogeneity of Variances," *Journal of the American Statistical Association*, **68**, 195–198 (1973).

Levin, J. R., "Comment: Misinterpreting the Significance of "Explained Variation," *American Psychologist*, **22**, 675–676 (1967).

Lewis, D., *Quantitative Methods in Psychology*, McGraw-Hill, New York, 1960.

Linn, R. L., and Slinde, J. A., "The Determination of the Significance of Change between Pre- and Posttesting Periods," *Review of Educational Research*, **47**, 121–150 (1977).

Linn, R. L., and Werts, C. E., "Errors of Inference Due to Errors of Movement," *Educational and Psychological Measurement*, **33**, 531–545 (1973).

Lord, F. M., "Large-sample Covariance Analysis when the Control Variable is Fallible," *Journal of the American Statistical Association*, **55**, 307–321 (1960).

Lord, F. M., "A Paradox in the Interpretation of Group Comparisons," *Psychological Bulletin*, **68**, 304–305 (1967).

Lord, F. M., "Testing if Two Measuring Procedures Measure the Same Dimension," *Psychological Bulletin*, **79**, 71–72 (1973).

Lord, F. M., and Novick, M. R., *Statistical Theories of Mental Test Scores*. Addison-Wesley, Reading, Mass., 1968.

Magidson, J., "Towards a Causal Model Approach for Adjustment for Preexisting Differences in the Nonequivalent Control Group Situation: A General Alternative to ANCOVA," *Evaluation Quarterly*, **1**, 399–420 (1977).

Maxwell, S., and Cramer, E. M., "A Note on Analysis of Covariance," *Psychological Bulletin*, **82**, 187–190 (1975).

McClaran, V. R., "An Investigation of the Effect of Violating the Assumption of Homogeneity of Regression Slopes in the Analysis of Covariance Model upon the F-Statistic," doctoral dissertation, North Texas State University, Denton, 1972.

McGaw, B., and Cummings, J., "Covariance Adjustments with Estimates of True Scores," *Research Series*, Research Branch, Planning and Services Division, Department of Education, Queensland, Australia, 1975.

McKinlay, S. M., "The Design and Analysis of the Observational Study—A Review," *Journal of the American Statistical Association*, 70, 503–520 (1975).

McLean, J. E., "An Empirical Examination of Covariance with and without Porter's Adjustment for a Fallible Covariate," doctoral dissertation, The University of Florida, Gainesville, 1975.

McNemar, Q., "A Critical Examination of the University of Iowa Studies of Environmental Influences upon I.Q., *Psychological Bulletin*, 37, 63–92 (1940).

McNemar, Q., "Attenuation and Interaction," *Psychometrika*, 23, 259–266 (1958).

McNemar, Q., *Psychological Statistics*, Wiley, New York, 1969.

McSweeney, M., and Porter, A. C., "Small Sample Properties of Nonparametric Index of Response and Rank Analysis of Covariance," paper presented at Annual Meeting of the American Educational Research Association, New York, 1971.

Meehl, Paul, E., "Nuisance Variables and the ex-post-facto Design, in (M. Radner and S. Winokur (Eds.),) *Minnesota Studies in the Philosophy of Science*, Vol. 4, University of Minnesota Press, Minneapolis, 1970.

Mendro, R. L., "A Monte Carlo Study of the Robustness of the Johnson–Neyman Technique," paper presented at Annual Meeting of the American Educational Research Association, Washington, D.C., 1975.

Misselt, A. D., "The Effects of Measurement Error on Analysis of Covariance with Multiple Covariates," doctoral dissertation, University of Illinois, Urbana, 1977.

Neill, J. J., and Dunn, O. J., "Equality of Dependent Correlation Coefficients," *Biometrics*, 31, 531–543 (1975).

Neter, J., and Wasserman, W., *Applied Linear Statistical Models*, Richard D. Irwin, Homewood, Ill., 1974.

Newman, D., "The Distribution of the Range in Samples from a Normal Population, Expressed in Terms of an Independent Estimate of Standard Deviation," *Biometrika*, 31, 20–30 (1939).

Nunnally, J. C., *Introduction to Psychological Measurement*, McGraw-Hill, New York, 1970.

Nunnally, J. C., *Psychometric Theory*, (2nd ed). McGraw-Hill, New York, 1978.

Oken, D., "The Psychophysiology and Psychoendocrinology of Stress and Emotion, in Appley, M. H., and Trumbell, R. (eds.), *Psychological Stress*. Appleton-Century-Crofts, New York, 1967.

Overall, J. E., Spiegel, D. K., and Cohen, J., "Equivalence of Orthogonal and nonorthogonal Analysis of Variance," *Psychological Bulletin*, 82, 182–186 (1975).

Overall, J. E., and Woodward, J. A., "Nonrandom Assignment and the Analysis of Covariance," *Psychological Bulletin*, 84, 588–594 (1977).

Paulson, A. S., private communication, April 1979.

Pearson, E. S., and Hartley, H. O., *Biometrika Tables for Statisticians*, Vol. 1, 2nd ed., Cambridge, University Press, New York, 1958.

Peckham, P. D., "The Robustness of the Analysis of Covariance to Heterogeneous Regression Slopes," paper presented at Annual Meeting of the American Educational Research Association, Minneapolis, 1970.

Peckham, P. D., Glass, G. V., and Hopkins, K. D., "The Experimental Unit in Statistical Analysis," research paper No. 28, Laboratory of Educational Research, University of Colorado, Boulder, 1969.

Pedhazur, E. J., "Coding Subjects in Repeated Measures Designs," *Psychological Bulletin*, 84, 298–305 (1977).

Porter, A. C., "The Effects of Using Fallible Variables in the Analysis of Covariance," doctoral dissertation, University of Wisconsin, Madison, 1967.

Porter, A. C., and McSweeney, M., "Comparison of Rank Analysis of Covariance and Nonparametric Randomized Blocks Analysis," paper presented at Annual Meeting of the American
Educational Research Association, New York, 1971.

Porter, A. C., "Analysis Strategies for Some Common Evaluation Paradigms," paper presented at
Annual Meeting of the American Educational Research Association, New Orleans, 1973.

Potthoff, R. F., "On the Johnson–Neyman Technique and Some Extensions Thereof," *Psychometrika*, **29**, 241–256 (1964).

Pravalpruk, K., and Porter, A., "The Effect of Multiple Fallible Covariables in Analysis of Covariance and Two Correction Methods," paper presented at the Annual Meeting of
the American Educational Research Association, Chicago, 1974.

Puri, M. L., and Sen, P. K., "Analysis of Covariance Based on General Rank Scores," *Annals of
Mathematical Statistics*, **40**, 610–618 (1969).

Quade, D., "Rank Analysis of Covariance," *Journal of the American Statistical Association*, **62**,
1187–1200 (1967).

Reeder, M. W., and Carter, M. W., "The Analysis of Covariance in the Presence of Heterogeneous Slopes," paper presented at Annual Meeting of the American Statistical Association,
Boston, 1976.

Reichardt, C. S., "The Design and Analysis of the Nonequivalent Group Quasi-Experiment,"
doctoral dissertation, Northwestern University, Evanston, Ill, 1979.

Richards, J. M., Jr., "A Simulation Study of the Use of Change Measures to Compare
Educational Programs," *American Educational Research Journal*, **12**, 299–311 (1975).

Rindskopf, D. M., "A Comparison of Various Regression-correlation Methods for Evaluating
Nonexperimental Research," doctoral dissertation, Iowa State University, Ames, 1976.

Rindskopf, D. M., "Using Structural Equation Models to Analyze Nonexperimental Data," in
R. F. Boruch and P. M. Wortman (Eds.), *Secondary Analysis*, Jossey-Bass, San Francisco,
1978.

Robinson, J., "The Analysis of Covariance under a Randomization Model," *Journal of the Royal
Statistical Society*, **35**, 368–376 (1973).

Robinson, J. J., "The Effect of Non-Homogeneous Within-Group Regression Coefficients and
Sample Size on the Distribution of the *F*-Statistic in the Analysis of Covariance," dissertation, University of Oregon, Eugene, 1969.

Robson, D. S., and Atkinson, G. F., "Individual Degrees of Freedom for Testing Homogeneity
of Regression Coefficients in a One-way Analysis of Covariance," *Biometrics*, **16**, 593–605
(1960).

Rogosa, D. R., "Some Results for the Johnson–Neyman Technique," doctoral dissertation,
Stanford University, Stanford, Calif., 1977.

Rubin, D. B., "Assignment to Treatment Group on the Basis of a Covariate," *Journal of
Educational Statistics*, **2**, 1–26 (1977).

Scheffé, H., *The Analysis of Variance*, Wiley, New York, 1959.

Searle, S. R., *Linear Models*, Wiley, New York, 1971.

Searle, S. R., "Alternative Covariance Models for the 2-Way Crossed Classification," *Communications in Statistics*, **A8**, 799–818 (1979).

Shapiro, S. S., Wilk, M. B., and Chen, H. J., "A Comparative Study of Various Tests for
Normality," *Journal of the American Statistical Association*, **63**, 1343–1372 (1968).

Shields, J. L., "An Empirical Investigation of the Effect of Heteroscedosticity and Heterogeneity
of Variance on the Analysis of Covariance and the Johnson–Neyman Technique, technical
paper No. 292, U.S. Army Research Institute for the Behavioral and Social Sciences,
Alexandria, Virginia, 1978.

Smith, F. W., "Interpretation of Adjusted Treatment Means and Regressions in Analysis of
Covariance," *Biometrics*, **13**, 282–307 (1957).

Snedecor, G. W., and Cochran, W. G., *Statistical Methods*, Iowa State University Press, Ames, 1967.

Sörbom, D., "An Alternative to the Methodology for Analysis of Covariance," *Psychometrika*, **43**, 381–396 (1978).

Speed, F. M., Hocking, R. R., and Hockney, O. P., "Methods of Analysis of Linear Models with Unbalanced Data," *Journal of the American Statistical Association*, **73**, 105–112 (1978).

Steel, R. G. D., and Federer, W. T., "Yield-stand Analyses," *Journal of the Indian Society of Agricultural Statistics*, **7**, 27–45 (1955).

Strenio, J. F., Bryk, A. S., and Weisberg, H. I., "An Individual Growth Model Perspective for Evaluating Educational Programs," *Proceedings of the Social Statistics Section*, American Statistical Association, 1977.

Stroud, T. W. F., "Comparing Conditional Means and Variance in a Regression Model with Measurement Errors of Known Variances," *Journal of the American Statistical Association*, **67**, 407–414 (1972).

Tatsuoka, M. M., *Multivariate Analysis*: *Techniques for Educational and Psychological Research*, Wiley, New York, 1971.

Thorndike, R. L., "Regression Fallacies in the Matched Groups Experiment," *Psychometrika*, **7**, 85–102 (1942).

Tukey, J. W., "Comparing Individual Means in the Analysis of Variance," *Biometrics*, **5**, 99–114 (1949).

Tukey, J. W., "The Problem of Multiple Comparisons, Ditto, Princeton University, 1953.

Villegas, C., "Confidence Region for a Linear Relation," *The Annals of Mathematical Statistics*, **35**, 780–788 (1964).

Weisberg, H. I., Strenio, J. F. and Bryk, A. S., "Estimation of Individual Growth Parameters and Their Relationship to Background Variables," paper presented at Annual Meeting of the American Statistical Association, New York, 1979.

Werts, C. E., and Linn, R. L., "Analyzing School Effects: ANCOVA with a Fallible Covariate," *Educational and Psychological Measurement*, **31**, 95–104 (1971).

Werts, C. E., Rock, D. A., Linn, R. L., and Jöreskog, K. G., "Comparison of Correlations, Variances, Covariances, and Regression Weights with or without Measurement Error," *Psychological Bulletin*, **83**, 1007–1013 (1976).

Williams, E. G., "The Comparison of Regression Variables," *Journal of the Royal Statistical Society*, Series B, **21**, 396–399 (1959).

Winer, B. J., *Statistical Principles in Experimental Design*, 2nd ed., McGraw-Hill, New York, 1971.

Wortman, P. M., Reichardt, C. S., and St. Pierre, R. G., "The First Year of the Education Voucher Demonstration," *Evaluation Quarterly*, **2**, 193–214 (1978).

Wunderlich, K. W., and Borich, G. D., "Curvilinear Extensions to Johnson–Neyman Regions of Significance and Some Applications to Educational Research," paper presented at Annual Meeting of the American Educational Research Association, Chicago, 1974.

Index